Computer-Operated Systems Control

ELECTRICAL ENGINEERING AND ELECTRONICS

A Series of Reference Books and Textbooks

EXECUTIVE EDITORS

Marlin O. Thurston
Department of
Electrical Engineering
The Ohio State University
Columbus, Ohio

William Middendorf
Department of
Electrical and Computer Engineering
University of Cincinnati
Cincinnati, Ohio

EDITORIAL BOARD

Maurice Bellanger
Télécommunications, Radioélectriques, et
 Téléphoniques (TRT)
Le Plessis-Robinson, France

J. Lewis Blackburn
Bothell, Washington

Sing T. Bow
Department of Electrical Engineering
The Pennsylvania State University
University Park, Pennsylvania

Norman B. Fuqua
Reliability Analysis Center
Griffiss Air Force Base, New York

Charles A. Harper
Westinghouse Electric Corporation
 and Technology Seminars, Inc.
Timonium, Maryland

Naim A. Kheir
Department of Electrical and
 Systems Engineering
Oakland University
Rochester, Michigan

Lionel M. Levinson
General Electric Company
Schenectady, New York

V. Rajagopalan
Department of Engineering
Université du Québec
 à Trois-Rivières
Trois-Rivières, Quebec, Canada

Earl Swartzlander
TRW Defense Systems Group
Redondo Beach, California

Spyros G. Tzafestas
Department of Electrical Engineering
National Technical University
 of Athens
Athens, Greece

Sakae Yamamura
Central Research Institute of
 the Electric Power Industry
Tokyo, Japan

1. Rational Fault Analysis, *edited by Richard Saeks and S. R. Liberty*
2. Nonparametric Methods in Communications, *edited by P. Papantoni-Kazakos and Dimitri Kazakos*
3. Interactive Pattern Recognition, *Yi-tzuu Chien*
4. Solid-State Electronics, *Lawrence E. Murr*
5. Electronic, Magnetic, and Thermal Properties of Solid Materials, *Klaus Schröder*
6. Magnetic-Bubble Memory Technology, *Hsu Chang*
7. Transformer and Inductor Design Handbook, *Colonel Wm. T. McLyman*
8. Electromagnetics: Classical and Modern Theory and Applications, *Samuel Seely and Alexander D. Poularikas*
9. One-Dimensional Digital Signal Processing, *Chi-Tsong Chen*
10. Interconnected Dynamical Systems, *Raymond A. DeCarlo and Richard Saeks*
11. Modern Digital Control Systems, *Raymond G. Jacquot*
12. Hybrid Circuit Design and Manufacture, *Roydn D. Jones*
13. Magnetic Core Selection for Transformers and Inductors: A User's Guide to Practice and Specification, *Colonel Wm. T. McLyman*
14. Static and Rotating Electromagnetic Devices, *Richard H. Engelmann*
15. Energy-Efficient Electric Motors: Selection and Application, *John C. Andreas*
16. Electromagnetic Compossibility, *Heinz M. Schlicke*
17. Electronics: Models, Analysis, and Systems, *James G. Gottling*
18. Digital Filter Design Handbook, *Fred J. Taylor*
19. Multivariable Control: An Introduction, *P. K. Sinha*
20. Flexible Circuits: Design and Applications, *Steve Gurley, with contributions by Carl A. Edstrom, Jr., Ray D. Greenway, and William P. Kelly*
21. Circuit Interruption: Theory and Techniques, *Thomas E. Browne, Jr.*
22. Switch Mode Power Conversion: Basic Theory and Design, *K. Kit Sum*
23. Pattern Recognition: Applications to Large Data-Set Problems, *Sing-Tze Bow*
24. Custom-Specific Integrated Circuits: Design and Fabrication, *Stanley L. Hurst*
25. Digital Circuits: Logic and Design, *Ronald C. Emery*
26. Large-Scale Control Systems: Theories and Techniques, *Magdi S. Mahmoud, Mohamed F. Hassan, and Mohamed G. Darwish*
27. Microprocessor Software Project Management, *Eli T. Fathi and Cedric V. W. Armstrong (Sponsored by Ontario Centre for Microelectronics)*
28. Low Frequency Electromagnetic Design, *Michael P. Perry*
29. Multidimensional Systems: Techniques and Applications, *edited by Spyros G. Tzafestas*
30. AC Motors for High-Performance Applications: Analysis and Control, *Sakae Yamamura*

31. Ceramic Materials for Electronics: Processing, Properties, and Applications, *edited by Relva C. Buchanan*
32. Microcomputer Bus Structures and Bus Interface Design, *Arthur L. Dexter*
33. End User's Guide to Innovative Flexible Circuit Packaging, *Jay J. Miniet*
34. Reliability Engineering for Electronic Design, *Norman B. Fuqua*
35. Design Fundamentals for Low-Voltage Distribution and Control, *Frank W. Kussy and Jack L. Warren*
36. Encapsulation of Electronic Devices and Components, *Edward R. Salmon*
37. Protective Relaying: Principles and Applications, *J. Lewis Blackburn*
38. Testing Active and Passive Electronic Components, *Richard F. Powell*
39. Adaptive Control Systems: Techniques and Applications, *V. V. Chalam*
40. Computer-Aided Analysis of Power Electronic Systems, *Venkatachari Rajagopalan*
41. Integrated Circuit Quality and Reliability, *Eugene R. Hnatek*
42. Systolic Signal Processing Systems, *edited by Earl E. Swartzlander, Jr.*
43. Adaptive Digital Filters and Signal Analysis, *Maurice G. Bellanger*
44. Electronic Ceramics: Properties, Configuration, and Applications, *edited by Lionel M. Levinson*
45. Computer Systems Engineering Management, *Robert S. Alford*
46. Systems Modeling and Computer Simulation, *edited by Naim A. Kheir*
47. Rigid-Flex Printed Wiring Design for Production Readiness, *Walter S. Rigling*
48. Analog Methods for Computer-Aided Circuit Analysis and Diagnosis, *edited by Takao Ozawa*
49. Transformer and Inductor Design Handbook, Second Edition, Revised and Expanded, *Colonel Wm. T. McLyman*
50. Power System Grounding and Transients: An Introduction, *A. P. Sakis Meliopoulos*
51. Signal Processing Handbook, *edited by C. H. Chen*
52. Electronic Product Design for Automated Manufacturing, *H. Richard Stillwell*
53. Dynamic Models and Discrete Event Simulation, *William Delaney and Erminia Vaccari*
54. FET Technology and Application: An Introduction, *Edwin S. Oxner*
55. Digital Speech Processing, Synthesis, and Recognition, *Sadaoki Furui*
56. VLSI RISC Architecture and Organization, *Stephen B. Furber*
57. Surface Mount and Related Technologies, *Gerald Ginsberg*
58. Uninterruptible Power Supplies: Power Conditioners for Critical Equipment, *David C. Griffith*
59. Polyphase Induction Motors: Analysis, Design, and Application, *Paul L. Cochran*

60. Battery Technology Handbook, *edited by H. A. Kiehne*
61. Network Modeling, Simulation, and Analysis, *edited by Ricardo F. Garzia and Mario R. Garzia*
62. Linear Circuits, Systems and Signal Processing: Advanced Theory and Applications, *edited by Nobuo Nagai*
63. High-Voltage Engineering: Theory and Practice, *edited by M. Khalifa*
64. Large-Scale Systems Control and Decision Making, *edited by Hiroyuki Tamura and Tsuneo Yoshikawa*
65. Industrial Power Distribution and Illuminating Systems, *Kao Chen*
66. Distributed Computer Control for Industrial Automation, *Dobrivoje Popovic and Vijay P. Bhatkar*
67. Computer-Aided Analysis of Active Circuits, *Adrian Ioinovici*
68. Designing with Analog Switches, *Steve Moore*
69. Contamination Effects on Electronic Products, *Carl J. Tautscher*
70. Computer-Operated Systems Control, *Magdi S. Mahmoud*

Additional Volumes in Preparation

Electrical Engineering-Electronics Software

1. Transformer and Inductor Design Software for the IBM PC, *Colonel Wm. T. McLyman*
2. Transformer and Inductor Design Software for the Macintosh, *Colonel Wm. T. McLyman*
3. Digital Filter Design Software for the IBM PC, *Fred J. Taylor and Thanos Stouraitis*

Computer-Operated Systems Control

Magdi S. Mahmoud

Kuwait University
Safat, Kuwait

Marcel Dekker, Inc.   New York • Basel • Hong Kong

Library of Congress Cataloging--in--Publication Data

Mahmoud, Magdi S.
  Computer-operated systems control/Magdi S. Mahmoud.
      p.   cm. -- -- (Electrical engineering and electronics; 70)
  Includes bibliographical references and index.
  ISBN 0-8247-8092-2
  1. Digital control systems. I. Title. II. Series.
TJ223.M53M34   1991
629.8'9-- --dc20   91-7683
                                                                 CIP

This book is printed on acid-free paper.

Copyright © 1991 by MARCEL DEKKER, INC. All Rights Reserved

Neither this book nor any part may be reproduced or transmitted in any form or by any means, electronic or mechanical, including photocopying, microfilming, and recording, or by any information storage and retrieval system, without permission in writing from the publisher.

MARCEL DEKKER, INC.
270 Madison Avenue, New York, New York 10016

Current printing (last digit):
10 9 8 7 6 5 4 3 2 1

PRINTED IN THE UNITED STATES OF AMERICA

To my loving wife
Salwa
who means the most in my life

Preface

Problems in systems control are pervasive in the modern world, appearing in social science, biology, business management, industry, and engineering, to name a few. Extensive use is made of mathematical models, computer-aided analysis and design, simulation, and optimization with respect to quantified goals. Although systems control obviously is not a panacea, it is clear that it is playing an increasingly important role in finding solutions to many significant problems.

Historically, systems control engineers in the 1960s had to master analog computing technology because analog computers were the major tools available. Systems control components were also made based on analog technology in mechanics, pneumatics, and electronics. Recent development in microprocessor technology coupled with a gradual evolution in control theory and techniques therefore have many important areas of application and promise to have even wider usage in the future. There has been a steady shift in using digital computers from components in complex processes to basic elements in individual processing units. Due to their small size and low price, modern digital computers are now outperforming their analog counterparts in several areas. The fundamental lesson learned is that systems control, computers, and information processing subjects should be closely integrated to ensure that a user (student or designer) can move fluently from systems control analysis and design to digital computer implementation. This book attempts to benefit from this lesson.

Therefore the purpose of this book is to present an integrated treatment of the techniques of systems control analysis, components, and design with emphasis on the role of computing elements for problems and examples drawn from diverse fields. These fields include chemical, mechanical, biological, ecological, and sociological systems with the intention of charging the reader with a greater awareness of the general applicability of computer-operated systems control methodologies to many facets of life. Specifically, the book is a guided tour through the growth pattern of computer technological use in control systems. It aims initially at providing a step-by-step analysis of dynamic systems that include analog elements, digital devices, or microprocessor or microcomputer (micros for brevity) chips. Another aim is to bring to systems control users relevant information about the features, characteristics, and architectures of micros and their potential applications.

We hope that the book will be found useful and stimulating by researchers, scientists, and engineers in industry as well as students in formal courses. Every attempt has been made to keep its distinctive features in covering both theoretical aspects and implementation issues.

While book-length studies exist for each topic covered here, it is felt that a book embracing a wide spectrum of coherent topics would be useful in stressing their important interrelationships and the logic for their development.

The material contained in this volume arose from lectures notes that have been compiled and organized for courses at Pittsburgh University (U.S.A.), UMIST (U.K.), Kuwait University (Kuwait) and Cairo University (Egypt). For convenience, the book is organized into three parts:

Part I   Analog Control Systems
Part II   Digital Control Systems
Part III Microcomputer-Based Control Systems

After a brief introduction to the development of digital computers in Chapter 1, Part I presents a balanced treatment of the main techniques of analog control systems in Chapters 2–5. Then follows methods of analysis and design of digital control systems in Chapters 6–8 to form Part II. In both parts, attention is directed to the role of the computing element and its capabilities. Part III with Chapters 9–12 covers the relevant information on microcomputer-based control systems, from the viewpoints of hardware architecture, software support,

Preface

and applications. Chapter 13 brings together the up-to-date information and technology about 16-bit and 32-bit microprocessor architectures.

Additionally, the book stresses the main ideas, concepts, and basic knowledge and follows a step-by-step development of the subject with an in-depth treatment of some major points when required. The treatment should suit first-year graduate or senior undergraduate students. Numerous examples are worked out and problems are given at the end of each chapter.

The author is indebted to many people for their various contributions. Foremost, I would like to thank Professor Abdel-Moniem Y. Belal (Cairo University, Egypt) for introducing me to the fascinating field of systems engineering. I would like to recognize most gratefully the support and encouragement of Professor M. G. Singh (UMIST, U.K.) and Professor A. P. Sage (GMU, U.S.A.). I would like to acknowledge the truly outstanding environment provided by Dr. Adel Assem (Director, Techno-Economic Division at KISR, Kuwait) during the early stages of writing this book. Discussions with colleagues and friends on various occasions have reflected beneficially on the style and content of the material. In particular, I should mention Professor M. I. Younis (AUC, Egypt), Professors S. Z. Eid and M. F. Hassan (Cairo University, Egypt), Professors W. G. Vogt and M. H. Mickle (Pitt University, U.S.A.), Professor N. Munro (UMIST, U.K.) and Drs. A. A. Bahnasawi, H. M. El-Sayed (Cairo University, Egypt) and Professor A. A. Hanafi (Kuwait University, Kuwait). The excellent artwork made by Mr. Abdel-Aziz Shaheen is greatly appreciated. Special thanks goes to Mr. Saju Kuriakose for expert typing and preparation of the initial draft of the manuscript. Last but not least, this book was made possible by the encouragement and continued support provided by the staff of Marcel Dekker, Inc.

Magdi S. Mahmoud

Contents

Preface   *v*

1 Introduction   1

  1.1 Some Concepts and Definitions   2
  1.2 Examples of Control Systems   3
  1.3 Analog Versus Digital Processing   8
      1.3.1 Analog Processing   9
      1.3.2 Digital Processing   9
  1.4 Microprocessors and Microcomputers   10
      1.4.1 Main Features   11
      1.4.2 Architectures   12
      1.4.3 Programming   14
      1.4.4 Applications   15
  1.5 Computer-Controlled Systems   16
      1.5.1 Basic Structure   16
      1.5.2 Some Remarks   16
  1.6 Outline of the Book   18
      Notes and References   20

Part I: Analog Control Systems
2 Systems Control Modeling and Representation   21

   2.1 Transfer Functions   21
   2.2 Block Diagram Algebra   24
       2.2.1 A Closed-Loop System   25
       2.2.2 Some Reduction Rules   26
       2.2.3 Linear Multivariable Systems   27
   2.3 Elementary Physical Systems Models   29
       2.3.1 Electrical Systems   32
       2.3.2 Hydraulic Systems   38
       2.3.3 Mechanical Systems   41
       2.3.4 Pneumatic Systems   45
       2.3.5 Thermal Systems   47
   2.4 Feedback   51
       2.4.1 Control Actions   52
       2.4.2 System Sensitivity   53
       2.4.3 Steady-State Error   55
       2.4.4 System Types   59
   2.5 System Response   60
   2.6 Examples   65
   2.7 Problems   68
       Notes and References   70

3 Systems Analysis Via State Variables   71

   3.1 Introduction   71
   3.2 State-Space Formulation   72
       3.2.1 Obtaining the State Equations   75
       3.2.2 Free Response   80
       3.2.3 Forced Response   81
       3.2.4 Properties of the State Transition Matrix   82
       3.2.5 Examples   86
       3.2.6 State Model Transformations   91
   3.3 Performance Indices   92
   3.4 Relationship of State Description to the Transfer
       Matrix   94
   3.5 Examples   95
   3.6 Problems   98
       Notes and References   100

Contents

4 Stability and Frequency Response Methods — 101

- 4.1 The Concept of Stability — 101
- 4.2 Stability Criteria — 102
  - 4.2.1 Characteristics Equation — 103
  - 4.2.2 Routh Criterion — 108
  - 4.2.3 Hurwitz Criterion — 113
  - 4.2.4 Relative Stability — 115
  - 4.2.5 Input-Output Stability — 116
  - 4.2.6 Root-Locus Method — 119
  - 4.2.7 Lyapunov Method — 129
- 4.3 Frequency Response — 132
  - 4.3.1 Polar Plots — 133
  - 4.3.2 Bode Logarithmic Plots — 137
  - 4.3.3 Stability Margins — 142
  - 4.3.4 The Nyquist Criterion — 145
- 4.4 Sensitivity Analysis — 149
- 4.5 Some Notes — 152
- 4.6 Problems — 154
- Notes and References — 157

5 Introduction to Design — 159

- 5.1 The Design Problem — 159
- 5.2 Some Guidelines — 160
- 5.3 Compensation Methods — 161
  - 5.3.1 Lead Compensator — 165
  - 5.3.2 Lag Compensator — 169
  - 5.3.3 Lead-Lag Compensator — 172
  - 5.3.4 Technical Considerations — 175
  - 5.3.5 A Synthesis Procedure — 176
  - 5.3.6 Examples — 177
- 5.4 State Variable Design — 181
  - 5.4.1 Concept of Controllability and Observability — 181
  - 5.4.2 State Feedback — 182
  - 5.4.3 Output Feedback — 185
  - 5.4.4 Decoupling Control — 187
  - 5.4.5 Observer-Based Feedback Control — 189
  - 5.4.6 Discussions — 193
- 5.5 Parameter Optimization — 194
  - 5.5.1 Quadratic Performance Criteria — 194

5.5.2 Optimization Using Parseval's Theorem	196
5.5.3 The Optimum State Regulator	199
5.5.4 The Optimum Output Regulator	203
5.5.5 Spectral Factorization	205
5.5.6 Design Considerations	208
5.6 Problems	209
Notes and References	212

Part II: Digital Control Systems

6 Methods of Analysis — 214

6.1 Introduction	214
6.2 The Z Transform	215
6.2.1 Definition and Properties	216
6.2.2 The Pulse Transfer Function	220
6.2.3 Inverse Transform	223
6.2.4 Poles and Zeros	227
6.3 Difference Equations	230
6.3.1 Solution by the Z-Transform	230
6.3.2 Numerical Solution	233
6.3.3 Computer Realization	234
6.4 Discretization	236
6.4.1 Principles and Issues	236
6.4.2 Different Schemes	238
6.5 Problems	244
Notes and References	245

7 Design Algorithms — 246

7.1 Stability Tests	246
7.1.1 Jury Test	246
7.1.2 The W-Plane Method	249
7.2 Algorithms of Digital Controller Design	250
7.2.1 Algorithm 1	250
7.2.2 Algorithm 2	252
7.2.3 Algorithm 3	254
7.2.4 Algorithm 4	255
7.2.5 Some Comments	256
7.2.6 On-Off Control	257

Contents

    7.3 The Root-Locus Diagram   257
        7.3.1 Transient Response and Relative Stability   258
        7.3.2 Illustrative Examples   258
    7.4 Design Realization   265
        7.4.1 Direct Programming   265
        7.4.2 Cascade Programming   266
        7.4.3 Parallel Programming   268
        7.4.4 Examples   268
    7.5 Problems   276
    Notes and References   278

8 Optimal Design Methods   **279**

    8.1 Introduction   279
    8.2 Linear Quadratic Control   280
        8.2.1 Problem Formulation   280
        8.2.2 Derivation of the Optimal Sequence   282
        8.2.3 Some Properties   284
        8.2.4 Examples   286
    8.3 Filtering and Prediction   287
        8.3.1 The Estimation Problem   289
        8.3.2 Principal Methods of Obtaining Estimates   290
        8.3.3 Development of the Kalman Filter Equations   294
        8.3.4 Examples   302
    8.4 Linear Quadratic Gaussian Control   309
        8.4.1 Problem Formulation   309
        8.4.2 Solution Strategy   310
        8.4.3 Concluding Remarks   312
    8.5 Problems   312
    Notes and References   314

Part III: Microcomputer-Based Control Systems

9 Basic Hardware and Peripherals   **315**

    9.1 Introduction   315
        9.1.1 Arithmetic and Logic Unit (ALU)   317
        9.1.2 Registers   318
        9.1.3 Stacks   320
        9.1.4 Control Unit   321
        9.1.5 Program Counter   322

	9.2 Microprocessor Characteristics	323
	9.2.1 Purpose	323
	9.2.2 Bit Width	324
	9.2.3 Bit-Slicing	325
	9.2.4 Processing Speed	325
	9.2.5 Common Microprocessor Families	326
	9.3 Memory	332
	9.3.1 Memory Hierarchies	333
	9.3.2 Working Store	334
	9.3.3 Medium Capacity Storage	336
	9.3.4 High Capacity Storage	337
	9.4 Input and Output Methods	337
	9.4.1 Data Transfer	339
	9.4.2 Serial I/O	339
	9.4.3 Common I/O Methods	341
	9.4.4 Communication Buses	341
	9.4.5 I/O Transfer	342
	9.5 Interface Components	343
	9.5.1 Driver Circuits	343
	9.5.2 Receivers	344
	9.5.3 Data Acquisition	344
	9.5.4 Some Remarks	346
	9.6 Problems	347
	Notes and References	347
10	**Software Support**	**348**
	10.1 Introduction	348
	10.2 Application Programs	349
	10.2.1 Machine Code	349
	10.2.2 Hex Code Programming	350
	10.2.3 Assembly Language Programming	352
	10.2.4 High-Level Languages	355
	10.3 Operating Systems	360
	10.3.1 Classification	361
	10.3.2 Single-User (Single-Job) Operating System	363
	10.3.3 Single Foreground/Background Operating System	373
	10.3.4 Real-Time Multitasking Operating Systems	377
	10.4 Software Development Tools	381
	10.4.1 Loaders	381

Contents

	10.4.2 Editors	382
	10.4.3 Linkers	382
	10.4.4 Debugger	383
	10.4.5 Monitor	383
	10.4.6 File Handling/Management	384
	10.4.7 Microcomputer Operating System	386
10.5	Interfacing	387
	10.5.1 Introduction	387
	10.5.2 Microprocessor I/O Ports	390
	10.5.3 Parallel I/O Ports	390
	10.5.4 Serial I/O Ports	392
	10.5.5 Data Acquisition Modules	393
10.6	Discussion	398
	Notes and References	399

11 Industrial Process Control Systems — 400

11.1	Introduction	400
11.2	Functions and Benefits of a Batch Process Control	402
11.3	Process Control Concepts and Actions	404
	11.3.1 Process Control Concepts	404
	11.3.2 Computer Control Actions	406
11.4	Supervisory Control	414
	11.4.1 Ratio Control	416
	11.4.2 Program Control	418
	11.4.3 Sequence Control	419
	11.4.4 Cascade Control	420
	11.4.5 Some Practical Considerations	422
	11.4.6 Software for Process Control	424
	11.4.7 Microprocessor Based Programmable Controllers	425
11.5	Advanced Control	427
	11.5.1 Dead Time Compensators	427
	11.5.2 Inferential Control	437
	11.5.3 Feedforward Control	445
	11.5.4 Constraint Control	451
	11.5.5 Adaptive Control	459
	11.5.6 Multivariable Control	461
	11.5.7 Self-Tuning Control	462
11.6	Sensors and Actuators	464
	11.6.1 Introduction	464

11.6.2 Sensors	464
11.6.3 Actuators	474
11.7 Discussions	477
Notes and References	478

12 Distributed Digital Control Systems — 479

12.1 Introduction	479
12.2 Distributed Processing	482
12.2.1 Elements	482
12.2.2 Hierarchical Processing Systems	483
12.2.3 Horizontal Processing Systems	485
12.2.4 Benefits	487
12.3 Multi-Processor Systems	490
12.3.1 Interconnection of Processors	491
12.3.2 Linking Techniques	495
12.3.3 Advantages	504
12.4 Integrated Systems Control	506
12.4.1 Plant	506
12.4.2 Controller	507
12.4.3 Information Processor	508
12.4.4 Functional Multi-layer Control Hierarchy	509
12.4.5 Multi-level Control Hierarchy	515
12.5 Discussions	522
Notes and References	523

13 Advanced Microprocessor Architecture — 524

13.1 Overview	524
13.2 16-Bit Microprocessors	525
13.2.1 Intel 8086/8088 Family	525
13.2.2 Motorola MC68000 Family	540
13.2.3 Zilog Z8000 Family	548
13.2.4 Comparisons and Evaluation	556
13.3 32-Bit Microprocessors	556
13.3.1 Introduction	556
13.3.2 Intel 80386	560
13.3.3 Motorola C68020	567
13.3.4 Zilog Z80000	575
13.3.5 Inmos T424 Transputer	581
13.4 Conclusions	583
Notes and References	586

Contents

Appendix A: Laplace Transforms and Properties	*587*
Appendix B: Elements of Matrix Algebra	*594*
B.1 Notation	*594*
B.2 Basic Operations	*595*
B.3 Special Types of Matrices	*596*
B.4 Determinants	*596*
B.5 Matrix Inversion	*597*
B.6 Partitioned Matrices	*598*
B.7 Range and Null Spaces	*598*
B.8 The Pseudo-inverse of a Matrix	*599*
B.9 Functions of a Square Matrix	*599*
B.10 The Cayley-Hamilton Theory	*600*
B.11 Differentiation and Integration	*600*
B.12 Eigenvalues and Eigenvectors	*602*
B.13 Norms	*602*
Appendix C: A Derivation of the Riccati Equation	*605*
Appendix D: Introduction to Random Variables and Gauss–Markov Processes	*609*
D.1 Basic Concepts of Probability Theory	*609*
D.2 Mathematical Properties of Random Variables	*612*
D.3 Stochastic Properties	*617*
D.4 Linear Discrete Models with Random Inputs	*622*
Appendix E: Instruction Sets	*628*
Bibliography	*635*
Index	*647*

Computer-Operated Systems Control

1
Introduction

Systems control concepts and operations are undoubtedly of interest to almost all engineering and physical disciplines. They have been in existence since the appearance of living creatures on the earth. In this sense, systems control has probably been going on forever in some part of the universe. By the same token, the use of control components and devices surely dates far back in history. On tracing historical records we find that the Babylonian practiced feedback control in land irrigation, the Romans invented a water-level control device and the Ancient Egyptians in the Hellenistic period produced the water clock which is believed to have employed the first float value regulator.

Thus, over the years, there has been a gradual evolution from manual control (in which there is an operator involved in the control system) to automatic control (which signifies the absence of the intervention of a human operator). In contemporary literature when we speak of control, we usually have in mind automatic control with all of its forms, namely classical, modern, and advanced.

Within the past several decades there has been an enormous growth in the application of controllers and instruments, not only in industrial plants, but generally throughout technology. One of the goals of this book is to follow that growth and document its advantages and limitations. We begin by introducing some basic terminology.

1.1 SOME CONCEPTS AND DEFINITIONS

The global domain in the subsequent chapters will be control systems, which we now define. A *control system* is an interconnection of components forming a system configuration by which any quantity of interest is maintained or altered in accordance with a desired manner. Our primary concern is an understanding of control system operations and the various terms used in their description. Consider the following terms.

Simply stated, *a dynamic variable* is any physical parameter that can change either spontaneously or from exogenous effects. The word *dynamic* conveys the concept of a time variable that can be produced by various ways; the word *variable* here relates to the capacity to vary in response to these ways. In systems control applications, we are interested in those dynamic variables that require regulation in some sense. Sometimes, we call the regulated dynamic variables *controlled variables*. Typical examples are temperature, pressure, flow rate, force, level, light intensity, and humidity.

The primary objective of systems control application is to force some dynamic variable to stay close to some preselected path or pattern. To achieve this, corrective action must be constantly provided to enable the dynamic variable to approach the desired path. The term *regulation* defines this operation of path-following. We can say that systems control regulates a dynamic variable. When the desired path reduces to a specified value, regulation will correspond to value-maintenance.

From the foregoing considerations, it should be clear that a control system for the time being, consists of two main blocks: a plant (process) to be controlled and a controller that exerts a proper action on the plant for an input (command) signal. The output is the controlled variable. Figure 1.1 shows a typical control system.

The input signal is processed in one forward path to yield the output variable. This is frequently called an *open-loop* system since there is no loop, but rather a direct transmission from the input to the output. By construction it is evident that the open-loop system is unregulated in view of the preceding definition.

Figure 1.1 A typical control system (open-loop).

Introduction

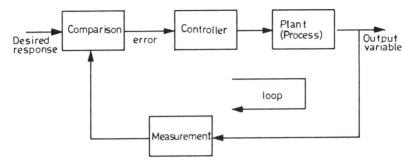

Figure 1.2 A closed-loop control system (single variable).

In contrast to an open-loop control system, a *closed-loop* control system utilizes an additional measure of the actual output in order to compare it with the desired response. This is shown in Fig. 1.2.

We emphasize two points. First, the feedback information is used to provide regulation, and the loop is closed. Second, after necessary adjustment of the controller, the system can run by itself without further intervention. It is for this reason that the term "automatic" was coined and used for a long time.

1.2 EXAMPLES OF CONTROL SYSTEMS

In this section, we shall examine several control systems with which the reader has some exposure and/or experience. The first example concerns a home-heating system, as a representative closed-loop system. With reference to Fig. 1.3, the output variable is the room temperature θ_o which is sensed by a thermostat. There is a reference temperature θ_s, sometimes called a "set point," against which the room temperature is compared. When the room temperature falls below the set point, the

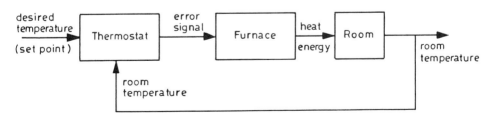

Figure 1.3 A home-heating system.

furnace (plant or process) is turned on. As a result, heat energy is released in the room causing the temperature to rise. This operation continues until the set point is reached when the furnace is turned off. Several important points should be clarified. In practice, there is a finite time for the rise-up or fall-down of room temperature. As we shall learn from Chapter 2, this is due to the finite capacity of the furnace as a thermal unit. Another point is that the thermostat has finite accuracy, and hence one should expect a nonzero value of the difference $2(\theta_s - \theta_o)$. Again, we shall analyze such behavior later.

As a second example let us consider the tank heater system shown in Fig. 1.4. A liquid enters the tank with a flow rate F_i (ft^3/min) and a temperature T_i (°F), where it is heated with steam having a flow rate F_s (lb/min). Let F and T be the flow rate and temperature of the stream leaving the tank. For simplicity, the tank is considered to be well stirred which implies that the temperature of the effluent is equal to the temperature of the liquid in the tank. One of the operational objectives of this heater could be to keep the liquid level h at a desired set point h_d when F_i changes. It is clear, though, that some form of control action is needed to alleviate the impact of the input disturbance. A possible configuration would be to measure the level of the liquid in the tank and then either

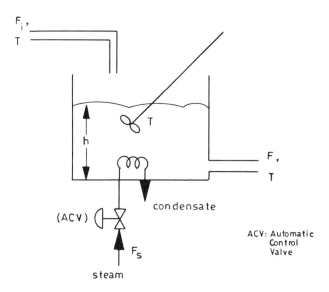

Figure 1.4 Stirred tank heater.

Introduction

1. Open or close the valve that affects the inlet flow rate F_i, or
2. Open or close the valve that affects the effluent flow rate F.

Both cases are drawn in Fig. 1.5. Note that case (1) represents an inlet control whereas case (2) stands for output control and in both cases the control law is driven by the error quantity $(h_d - h)$.

Another example concerns the conventional Ward-Leonard system, which is a configuration with a direct current (dc) generator driving an armature-controlled dc motor coupled to the load. A schematic diagram of the system is shown in Fig. 1.6(a).

Focusing on the motor, we can easily see that the armature circuit provides a closed-loop for the developed emf which means that a local feedback action has arisen quite naturally. This case will be explained later on in detail. In application, it is required to reduce the time constant (L/R); or equivalently speed up the system response. One way to accomplish this is through appropriate feedback signal formed by inserting a resistance R_a in series with the generator-armature circuit as shown in Fig. 1.6(b).

Note that the time constants of the field circuit and the motor generator loop are now $L_f/(R_f + R_a)$ and $(L_g + L_m)/(R_m + R_g + R_a)$, respectively.

A final point to stress is that the closed-loop control system in Fig. 1.2 is a basic disturbance-free configuration and therefore can be expanded to consider the following cases:

1. Multivariable control system in which more complex systems are considered and many controlled variables are analyzed; see Fig. 1.7
2. Feedforward control in which direct measurement of the disturbances is used to adjust the values of the manipulated variables; see Fig. 1.8
3. Estimator-based feedback control (inferential control) in which secondary measurements are used to adjust the values of the manipulated variables via feedback; see Fig. 1.9.

Many possible combinations of these cases occur in various system applications. It is significant to emphasize that, in all of the foregoing configurations, the controller is the active component that receives the information from the measurements and collects the data before taking appropriate control actions towards adjusting the values of the manipulated variables. More will be mentioned on every portion of the above discussions in the subsequent chapters.

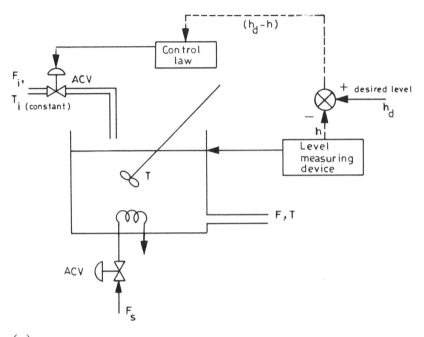

(a)

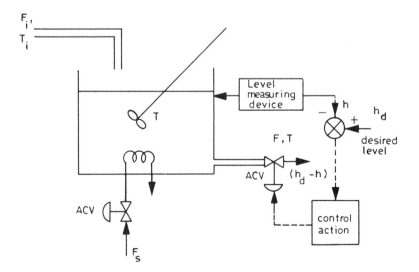

(b)

Figure 1.5 Liquid-level control schemes. (a) Inlet control. (b) Outlet control.

Introduction

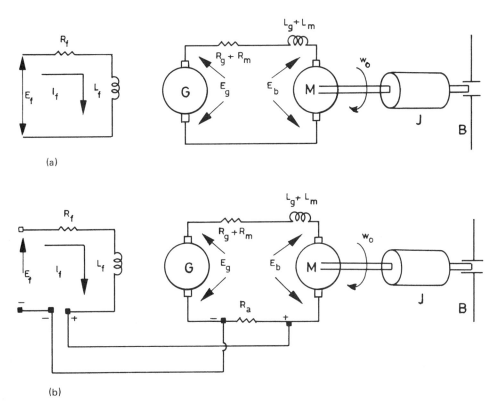

Figure 1.6 (a) Ward-Leonard system. (b) A Ward-Leonard system with feedback.

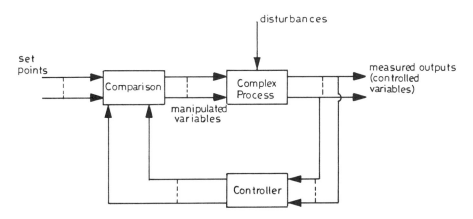

Figure 1.7 Multivariable control system.

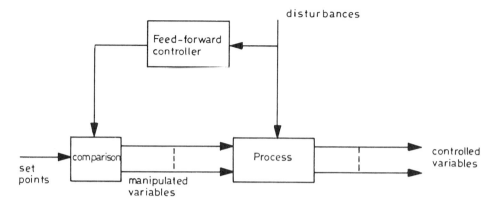

Figure 1.8 Feed-forward control system.

1.3 ANALOG VERSUS DIGITAL PROCESSING

The evolution of process and systems control has been the infusion of electronics technology into almost every facet because of

1. Low cost and reliability
2. Miniaturization, and
3. Ease of interface

The development of microelectronics and associated digital computer

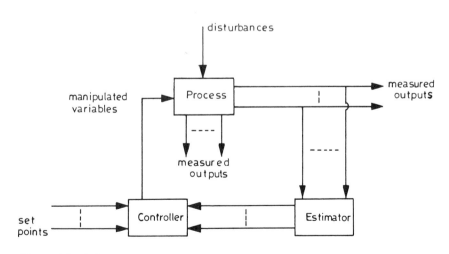

Figure 1.9 Inferential control.

Introduction

technology has brought about the rapid introduction of digital techniques in the field of industrial and process control. With reference to Fig. 1.2, this technological change has caused the transition from analog components and processing to digital components and processing. Just recall some aspects of control systems such as the initial transduction of a dynamic variable into electrical information which might always be of analog nature. It is inevitable, however, with the continued development of computing power, miniaturized digital electronics, and associated technology, that the evaluation and controller phase of systems control may now be digitally performed.

1.3.1 Analog Processing

In the sequel, we shall contrast the two methods of processing. An important feature of analog systems (components and processing) is to maintain continuous flow of energy through proper use of physical elements, transducers and actuators. The form of such components is system-dependent. For example, in thermal systems the use of thermistor (whose resistance is proportional to and an analog of temperature) and differential (operational) amplifiers provides a suitable way to regulate the temperature. The energy flow is of electric-type. In hydraulic systems, the use of hydraulic transmission (consists of a variable stroke hydraulic pump and a fixed stroke hydraulic motor) guarantees the development of large output torque and short response time. Here the flow is incompressible fluid. As distinct from hydraulic fluid, air medium (or other gases in special situations) is used in pneumatic control systems. Air medium has the advantage of being noninflammable and having almost negligible viscosity. The use of pneumatic bellows, relays and flapper valves are common in guided missiles and aircraft systems.

1.3.2 Digital Processing

We now turn attention to digital processing, in which all information carried in the control loop is *encoded* into an electrical signal that is binary in nature. Binary refers to a numbering system with a base of two, that is, 0 and 1 are the only possible counting states. The 0 state is often represented by a low level of voltage and the 1 state is a high level of voltage, both according to transistor-transistor logic (TTL) terminology. In application, the value of the dynamic variable is transformed into some *encoding* of the binary levels. The encoding itself is a correspondence between a set of binary numbers and the analog signal to be encoded. The set of binary numbers is commonly referred

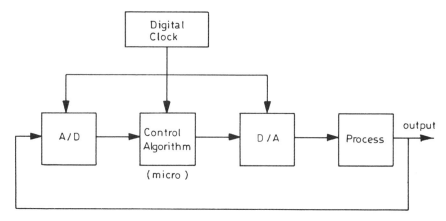

Figure 1.10 A schematic diagram of digital-processed control system.

to as a *word*, which may contain many binary counts, called *bits*. A prerequisite to encoding is the *sampling* of a continuous-time signal over discrete instants, which most of the time are equally spaced and hence called *periodic sampling*. To further enable the digital processing of control system, two main operations are needed: analog to digital converter (A/D) performing the function of conversion of an analog input to a digitally encoded signal, and digital to analog converter (D/A) providing the reverese action of converting the digitally encoded word into an appropriate analog output, that is, *decoding*. Here again, each bit, by design, will correspond to a certain level of output. A schematic diagram of digital processing in systems control is shown in Fig. 1.10.

Note that the use of a digital clock ensures synchronous operation of the converters. In general, there are two methods by which a digitally encoded signal can be transmitted through the systems control loop. One method, referred to as the *parallel transmission mode*, provides a separate wire (path or channel) for each binary number in the word. The alternate method is *serial transmission mode* of binary numbers over a single wire where the binary levels are provided in a time sequence over the wire.

1.4 MICROPROCESSORS AND MICROCOMPUTERS

The advent of microprocessors has enabled systems control methodologies to be widely applied in the process and manufacturing industries.

Introduction

Simply stated, microprocessors provide a unique opportunity for implementing control algorithms as economic alternatives to conventional analog controllers. This section briefly provides motivation, and outlines some aspects of microprocessors and microcomputers (micros for brevity).

1.4.1 Main Features

In the early 1970s, advances in semiconductor technology produced micros using large-scale integration (LSI). Recently, LSI technology has made it possible to fabricate a central-processing unit (CPU) or even an entire computer on a single integrated circuit (IC) or chip.

A *microprocessor* is a CPU that is usually implemented on the IC package. On combining micros with memory chips and input/output (I/O) devices, we obtain a *microcomputer*. Usually, micros operate in parallel transmission mode and are categorized as 4-bit, 8-bit, 16-bit, and 32-bit units. In view of its construction, the microprocessor is a digital device capable of receiving binary information, processing it in accordance with stored instructions and delivering output binary information. Micros contain a number of storage locations that they use for holding current data, current instructions, the address of any peripheral device with which they are due to communicate and the address of the next line of the program. These storage locations constitute a device, usually called *memory*, of which the following types are identified:

1. *Random access memory* (RAM) is used as general-purpose working space during computation and data transfer. It is volatile in the sense that the information that it contains is lost when the electrical supply is removed. Contents of words can be read and also altered at specific addresses.
2. *Read only memory* (ROM) is a rapid source of information of the type that never needs to be modified. Contents of words, which cannot normally be altered, may be also required for storing programs or data that do not change and must always be available to the microprocessor.
3. *Programmable read only memory* (PROM) serves rather similar purposes and it is suitable for on-off applications. This is due to the possibility of programming it in a desirable manner.
4. *Erasable programmable read only memory* (EPROM). The major disadvantage of PROMs is that they can be programmed only once. If program/data values have to be altered, a new PROM must be used. This difficulty can be overcome using EPROM in which the

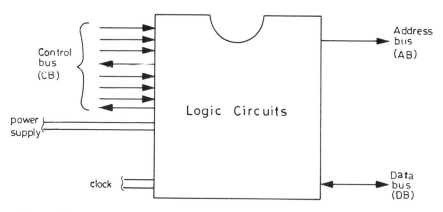

Figure 1.11 A generalized microprocessor.

contents of each memory location can be erased and reprogrammed.

All memory types are IC packages. In order to facilitate the transfer of data and instructions between the micro, memory and I/O devices, the following major buses (groups of lines or pins sharing information) are generally provided:

1. Address bus
2. Data bus
3. Control bus

A schematic of a generalized micro is given in Fig. 1.11 to show these buses.

The clock is to ensure synchronization among the components of the micro using suitable circuits and cycle generation. The size of DB, AB, and CB and the type of control signals depend on the method of fabricating the micro. Bidirectional buses are sometimes used to combine data input and data output buses.

1.4.2 Architectures

A generalized micro consists of several, digital building blocks, the most common of which are:

1. Arithmetic and logic unit (ALU)
2. Accumulator(s) (ACC)
3. Instruction register (IR) and memory address buffer register (MAR)

Introduction

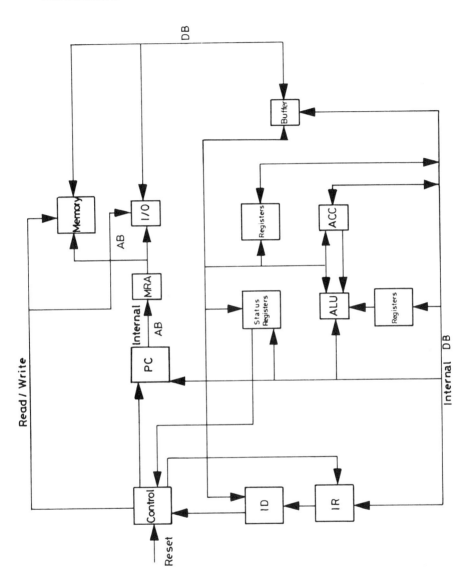

Figure 1.12 Internal architecture of a generalized micro.

4. Instruction decoder (ID)
5. Program counter (PC)
6. Registers (status and general-purpose)

Organization of these blocks differs from one micro to another. An internal architecture of a generalized micro is displayed in Fig. 1.12. Briefly, data processing is performed by the ALU and correct information routing is timely provided by the control unit (CU). The ID decodes instructions, which are temporarily stored in high-speed registers. The main memory contains instructions and data.

An *instruction set* is the list of operations that the micro is designed to execute. Common types of instructions are arithmetic and logic, data transfer, machine control, stock, and transfer of program control. Data-transfer instructions cause data to be transfered among CPU registers and memory locations. Stack operations, push and pop instructions, are included to facilitate the transfer of program control.

Instructions are usually fetched from the main memory and executed in the micro in two separate consecutive steps called the *fetch cycle* and *execution cycle*. Together they form the *instruction cycle*. The fetch cycle is common to all instructions whereas in the execution cycle, the actions inside the micro depend on the particular instruction. More details about the various aspects will be presented in Chapter 9.

1.4.3 Programming

The interfacing of the micro to real world devices is only the first stage in the development of a microprocessor-based or microcomputer-based system. Once the micro and the devices have been correctly assembled and interfaced, the microprocessor must be programmed to carry out the necessary functions. Four different types of programs (codes) are used:

1. *Machine code* is the actual code executed on the microprocessor and is expressed in a sequence of bits.
2. *Hex code* is a possible alternative to machine code using hexadecimal numbers (base 16 using numeric digits 0 to 9 and letters A to F). In this case, every four bits can be combined and represented by an equivalent hex code.
3. *Assembly language* further simplifies the task of writing application programs by using mnemonics to represent each basic instruction.
4. *High-level languages* are general purpose, application-oriented programs that can be used in conjunction with translator programs (compilers). In general, the preparation of an application program

Introduction

is easy and less time consuming. Popular examples of compiled high-level languages are FORTRAN, COBOL, PASCAL and C. In normal practice only the latter two, that is, assembly and high-level languages are used, particularly for microcomputer environment.

1.4.4 Applications

The evolution of micros has taken place in a sequence of generations:

1. First generation (1971–1972) were 4-bit and 8-bit p-channel MOS microprocessors.
2. Second generation (1973–1978) were mainly 8-bit n-channel, with increased instruction execution speed, comprehensive hardware/software, and compatible peripheral chips.
3. Third generation (1978–1981) marked the beginning of 16-bit microcomputers (supermicros) by speeding up the memory cycle and increasing the data bus width. Several sophisticated architectural features have been also added.
4. Fourth generation (1981–now) have incorporated hardware and software facilities that yield more powerful machines. Some manufacturers use 32-bit CPUs.

The key to successfully using micros in practical applications is via the software. These include development facilities for standard programs such as editor, assembler/compiler/interpreter, simulator, and debugger. In view of their small size, low cost, flexibility, and fast processing of data, microprocessors have been used in microcomputer applications that previously employed hardwired logic. The wide spectrum of applications ranges as follows:

1. *Industrial process systems.* Micros are now used for implementing process-control algorithms as economic alternatives to conventional analog controllers. Examples include Intel 8748, Rockwell R6500/1, Honeywell TDC3000, MICON MDC-200, KENT P4000 and Motorola 6800.
2. *Manufacturing systems.* The inherent advantages of micros are involved in recent manufacturing operations such as numerical control of flexible automation, machine tools, on-line inspection, automatic testing and quality control.
3. *Distributed processing systems.* Micros are ideal components for distributed processing systems which play an important role in production organization. An alternative to centralized computer process control systems is the increasingly popular concept of dis-

tributed processing, which aims at providing the required computing power at the point of use. In such systems, the control functions are divided and distributed to a number of processors linked in a network.

These applications are discussed in detail in Chapters 11 and 12.

1.5 COMPUTER-CONTROLLED SYSTEMS

A computer-controlled system (CCS) essentially considers a microcomputer as an element in the control loop. One tentative structure of the CCS would resemble Fig. 1.10 by replacing the control algorithm by a micro. This micro allows compensation and control improvement.

1.5.1 Basic Structure

A basic configuration of a CCS is shown in Fig. 1.13. It contains the following main functional elements.

1. *Sensors/transducers* for measuring physical parameters such as pressure, temperature, velocity, position and torque; then converting any of them into an appropriate electrical signal.
2. *Actuators* to deliver the proper output signal to the physical plant
3. *A/D and D/A converters* for converting the analog data into digital format (A/D) and digital data into analog format (D/A)
4. *Multiplexers and demultiplexers* to manipulate large quantities of data and organize the flow of information from the sensors without incurring heavy system overheads
5. *Miscellaneous elements* such as amplifiers, signal conditioning, and filters.

1.5.2 Some Remarks

The inclusion of a digital computer (micros, mini or mainframe) in the systems control loop is expected to introduce some errors, delays and discontinuities. It is the task of the system designer to alleviate the sources of such problems and provide adequate solutions to them. Here, we shed some light on the difficulties encountered.

One of the problems results from the finite operation of an A/D converter, finite word length, limited conversion time, and a particular sampling rate. The second problem comes from the need to discretize known continuous functions. Some tools have to be worked out to eliminate several unpleasant factors that tend to lower the system performance.

Introduction

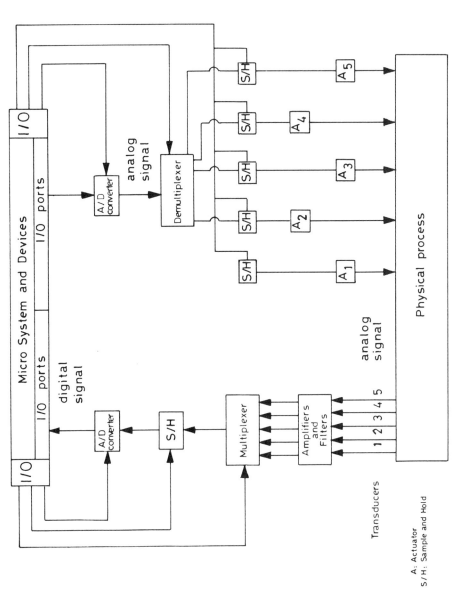

Figure 1.13 A basic configuration of a computer-control system.

In addition to these considerations, delays and quantization noise brought about by digital devices have to be carefully examined. Both can be reduced in return for additional hardware costs. The issue becomes that of system performance versus complexity. Note that there is a mix of signals (digital and analog) for which twin problems always arise.

Another significant point is that of real-time processing. This demands constraints on the order of computation and the organization of processing. Techniques for multitasking and parallel-processing are relevant with regard to sequential algorithms.

Indeed the method of tackling the design problems and issues (whether global or component-wise) deserves systematic analysis and evaluation to save subsequent efforts and/or adaptation. These aspects are considered in subsequent chapters.

1.6 OUTLINE OF THE BOOK

The present volume aims to cover the material necessary for

1. Establishing a strong theoretical background about systems control modeling and dynamics
2. Understanding the internal architecture of micros
3. Developing microcomputer-based systems for use in interactive design and simulation

The book is organized into three parts: Part I on analog control systems; Part II on digital control systems; Part III on microcomputer-based control systems.

Part I includes Chapters 2 through 5. In Chapter 2, concepts of modeling and dynamics of analog control systems are presented. Then follows the state variable analysis and assessment of plant response by selected performance indices in Chapter 3. We continue the study on analog systems in Chapter 4 by examining stability criteria and the associated frequency response methods. Chapter 5 is devoted to the design problem using different techniques.

Part II includes Chapters 6 through 8. Chapter 6 discusses the methods of converting continuous models to discrete ones and studies the various factors involved. In Chapter 7 control design algorithms are developed. Optimization methodologies are the subject of Chapter 8 for both deterministic and stochastic cases.

Part III includes Chapters 9 through 12. Hardware principles and elements are described in Chapter 9, and software support and interfac-

Introduction

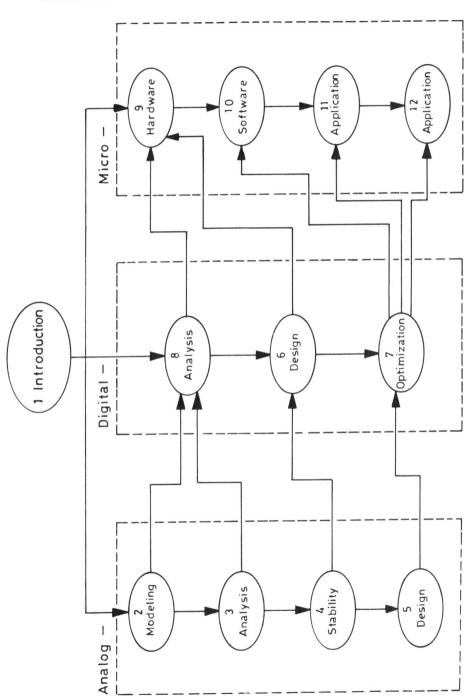

Figure 1.14 Structure of the book.

ing are contained in Chapter 10. Chapters 11 and 12 document case studies and practical characterization of microcomputer control systems.

Structuring the book into three constructive parts serves many objectives:

1. It keeps the identity of every stage in the development of control systems.
2. Each part constitutes a concise and self-contained course in the selected subject.
3. It focuses on the important design issues and supplies implementable solutions.

A block diagram showing the interrelationships among different parts and/or chapters of the book is displayed in Fig. 1.14. It helps in drawing some useful chapter-dependence to form specific domains of knowledge within the systems control arena.

For convenience, a list of references and textbooks organized in alphabetical order is included at the end of this volume. However, at the end of each chapter some notes are presented to assist the reader in selecting further reading.

NOTES AND REFERENCES

Recent orientation of control engineering information and basic tools can be found in several books. Kuo (1982) contains an updated version of the fundamental techniques. With a wider scope of systems engineering applications, Dorf (1980) presents an excellent summary of modern control concepts. For a first one-semester course on control engineering, both Leigh (1982) and El-Hawary (1984) are evenly recommended. Given the direction of systems control focus on process and chemical engineering subjects, basic knowledge can be obtained from Weber (1973) and advance treatment can be found in Stephanapoulos (1984). The subject of fundamentals and applications of micros is quite large and can be accommodated in more than one volume. However, for a modest coverage of basic facts and illustrative case studies, the reader is advised to consult Clements (1982), Fletcher (1984), Gibson and Lin (1981), Ashon (1984), and Kochhar and Burns (1983), among many others. By the same token, subjects related to computer-based systems control are quite vast and can be exposed in different ways. Major titles dealing with these subjects are Katz (1981), Mahmoud and Singh (1984), Iserman (1981), Astrom and Wittenmark (1984), Benett (1988), and Leigh (1985).

2
Systems Control Modeling and Representation

In our endeavor to understand, manage, and control real-life systems, an initial and important step would be to obtain quantitative pictures about these systems. The most common method of representing systems behavior has been through *mathematical models*. These models should describe the relationships among the various system variables under some assumptions concerning the system operation. In the modeling of lumped continuous-time dynamic systems, the descriptive equations are usually *differential equations*. Furthermore, if these equations can be *linearized*, then the *Laplace transform** can be utilized in order to simplify the method of solution. System characterization via Laplace transform is referred to as the *transform function* and it is the subject of the next section.

2.1 TRANSFER FUNCTIONS

Suppose that the input and output of a system are related by a linear ordinary differential equation (ODE) with constant coefficients. An

---

*Readers unfamiliar with the Laplace Transform should consult Appendix A where properties of the transform needed for this chapter are developed.

equivalent statement is that the system is linear and time-invariant (LTI) which corresponds to lumped characterization of physical engineering models. On taking the Laplace transform of both sides of the ODE, in the manner developed in Appendix A, the result is that the transform of the output is linearly related to the transform of the input and linearly related to the initial conditions also.

Let $U(s)$ and $C(s)$ be the transform of the input to and the output of the LTI, respectively. Then $C(s)$ is

$$C(s) = G(s)U(s) + \text{(initial condition terms)} \tag{1}$$

The initial condition terms are absent if the initial conditions are zero. In this case we define

$$G(s) \triangleq \frac{C(s)}{U(s)}, \quad \text{initial conditions} = 0 \tag{2}$$

as the *transfer function* of the LTI system. It is customary to use (2) in evaluating $G(s)$ and then incorporating (1), one obtains the output response.

Example 2.1. For the mechanical translational system shown in Fig. 2.1, which consists of a spring-mass-dashpot combination, the force balance equation is

$$M\ddot{x} + B\dot{x} + Kx = F$$

where M is the mass, K is the stiffness (elastance) of the spring, B is the viscous friction (dumping) and F is the applied force. This is a

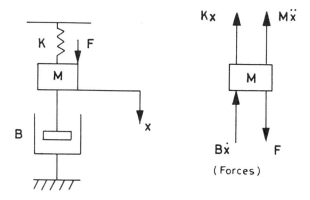

Figure 2.1 A spring-mass-dashpost combination.

Systems Control Modeling and Representation

second-order ODE. Applying Laplace transform we obtain

$$M[s^2X(s) - sx(0) - \dot{x}(0)] + B[sX(s) - x(0)] + KX(s) = F(s)$$

On collecting terms and rearranging

$$X(s)(s^2M + sB + K) + (-sM - B)x(0) - M\dot{x}(0) = F(s)$$

or

$$F(s) = (s^2M + sB + K)X(s) - [(sM + B)x(0) + M\dot{x}(0)]$$

A simple comparison with (1) gives

$$G(s) \triangleq \frac{X(s)}{F(s)} = \frac{1}{s^2M + sB + K}$$

as the transfer function of the (spring-mass-dashpot) combination. In practice, this combination represents a mechanical system on which we will elaborate later on.

Example 2.2. Consider the RC network of Fig. 2.2. We can write the impendance of the series arm

$$Z_1 = \frac{R_1/sC_1}{R_1 + (1/sC_1)}$$

$$= \frac{R_1}{1 + sC_1R_1}$$

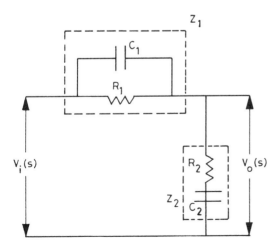

Figure 2.2 A lead-lag RC network.

Also the impedance of the shunt arm

$$Z_2 = R_2 + (1/sC_2)$$
$$= \frac{1 + sC_2R_2}{sC_2}$$

As an impedance divider, the ratio $V_o(s)/V_i(s)$ is given by

$$\frac{V_o(s)}{V_i(s)} = \frac{Z_2}{Z_1 + Z_2}$$

In terms of the time constants $\tau_1 = R_1C_1$, $\tau_3 = R_1C_2$, $\tau_2 = R_2C_2$, the above ratio can be simplified into the form

$$G(s) = \frac{V_o(s)}{V_i(s)}$$
$$= \frac{(1 + \tau_1 s)(1 + \tau_2 s)}{1 + (\tau_1 + \tau_2 + \tau_3)s + \tau_1\tau_2 s^2}$$

which, as will be shown later, represents a lead-lag compensator.

The foregoing examples illustrate the concept of transfer function. We should stress the fact that Laplace-transforming the linear ODEs converts them into algebraic equations in the complex variable s. This provides a very convenient representation of system dynamics. For the time being, we should keep in mind that the input-output behavior of a linearized system, or element of a linearized system, is given by $G(s) = C(s)/R(s)$.

2.2 BLOCK DIAGRAM ALGEBRA

A convenient graphical representation of the relationships between linear system variables is offered by the *block diagram* as shown in Fig. 2.3(a), wherein the signal into the block represents the input $R(s)$ and the signal out of the block represents the output $C(s)$, while the block itself stands for the transfer function $G(s)$. The information (signal) flows unidirectionally from the input to the output. In view of Example 2.2, the output is always equal to the input multiplied by the transfer function. When dealing with interconnected or multivariable systems, we expect the resulting block diagram to be a complex configuration of individual blocks. These blocks are connected by lines with arrows indicating the signal flow. In addition, *summing* points and *take-off*

Systems Control Modeling and Representation

(a)

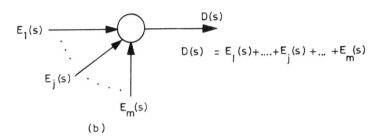

$$D(s) = E_1(s) + \ldots + E_j(s) + \ldots + E_m(s)$$

(b)

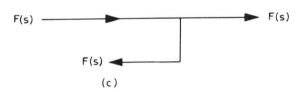

(c)

Figure 2.3 Ingredients of a block diagram. (a) Block. (b) Summing point. (c) Take-off point.

points are used for signals manipulation as shown in Figs. 2.3(b) and (c), respectively.

2.2.1 A Closed-Loop System

A fundamental block diagram configuration is the single-variable closed-loop system shown in Fig. 2.4(a), from which we can write

$$C(s) = G(s)E(s) \tag{3}$$

$$E(s) = R(s) - B(s) = R(s) - H(s)C(s) \tag{4}$$

Eliminating $E(s)$ using (3), we have

$$C(s) = G(s)[R(s) - H(s)C(s)]$$

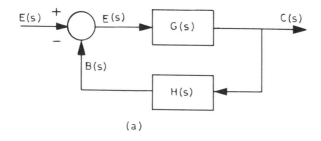

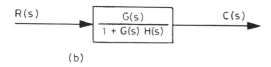

Figure 2.4 A closed-loop system. (a) Single-variable. (b) Equivalent block.

or

$$T(s) = \frac{C(s)}{R(s)} = \frac{G(s)}{1 + G(s)H(s)} \tag{5}$$

Therefore, the system shown in Fig. 2.4(a) can be reduced to a single block shown in Figure 2.4(b).

2.2.2 Some Reduction Rules

In this section, we introduce some rules that can be used to simplify block diagram configurations. A summary of the important rules is given in Table 2.1. We take note that all these rules are derived by simple algebraic manipulations of the relations associated with the blocks. The following example illustrates the use of the reduction rules.

Example 2.3. It is required to obtain the ratio C/R for the system shown in Fig. 2.5. We first move point B to coincide with A by applying rule 5. Then we eliminate the feedback loop at the left (input) and the summing point at the right by applying rules 6 and 7, respectively. Finally, we apply rule 1 to obtain the desired results. The steps are shown in Fig. 2.6.

Systems Control Modeling and Representation

Table 2.1 Reduction Rules of Block Diagram

Rule	Original Diagram	Equivalent Diagram
1. Combining blocks in cascade	$R_1 \to [G_1] \to R_1G_1 \to [G_2] \to R_1G_1G_2$	$R_1 \to [G_1G_2] \to R_1G_1G_2$
2. Moving a summing point ahead	$R_1 \to \oplus \to [G] \to G(R_1-R_2)$, with $-R_2$ input	$R_1 \to [G] \to \oplus \to G(R_1-R_2)$, with $R_2 \to [G]$ input
3. Moving a summing point backward	$R_1 \to [G] \to \oplus \to (R_1G-R_2)$, with R_2 input	$R_1 \to \oplus \to [G] \to (R_1G-R_2)$, with $R_2 \to [1/G]$ input
4. Moving a take-off point ahead	$R_1 \to [G] \to R_1G$, branch R_1	$R_1 \to [G] \to R_1G$, branch $R_1G \to [1/G] \to R_1$
5. Moving a take-off point backward	$R_1 \to [G] \to R_1G$, branch R_1G	$R_1 \to [G] \to R_1G$, branch $R_1 \to [G] \to R_1G$
6. Eliminating a feedback loop	$R_1 \to \oplus \to [G] \to R_2$, feedback $[H]$	$R_1 \to [\frac{G}{1-GH}] \to R_2$
7. Eliminating a summing point	$R_1 \to [G_1] \to \oplus \to R_2$, with $[G_2]$ branch	$R_1 \to [G_1+G_2] \to R_2$

2.2.3 Linear Multivariable Systems

When multiple inputs are present in a linear system, each input can be treated independently and the complete output can then be obtained by the principle of superposition. In the case of the linear multivariable system depicted in Fig. 2.7 (r inputs and m outputs), the jth output $C_j(s)$ is given by

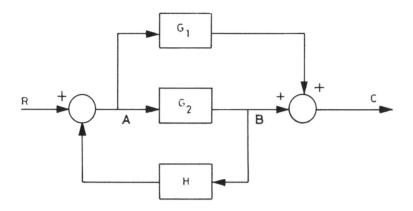

Figure 2.5 System of Example 2.3.

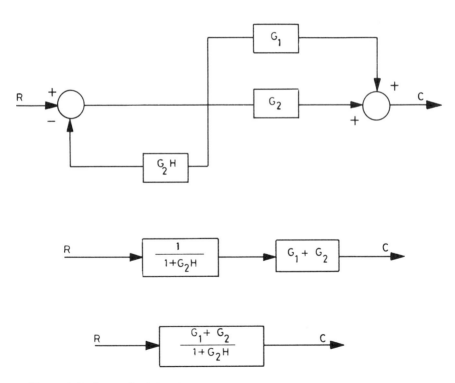

Figure 2.6 Steps of solving Example 2.3.

Systems Control Modeling and Representation

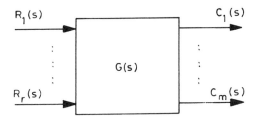

Figure 2.7 A linear multivariable system.

$$C_j(s) = \sum_{k=1}^{r} G_{jk}(s)R_k(s), \qquad j = 1, \ldots, m \qquad (6)$$

where $R_k(s)$ is the kth input and G_{jk} is the transfer function between the kth input and the jth output with all other inputs set to zero. Indeed, when calculating G_{jk}, we are generally going to employ the reduction rules of Table 2.1.

2.3 ELEMENTARY PHYSICAL SYSTEMS MODELS

As indicated in Chapter 1, a closed-loop control system is composed of devices and components that suit the application at hand. To help in describing the dynamic performance of a control system based on a physical process, one usually utilizes the physical laws of the process. This approach applies well to electrical, fluid, mechanical, and thermal systems. In Table 2.2, a summary of the variables of dynamic systems is given. We note the variables are classified into a *through* variable (corresponding to an intensive quantity giving the flux of energy flow) and an *across* variable (corresponding to an extensive quantity giving the pitch of the energy flow). Sometimes, *through* and *across* variables are termed *flow* and *effort* variables, respectively. As we shall see, mathematical models can be found for a system by the application of one or more fundamental laws peculiar to the physical nature of the system or component. For example, electrical systems use Kirchoff's and Ohm's laws or equivalent laws; mechanical (translational and rotational) systems use Newton's law and the d'Alembert principle; and finally, hydraulic systems use flow and continuity equations. Regardless of the nature of the system, however, the application of any of these laws yields ODEs that have the same basic form. This leads us to consider them as analogous systems; a comparison of the variables involved is given in Table 2.3.

Table 2.2 Summary of System Variables in Physical Processes

System	Through Variable (flow)	Integrated Through Variable	Across Variable (effort)	Integrated Across Variable
Electrical	Current	Charge	Voltage difference	Flux linkage
Mechanical (translational)	Force	Translational momentum	Velocity difference	Displacement difference
Mechanical (rotational)	Torque	Angular momentum	Angular velocity difference	Angular displacement difference
Fluid	Volumetric flow rate	Volume	Pressure difference	Pressure momentum
Thermal	Heat flow rate	Heat energy	Temperature difference	

Table 2.3 Comparison of Variables in Analogous Systems

System	Effort	Flow	Dissipation	Capacitive	Inductive
Electrical	Voltage v	Current i	Resistance R	Capacitance C	Inductive L
Mechanical (translation)	Velocity v	Force F	Viscous friction B	Mass M	Stiffness K
Mechanical (rotation)	Angular velocity ω	Torque T	Viscous friction B	Moment of inertia J	Stiffness K
Hydraulic	Pressure P	Volumetric flow q	Fluid resistance R	Fluid capacitance C	Fluid inertance I
Thermal	Temperature T	Heat flow q	Thermal resistance R	Thermal capacitance C	–

Most of the time, linearized relationships are invoked to facilitate further development. In the following sections, we describe mathematical models of some physical systems with particular focus on components and elements for use in control engineering.

2.3.1 Electrical Systems

Some of the important electrical components commonly used in systems control applications are as follows.

Servomotors

In electrical control systems, the power devices are either ac or dc servomotors. Ac servomotors are best suited for low power applications. They are rugged, light in weight and have no brush contacts as is the case with dc servomotors. In practice, the error signal derived from transducers (sometimes called synochro) can be amplified by ac amplifiers to produce a control signal for servomotors.

Ac Servomotors. An ac servomotor is basically a two-phase induction motor with high rotor resistance to guarantee a linear torque-speed relationship. One of the phases, known as the reference phase, is excited by a constant voltage and the other phase, known as the control phase, is energized by a voltage which is 90° out of phase with respect to the voltage of the reference phase. A circuit diagram is shown in Fig. 2.8.

The control phase voltage is supplied from a servo-amplifier and it

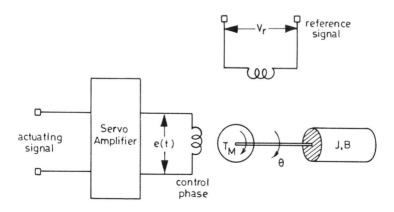

Figure 2.8 A schematic diagram of two-phase servomotor.

Systems Control Modeling and Representation

has a variable amplitude and polarity ($\pm 90°$ phase angle with respect to the reference phase). The direction of rotation of the motor reverses as the polarity of the control phase signal changes sign. From symmetrical components of electrical machines, it can be verified that the starting torque of a servomotor under unbalanced operation and low speed, is a function of both the speed $\dot{\theta}_M$ and the rms control voltage E_c

$$T_M = f_M(\dot{\theta}_M, E_c) \qquad (7)$$

Expanding (7) in a Taylor's series and retaining first-order terms with $\Delta T_M = T_m$, $\Delta E_c = E_m$, $\Delta \dot{\theta}_M = \dot{\theta}_m$ we can write

$$T_m = K E_m - f \dot{\theta}_m \qquad (8)$$

where

$$K \triangleq \frac{\partial T_M}{\partial E_c} \quad \text{and} \quad f \triangleq \frac{-\partial T_M}{\partial \dot{\theta}_M} \qquad (9)$$

If the load consists of inertia J and viscous friction B, we can write the torque relation in the form

$$T_m = J\ddot{\theta}_m + B\dot{\theta}_m = KE_m - f\dot{\theta}_m \qquad (10)$$

Taking the Laplace transform of (10) and dropping the effect of initial conditions, we arrive at

$$(s^2 J + sB)\theta_m(s) = KE_m(s) - fs\,\theta_m(s)$$

from which we define the motor transfer function as

$$G_m(s) = \frac{\theta_m(s)}{E_m(s)}$$

$$= \frac{K}{s^2 J + s(f + B)}$$

$$= \frac{K_m}{s(s\tau_m + 1)} \qquad (11)$$

where $\tau_m = J/(f + B) =$ motor time constant and $K_m = K/(f + B) =$ motor gain constant.

Dc Servomotors. In servo applications, a dc motor is required to produce rapid accelerations from standstill. This demands low inertia and high starting torque. Low inertia is attained with reduced armature diameter and an increase in armature length. In systems control, dc motors are used in either of two control modes, armature control mode (fixed field) and field control mode (fixed armature current).

Armature Control. In terms of the armature-controlled dc motor shown in Fig. 2.9, we express the air gap flux Φ, assuming linear portion of the magnetization curve, as

$$\Phi = K_f i_f, \qquad K_f \text{ is constant} \tag{12}$$

The torque T_m developed by the motor is proportional to the product of the armature current and air gap flux

$$T_m = K_i K_f i_f i_a, \qquad K_i \text{ is constant} \tag{13}$$

Keeping i_f constant (fixed field), (13) reduces to

$$T_m = K_t i_a \tag{14}$$

where K_t is the motor torque constant. The remaining electromechanical relations are

$$e_b = K_b \frac{d\theta}{dt} \tag{15}$$

$$L_a \frac{di_a}{dt} + R_a i_a + e_b = e \tag{16}$$

$$J \frac{d^2\theta}{dt^2} + B \frac{d\theta}{dt} = T_m \tag{17}$$

where K_b is the back emf constant.

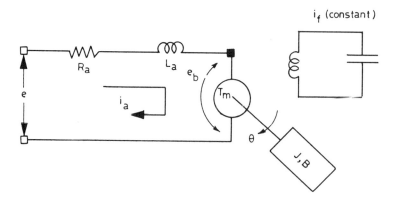

Figure 2.9 Armature-controlled dc motor.

Systems Control Modeling and Representation

Assuming zero initial conditions, and taking the Laplace transform of (14)–(17), we obtain the motor transfer function

$$G_m(s) \triangleq \frac{\theta(s)}{E(s)}$$

$$= \frac{K_t}{s[(R_a + sL_a)(sJ + B) + K_t K_b]}$$

Usually L_a is negligible and hence we can put $G_m(s)$ in the form

$$G_m(s) = \frac{K_t/R_a}{s\{sJ + [B + (K_t K_b/R_a)]\}} \tag{18}$$

which is identical to (11) where $K_m = K_t/[R_a B + K_t K_b]$, $\tau_m = J/[B + (K_t K_b/R_a)]$ signify the motor gain and motor time constants, respectively.

Field Control. A field-controlled dc motor is shown in Fig. 2.10(a) where the armature current in this case is fed from a constant current source.

With reference to (13), we can write

$$T_m = K_d i_f \tag{19}$$

where K_d is constant. It is then easy to write

$$e = R_f i_f + L \frac{di_f}{dt} \tag{20}$$

from the field circuit and

$$J \frac{d^2\theta}{dt^2} + B \frac{d\theta}{dt} = T_m = K_d i_f \tag{21}$$

Taking the Laplace transform of (20) and (21), assuming zero initial conditions, we obtain

$$(sL_f + R_f)I_f(s) = E(s)$$

$$(s^2 J + sB)\theta(s) = K_d I_f(s)$$

from which the transfer function of the motor is obtained as

$$G_m(s) = \frac{\theta(s)}{E(s)}$$

$$= \frac{K_d}{s(sL_f + R_f)(sJ + B)}$$

$$= \frac{K_m}{s(s\tau_f + 1)(s\tau_{me} + 1)} \tag{22}$$

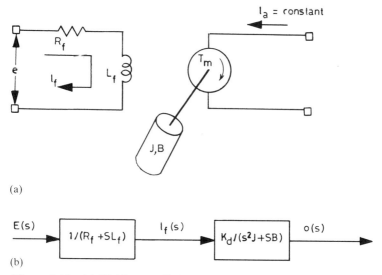

(a)

(b)

Figure 2.10 (a) Field-controlled dc motor. (b) Block diagram.

and shown in Fig. 2.10(b). The constants $K_m = K_d/BR_f$, $\tau_f = L_f/R_f$ and $\tau_{me} = J/B$ are of the motor gain, the field circuit, and the mechanical load.

In practice, field control is used advantageously for small-size motors while armature control can be economically employed for large-size motors to benefit from back emf effects on damping.

AC Tachometer

Here also, the field stator coils are mounted at right angles to each other (space quadrature). The rotor is of light inertia and high conductivity. From the theory of machines (reference to Fig. 2.11), it can be shown that the tachometer action output voltage proportional to angular velocity is expressed as

$$V_t = K_t \frac{d\theta}{dt} = K_t \omega \tag{23}$$

where K_t is the tachometer constant.

There are other arrangements of components which are frequently used in control applications. These include *stepper motors* which are electromechanical devices that actuate a train of step angular (or linear) movements in response to a train of input pulses on a one to one basis;

Systems Control Modeling and Representation

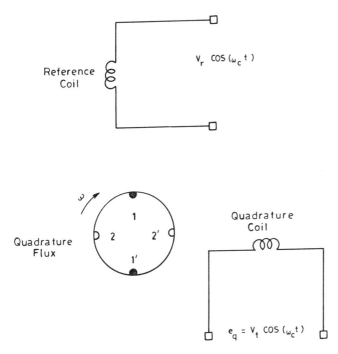

Figure 2.11 AC tachometer.

that is one step actuation for each pulse input. In practice, a stepper motor is the actuator element of incremental-motion control systems such as computer peripherals (printers, tap drives), machine tool, and digital process control systems. The two most widely used types of stepper motors are

1. Variable reluctance motor
2. Permanent magnet motor

In applications where a control signal is required to have a power level higher than the capability of linear electronic amplifiers, *rotating amplifiers* or *cross field machines* are the ideal choice. An ordinary dc generator, in fact, is such an amplifier where a field voltage at a lower power level ($V_f i_f$) controls a large armature power ($V_a i_a$). An *amplidyne* is such a cross field machine, specially suited for control applications. Briefly stated, the amplidyne is equivalent to accommodating two stages of power amplification in a single machine frame.

2.3.2 Hydraulic Systems

Hydraulic feedback systems utilize components that operate on hydraulic principles. In these components, power is transmitted through the action of fluid flow under pressure.

Hydraulic Transmission

When a large torque is required in a control device, it is possible to use a hydraulic transmission, shown in Fig. 2.12. It consists of a variable stroke hydraulic pump and a fixed stroke hydraulic motor. Control of the motor is exercised by varying the amount of oil delivered by the pump. This is carried out by mechanically changing the pump stroke. We note that hydraulic pump and motor are dual in operation; that is, in a pump, the input is mechanical power (torque at a certain speed) and output is hydraulic power (flow at a certain pressure) and in a motor, the opposite is true.

To derive the transfer function of the hydraulic pump-motor system, we define q_p as the total volumetric flow rate from the pump. This equals the sum of the flow rates in the system. Thus

$$q_p = q_m + q_k + q_c \qquad (24)$$

where q_m = volumetric flow rate through the motor, q_k = volumetric leakage flow rate of both pump and motor, and q_c = compressibility flow rate. These rates are given by

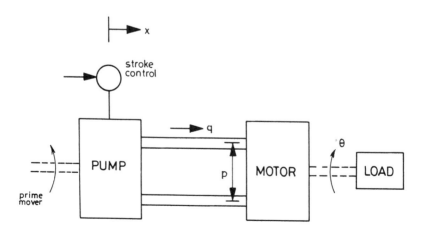

Figure 2.12 Hydraulic transmission.

Systems Control Modeling and Representation

$$q_p = K_p x \quad (25)$$

$$q_m = K_m \frac{d\theta}{dt} \quad (26)$$

$$q_k = K_k p \quad (27)$$

$$q_c = K_c \frac{dp}{dt} \quad (28)$$

where K_p = a constant (the rate of volumetric flow per unit stroke angle), x = the stroke angle, K_m = motor displacement constant, K_k = a constant (leakage coefficient of the complete system), and K_c = coefficient of compressibility. From (24)–(28) we obtain

$$K_p x = K_m \frac{d\theta}{dt} + K_k p + K_c \frac{dp}{dt} \quad (29)$$

For an inertial load within viscous friction, the developed torque T_m is

$$T_m = K_m p$$

$$= J \frac{d^2\theta}{dt^2} + B \frac{d\theta}{dt} \quad (30)$$

Combining (29) and (30) and eliminating p we obtain

$$K_p x = \frac{K_c J}{K_m} \frac{d^3\theta}{dt^3} + \frac{[K_c B + K_k J] \frac{d^2\theta}{dt^2}}{K_m} + \frac{K_m + K_k B}{K_m} \frac{d\theta}{dt} \quad (31)$$

whose Laplace transform results in the transfer function

$$G(s) = \frac{\theta(s)}{X(s)}$$

$$= \frac{K_p}{s[s^2(K_c J/K_m) + s(K_c B + K_k J)/K_m + (K_m + K_k B/K_m)]} \quad (32)$$

Normally, $K_c \ll K_m$ and therefore (32) may be simplified into

$$G(s) = \frac{\theta(s)}{X(s)}$$

$$= \frac{K_p}{s[s(K_k J/K_m) + (K_m + K_k B/K_m)]}$$

$$= \frac{K}{s(\tau_p s + 1)} \quad (33)$$

with $K = K_p/[K_m + (K_k B/K_m)]$ and $\tau_p = JK_k/(K_k B + K_m^2)$.

It is interesting to observe the analogy of (33) with (11), as both have the same structure and hence same properties.

Hydraulic Linear Actuator

The linear actuators are piston devices. A simple hydraulic actuator is shown in Fig. 2.13 in which the motion of the spool regulates the flow of fluid (mainly oil) to either side of the power cylinder. Note that when the spool moves to the right, the oil from the high pressure source enters into the power cylinder, on the left of the power piston. This in turn creates a differential pressure across the piston which causes the power piston to move to the right, pushing the oil in front of it to the sump. The oil is pressurized by a pump and is recirculated in the system. The load rigidly coupled to the piston moves a distance y from its reference position in response to the displacement x of the valve spool from its neutral position.

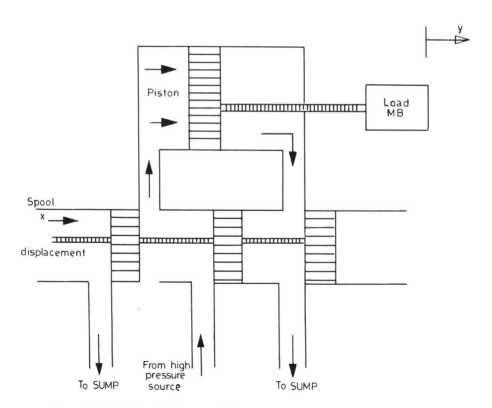

Figure 2.13 Hydraulic linear actuator.

Under the assumption that the relationship between the volumetric oil flow rate q into the power piston and the differential pressure p for small values of spool displacement x is linear, then

$$q = K_1 x - K_2 p \qquad (34)$$

On neglecting leakage and compressibility effects, the oil flow rate into the piston is proportional to the rate at which the piston moves

$$q = A \frac{dy}{dt} \qquad (35)$$

where A is the cross-sectional area of the piston. But the force on the piston is Ap, which moves the load consisting of mass M and viscous friction B according to

$$Ap = M \frac{d^2 y}{dt^2} + B \frac{dy}{dt} \qquad (36)$$

Taking the Laplace transform of (34)–(36), we arrive at the transfer function

$$G(s) = \frac{Y(s)}{X(s)}$$

$$= \frac{A(K_1/K_2)}{s\{sM + [B + (A^2/K_2)]\}}$$

$$\equiv \frac{K}{s(\tau_h s + 1)} \qquad (37)$$

where $K = AK_1/(BK_2 + A^2)$ and $\tau_h = MK_2/(BK_2 + A^2)$. By the same token, (37) is similar to (11) and contains the term $1/s$ which accounts for the integrating action of the device.

2.3.3 Mechanical Systems

There are two types of mechanical systems, one is translation and the other is rotational. In the case of translation, the basic law (known as Newton's law) is that the sum of the applied forces must be equal to the sum of the reactive forces. We restrict ourselves to linear functions.

Accelerometer

In its simplest form, an accelerometer consists of a spring-mass-dashpot system as shown in Fig. 2.14. Usually the frame of the acceler-

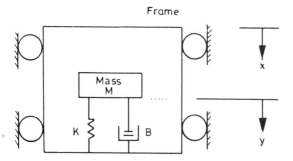

Figure 2.14 Accelerometer.

ometer is attached to a moving vehicle. In terms of x = displacement of the moving vehicle with respect to a fixed reference frame, y = displacement of the mass M with respect to the accelerometer frame, define

$$x_0 = y - x \qquad (38)$$

as the relative displacement. Direct application of the balance of forces gives

$$M\frac{d^2y}{dt^2} + K(y - x) + B\left(\frac{dy}{dt} - \frac{dx}{dt}\right) = 0$$

which after rearrangement and using (38), reduces to

$$-M\frac{d^2x}{dt^2} = M\frac{d^2x_0}{dt^2} + B\frac{dx_0}{dt} + Kx_0 \qquad (39)$$

Taking the Laplace transform of (39), the transfer function $X_0(s)/X(s)$ is given by

$$G(s) = \frac{X_0(s)}{X(s)}$$

$$= -\frac{s^2 M}{s^2 M + sB + K}$$

$$= -\frac{s^2}{s^2 + (B/M)s + (K/M)} \qquad (40)$$

where $\sqrt{(K/M)} = \omega_n$ is known as the natural frequency of mechanical oscillation.

Systems Control Modeling and Representation

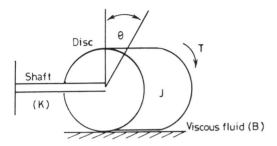

Figure 2.15 A simple rotational system.

For rotational systems, the torque equations parallel the writing of force equations in the translation case. Together with the displacement, velocity and acceleration terms are now being replaced by angular quantities. To illustrate this, consider the simple rotational system in Fig. 2.15. The rotating disc has a moment of inertia J, a shaft of stiffness K and moves against a medium with a viscous friction coefficient B. The balance of torque readily gives

$$T - B\frac{d\theta}{dt} - K\theta = J\frac{d^2\theta}{dt^2}$$

or

$$T = J\frac{d^2\theta}{dt^2} + B\frac{d\theta}{dt} + K\theta \tag{41}$$

which is a linear constant coefficient differential equation.

Gear Trains

To attain mechanical matching of motor to load in control systems, gear trains are used. Typically, a servomotor operates at high speed but low torque and to drive a load with high torque and low speed, gear trains are the proper device to use. It gives torque magnification and speed reduction. Consider a simple system as shown in Fig. 2.16. For the first shaft, direct application of torque balance gives

$$T_m = J_1 \frac{d^2\theta_1}{dt^2} + B_1 \frac{d\theta_1}{dt} + T_1 \tag{42}$$

where T_1 is the load torque exerted on gear 1 due to the rest of the

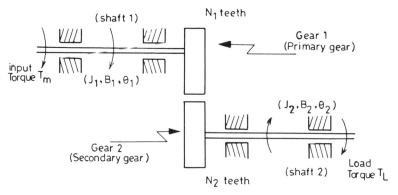

Figure 2.16 Gear train system.

gear train. For the second shaft

$$T_2 = J_2 \frac{d^2\theta_2}{dt^2} + B_2 \frac{d\theta_2}{dt} + T_L \qquad (43)$$

where T_2 is the torque transmitted to gear 2. Let R_1 be the radius of gear 1 and R_2 be that of gear 2. Since the linear distance travelled along the surface of each gear is the same, $\theta_1 R_1 = \theta_2 R_2$. The number of teeth on the gear surface being proportional to the gear radius, we obtain

$$\frac{\theta_2}{\theta_1} = \frac{N_1}{N_2} \qquad (44)$$

Ignoring power losses, we must have

$$T_1 \theta_1 = T_2 \theta_2 \qquad (45)$$

Using (44) and (45) to eliminate T_1 and T_2 from (42) and (43) and rearranging, we obtain

$$\left[J_1 + \left(\frac{N_1}{N_2}\right)^2 J_2 \right] \frac{d^2\theta_1}{dt^2} + \left[B_1 + \left(\frac{N_1}{N_2}\right)^2 B_2 \right] \frac{d\theta_1}{dt} + \frac{N_1}{N_2} T_L = T_m \qquad (46)$$

which characterizes the dynamics of gear 1. Define

$$J_{e1} = J_1 + \left(\frac{N_1}{N_2}\right)^2 J_2$$

$$B_{e2} = B_1 + \left(\frac{N_1}{N_2}\right)^2 B_2 \qquad (47)$$

Systems Control Modeling and Representation

as the equivalent moment of inertia and viscous friction of gear train referring to shaft 1; we rewrite (46) in the form

$$J_{e1}\frac{d^2\theta_1}{dt^2} + B_{e2}\frac{d\theta_1}{dt} + \frac{N_1}{N_2}T_L = T_m \qquad (48)$$

Had we made the equivalence with respect to gear 2, we would have arrived at

$$J_{e2}\frac{d^2\theta_2}{dt^2} + B_{e2}\frac{d\theta_2}{dt} + T_L = \frac{N_2}{N_1}T_m \qquad (49)$$

where

$$J_{e2} = J_2 + \left(\frac{N_2}{N_1}\right)^2 J_1$$

$$B_{e2} = B_2 + \left(\frac{N_2}{N_1}\right)^2 B_1 \qquad (50)$$

In the foregoing analysis, the stiffness of the shafts of the gear train is assumed to be infinite. It is a simple exercise to show that its effect in the case of finite value would be an additive term in either (46) or (49). More importantly, the case of multiple gear systems can be dealt with by extending and repeating the application of (47) and (50) in the desirable manner.

2.3.4 Pneumatic Systems

Air medium is used advantageously in pneumatic control systems since it is noninflammable and has very small viscosity. Pneumatic systems find considerable application in process contol, guided missiles, and aircraft systems.

Pneumatic Flapper Valve

This is an important component of many pneumatic systems. The power source for this device is the supply of air at constant pressure. A schematic diagram of the valve is shown in Fig. 2.17. Pressurized air P_1 is fed through the orifice A_0 and is ejected from the nozzle towards the flapper. The flapper is positioned against the nozzle opening and the nozzle back pressure P_2 is used to control the movement of a spring-loaded piston with K_2 and K_2 being the spring stiffness and piston cross-sectional area, respectively.

In what follows, we shall carry out approximate analysis to disclose

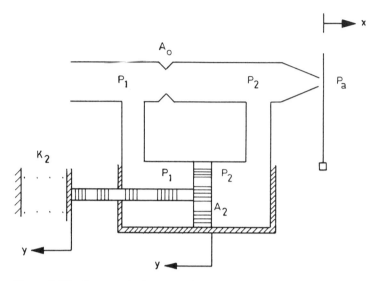

Figure 2.17 Pneumatic flapper valve.

the functional operation of the flapper valve. The approximation is based on linearized relationships and constant supply and ambient pressures P_1 and P_a. For a fixed inlet orifice, a linearized expression of the mass rate of flow into the chamber m_i can be put in the form

$$m_i = -C_1 p_2 \tag{51}$$

where small letters indicate perturbation in the corresponding variables ($m_i = \Delta M_i$ and $p_2 = \Delta P_2$) and the minus sign stresses the fact that as p_2 increases, m_i decreases. On the other hand, the mass rate of flow out from the chamber M_0 is a function of X and P_2. Thus (with $m_0 = \Delta M_0$, $x = \Delta X$)

$$m_0 = C_2 x + C_3 p_2 \tag{52}$$

The change in mass w of air in the chamber can then be expressed using (51) and (52) as

$$\frac{dw}{dt} = m_i - m_0$$

$$= -C_1 p_2 - C_2 x - C_3 p_2 \tag{53}$$

From the equation of state, the total mass W of air in the chamber is $W = P_2 V_2 / R T_2$ which when linearized for constant stagnation tempera-

ture T_2, gives
$$w = C_4 V_2 + C_5 p_2 \tag{54}$$
where the term $(C_4 V_2)$ represents the change in mass due to change in volume whereas the term $(C_5 p_2)$ represents the change in mass due to pressure. But from Fig. 2.17 we have
$$v_2 = A_2 y \tag{55}$$
and the summation of forces acting on the piston yield the perturbed relation
$$y = \frac{A_2}{K_2} p_2 \tag{56}$$
Taking the Laplace transform of (53), eliminating w from the foregoing relations, with some algebraic manipulations we arrive at
$$p_2 = -\left(\frac{K_1}{1 + \tau s}\right) x \tag{57}$$
$$y = -\left(\frac{A_2}{K_2} \frac{K_1}{1 + \tau s}\right) x \tag{58}$$
where
$$K_1 = \frac{C_2}{C_1 + C_3} \tag{59}$$
$$\tau = \frac{C_5 + A_2^2 C_4 K_2}{C_1 + C_3} \tag{60}$$

For most flapper valves, the time constant τ may generally be regarded as negligible, so y/x represents a constant ratio greater than 1 and thus we achieve pneumatic amplifier result.

2.3.5 Thermal Systems

Modeling of thermal systems tends to be generally nonlinear and therefore we shall use linearized relationships to facilitate subsequent analysis. The fundamental concept used for deriving the termal system equation is that the difference in heat coming into and leaving a body is equal to the increase of the thermal energy of the system. The physical properties used are mass, specific heat, thermal capacitance, conductance, and resistance. Temperature is the driving potential and heat is the quantity which flows.

Define T = temperature, q = heat flow, R = thermal resistance, and C = thermal capacitance. Then

$$\Delta T = qR \tag{61}$$

$$q = C\frac{dT}{dt} \tag{62}$$

Specify the two basic thermal relationships, where R and C may result from thermal conduction, convection, and radiation.

Next we illustrate the various system components by some examples.

Example 2.4. A small sphere of mass m, volume V, surface area A, and at a temperature t_0 is dropped in a large oil bath kept at constant temperature t_B. If h specifies the heat-transfer coefficient of convection, it is desired to express the sphere temperature in the Laplace domain. To do this, we use the fact that the heat gained by the sphere due to convection is equal to the internal energy increase. Thus

$$c\rho V \frac{dt_s}{dt} = Ah(t_B - t_S) \tag{63}$$

where c is the specific heat, ρ is the density and t_S is the temperature of the sphere. Taking the Laplace transform we have

$$sT_S = \alpha(T_B - T_S)$$

or

$$T_S = \frac{\alpha T_B}{s + \alpha}$$

where $\alpha = Ah/c\rho V$. Note that the describing relation is of first-order, which is the case in almost all thermal systems.

Example 2.5. Obtain the transfer function $Q_0(s)/Q_i(s)$ for the fluid system shown in Fig. 2.18. Assuming incompressible liquids, the conservation of mass yields for tank 1

$$q_i - q_2 = A_1 \frac{dh_1}{dt} \tag{64}$$

and for tank 2

$$q_2 - q_0 = A_2 \frac{dh_2}{dt} \tag{65}$$

where A is the cross-sectional area of the tank. By considering the flow

Systems Control Modeling and Representation

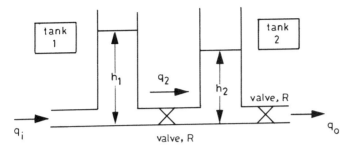

Figure 2.18 Double liquid-level system.

through the valves, we obtain

$$q_0 = \frac{\rho h_2}{R}, \qquad q_2 = \frac{\rho(h_1 - h_2)}{R} \tag{66}$$

Letting $C_1 = A_1/\rho$, $C_2 = A_2/\rho$ in (64), (65) and taking the Laplace transform, using (66) we reach

$$\frac{Q_0(s)}{Q_i(s)} = \frac{1}{[R^2 C_1 C_2 s^2 + (2C_1 + C_2)Rs + 1]}$$

as the desired transfer function.

To conclude this section, it remains to shed some light on two types of devices frequently used in control systems, *error detecting devices* and *transducers*. One device of common use in large electrical equipment is the *synchro* whose main function is to transmit torques and/or as position indicator. Another device is the *linear variable differential transfer* (*LVDT*) whose main function is convert displacement into an output voltage. The *potentiometer* is a classical error sensing device but can be used advantageously in conunction with high gain (operational) amplifiers. *Hydraulic valves* are commonly used as error sensors in hydraulic and gear train systems.

Transducers are a very important part of a control system since they provide a usable signal which measures a variable that must be either controlled or is useful as a control parameter. There are a large variety of transducers currently available. Their accuracy and cost is dependent upon the intended use. Most of them have nonlinear characteristics that can be satisfactorily linearized around their point of operation. A list of some common transducers, their use, and method of operation is given in Table 2.4.

Table 2.4 Classification of Transducers

Transducer	Use	Method of Operation
Capacitive probe	Liquid level	Capacitance changes between electrodes due to variable dielectric level change
Diaphragm	Pressure	Deflection of a circular plate that is proportional to the pressure
Geiger counter	Nuclear radiation	Ionization current produced by iron pair in gas or solid subjected to incident radiation
Gyroscope	Orientation and guidance	Change causes displacement relative to fixed axis of rotating wheel
Photodiode	Light detection	Resistance change in semi-conductor device junction due to light
Photovoltaic cell	Light detection	Output voltage when a junction of two dissimilar metals is illuminated
Piezoelectric crystal	Pressure	Mechanical distortion of crystal produces voltage
Potentiometer	Displacement	Change in voltage due to variable resistance or magnetic coupling
Pyranometer	Solar radiation	Thermopiles measure temperature of black and white surfaces to yield a temperature difference
Resistance thermometer	Temperature	Change in temperature causes change in electrical resistance of material
Solar cell	Orientation relative	Provides a signal proportional to cosine of angle between cell and normal sun vector
Strain gage	Strain	Electrical resistance change due to material deformation
Tachometer	Velocity	Voltage that is proportional to speed of the armature rotating in magnetic field
Thermocouple	Temperature	Emf proportional to temperature

Systems Control Modeling and Representation

We now move ahead to examine features and notions of control systems, starting with the fundamental notion of feedback around which much of control theory evolves.

2.4 FEEDBACK

The term *feedback* refers to the process of returning a fraction of the controlled variable (output) to the input. A *feedback control system* is a control system that operates to achieve prescribed relationships between selected system variables by comparing functions of these variables and utilizing the result to enforce control. A standard block diagram of a feedback control system is displayed in Fig. 2.19.

Our purpose here is to analyze the feedback control system and to delineate its properties. We first focus on the controlling system.

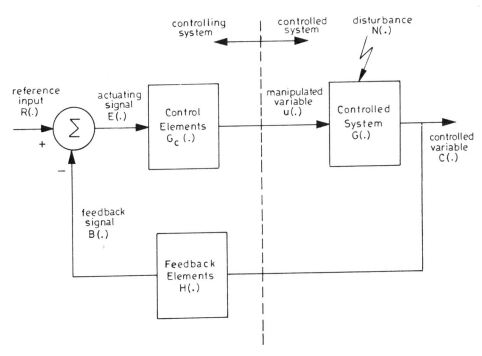

Figure 2.19 A standard feedback control system.

2.4.1 Control Actions

The term *control action* refers to the nature of the change of the element's (or controlling system's) output affected by the input. In this regard, the input may be the feedback signal $b(t)$, when the command is constant, the actuating signal, or the output of another control element. The output may be a signal to another element or the value of a manipulated variable $m(t)$. Briefly stated, control action can be used to produce regulatory effects for the purpose of improving system performance. We describe below basic types of control actions (reference is made to Fig. 2.18):

1. *Proportional action*, that is the output of the control element is linearly related to the input

$$m(t) = K_p e(t)$$

or equivalently

$$M(s) = K_p E(s) \qquad (67)$$

2. *Derivative action*, that is the output of the control element is proportional to the time rate of change of the input

$$m(t) = K_d \frac{de(t)}{dt}$$

or equivalently

$$M(s) = (sK_d) E(s) \qquad (68)$$

3. *Integral action*, that is the output of the control element is proportional to the integral of the input

$$m(t) = K_i \int_0^t e(t)\, dt$$

or equivalently

$$M(s) = \frac{K_i}{s} E(s) \qquad (69)$$

The following combinations of control actions are in common use.

1. *Proportional plus derivative action* combines (67) and (68) in the form

$$M(s) = K_p \left(1 + \frac{sK_d}{K_p}\right) E(s)$$

$$= K_p (1 + s\tau_d) E(s) \qquad (70)$$

Systems Control Modeling and Representation

2. *Proportional plus integral action* combines (67) and (69) in the form

$$M(s) = K_p\left(1 + \frac{K_i}{sK_p}\right)E(s)$$

$$= K_p\left(1 + \frac{1}{s\tau_i}\right)E(s) \quad (71)$$

3. *Proportional plus derivative plus integral action* combines (67) through (69) in the form

$$M(s) = K_p\left(1 + \frac{sK_d}{K_p} + \frac{K_i}{sK_p}\right)E(s)$$

$$= K_p\left(1 + s\tau_d + \frac{1}{s\tau_i}\right)E(s) \quad (72)$$

which is the most general form of control action.

Going back to Fig. (2.19), it is easy to write

$$C(s) = G_c(s)G(s)E(s)$$

$$E(s) = R(s) - H(s)C(s)$$

which when combined yields

$$\frac{C(s)}{R(s)} = \frac{G_c(s)G(s)}{1 + G_c(s)G(s)H(s)} \quad (73)$$

which characterizes the feedback transfer function.

2.4.2 System Sensitivity

One of the primary purposes of using feedback in control systems is to reduce the sensitivity of the system to parameter variations. The parameters of a system may vary with age, with changing environment or the like. Conceptually, *sensitivity* is a measure of the effectiveness of feedback in reducing the influence of these variations on system performance. Suppose due to parameter variations, $G(s)$ changes to $G(s) + \Delta G(s)$ where $|G(s)| \gg |\Delta G(s)|$. From (73) the output changes to

$$C(s) + \Delta C(s) = \frac{G_c(s)[G(s) + \Delta G(s)]R(s)}{[1 + G_c(s)G(s)H(s) + \Delta G(s)G_c(s)H(s)]} \quad (74)$$

Since $|\Delta G(s)| \ll |G(s)|$, we have from (73) and (74)

$$\Delta C(s) = \frac{\Delta G(s) G_c(s) R(s)}{1 + G_c(s) G(s) H(s)} \tag{75}$$

Setting $H(s) = 0$ gives the corresponding result in the open-loop case

$$\Delta C_0(s) = \Delta G(s) G_c(s) R(s) \tag{76}$$

Simple comparison of (75) and (76) reveals that the change in the output of the feedback system due to variation in $G(s)$ is reduced by a factor of $1 + G_c G(s) H(s)$ which is much greater than unity in most practical cases.

The term *system sensitivity* is used to describe the relative variation in the overall transfer function $T(s) = C(s)/R(s)$ due to variation in a parameter p and is defined as

$$S_p^T \triangleq \frac{\partial T/T}{\partial p/p}$$

$$\equiv \frac{\partial T}{\partial p} \frac{T}{p} \tag{77}$$

We now discuss some typical cases.

1. Parameter p in $G(s)$ only. Use the chain rule of differentiation to obtain

$$\frac{\partial T}{\partial p} = \frac{\partial T}{\partial G} \frac{\partial G}{\partial p}$$

but from (73) we obtain

$$\frac{\partial T}{\partial G} = \frac{G_c(s)}{[1 + G_c(s) G(s) H(s)]^2}$$

and finally from (77)

$$S_p^T = \frac{p(\partial G/\partial p)}{G(s)[1 + G_c(s) G(s) H(s)]} \tag{78}$$

A special case of (78) is when $p \equiv G(s)$ in which case

$$S_G^T = \frac{1}{1 + G_c(s) G(s) H(s)} \tag{79}$$

which is in accordance with the foregoing analysis.

2. Parameter p in $H(s)$ only. In this case we invoke

Systems Control Modeling and Representation

$$\frac{\partial T}{\partial p} = \frac{\partial T}{\partial H}\frac{\partial H}{\partial p}$$

but from (73) we have

$$\frac{\partial T}{\partial H} = -\frac{G_2^2(s)G^2(s)}{[1 + G_c(s)G(s)H(s)]^2}$$

and finally from (77) we arrive at

$$S_p^T = -\frac{pG_c(s)G(s)/(\partial H/\partial p)}{1 + G_c(s)G(s)H(s)} \tag{80}$$

An interesting result occurs when $p \equiv H(s)$ in which (80) reduces to

$$S_H^T = -\frac{G_c(s)G(s)H(s)}{1 + G_c(s)G(s)H(s)} \tag{81}$$

which indicates that for large values of $G_c(s)G(s)H(s)$, sensitivity of the feedback system with respect to $H(s)$ approaches unity. Other cases can be studied in the same manner. This emphasizes the need to use feedback elements which do not vary with environmental changes or can be maintained constant. The use of feedback in reducing sensitivity to parameter variations is an important advantage of feedback control systems. To have a highly accurate feedback system, the selection of control elements $G_c(s)$ and feedback elements $H(s)$ requires careful consideration.

The price for improvement in sensitivity by use of feedback is paid in terms of *loss of system gain*. It is clear that the open-loop system $H(s) = 0$ has a gain $G_c(s)G(s)$, while the gain of the feedback control systems is $G_c(s)G(s)/[1 + G_c(s)G(s)H(s)]$. The reduction factor is the same as for system sensitivity to parameter variations.

2.4.3 Steady-State Error

For brevity we let $G_c(s)G(s) \equiv G_p(s)$. With reference to Fig. 2.18, the error signal can be written in the form

$$E(s) = R(s)F(s) \tag{82}$$

with the error transfer function defined as

$$F(s) = \frac{1}{1 + G_p(s)H(s)} \tag{83}$$

The inverse Laplace transform of $E(s)$ produces the error signal $e(t)$ as convolution integral

$$e(t) = \int_0^t f(\tau) r(t - \tau) \, d\tau \qquad (84)$$

where the lower integration limit is set zero to preserve the physical character of reference signal $r(t)$ and $f(\tau)$ is the inverse Laplace transform of $F(s)$. On expanding the function $r(t - \tau)$ in a Taylor series, we can express (84) as

$$e(t) = \int_0^t f(\tau) \left[r(t) + \sum_{k=1}^{\infty} \frac{(-\tau)^k}{k!} \frac{d^k r(t)}{dt^k} \right] d\tau \qquad (85)$$

At this stage we define the generalized error coefficient $C_k(t)$ by

$$C_k(t) = (-)^k \int_0^t \tau^k f(\tau) \, d\tau, \qquad K = 0, 1, 2 \qquad (86)$$

which when used in (85) yields

$$e(t) = \sum_{k=0}^{\infty} \frac{C_k(t)}{k!} \frac{d^k r(t)}{dt^k} \qquad (87)$$

The interest in (87) stems from the fact that it expresses the error signal in terms of the test signal $r(t)$ and its derivatives.

The steady-state error e_{ss} is defined by

$$e_{ss} \triangleq \lim_{t \to \infty} e_s(t) \qquad (88)$$

where

$$e_s(t) = \sum_{k=0}^{\infty} \frac{C_k}{k!} \frac{d^k r_s(t)}{dt^k} \qquad (89)$$

$$C_k = \lim_{t \to \infty} C_k(t)$$

$$= (-1)^k \int_0^{\infty} \tau^k f(\tau) \, d\tau \qquad (90)$$

Here $e_s(t)$ denotes the steady-state part of $e(t)$, $r_s(t)$ represents the steady-state part of $r(t)$, and C_k is the steady-state generalized error coefficient. Instead of computing (86) directly, we can use the properties of Laplace transform. By definition

$$F(s) \triangleq \int_0^{\infty} f(\tau) e^{-s\tau} \, d\tau \qquad (91)$$

Systems Control Modeling and Representation 57

Taking the limit of both sides as $s \to 0$ yields

$$\lim_{s \to 0} F(s) = \lim_{s \to 0} \int_0^\infty f(\tau) e^{-s\tau} d\tau$$

$$= \int_0^\infty f(\tau) d\tau$$

A comparison of the right-hand side of the above relation with (86) for $k = 0$ indicates that

$$C_0 = \lim_{s \to 0} F(s) \tag{92}$$

Taking the derivative of (91) with respect to s, we arrive at

$$\lim_{s \to 0} [dF(s)/ds] = \lim_{s \to 0} \int_0^\infty (-\tau) f(\tau) e^{-\tau s} d\tau$$

$$= -\int_0^\infty \tau f(\tau) d\tau$$

which in the light of (86) shows that

$$C_1 = \lim_{s \to 0} [dF(s)/ds] \tag{93}$$

Repeating the foregoing process, we reach

$$C_k = \lim_{s \to 0} [d^k F(s)/ds^k] \tag{94}$$

Note that (94) can be used in conjunction with (88) and (89) to obtain the steady state error information.

Had we started from (82) and used a Maclaurin series expansion, we could have obtained:

$$E(s) = F(0)R(s) + dF(s)/ds|_{s=0} sR(s)$$
$$+ d^2F(s)/ds^2|_{s=0} s^2 R(s)/2! + \cdots$$

$$= \sum_{k=0}^\infty (C_k/k!) s^k R(s) \tag{95}$$

which is the Laplace transform of (89). C_k is given by (90) as before.

An alternative route would be to invoke the final value theorem of Laplace transform in (88) and incorporate (82) and (83) to yield

$$e_{ss} = \lim_{t \to 0} e_s(t)$$

$$= \lim_{s \to 0} sE(s)$$

$$= \lim_{s \to 0} \frac{sR(s)}{1 + G_p(s)H(s)} \tag{96}$$

which shows the dependence of the steady-state error on the loop transfer function $G_p(s)H(s)$ and the input signal $R(s)$. The expression for steady-state error for various types of standard test signals are derived below.

1. *Unit-step input.* In this case $r(t) = 1$ for $t \geq 0$ and thus $R(s) = 1/s$. From (96)

$$e_{ss} = \lim_{s \to 0} \frac{1}{1 + G_p(s)H(s)}$$

$$= \frac{1}{1 + G_p(0)H(0)}$$

$$\triangleq \frac{1}{1 + K_p} \tag{97}$$

where $K_p = G_p(0)H(0)$ is defined as the *steady-state step (position) error constant.* An equivalent definition is that

$$K_p \triangleq \frac{C_{ss}}{e_{ss}}$$

$$= \frac{\lim_{s \to 0} s[R(s) - E_s]}{\lim_{s \to 0} sE_s(s)}$$

$$= \frac{1 - C_0}{C_0}$$

since $E_s(s) = C_0/s$.

2. *Unit-ramp input.* In this case $r(t) = t$ for $t \geq 0$ which means that $dr/dt = 1$, $d^k r/dt^k = 0$ for $k > 1$ and subsequently $R(s) = 1/s^2$. From (96)

$$e_{ss} = \lim_{s \to 0} \frac{1}{s + sG_p(s)H(s)}$$

$$= \lim_{s \to 0} \frac{1}{sG_p(s)H(s)}$$

$$\triangleq \frac{1}{K_v} \tag{98}$$

Systems Control Modeling and Representation

where $K_v = \lim_{s \to 0} sG_p(s)H(s)$ is defined as the *steady-state ramp (velocity) error constant*. An equivalent definition is that

$$K_v = \frac{\lim_{t \to \infty} \dot{c}(t)}{\lim_{t \to \infty} e(t)}$$

Since $c(t) = r(t) - e(t)$ where $H(s) = 1$ and $C_0 = 0$, then

$$\lim_{t \to \infty} \dot{c} = \lim_{s \to 0} s[sR(s) - C_1]$$

$$= 1$$

$$\lim_{t \to \infty} e(t) = \lim_{s \to 0} sE_s(s) \lim_{s \to 0} sC_1\left(\frac{1}{s}\right) = C_1$$

and finally $K_v = 1/C_1$.

3. *Unit-parabolic input*. In this case $r(t) = t^2/2$ for $t \geq 0$ and thus $dr/dt = t$, $d^2r/dt^2 = 1$, $d^k r/dt^k = 0$ for $K > 2$ and $R(s) = 1/s^3$. From (96)

$$e_{ss} = \lim_{s \to 0} \frac{1}{s^2 + s^2 G_p(s)H(s)}$$

$$= \lim_{s \to 0} \frac{1}{s^2 G_p(s)H(s)}$$

$$= \frac{1}{K_a} \tag{99}$$

where $K_a = \lim_{s \to 0} s^2 G_p(s)H(s)$ is defined as the *steady-state parabolic (acceleration) error constant*. An equivalent definition is that

$$K_a = \frac{\lim_{t \to \infty} \ddot{c}(t)}{\lim_{t \to \infty} e(t)} \tag{100}$$

Using similar arguments with $H(s) = 1$, $C_0 = 1$, $C_1 = 0$ we can conclude that $K_a = 1/C_2$. The error constants K_p, K_v, and K_a describe the ability of a system to reduce or eliminate steady-state errors. In order to assess the dynamic behavior of the error function with time, one has to rely on the generalized relations (85)–(87).

2.4.4 System Types

From now onwards we shall consider the *loop transfer function* $G_p(s)H(s)$ as given by

Table 2.5 Steady-State Error for Various Inputs

Input Type	Steady-State Error		
	Type-0 System	Type-1 System	Type-2 System
Unit-step	$1/(1 + K_p)$	0	0
Unit-ramp	∞	$1/K_v$	0
Unit-parabola	∞	∞	$1/K_a$

$$G_p(s)H(s) = \frac{K_m N(s)}{s^m D(s)} \qquad (101)$$

where the numerator polynomial $N(s)$ and the denominator polynomial $D(s)$ have the form

$$N(s) = (T_{z1}s + 1)(T_{z2}s + 1)T_{z3}s + 1) \cdots$$
$$= 1 + b_1 s + b_2 s^2 + b_3 s^3 + \cdots \qquad (102)$$
$$D(s) = (T_{p1}s + 1)(T_{p2}s + 1)(T_{p3}s + 1) \cdots$$
$$= 1 + a_1 s + a_2 s^2 + a_3 s^3 + \cdots \qquad (103)$$

where K_m is the open-loop gain, $-1/T_{zk}$ is the position of the kth zero and $-1/T_{ph}$ is the position of the hth poles. Note in (101) that there is an mth order pole at $s = 0$ (origin) which corresponds to the number of integrations in the system. Since for small values of s (tends to zero), the term s^m dominates in determining the steady-state error, control systems are classified in accordance with the number of integrations in the loop transfer function. Steady-state error for various inputs and systems are summarized in Table 2.5. Note that the entries in Table 2.5 have been computed using (96) through (101). Although it may appear that there would be no limit to the number of integrations, types-0, -1, -2, are thus for the most commonly employed systems.

2.5 SYSTEM RESPONSE

Having represented control systems by transfer functions or block diagrams, we turn our attention to system response; that is how does a linear system respond as a function of time when subjected to various types of stimuli? In the sequel we are interested in the system output without regard to the behavior of variables inside the control system. Specifically if we desire $C(s)$, we can work with $C(s)/R(s)$, which represents the transfer function, and specify $R(s)$ and obtain the output. On

Systems Control Modeling and Representation

the other hand, if we need $E(s)$, we would work with $E(s)/R(s)$ and specify $R(s)$. In any event it is important to recognize that we use the block diagram algebra in our course to characterize the system behavior.

In studying the system response of feedback control systems there are three things we wish to know, namely the *transient response*, the *steady-state* or forced response, and the *stability* of the system. The transient solution yields information on how much the system deviates from the input and the time necessary for the system response to settle to within certain limits. The steady-state or forced response gives an indication of the accuracy of the system. Whenever the steady-state output does not agree with the input, the system is said to have a steady-state error. By stability, we mean in short that the output does not become uncontrollably large; about which we shall say a great deal. With reference to Fig. 2.18, the output transfer function is written in the form

$$\frac{C(s)}{R(s)} = \frac{G_p(s)}{1 + G_p(s)H(s)} \tag{104}$$

The system transients depend only on the internal properties of the system which are mainly determined by the poles of (101). These poles or characteristic roots are the values that solve the *characteristic equation*

$$1 + G_p(s)H(s) = 0 \tag{105}$$

For convenience we rewrite (104) in the form

$$\frac{C(s)}{R(s)} = K \frac{\Pi_{j=1}^{r}(s + z_j)}{\Pi_{j=1}^{n}(s + p_j)} \tag{106}$$

with r zeros at $z_1, \ldots, z_r$ and n poles at $p_1, \ldots, p_n$. In terms of (101), we can state

$$K_m = K \frac{\Pi_{j=1}^{r} z_j}{\Pi_{j=1}^{n}(p_j)} \tag{107}$$

To obtain the output from (104) we usually set $R(s) = 1$ (corresponding to $r(t) = \delta_\epsilon$ an impulse input) and seek the partial fractions expansion to yield

$$C(s) = \sum_i \frac{K_i}{s + s_i} + \sum_i \frac{K_i^+}{s + s_i^+} + \sum_l \frac{K_l}{s + s_l} \tag{108}$$

where

$$K_i = \lim_{s \to -s_i} (s + s_i)C(s) \tag{109}$$

$$K_l = \lim_{s \to -s_l} (s + s_l)C(s) \tag{110}$$

and s_i, s_i^+ are complex conjugates but s_l is real. If we denote $s_i = -\alpha_i + j\omega_r$, then the output in the time domain is obtained by taking the inverse Laplace transform of (108)

$$\begin{aligned}c(t) &= \sum_i K_i \exp[-(-\sigma_i + j\omega_i)t] \\ &+ \sum_i K_i^+ \exp[-(-\sigma_i - j\omega_i)t] + \sum_l K_l \exp(s_l t) \\ &= \sum_i |K_i| \exp(+\sigma_i t) \cos(-\omega_i t + \phi_m) + \sum_l K_l \exp(s_l t) \quad (111)\end{aligned}$$

where ϕ_m is the phase contribution of the constant K_i. Note that the second term of (110) is obtained by combining two terms. The important fact here is that the form of the transient response is a function of the location of the closed-loop poles, which are identical to the roots of the characteristic equation, on the s-plane. For those cases where the closed-loop poles are in the right half-plane, growing oscillations occur; otherwise decaying (poles with negative real parts) or sustained (poles with zero real parts) oscillations take place.

Turning to the forced response, it is sufficient to know that (104) can be manipulated for given $R(s)$, corresponding to input $r(t)$, using Laplace transform techniques. The case of the second-order system is most interesting in the physical world and for which (104) can be reduced into the normalized form

$$\frac{C(s)}{R(s)} = \frac{\omega_n^2}{s^2 + 2\delta\omega_n s + \omega_n^2} \tag{112}$$

where ω_n is the *natural frequency* and δ is the *damping ratio* of the control system. The poles of (112) are positioned in the s-plane as a function of δ as follows:

$p_1, p_2 = \pm j\omega_n$	$\delta = 0$, underdamping
$p_1, p_2 = -\delta\omega_n \pm j\omega_n\sqrt{1-\delta^2}$	$\delta < 1$, underdamping
$p_1, p_2 = -\omega_n$	$\delta = 1$, damping critical
$p_1, p_2 = -\delta\omega_n \pm \omega_n\sqrt{\delta^2 - 1}$	$\delta > 1$, overdamping

Systems Control Modeling and Representation

The migration for the closed-loop poles as a function of δ is shown in Fig. 2.20 for constant natural frequency.

The results of evaluating (112) when $R(s) = 1/s$ (unit step response) are summarized in Table 2.6. It is left as an exercise to the reader to verify these results. The most appealing case is that of underdamping ($\delta < 1$) which has the following important properties (all can be derived by inspection of Table 2.6).

1. Frequency of damped oscillations is $\omega_d = \omega_n \sqrt{1 - \delta^2}$.
2. Time to first overshoot is

$$t_p = \frac{\pi}{\omega_n \sqrt{1 - \delta^2}} = \frac{\pi}{\omega_d}$$

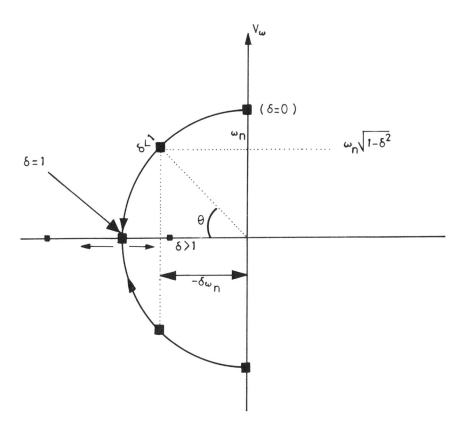

Figure 2.20 Pattern of closed-loop poles of second-order system.

Table 2.6 Unit Step Response of Second-Order Systems

Damping Ratio	Response $C(t)$ $(r(t) = 1)$
Undamping ($\delta = 0$)	$C(t) = 1 - \cos(\omega_n t)$
Underdamping ($\delta < 1$)	$C(t) = 1 + \dfrac{\exp(-\delta\omega_n t)}{\sqrt{1-\delta^2}} \sin[\omega_n \sqrt{1-\delta^2}(t-\theta)]$
	$\cos\theta = -\delta, \quad \sin\theta = \sqrt{1-\delta^2}$
Critical damping ($\delta = 1$)	$C(t) = 1 - (1+\omega_n t)\exp(-\omega_n t)$
Overdamping ($\delta > 1$)	$C(t) = 1 + \left(\dfrac{\omega_n}{2\sqrt{\delta^2-1}}\right)\left(\dfrac{e^{p_1 t}}{p_1} - \dfrac{e^{p_2 t}}{p_2}\right)$
	$p_1 = (-\delta + \sqrt{\delta^2-1})\omega_n$
	$p_2 = (-\delta - \sqrt{\delta^2-1})\omega_n$

3. The peak value of the output is

$$c_p = 1 + \exp\left(\frac{-\delta\pi}{\sqrt{1-\delta^2}}\right)$$

4. Rise time from 0 to 100%

$$t_r = \frac{\theta}{\omega_d}, \quad \cos\theta = -\delta$$

5. Time required for the response to reach and stay with 2% of the final value ($C(t) = 1$)) is

$$t_s = \frac{4}{\delta\omega_n}$$

When dealing with high-order systems, we usually rely on the partial fraction expansion method to convert the problem at hand into summation first- and second-order terms. The use of the inverse Laplace transform is thus straightforward. Note in general that

$$C(s) = \left(\frac{G_p(s)}{1 + G_p(s)H(s)}\right) R(s)$$

$$= G_f(s) R(s)$$

Systems Control Modeling and Representation

So that

$$C(t) = \mathscr{L}[G_f(s)R(s)] \tag{113}$$

where $G_f(s)$ is the feedback transfer function. When $R(s) = 1$, $C(s) = G_f(s)$, and $C(t) = \mathscr{L}[G_f(s)] = g(t)$ which is frequently known as the *impulse response*. The interpretation is that when the input to a linear control system is a unit impulse, the output is thus the impulse response. Inversely, the Laplace transform of the impulse response gives the transfer function $G_f(s)$ (or $G_p(s)$ if $H(s) = 0$) of the system under consideration. From a theoretical standpoint, the measurement of the impulse response of a system provides a complete description of the control system. This is obvious since from (105) the output is defined by the convolution from

$$C(t) = g(t)r(t)$$
$$= \int_0^\infty g(\tau)r(t-\tau)\,d\tau \tag{114}$$

which is implemented in practice by various discretization schemes.

2.6 EXAMPLES

In this section, we illustrate the foregoing concepts by different examples.

Example 2.6. The forward loop of a unity feedback control system $H(s) = 1$ is given by $G_p(s) = K/s(s^2 + 19s + 118)$. It is desired to obtain $e(t)$ if $K = 240$ and $r(t) = t$. Since $E(s) = \{1/[1 + G_p(s)]\}R(s)$, $R(s) = 1/s^2$, and $K = 240$. Thus with some manipulation we reach

$$E(s) = \frac{s^2 + 19s + 118}{s(s+5)(s+6)(s+8)}$$

by partial fraction expansion, this takes the form

$$E(s) = \frac{R_1}{s} + \frac{R_2}{s+5} + \frac{R_3}{s+6} + \frac{R_4}{s+8}$$

where the residues $R_1, \ldots, R_4$ are calculated as follows

$$R_1 = \lim_{s \to 0} sE(s) = 0.492$$

$$R_2 = \lim_{s \to -5} (s+5)E(s) = -3.2$$

$$R_3 = \lim_{s \to -6} (s + 6)E(s) = 3.33$$

$$R_4 = \lim_{s \to -8} (s + 8)E(s) = -0.625$$

Finally the inverse Laplace transform yields

$$e(t) = 0.492 - 3.2e^{-5t} + 3.33e^{-6t} - 0.625e^{-8t}.$$

Example 2.7. It is required to derive the error series for the system with

$$F(s) = \frac{s^2}{s^2 + 2\delta\omega_m s + \omega_n^2}$$

From the definition of the generalized coefficients, it is easy to see that

$$C_0 = \lim_{s \to 0} F(s) = 0$$

$$C_1 = \lim_{s \to 0} \frac{dF(s)}{ds} = 0$$

$$C_2 = \lim_{s \to 0} \frac{d^2 F(s)}{ds^2} = \frac{2}{\omega_n^2}$$

$$C_3 = \lim_{s \to 0} \frac{d^3 F(s)}{ds^3} = -\frac{12\delta}{\omega_n^3}$$

Substituting the above error coefficients in (97), the steady-state error becomes

$$e_s(t) = \frac{1}{\omega_n^2} \frac{d^2 r(t)}{dt^2} - \frac{2\delta}{\omega_n^3} \frac{d^3 r(t)}{dt^3}$$

Since the order of the numerator equals that of the denominator, we expect $C_1 = C_2 = 0$. Note also that to cancel out the steady-state error or to keep it constant, the input must be limited to parabolic at the most.

Example 2.8. Consider the standard feedback control system with $G(s) = K/s(s + 1)$, $G_c(s) = 25(s + 1)/(s + 5)$ and $H(s) = 1$. The nominal value of process parameter $K = 1$. Let us evaluate the sensitivity of transfer function $C(s)/R(s)$ to variation in parameter K. First we calculate

$$\frac{C(s)}{R(s)} = \frac{25K}{s^2 + 5s + 25K}$$

Systems Control Modeling and Representation

Therefore

$$S_k^T = \frac{\partial T}{\partial K}\frac{K}{T}$$

$$= \frac{s(s+5)}{s^2 + 5s + 25k}$$

Since the normal value of $K = 1$, then

$$S_k^T = \frac{s(s+5)}{s^2 + 5s + 25}$$

which has the magnitude of 1.41 at $\omega = 5$.

Example 2.9. Consider the mechanical system shown in Fig. 2.21. It is required to obtain the transfer function $X_2(s)/F(s)$.

Let the displacements of masses M_1 and M_2 be x_1 and x_2, respectively. Consider first mass M_2 and study the balance of forces to produce

$$F(t) - B_2\left(\frac{dx_2}{dt} - \frac{dx_1}{dt}\right) - K_2(x_2 - x_1) = M_2\frac{d^2x_2}{dt^2}$$

For mass M_1 we obtain

$$B_2\left(\frac{dx_2}{dt} - \frac{dx_1}{dt}\right) + K_2(x_2 - x_1) - B_1\frac{dx_1}{dt} - K_1x_1 = M_1\frac{d^2x_1}{dt^2}$$

Rearranging the foregoing relations we obtain

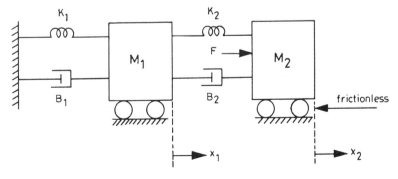

Figure 2.21 Consider the mechanical system shown below.

$$M_1 \frac{d^2 x_1}{dt^2} + B_1 \frac{dx_1}{dt} - B_2\left(\frac{dx_2}{dt} - \frac{dx_1}{dt}\right) + K_1 x_1 - K_2(x_2 - x_1) = 0$$

$$M_2 \frac{d^2 x_2}{dt^2} + B_2\left(\frac{dx_2}{dt} - \frac{dx_1}{dt}\right) + K_2(x_2 - x_1) = F(t)$$

These are two simultaneous second-order linear differential equations. Laplace transforming these equations with some manipulations yields

$$V_1(s)X_1(s) - V_3(s)X_2(s) = 0$$
$$V_2(s)X_2(s) - V_3(s)X_1(s) = F(s)$$

with

$$V_1(s) = s^2 M_1 + s(B_1 + B_2) + (K_1 + K_2)$$
$$V_2(s) = s^2 M_2 + sB_2 + K_2$$
$$V_3(s) = (sB_2 + K_2)$$

eliminating $X_1(s)$ we obtain

$$\frac{X_2(s)}{F(s)} = \frac{V_1(s)}{V_2(s)V_1(s) - V_3^2(s)}$$

which can be simplified into the form

$$\frac{X_2(s)}{F(s)} = s^2 M_1 + s(B_1 + B_2) + (K_1 + K_2)/[s^4 M_1 M_2$$
$$+ s^3(M_2(B_1 + B_2) + M_1 B_2) + s^2(M_2(K_1 + K_2)$$
$$+ B_1 B_2 + M_1 K_2) + s(B_2 K_1 + K_2 B_1) + K_1 K_2]$$

2.7 PROBLEMS

1. An amplidyne and a servomotor are connected as shown in Fig. 2.22. Obtain the transfer function $\theta(s)/V_c(s)$.
2. The forward loop of a unity feedback control system is $G_p(s) = K/s(s + 6)$. Obtain the output response for $K = 9$ and the input is unit step.
3. A standard feedback control system has $G_c(s) = (1 + \tau_c s)/(1 + \tau_b s)$, $G(s) = K_0/(1 + \tau_0 s)$, and $H(s) = Ks/(1 + \tau_p s)$. Find the closed-loop sensitivity to variations in K_0, τ_0, τ_p, and τ_b.
4. Consider the normalized from of second-order system

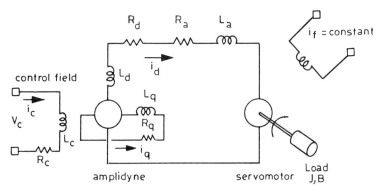

Figure 2.22 An amplidyne and a servomotor are connected as shown.

$$\frac{C(s)}{R(s)} = \frac{\omega_n^2}{s^2 + 2\delta\omega_n + \omega_n^2}$$

with $\delta < 1$, $r(t) = A$, $0 \leq t < t_1$, $r(t) = (A + B)$, $5t_1 \leq t$. It is desired to calculate the factors t_1, A, and B if the output $C(t)$ must settle at $A + B$ for $t > t_1$.

5. A pitch-attitude-control system for a missile is approximated by the standard feedback configuration with

$$G(s) = \frac{K_2}{(s^2 5 - b)}, \qquad G(s) = G_a G_b$$

$$G_a = \frac{a}{s + a}, \qquad G_b = \frac{3(s + 0.2)}{s(1 + 0.04s)}$$

$H(s) = 1 + h_1 s$ and the nominal values are

$a = 15$, $\quad K_2 = 6.45$, $\quad b = 2.14$, $\quad h_1 = 0.33$

Find
  a. The closed-loop transfer function
  b. The type of the system
  c. The error constants K_p, K_v, K_a
  d. The sensitivity to changes in K_2, a, h_1, and b

6. Consider the standard feedback configuration with $G(s) = 5/(s + 2)$, $H(s) = 0.1$ and $G_c(s)$ may take one of the forms
  a. $G_c(s) = 2.0$ (proportional)
  b. $G_c(s) = 2.0/s$ (integral)
  c. $G_c(s) = 2.0\, s$ (derivative)
  For each case find

a. The closed-loop transfer function
b. The output when $r(t) = 4.0$

Comment on the results when compared to each other.

NOTES AND REFERENCES

The material covered in this chapter is a fundamental prerequisite to the understanding of analog (continuous-time) control systems. Our purpose has been to build a common spectrum of tools and modeling methods to deal with a variety of physical systems. Selected references for collateral reading include the books of Ogata (1979), Anand (1984), Raven (1978), Dazzo and Houpis (1981), and Nagrath and Gopal (1982), are recommended for their emphasis on the identification of systems functions and operations. The paper by Kalman (1963) is seminal and contains a wealth of knowledge. Wellstead (1979) provides a unique modeling approach based on the concept of energy manipulations. General segments of the material discussed here are found in Perkins and Cruz (1969), DeRusso (1965), Clare and Frederick (1978), and Karanopp and Rosenberg (1975) where the focus has been on the mathematical treatment, sensitivity analysis, and system simulation. Interesting process engineering models and examples have been dealt with in Weber (1973), Luyben (1974), and Shinkskey (1981). Classical references on similar topics include Cannon (1967), Shearer et al. (1967), Mayr (1970), Popov (1962), and Newton et al. (1957). The book by Dorf (1980) contains several interesting practical problems and real-life models drawn from various fields.

3
Systems Analysis Via State Variables

3.1 INTRODUCTION

In the preceding chapter, we have developed and studied several useful control engineering models of physical systems. The Laplace transform was utilized to transform the differential equations representing the system into an algebraic equation expressed in terms of the complex variable s. Utilizing this algebraic equation, we were able to obtain a transfer function representation of the input-output relationship. Though the transfer function model provides us with simple and powerful analysis technique, it suffers from certain drawbacks. First, a transfer function is only defined under zero initial conditions. Second, it is only applicable to linear time-invariant systems and there too it is generally restricted to single-input single-output systems. Multivariable transfer functions can be developed and studied via a relatively complex linear computational algebra. Another limitation of the transfer function technique is that it reveals only the system output for a given input. No information is provided regarding the internal status of the systems. There may be situations where the output of a system is stable and yet some of the system elements may have a tendency to exceed their specified ratings. It may sometimes also be necessary and advantageous to provide feedback information about some of the internal variables of a system, rather than the output alone, for the purpose of stabilizing and improving the performance of a system.

Motivated by the preceding discussions, a more general mathematical representation of a system in terms of its internal dynamics was developed and termed the *state-variables* or *state-space approach*. It is a very powerful technique for the analysis and design of linear, nonlinear, time-invariant or time-varying multi-input multi-output systems. The organization of the state variable approach is such that it is easily amendable to analog and digital computers solution.

Before proceeding further, we should assert that both the frequency-domain (transfer function) and the time-domain (state variables) approaches are complementary and one approach cannot override the other.

3.2 STATE-SPACE FORMULATION

In this section our aim is to introduce the state variable approach for LTI systems as a powerful tool for time-domain analysis of control systems. This approach is based on the concept of the *state* of a system; which is defined as a set of numbers such that the knowledge of these numbers and the input functions will, with the equations describing the dynamics, provide the future state and the output of the system.

For a dynamic system, the state of a system is described in terms of a set of *state variables* $\{x_1(t), x_2(t), \ldots, x_n(t)\}$. The state variables are those minimal set of variables which determine the future behavior of a system when the present state of the system and the excitation signals are known. The n state variables are the components of vector **x** termed the *state vector*. The space defined by the state variables as coordinates is called the *state space*.

The idea central to state representation can be illustrated in terms of the simple pendulum system shown in Fig. 3.1. We can ascertain the state of the system uniquely at any time if we know the angular position and angular velocity of the pendulum, that is $\theta(t)$ and $d\theta(t)/dt$. Let us define

$$x_1(t) = \theta(t), \qquad x_2(t) = \frac{d\theta(t)}{dt}$$

Systems Analysis Via State Variables

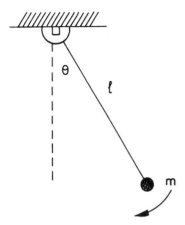

Figure 3.1 A simple pendulum.

where $x_1(t)$ and $x_2(t)$ are the state variables of this system. If we consider θ to be small so that the governing equations are linear, then we can easily establish

$$\frac{dx_1(t)}{dt} = x_2(t)$$

$$\frac{dx_2(t)}{dt} = (-g/l)x_1(t)$$

which are the state equations. We have already shown in Chapter 2, albeit with different terminology, how the mathematical model of various systems can be written in this way using physical variables.

It should be emphasized that although the state of a dynamic system is uniquely determined from a specified set of state variables, that set itself is not unique. We can therefore select a variety of such sets. The particular one chosen is dictated often by convenience or the type of information desired.

In general, then, when we model a system using state variables we do so by appropriately writing down a set of first-order linear differential equations of the form

$$\dot{x}_1(t) = a_{11}x_1(t) + a_{12}x_2(t) + \cdots + a_{1n}x_n(t) + b_1 r(t)$$
$$\dot{x}_2(t) = a_{21}x_1(t) + a_{22}x_2(t) + \cdots + a_{2n}x_n(t) + b_2 r(t)$$
$$\vdots \tag{1}$$
$$\dot{x}_n(t) = a_{n1}x_1(t) + a_{n2}x_2(t) + \cdots + a_{nn}x_n(t) + b_n r(t)$$

and

$$y(t) = C_1 x_1(t) + C_2 x_2(t) + \cdots + c_n x_n(t) \tag{2}$$

where $\{x_1(t), x_2(t), \ldots, x_n(t)\}$ are the state variables, $r(t) = u(t)$ is the forcing function or system input and $y(t)$ is the system output. For notational economy, we rewrite (1) and (2) using the vectors

$$\mathbf{x}(t) = \begin{bmatrix} x_1(t) \\ x_2(t) \\ \vdots \\ x_n(t) \end{bmatrix}, \quad \mathbf{b} = \begin{bmatrix} b_1 \\ b_2 \\ \vdots \\ b_n \end{bmatrix}, \quad \mathbf{c} = \begin{bmatrix} c_1 \\ c_2 \\ \vdots \\ c_n \end{bmatrix} \tag{3}$$

and

$$A = \begin{bmatrix} a_{11} & a_{12} & \cdots & a_{1n} \\ a_{21} & a_{22} & \cdots & \cdot \\ \cdot & \cdot & & \cdot \\ \cdot & \cdot & & \cdot \\ a_{n1} & & & a_{nn} \end{bmatrix} \tag{4}$$

in the compact form

$$\dot{\mathbf{x}}(t) = A\mathbf{x}(t) + \mathbf{b}u(t) \tag{5}$$
$$y(t) = \mathbf{c}^t \mathbf{x}(t) \tag{6}$$

where the superscript t denotes vector (matrix) transpose. The $n \times n$ matrix A, the $n \times 1$ matrix $\mathbf{b}$ and the $1 \times n$ matrix $\mathbf{c}^t$ are termed the system (coefficient), input, and output matrices, respectively. Next, we look into methods of deriving the model (1), (2) or equivalently (5),

Systems Analysis Via State Variables

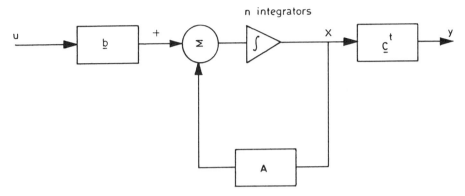

Figure 3.2 Block diagram of the state model (single-variable).

(6). The block diagram representation of the state model (5), (6) is shown in Fig. 3.2.

3.2.1 Obtaining the State Equations

The selection of state variables is an important aspect of state-space formulation. We will mention here some of the standard methods to obtain the state equations. The first method is called the *phase-variable method* in which the first dependent variable in the ODE of the system is selected to be the first state, while its first derivative is taken as the second state, and so on.

Example 3.1. Consider the ODE

$$3\frac{d^3y}{dt^3} + 4\frac{d^2y}{dt^2} + \frac{dy}{dt} = 7t$$

To obtain the state-space representation in terms of the phase variables, we define

$$x_1 = y, \qquad x_2 = \frac{dy}{dt} = \frac{dx_1}{dt}$$

$$x_3 = \frac{d^2y}{dt^2} = \frac{dx_2}{dt}, \qquad u = t$$

Thus

$$\frac{dx_1}{dt} = x_2$$

$$\frac{dx_2}{dt} = x_3$$

and from the ODE

$$\frac{dx_3}{dt} = \frac{-4}{3}x_3 + \frac{-1}{3}x_2 + \frac{7}{3}u(t)$$

which correspond to (1) and in terms of (5) we write

$$\frac{d\mathbf{x}}{dt} = \begin{bmatrix} 0 & 1 & 0 \\ 0 & 0 & 1 \\ 0 & \frac{-1}{3} & \frac{-4}{3} \end{bmatrix} \mathbf{x} + \begin{bmatrix} 0 \\ 0 \\ \frac{7}{3} \end{bmatrix} u(t),$$

Note that the nth-order form of ODE

$$\frac{d^n y}{dt^n} + a_{n-1}\frac{d^{n-1} y}{dt^{n-1}} + \cdots + a_2\frac{d^2 y}{dt^2} + a_1\frac{dy}{dt} + a_0 y = u(t)$$

can be cast into the standard form

$$\frac{d\mathbf{x}}{dt} = \begin{bmatrix} 0 & 1 & 0 & 0 & \cdots & 0 \\ 0 & 0 & 1 & 0 & \cdots & 0 \\ 0 & 0 & 0 & 1 & \cdots & 0 \\ \cdot & \cdot & \cdot & \cdot & \cdot & \cdot \\ \cdot & \cdot & \cdot & \cdot & \cdot & \cdot \\ \cdot & \cdot & \cdot & \cdot & \cdot & \cdot \\ 0 & 0 & 0 & 0 & \cdots & 1 \\ -a_0 & -a_1 & -a_2 & -a_3 & \cdots & -a_{n-1} \end{bmatrix} \mathbf{x} + \begin{bmatrix} 0 \\ 0 \\ 0 \\ \cdot \\ \cdot \\ \cdot \\ 0 \\ 1 \end{bmatrix} u(t) \quad (7)$$

$$\mathbf{y} = [1 \ 0 \ 0 \ 0 \ 0]\mathbf{x}$$

where $x_1 = y$, $x_2 = dy/dt$, $x_2 = d^2y/dt^2, \ldots$ are the state variables and (7) is now called the *phase-variable canonical form*. Figure 3.3 depicts a simulation diagram of (7) in the manner of the state model of Figure 3.2.

Next, suppose the system is represented by the transfer function

$$\frac{Y(s)}{U(s)} = \frac{\sum_{j=0}^n b_j s^j}{\sum_{j=0}^n a_j s^j}$$

Systems Analysis Via State Variables

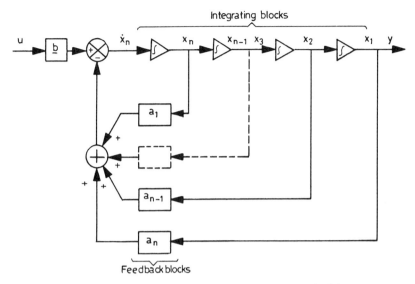

Figure 3.3 A simulation diagram of phase-variable canonical form.

This can be easily converted to

$$\sum_{j=0}^{n} s^j [b_j U(s) - a_j Y(s)] = 0 \tag{8}$$

By properly adding and subtracting certain terms and expanding the inverse Laplace transform of (8) we obtain

$$\left[b_0 u(t) - a_0 y(t) - \frac{dx_n}{dt} \right]$$

$$+ \frac{d}{dt}\left[b_1 u(t) - a_1 y(t) + x_n - \frac{dx_{n-1}}{dt} \right]$$

$$+ \frac{d^2}{dt^2}\left[b_2 u(t) - a_2 y(t) + x_{n-1} - \frac{dx_{n-2}}{dt} \right]$$

.
.
.

$$+ \frac{d^{n-1}}{dt^{n-1}}\left[b_{n-1} u(t) - a_{n-1} y(t) + x_2 - \frac{dx_1}{dt} \right]$$

$$+ \frac{d^n}{dt^n}[b_n u(t) - a_n y(t) + x_1(t)] = 0$$

Since each term of the foregoing expansion is algebraically equal to zero, we obtain by solving backwards and rearranging

$$x_1(t) = a_n y(t) - b_n u(t) \tag{9}$$

$$\frac{dx_1}{dt} = x_2 + b_{n-1} u(t) - a_{n-1} y(t)$$

$$\frac{dx_2}{dt} = x_3 + b_{n-2} u(t) - a_{n-2} y(t)$$

$$\cdot \quad \cdot$$
$$\cdot \quad \cdot \tag{10}$$
$$\cdot \quad \cdot$$

$$\frac{dx_{n-1}}{dt} = x_n + b_1 u(t) - a_1 y(t)$$

$$\frac{dx_n}{dt} = b_0 u(t) - a_0 y(t)$$

Substituting (9) into (10) and rearranging to obtain the compact form (5), (6)

$$\frac{d\mathbf{x}}{dt} \triangleq \begin{bmatrix} \dfrac{dx_1}{dt} \\ \dfrac{dx_2}{dt} \\ \dfrac{dx_3}{dt} \\ \cdot \\ \cdot \\ \cdot \\ \dfrac{dx_{n-1}}{dt} \\ \dfrac{dx_n}{dt} \end{bmatrix} = \begin{bmatrix} \dfrac{-a_{n-1}}{a_n} & 1 & 0 & 0 & \cdots & 0 \\ \dfrac{-a_{n-2}}{a_n} & 0 & 1 & 0 & \cdots & 0 \\ \dfrac{-a_{n-3}}{a_n} & 0 & 0 & 1 & \cdots & 0 \\ \cdot & \cdot & \cdot & \cdot & \cdot & \cdot \\ \cdot & \cdot & \cdot & \cdot & \cdot & \cdot \\ \cdot & \cdot & \cdot & \cdot & \cdot & \cdot \\ \dfrac{-a_1}{a_n} & 0 & 0 & 0 & \cdots & 1 \\ \dfrac{-a_0}{a_n} & 0 & 0 & 0 & \cdots & 0 \end{bmatrix} \begin{bmatrix} x_1(t) \\ x_2(t) \\ x_3(t) \\ \cdot \\ \cdot \\ \cdot \\ x_{n-1}(t) \\ x_n(t) \end{bmatrix}$$

Systems Analysis Via State Variables

$$+ \begin{bmatrix} \dfrac{b_{n-1} - b_n a_{n-1}}{a_n} \\ \dfrac{b_{n-2} - b_n a_{n-2}}{a_n} \\ \dfrac{b_{n-3} - b_n a_{n-3}}{a_n} \\ \cdot \\ \cdot \\ \cdot \\ \dfrac{b_0 - b_n a_0}{a_n} \end{bmatrix} u(t) \quad (11)$$

$$y = \begin{bmatrix} \dfrac{1}{a_n} & 0 & 0 & \cdots & 0 \end{bmatrix} \mathbf{x} + \begin{bmatrix} \dfrac{b_n}{a_n} \end{bmatrix} u(t) \quad (12)$$

The case in which $a_n = 1$ is frequently used in linear systems analysis and we shall use it from now onwards.

Other state representations that are frequently encountered in linear systems analysis are

1. *Controller canonical form* in which

$$A = \begin{bmatrix} -a_0 & \cdots & -a_{n-1} \\ 1 & \cdot & \cdot \\ \cdot & \cdot & \cdot \\ \cdot & \cdot & \cdot \\ \cdot & \cdot & \cdot \\ 0 & 1 & 0 \end{bmatrix}, \quad \mathbf{b} = \begin{bmatrix} 1 \\ \cdot \\ \cdot \\ \cdot \\ \cdot \\ 0 \end{bmatrix} \quad (13)$$

$$\mathbf{c}' = \begin{bmatrix} b_{n-1} & b_{n-2} & \cdots & b_0 \end{bmatrix} \quad (14)$$

2. *Controllability canonical form* in which

$$A = \begin{bmatrix} 0 & 0 & \cdots & -a_{n-1} \\ 1 & 0 & \cdots & -a_{n-2} \\ 0 & 1 & \cdot & \cdot \\ \cdot & \cdot & \cdot & \cdot \\ \cdot & \cdot & \cdot & \cdot \\ 0 & 0 & 1 & -a_0 \end{bmatrix}, \quad \mathbf{b} = \begin{bmatrix} 1 \\ 0 \\ \cdot \\ \cdot \\ \cdot \\ 0 \end{bmatrix} \quad (15)$$

$$\mathbf{c}' = [\beta_{n-1} \quad \beta_{n-2} \quad \cdots \quad \beta_0] \tag{16}$$

3. *Observability canonical form* in which

$$A = \begin{bmatrix} 0 & 1 & \cdots & \\ \cdot & \cdot & & \cdot \\ \cdot & \cdot & & \cdot \\ \cdot & \cdot & & \cdot \\ \cdot & \cdot & & 1 \\ -a_{n-1} & -a_{n-2} & & -a_0 \end{bmatrix}, \quad \mathbf{b} = \begin{bmatrix} \beta_{n-1} \\ \beta_{n-2} \\ \cdot \\ \cdot \\ \cdot \\ \beta_0 \end{bmatrix} \tag{17}$$

$$\mathbf{c}' = [1 \quad 0 \quad \cdots \quad 0] \tag{18}$$

4. *Observer canonical form* in which

$$A = \begin{bmatrix} -a_{n-1} & 1 & \cdots & 0 \\ -a_{n-2} & 0 & \cdots & \cdot \\ \cdot & \cdot & & \cdot \\ \cdot & \cdot & & \cdot \\ \cdot & \cdot & & 1 \\ -a_0 & 0 & \cdots & 0 \end{bmatrix}, \quad \mathbf{b} = \begin{bmatrix} b_{n-1} \\ b_{n-2} \\ \cdot \\ \cdot \\ \cdot \\ b_0 \end{bmatrix} \tag{19}$$

$$\mathbf{c}' = [1 \quad 0 \quad \cdots \quad 0] \tag{20}$$

where in (16), (17) the β are usually termed the *Markovian parameters* given by

$$\begin{bmatrix} \beta_0 \\ \beta_1 \\ \beta_2 \\ \cdot \\ \cdot \\ \cdot \\ \beta_{n-1} \end{bmatrix} = \begin{bmatrix} 1 & 0 & \cdots & 0 \\ a_0 & 1 & \cdots & 0 \\ a_1 & a_0 & \cdots & 0 \\ \cdot & \cdot & & \cdot \\ \cdot & \cdot & & \cdot \\ \cdot & \cdot & & \cdot \\ a_{n-1} & \cdots & a_1 a_0 & 1 \end{bmatrix}^{-1} \begin{bmatrix} b_{n-1} \\ b_{n-2} \\ b_{n-3} \\ \cdot \\ \cdot \\ \cdot \\ b_0 \end{bmatrix} \tag{21}$$

3.2.2 Free Response

We now consider the solution of the state equations represented by (5) and (6). By similarity to the ODE, the solution consists of two parts: free response corresponding to the homogeneous system ($u(t) = 0$ or

Systems Analysis Via State Variables

no control forces being applied to the system) and forced response corresponding to the nonhomogeneous system (control forces are applied to the system, $u(t) \neq 0$). We shall consider first the free response. The homogeneous system of (4) is given by

$$\dot{\mathbf{x}} = A\mathbf{x}, \quad \mathbf{x}(0) = \mathbf{x}_0 \tag{22}$$

where $\mathbf{x}_0$ is the initial condition. By way of analogy to the scalar case ($\dot{x} = ax, x(0) = x_0$ has a solution $x(t) = e^{at}x_0$), the solution of (22) takes the form

$$\mathbf{x}(t) = e^{At}\mathbf{x}_0 \tag{23}$$

Note that if the initial condition was given as $\mathbf{x}(t_0)$ at initial time t_0, then (23) would be replaced by

$$\mathbf{x}(t) = e^{A(t-t_0)}\mathbf{x}(t_0) \tag{24}$$

Again, by similarity to the scalar case, we can write (23) as

$$\mathbf{x}(t) = \left[I + At + \frac{(At)^2}{2!} + \frac{(At)^3}{3!} + \cdots + \frac{(At)^j}{j!} + \cdots \right] \mathbf{x}_0 \tag{25}$$

An interpretation of (24), and the same applies to (23), is that the initial state $\mathbf{x}(t_0)$ at $t = t_0$ is driven to a state $\mathbf{x}(t)$ at time t. This transition in state is carried out by the matrix exponential $e^{A(t-t_0)}$; in which the time elapsed in transition $(t - t_0)$ appears clearly in the exponent. Because of this property, the term $e^{A(t-t_0)}$ is known as the *state transition matrix* and is denoted by $\Phi(t, t_0)$.

3.2.3 Forced Response

Now, we turn attention to the determination of the solution of the nonhomogeneous state equation (forced system)

$$\dot{\mathbf{x}} = A\mathbf{x}(t) + B\mathbf{u}(t), \quad \mathbf{x}(t_0) = \mathbf{x}_0 \tag{26}$$

where a control vector $\mathbf{u}(t)$ is used for generality. Rewriting (26) in the form

$$\dot{\mathbf{x}} - A\mathbf{x}(t) = B\mathbf{u}(t)$$

and multiplying both sides by $\bar{e}^{At}$ we have

$$\bar{e}^{At}[\dot{\mathbf{x}} - A\mathbf{x}(t)] = \bar{e}^{At}B\mathbf{u}(t) \tag{27}$$

But from matrix algebra in Appendix B, the left-hand side of (27) is

$$\frac{d}{dt}[\tilde{e}^{At}\mathbf{x}(t)] = \tilde{e}^{At}\mathbf{Bu}(t)$$

which we now integrate with respect to t between the limits t_0 and t to arrive at

$$\tilde{e}^{At}\mathbf{x}(t)\big|_{t_0}^{t} = \int_{t_0}^{t} \tilde{e}^{A\tau}\mathbf{Bu}(\tau)\,d\tau$$

$$\tilde{e}^{At}\mathbf{x}(t) - \tilde{e}^{At_0}\mathbf{x}(t_0) = \int_{t_0}^{t} \tilde{e}^{A\tau}\mathbf{Bu}(\tau)\,d\tau \tag{28}$$

Premultiplying (28) by e^{At} and rearranging, we reach

$$\mathbf{x}(t) = e^{A(t-t_0)}\mathbf{x}(t_0) + \int_{t_0}^{t} e^{A(t-\tau)}\mathbf{Bu}(\tau)\,d\tau \tag{29}$$

in which it is readily evident that the first term constitutes the homogeneous solution in view (24) and the second term represents the forced response. We thus call $\mathbf{x}(t)$ in (29) as the complete response which is subsumed of two parts:

Zero-input response (fee response) resulting from the initial conditions or equivalently $u(t) = 0$ for $t \geq 0$.
Zero-state response (forced response) resulting from the effect of input excitation only ($\mathbf{x}(t_0) = \mathbf{0}$).

3.2.4 Properties of the State Transition Matrix

In the above discussion, the state transition matrix has been defined as

$$\Phi(t, t_0) = e^{A(t-t_0)}$$

For a LTI system, it is customary to write

$$\Phi(t, t_0) = \Phi(t - t_0) \tag{30}$$

since in this case the time difference $(t - t_0)$ is of interest irrespective of the initial time t_0 and the current time t. Certain useful properties of the state transition matrix $\Phi(\ )$ are given below:

1. $\Phi(0) = e^{A0} = I$
2. $\Phi(\tau) = e^{A\tau} = [\tilde{e}^{A\tau}]^{-1} = [\Phi(-\tau)]^{-1}$ or $\Phi^{-1}(\tau) = \Phi(-\tau)$
3. $\Phi(\tau_1 + \tau_2) = e^{A(\tau_1 + \tau_2)} = e^{A\tau_1}e^{A\tau_2} = \Phi(\tau_1)\Phi(\tau_2) = e^{A\tau_2}e^{A\tau_1} = \Phi(\tau_2)\Phi(\tau_1)$
4. $d\Phi(t)/dt = Ae^{At} = e^{At}A = A\Phi(t) = \Phi(t)A$

Systems Analysis Via State Variables

In terms of the state transition matrix $\Phi(t)$, the solution of the forced system given by (29) can be written as

$$\mathbf{x}(t) = \Phi(t - t_0)\mathbf{x}(t_0) + \int_{t_0}^{t} \Phi(t - \tau)\mathbf{Bu}(\tau)\,d\tau \quad (31)$$

where, for emphasis, we can express $\Phi(t - \tau)$ in the form

$$\Phi(t - \tau) = I + A(t - \tau) + \frac{A^2(t - \tau)^2}{2!} + \frac{A^3(t - \tau)^3}{3!} + \cdots \quad (32)$$

as infinite series that can be truncated after a finite number of terms to obtain an approximation for the transition matrix.

To complete this section, we describe below some of the methods for computing the transition matrix.

Laplace transform method

On taking the Laplace transform of (22) we obtain

$$sX(s) - \mathbf{x}(0) = AX(s) \quad (33)$$

where $\{\mathbf{x}(t), X(s)\}$ are the Laplace transform pair. Rearranging (33) and inverting we reach

$$(sI - A)X(s) = \mathbf{x}(0)$$

$$X(s) = (sI - A)^{-1}\mathbf{x}(0)$$

$$\mathbf{x}(t) = \mathscr{L}^{-1}[(sI - A)^{-1}]\mathbf{x}(0) \quad (34)$$

A comparison of (23) and (34) reveals that

$$\Phi(t) = e^{At} = \mathscr{L}^{-1}[(sI - A)^{-1}] = \mathscr{L}^{-1}[\Phi(s)] \quad (35)$$

where $\Phi(s) = (sI - A)^{-1}$ is called the *resolvent matrix*.

Canonical transformation method

We start by the state model (22) or its solution (23) and introduce a new state vector $\mathbf{z}$ such that

$$\mathbf{x}(t) = M\mathbf{z}(t) \quad (36)$$

where M is called the *modal matrix*. For a concise account of matrix algebra, the reader is referred to Appendix B. It suffices here to consider the columns of M as the right-eigenvectors corresponding to the distinct eigenvalues of A. The case of repeated eigenvalues is treated similarly with extra manipulations. Under the transformation (36), the original state model (22) is modified to

$$\dot{\mathbf{z}} = M^{-1}AM\mathbf{z}(t) = J\mathbf{z}(t) \tag{37}$$

where J is a diagonal matrix whose diagonal elements are the distinct eigenvalues of A, frequently called the *Jordan form*. The solution of (37) is

$$\mathbf{z}(t) = e^{Jt}\mathbf{z}(0) \tag{38}$$

In view of (25) we write the matrix function e^{Jt} in the form

$$e^{Jt} = I + Jt + \frac{J^2 t^2}{2!} + \frac{J^3 t^3}{3!} + \cdots$$

Since J is diagonal so is the term $J^i t^i/i!$ for all $i = 1, \ldots, \infty$. More importantly, the diagonal elements of $J^i t^i/i!$ have the general form $\lambda_j^i/i!$ where λ_j is the jth eigenvalue of A, $j = 1, \ldots, n$. Looked at in this light, it is readily evident that e^{Jt} will be a diagonal matrix, being the sum of diagonal matrices, with the jth term as the series

$$1 + \lambda_j t + \frac{\lambda_j^2 t^2}{2!} + \frac{\lambda_j^3 t^3}{3!} + \cdots$$

which in view of (25) corresponds to $e^{\lambda_j t}$. Thus

$$e^{Jt} = \text{diag}[e^{\lambda_1 t} \quad e^{\lambda_2 t} \quad \cdots \quad e^{\lambda_n t}] \tag{39}$$

From (36) and (38) we have

$$\mathbf{x}(t) = M e^{Jt} M^{-1} \mathbf{x}(0) \tag{40}$$

Comparison of (40) with (23) yields

$$e^{At} = M e^{Jt} M^{-1} \tag{41}$$

which is the desired relation. Note that e^{Jt} is given by (39).

Cayley–Hamilton method

As explained in Appendix B, the *Cayley–Hamilton theorem* states that every square matrix satisfies its own characteristic equation. Thus if the characteristic equation of A is

$$\begin{aligned} q(\lambda) &= \det[\lambda I - A] \\ &= \lambda^n + a_1 \lambda^{n-1} + a_2 \lambda^{n-2} + \cdots + a_{n-1}\lambda + a_n \\ &= 0 \end{aligned} \tag{42}$$

Systems Analysis Via State Variables

then
$$q(A) = A^n + a_1 A^{n-1} + \cdots + a_{n-1} A + a_n I = 0 \tag{43}$$

To use this result, we consider that A has the characteristic equation (42) and a set of distinct eigenvalues $\lambda_1, \ldots, \lambda_n$. We focus first on the scalar situation. The matrix function
$$f(A) = \sum_{j=0} k_j A^j$$
can be computed by consideration of the corresponding scalar function
$$f(\lambda) = \sum_{j=0} k_j \lambda^j$$

Dividing $f(\lambda)$ by $q(\lambda)$ in (42), we then obtain
$$\frac{f(\lambda)}{q(\lambda)} = Q(\lambda) + \frac{R(\lambda)}{q(\lambda)}$$
or
$$f(\lambda) = Q(\lambda) q(\lambda) + R(\lambda) \tag{44}$$

where $Q(\lambda)$ is a quotient polynomial and $R(\lambda)$ is the remainder polynomial expressed in the form
$$R(\lambda) = \alpha_0 + \alpha_1 \lambda + \alpha_2 \lambda^2 + \cdots + \alpha_{n-1} \lambda^{n-1} \tag{45}$$

For the particular points at which $\lambda = \lambda_j$, $j = 1, \ldots, n$ then $q(\lambda_j) = 0$ by definition and subsequently (44) reduces to
$$f(\lambda_j) = R(\lambda_j), \quad j = 1, \ldots, n \tag{46}$$

Repeated application of (46) will yield the n coefficients $(\alpha_0, \alpha_1, \ldots, \alpha_{n-1})$. Now, we produce the matrix result. An equivalent matrix expression of (44) is
$$F(A) = Q(A) q(A) + R(A)$$

Since from (43), $q(A)$ is identically zero at the eigenvalues $\lambda_1, \ldots, \lambda_n$, thus we obtain
$$f(A) = R(A)$$
$$= \alpha_0 I + \alpha_1 A + \cdots + \alpha_{n-1} A^{n-1} \tag{47}$$

which is the desired result. Formally we proceed as follows (for the case of distinct eigenvalues)

1. Determine the eigenvalues of matrix A.

2. Solve n simultaneous equations given by (46) for the coefficients $\alpha_0, \ldots, \alpha_{n-1}$.
3. Substitute the results in (47).

The case of repeated eigenvalues is treated in Appendix B.

To conclude this section, we should recall that (5) and (6) have been posed as a linear, general description of LTI. This can be related back to transfer functions by Laplace transforming (5), (6) to yield

$$sX(s) - \mathbf{x}_0 = AX(s) + \mathbf{b}U(s) \tag{48}$$

$$Y(s) = \mathbf{c}'X(s) \tag{49}$$

From (48) we obtain

$$(sI - A)X(s) = \mathbf{x}_0 + \mathbf{b}U(s)$$

or

$$X(s) = [(sI - A)^{-1}]\mathbf{x}_0 + [(sI - A)^{-1}\mathbf{b}U(s)]$$

By inverse Laplace transformation

$$\mathbf{x}(t) = \mathcal{L}^{-1}[(sI - A)^{-1}]\mathbf{x}_0 + \mathcal{L}^{-1}[(sI - A)^{-1}\mathbf{b}U(s)]$$
$$= \Phi(t)\mathbf{x}_0 + \mathcal{L}^{-1}[\Phi(s)\mathbf{b}U(s)] \tag{50}$$

which ties the transition matrix and the resolvent matrix to the state vector.

3.2.5 Examples

We now consider some illustrative examples.

Example 3.2. A system is characterized by

$$\dot{x}_1 = 0, \qquad \dot{x}_2 = 3x_1(t) - 3x_2(t), \qquad y(t) = x_1(t) + 2x_2(t)$$

If $x_1(0) = 1$, $x_2(0) = 2$ determine $y(t)$ for $t > 0$.

It is clear that the system is in the form (5) and (6) with $u(t) = 0$. The system matrix A is given by

$$A = \begin{bmatrix} 0 & 0 \\ 3 & -3 \end{bmatrix}$$

and the associative resolvent matrix $\Phi(s)$

Systems Analysis Via State Variables

$$\Phi(s) = (sI - A)^{-1}$$

$$= \frac{1}{s(s+3)} \begin{bmatrix} s+3 & 0 \\ 3 & s \end{bmatrix}$$

$$= \begin{bmatrix} \dfrac{1}{s} & 0 \\ \dfrac{3}{s(s+3)} & \dfrac{1}{s+3} \end{bmatrix}$$

By inverse Laplace transformation

$$\Phi(t) = \begin{bmatrix} 1 & 0 \\ 1 - e^{-3t} & e^{-3t} \end{bmatrix}$$

From (23)

$$\mathbf{x}(t) = \Phi(t)\mathbf{x}_0$$

$$= \begin{bmatrix} 1 & 0 \\ 1 - e^{-3t} & e^{-3t} \end{bmatrix} \begin{bmatrix} 1 \\ 2 \end{bmatrix} = \begin{bmatrix} 1 \\ 1 + e^{-3t} \end{bmatrix}$$

and finally

$$y(t) = x_1(t) + 2x_2(t) = 3 + 2e^{-3t}$$

Example 3.3. Consider the mechanical system shown in Fig. 3.4 wherein mass M is acted upon by the force $F(t)$. Simple dynamics shows that the system is characterized by the relations

$$\frac{d}{dt}[v(t)] = \frac{1}{M} F(t)$$

$$\frac{d}{dt}[x(t)] = v(t)$$

One possibility to obtain the state model is to define the following variables as

$$x_1(t) = x(t), \quad x_2(t) = v(t)$$
$$u(t) = F(t), \quad y(t) = x_1(t)$$

The dynamical model can then be rewritten in the form

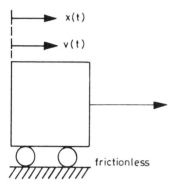

Figure 3.4 A simple mechanical system.

$$\frac{d}{dt}\begin{bmatrix} x_1 \\ x_2 \end{bmatrix} = \begin{bmatrix} 0 & 1 \\ 0 & 0 \end{bmatrix}\begin{bmatrix} x_1 \\ x_2 \end{bmatrix} + \begin{bmatrix} 0 \\ \frac{1}{M} \end{bmatrix} u$$

$$y = \begin{bmatrix} 1 & 0 \end{bmatrix}\begin{bmatrix} x_1 \\ x_2 \end{bmatrix}$$

which corresponds to (5) and (6) with

$$A = \begin{bmatrix} 0 & 1 \\ 0 & 0 \end{bmatrix}, \quad \mathbf{b} = \begin{bmatrix} 0 \\ \frac{1}{M} \end{bmatrix}, \quad \mathbf{c}' = \begin{bmatrix} 1 & 0 \end{bmatrix}$$

It should be noted that the state of a system is not uniquely specified. To show this, define

$$z_1(t) = 2x(t) + v(t), \quad z_2(t) = x(t) + v(t), \quad y(t) = z_1(t) - z_2(t)$$

A little algebra indicates that

$$x(t) = z_1(t) - z_2(t), \quad v(t) = -z_1(t) + 2z_2(t)$$

But

$$\frac{dz_1(t)}{dt} = 2\frac{dx(t)}{dt} + \frac{dv(t)}{dt}$$

$$= 2v(t) + \frac{1}{M}F(t)$$

Systems Analysis Via State Variables

$$= -2z_1(t) + 4z_2(t) + \frac{1}{M}F(t)$$

The last two relations can be written in the standard form

$$\frac{d}{dt}\begin{bmatrix} z_1 \\ z_2 \end{bmatrix} = \begin{bmatrix} -2 & 4 \\ -1 & 2 \end{bmatrix}\begin{bmatrix} z_1 \\ z_2 \end{bmatrix} + \begin{bmatrix} \frac{1}{M} \\ \frac{1}{M} \end{bmatrix} u$$

$$y(t) = \begin{bmatrix} 1 & -1 \end{bmatrix}\begin{bmatrix} z_1 \\ z_2 \end{bmatrix}$$

which has the components

$$A = \begin{bmatrix} -2 & 4 \\ -1 & 2 \end{bmatrix}, \quad \mathbf{b} = \begin{bmatrix} \frac{1}{M} \\ \frac{1}{M} \end{bmatrix}, \quad \mathbf{c}' = \begin{bmatrix} 1 & -1 \end{bmatrix}$$

It is immediately observed that we have two state models for the same mechanical system. This, at least, leads us to conclude that the state variables are nonunique and that this example further brings about the fact that the state variables need not necessarily be physical variables of the system.

Though the state model is not unique, all such models have one characteristic in common for a given system, namely, the number of elements in the state vector is equal and minimal. This number n is referred to as the *order* of the system.

Example 3.4. An RLC network is shown in Fig. 3.5. It is desired to derive a state variable model of this network. With reference to Chapter 2 on electrical elements, this network possesses three energy storage elements, a capacitor C and two inductors L_1 and L_2. In turn, knowledge of $v(0)$, $i_1(0)$, $i_2(0)$, and $e(t)$ for $t \geq 0$ is sufficient to completely specify the behavior of the network. Thus we choose

$$x_1 = v(t), \quad x_2(t) = i_1(t), \quad x_3(t) = i_2(t)$$

The differential equations governing the behavior of the RLC network are

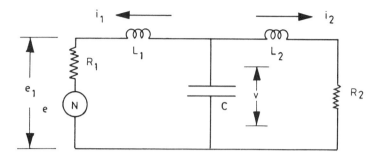

Figure 3.5 An RLC network.

$$i_1(t) + i_2(t) + C\frac{dv}{dt} = 0$$

$$L_1\frac{di_1}{dt} + R_1 i_1(t) + e(t) - v(t) = 0$$

$$L_2\frac{di_2}{dt} + R_2 i_2(t) - v(t) = 0$$

Rearranging the above equations to retain the rates of state variables in terms of the other quantities, we obtain

$$\frac{dv}{dt} = -\frac{1}{C}i_1 - \frac{1}{C}i_2$$

$$\frac{di_1}{dt} = \frac{1}{L_1}v - \frac{R_1}{L_1}i_1 - \frac{1}{L_1}e$$

$$\frac{di_2}{dt} = \frac{1}{L_2}v - \frac{R_2}{L_2}i_2$$

Let $u(t) = e(t)$; the state model is given by

$$\frac{d}{dt}\begin{bmatrix} x_1 \\ x_2 \\ x_3 \end{bmatrix} = \begin{bmatrix} 0 & -\frac{1}{C} & -\frac{1}{C} \\ \frac{1}{L_1} & -\frac{R_1}{L_1} & 0 \\ \frac{1}{L_2} & 0 & -\frac{R_2}{L_2} \end{bmatrix} \begin{bmatrix} x_1 \\ x_2 \\ x_3 \end{bmatrix} + \begin{bmatrix} 0 \\ -\frac{1}{L_1} \\ 0 \end{bmatrix} u$$

Assume that the voltage across R_2 as the output variable y_1; thus

$$y_1 = R_2 i_2$$

$$= \begin{bmatrix} 0 & 0 & R_2 \end{bmatrix} \begin{bmatrix} x_1 \\ x_2 \\ x_3 \end{bmatrix}$$

$$A = \begin{bmatrix} 0 & -\dfrac{1}{C} & -\dfrac{1}{C} \\ \dfrac{1}{L_1} & -\dfrac{R_1}{L_1} & 0 \\ \dfrac{1}{L_2} & 0 & -\dfrac{R_2}{L_2} \end{bmatrix}, \quad \mathbf{b} = \begin{bmatrix} 0 \\ -\dfrac{1}{L_1} \\ 0 \end{bmatrix}$$

$$\mathbf{c}' = \begin{bmatrix} 0 & 0 & R_2 \end{bmatrix}$$

3.2.6 State Model Transformations

Suppose that we have the state model

$$\dot{\mathbf{x}} = A\mathbf{x} + \mathbf{b}u, \quad \mathbf{x}(0) = \mathbf{x}_0 \tag{51}$$

$$y = \mathbf{c}'\mathbf{x} \tag{52}$$

of a given system according to certain choice of state variables. It is possible to obtain an alternative state-space representation of the system of desired properties by using appropriate change of coordinates. In general, we can write the alternative state model as

$$\dot{\mathbf{z}} = E\mathbf{z} + \mathbf{d}u, \quad \mathbf{z}(0) = \mathbf{z}_0 \tag{53}$$

$$y = \mathbf{g}'\mathbf{z} \tag{54}$$

It must be emphasized that both state models are subject to the same input u and yield the same output y. To relate the two state vectors, we introduce the linear transformation

$$\mathbf{z} = T\mathbf{x} \tag{55}$$

where the $n \times n$ matrix T, assumed invertible, is called the *similarity transformation*. We now link the state models via T by following either of two approaches: the first is to manipulate (51), (52) using (55) to reach (53), (54) and the second is the reverse route by manipulating (53), (54) using (55) to arrive at (51), (52). We follow the first approach. Multiplying (51) by T and making use of (55) we obtain

$$T\dot{\mathbf{x}} = TA\mathbf{x} + T\mathbf{b}u$$
$$\dot{\mathbf{z}} = (TAT^{-1})\mathbf{z} + T\mathbf{b}u \tag{56}$$

From (52) and (55) we have

$$y = \mathbf{c}'\mathbf{x} = (\mathbf{c}'T^{-1})\mathbf{z} \tag{57}$$

A comparison of (56), (57) with (53), (54) reveals that

$$E = TAT^{-1}, \qquad \mathbf{d} = T\mathbf{b}, \qquad \mathbf{g}' = \mathbf{c}'T^{-1} \tag{58}$$

which can be interpreted as, given the triplet $(A, \mathbf{b}, \mathbf{c}')$ and the similarity transformation T, we can use (58) to produce the triplet $(E, \mathbf{d}, \mathbf{g}')$.

Had we followed the second approach we would have arrived at

$$A = T^{-1}ET, \qquad \mathbf{b} = T^{-1}\mathbf{d}, \qquad \mathbf{c}' = \mathbf{g}'T \tag{59}$$

which can now be interpreted as, given the triplet $(E, \mathbf{d}, \mathbf{g}')$ and the similarity transformation T, we can use (59) to produce the triplet $(A, \mathbf{b}, \mathbf{c}')$.

Alternatively, given the two triplets $(A, \mathbf{b}, \mathbf{c}')$, $(E, \mathbf{d}, \mathbf{g}')$ (sometimes called *realizations*) the question asked is: What is T that relates the two realizations? The answer comes directly from either (58) or (59). One needs to solve

$$AT = ET, \qquad T\mathbf{b} = \mathbf{d}, \qquad \mathbf{c}' = \mathbf{g}'T \tag{60}$$

which are $(n \times n + n + n = n^2 + 2n)$ equations in n^2 unknowns. This means that there are $2n$ redundant equations in (60) which usually take the form of repeated equations.

3.3 PERFORMANCE INDICES

As already discussed, the design of a control system is an effort to satisfy a set of specifications that defines the overall performance of the system in terms of certain measurable quantities. A number of performance measures have been introduced so far in respect of dynamic response to step input (rise time, time to peak, percentage overshoot, damping factor, settling time, etc.) and the steady-state error e_{ss}, to both step and higher-order inputs. Unfortunately, these

Systems Analysis Via State Variables

measures have to be satisfied simultaneously in design, which in turn makes the design procedure intractable. If, however, a single performance index could be established on the basis of which one may evaluate the goodness of the system response, then the design procedure will become logical and systematic.

A number of such performance indices are used in practice, the most common being the *integral square error* (ISE), given by

$$\text{ISE} = \int_0^T e^2(t)\, dt \tag{61}$$

where the upper limit T is chosen sufficiently large so that $e(t)$ is negligible for $t > T$; say a finite multiple of the settling time. Note that criterion (63) will discriminate between excessively overdamped systems and excessively underdamped systems.

Another easily instrumented performance criterion is the *integral of the absolute magnitude of error* (IAE), which is written as

$$\text{IAE} = \int_0^T |e(t)|\, dt \tag{62}$$

The usefulness of (62) is found in analog computer simulation studies. In order to reduce the weighting of the large initial error and to penalize the small errors occurring later in the response more heavily, the following indices are proposed: *Integral time-absolute error* (ITAE) and *integral time-square error* (ITSE) where

$$\text{ITAE} = \int_0^T t|e(t)|\, dt \tag{63}$$

$$\text{ITSE} = \int_0^t te^2(t)\, dt \tag{64}$$

The performance index ITAE provides the best selectivity of the performance indices, that is, the minimum value of the integral is readily discernible as the system parameters are varied. The general form of the performance integral is

$$I = \int_0^T f[\mathbf{e}(t), r(t), c(t), t]\, dt \tag{65}$$

where $f[.,.,.,.]$ is a function of the error, input, output, and time.

3.4 RELATIONSHIP OF STATE DESCRIPTION TO THE TRANSFER MATRIX

The most general state-space description of a linear, time-invariant system with r inputs $\mathbf{u}(t)$, m outputs $\mathbf{y}(t)$, and n state variables $\mathbf{x}(t)$ is given by

$$\dot{\mathbf{x}}(t) = A\mathbf{x}(t) + B\mathbf{u}(t) \qquad (66)$$

$$\mathbf{y}(t) = C\mathbf{x}(t) + D\mathbf{u}(t) \qquad (67)$$

where A, B, C, and D are constant matrices of dimensions $n \times n$, $n \times r$, $m \times n$, and $m \times r$, respectively. It is easily evident that (66), (67) is a natural extension of (5), (6) as shown in Fig. 3.6 when compared to Fig. 3.2. The Laplace transform of (63) yields

$$sX(s) = AX(s) + BU(s) \qquad (68)$$

or

$$X(s) = (sI - A)^{-1}BU(s) \qquad (69)$$

Using (70) in the Laplace transform of (69) gives the desired input-output description

$$Y(s) = [C(sI - A)^{-1}B + D]U(s) \qquad (70)$$

From (70) we define the $m \times r$ matrix which premultiplies $U(s)$ as the *transfer matrix* $H(s)$,

$$H(s) = [C(sI - A)^{-1}B + D] \qquad (71)$$

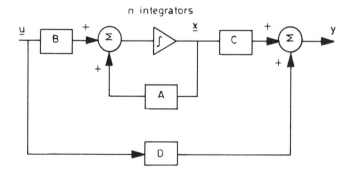

Figure 3.6 Block diagram of a multivariable system.

Systems Analysis Via State Variables

In terms of $\Psi(s) = \text{Adj}[sI - A]$ as the *adjoint matrix*, $\Delta(s) = \det[sI - A]$ as the *characteristic polynomial* and $(sI - A)^{-1} = \Psi(s)/\Delta(s)$, we write (71) as

$$H(s) = \frac{[C\Psi(s)B + D\Delta(s)]}{\Delta(s)} \tag{72}$$

It must be emphasized that each element of the adjoint matrix is a polynomial in s of degree less than or equal to $n - 1$. Since $\Delta(s)$ is an nth degree polynomial in s, each element $H_{ij}(s)$ of the H-matrix is a ratio of polynomials in s, with the degree of the denominator at least as great as the degree of the numerator. Such an $H(s)$ matrix is called a *proper* rational matrix (loosely, at least as many poles as zeros). If $D = 0$, meaning no direct coupling between input and output, then the numerator of every element in $H(s)$ will be of degree less than the denominator. In this case, $H(s)$ is a *strictly proper* rational matrix (loosely, more poles than zeros). It is clear from (71) that $D = \lim_{s \to \infty} H(s)$.

3.5 EXAMPLES

We conclude this chapter by solving typical examples.

Example 3.5. Find $f(A) = e^{At}$ for

$$A = \begin{bmatrix} 0 & 1 \\ -1 & -2 \end{bmatrix}$$

The characteristic equation $q(\lambda) = \det[\lambda I - A] =$ has the form

$$\det \begin{bmatrix} \lambda & -1 \\ 1 & \lambda + 2 \end{bmatrix} = (\lambda + 1)^2 = 0$$

Matrix A has eigenvalues $\lambda_1, \lambda_2 = -1$. Since A is of second-order, the polynomial $R(\lambda)$ will have the form

$$R(\lambda) = \alpha_0 + \alpha_1 \lambda$$

The coefficients α_0 and α_1 are evaluated using (46) as

$$f(-1) = e^{-t} = \alpha_0 - \alpha_1$$

$$\frac{d}{d\lambda} f(\lambda)|_{\lambda=-1} = te^{-t} = \frac{d}{d\lambda} R(\lambda)|_{\lambda=-1} = \alpha_1$$

The result is $\alpha_1 = te^t$, $\alpha_0 = (1+t)e^t$. From (47) we obtain

$$e^{At} = \alpha_0 I + \alpha_1 A$$

$$= \begin{bmatrix} (1+t)e^t & te^t \\ -te^t & (1-t)e^t \end{bmatrix}$$

Example 3.6. Obtain the state model for a system described by

$$\ddot{h}(t) + k_1 h(t) + B_1 \dot{g}(t) = f_1(t)$$
$$\ddot{g}(t) + k_2 g(t) + B_2 \dot{h}(t) = f_2(t)$$

A straightforward definition of state variables would be

$$x_1(t) = h(t), \qquad x_2(t) = \dot{h}(t)$$
$$x_3(t) = g(t), \qquad x_2(t) = \dot{g}(t)$$

and the state equations are

$$\dot{x}_1 = x_2(t)$$
$$\dot{x}_2 = -K_1 x_1(t) - B_1 x_4(t) + f_1(t)$$
$$\dot{x}_3 = x_4(t)$$
$$\dot{x}_4 = -K_2 x_3(t) - B_2 x_2(t) + f_2(t)$$

which can be put in the standard form (5), (6) with

$$A = \begin{bmatrix} 0 & 1 & 0 & 0 \\ -K_1 & 0 & 0 & -B_1 \\ 0 & 0 & 0 & 1 \\ 0 & -B_2 & -K_2 & 0 \end{bmatrix}, \quad bu(t) = \begin{bmatrix} 0 \\ f_1(t) \\ 0 \\ f_2(t) \end{bmatrix}$$

Example 3.7. A candidate performance index is

$$I = \int_0^\infty [\mathbf{x}' \quad \mathbf{x}] \, dt$$

Find the minimum value of I for the two cases

1. $\mathbf{x}(t) = \begin{bmatrix} x_1(t) \\ x_2(t) \end{bmatrix} = \begin{bmatrix} \sigma e^{\sigma t} \\ 2 e^{\sigma t} \end{bmatrix}$

   and σ is a parameter

Systems Analysis Via State Variables

2. $\mathbf{x}(t) = \begin{bmatrix} x_1(t) \\ x_2(t) \end{bmatrix} = \begin{bmatrix} \dfrac{\delta}{\sqrt{1-\delta^2}} e^{-\delta t} \sin(\sqrt{1-\delta^2}\,)t \\ \left(\dfrac{1}{2}\right) e^{\delta t} \end{bmatrix}$

and δ is a parameter.

For case (1), I is given by

$$I = \int_0^\infty [\sigma^2 e^{2\sigma t} + re^{2\sigma t}]\, dt$$

carrying out the integration yields $(\sigma^2 + 4)/2\sigma$. Setting $dI/d\sigma = 0$ results in $\sigma^* = 2$, and $I^* = 2$.

For case (2), I takes the form

$$I = \int_0^\infty \left[\delta^2 e^{2\delta t} \sin^2\dfrac{(\sqrt{1-\delta^2}\,)t}{(1-\delta^2)} + \dfrac{1}{4} e^{2\delta t}\right] dt$$

Simple algebra gives

$$I = \dfrac{\delta^2 + 0.5}{4\delta}$$

For minimizing J we obtain $dJ/d\delta = 0$, which yields $\delta^* = 1/\sqrt{2} = 0.707$, and $I^* = 0.3536$.

Example 3.8. Determine the transfer matrix for a system described by

$$\dot{\mathbf{x}} = \begin{bmatrix} -1 & 1 & 0 \\ 0 & -1 & 0 \\ 0 & 0 & -1 \end{bmatrix} \mathbf{x} + \begin{bmatrix} 2 & 0 \\ 1 & -1 \\ -1 & 1 \end{bmatrix} \mathbf{u}$$

$$\mathbf{y} = \begin{bmatrix} 2 & -1 & 3 \\ 0 & 1 & 0 \\ 3 & -1 & 6 \end{bmatrix} \mathbf{x} + \begin{bmatrix} 1 & 1 \\ 0 & 0 \\ 0 & 0 \end{bmatrix} \mathbf{u}$$

Using (71) and carrying out the matrix inversion and multiplications, the result is

$$H(s) = \begin{bmatrix} \dfrac{s^2 + 2s + 3}{(s + 1)^2} & \dfrac{s^2 + 6s + 3}{(s + 1)^2} \\ \dfrac{1}{s + 1} & \dfrac{-1}{(s + 1)} \\ \dfrac{-s + 2}{(s + 1)^2} & \dfrac{7s + 4}{(s + 1)^2} \end{bmatrix}$$

3.6 PROBLEMS

1. Using the fundamental relation $d\Phi/dt = \Phi(t)$ with $\Phi(0) = I$ and applying the Laplace transform techniques show that
$$A = \lim_{s \to \infty}[s^2\Phi(s) - sI]$$
where I is the unit matrix.

2. Given that the transfer matrix of a second-order system has the form
$$\Phi(s) = \begin{bmatrix} s + 6 & 1 \\ -5 & s \end{bmatrix}\left(\dfrac{1}{s^2 + 6s + 5}\right)$$
obtain A.

3. If $(A, \mathbf{b}, \mathbf{c}')$ and $(E, \mathbf{d}, \mathbf{g}')$ are state-space representations and are related by a constant similarity transformation, show that they have the same transfer function.

4. Find state representations in controller, observer, controllability, observability, and Jordan canonical forms of the transfer function
$$G(s) = \dfrac{s^2 + 5s + 4}{s^3 + 8s^2 + 17s + 10}$$

5. Suppose that
$$\Delta(s) = \det(sI - A) = s^n + a_1 s^{n-1} + a_2 s^{n-2} + \cdots + a_{n-1}s + a_n$$
Verify that
$$\begin{aligned}
\text{Adj}(sI - A) &= [Is^{n-1} + (A + a_1 I)s^{n-2} + \cdots \\
&\quad + (A^{n-1} + a_1 A^{n-2} + \cdots + a_{n-1}I)] \\
&= [A^{n-1} + (s + a_1)A^{n-2} + \cdots \\
&\quad + (s^{n-1} + a_1 s^{n-2} + \cdots + a_{n-1})I]
\end{aligned}$$

Systems Analysis Via State Variables

6. Let the state representation of a system be expressed in controller form. Define $e'_j = [0\ 0\ \cdots\ 0\ 1\ 0\ \cdots\ 0]$ with the entry 1 in the jth place. Show that

$$e'_j A = e'_{j-1}, \quad 2 \leq j \leq n$$

$$e'_1 A = [-a_1\ -a_2\ \cdots\ -a_n]$$

Verify that the above two relations are applicable to other canonical forms with appropriate manipulation of e_j and A.

7. A system is described by

$$\frac{d}{dt}\begin{bmatrix} x_1 \\ x_2 \\ x_2 \end{bmatrix} = \begin{bmatrix} -2 & -2 & 0 \\ 0 & 0 & 1 \\ 0 & -3 & -4 \end{bmatrix}\begin{bmatrix} x_1 \\ x_2 \\ x_3 \end{bmatrix} + \begin{bmatrix} 1 & 1 \\ 0 & 0 \\ 1 & 1 \end{bmatrix}\begin{bmatrix} u_1 \\ u_2 \end{bmatrix}$$

Find the linear transformation $\mathbf{x} = M\mathbf{z}$ which uncouples this system.

8. The state equations of a production model are

$$\dot{\mathbf{x}} = \begin{bmatrix} -1 & -K \\ 1 & 0 \end{bmatrix}\mathbf{x} + \begin{bmatrix} K & 0 \\ 0 & -1 \end{bmatrix}\begin{bmatrix} u_1 \\ u_2 \end{bmatrix}$$

Find the transition matrix $\Phi(t, 0)$ when $k = 0.1875$ and 2.5.

9. Determine the state equations of a constant-field dc motor feeding a load J when the state variables are
    a. $\mathbf{x}' = [\theta\ \ \dot\theta\ \ \ddot\theta]$
    b. $\mathbf{x}' = [\theta\ \ \dot\theta\ \ i_a]$

10. For a system represented by the state model $\dot{\mathbf{x}} = A\mathbf{x}$, the response is $[e^{2t}\ -2e^{2t}]$ when $\mathbf{x}'(0) = [1\ -2]$ and $\mathbf{x}'(t) = [e^t\ -e^t]$ when $\mathbf{x}'(0) = [1\ -1]$. Determine the system matrix A and the state transition matrix.

11. A model for an aircraft pitch-rate control system.
    The nominal parameters are:

$M = 0.7$	$h_0 = 1$
$w_n^2 = 0.65$	$a = 6.67$
$b = 0.008$	$K_1 = 800$
$2\xi w_n = 0.0165$	$h_1 = 0.05$
$K_2 = 0.033$	

    Find a state-space representation using the output $c(t)$ and its derivatives are the state variables.

NOTES AND REFERENCES

We assume that the reader has had some exposure to ordinary differential equations with constant coefficients, at least to the extent of knowing that the complete solution is composed of two parts: a complementary function (free response, solution of homogeneous DE) and a particular integral (forced response, solution of nonhomogeneous DE). Among the good texts covering the analog computation and simulation are Levine (1964), Rohrer (1970), Dorf (1965), and Blackman (1977). The seminal paper by Kalman (1963) provides much of the basic foundations of linear systems and their matheamtical properties. Derivation of canonical forms for state variable representations can be found in several standard books; chief among them are Kailath (1980), Wiberg (1973), Chen (1970), Brogan (1982), and D'Azzo and Houpis (1981). Some advanced information and in-depth treatment of the algebraic properties of linear systems can be found in Kailath (1980), De Russo et al. (1965), Desoer (1970), and Rosenbrock (1970). The subject of multi-input multi-output systems and the associated transfer matrices can be found in Perkins and Cruz (1969), Brogan (1982), Kailath (1980), Anderson and Vongpanitlerd (1973), and MacFarlane (1979). The subjects discussed here will be continued in the subsequent chapters with emphasis on stability and frequency response methods (Chapter 4) and design techniques (Chapter 5).

4
Stability and Frequency Response Methods

4.1 THE CONCEPT OF STABILITY

This chapter is devoted to the study of one of the most important aspects of systems control engineering analysis and design. It is generally referred to as *system stability*. Roughly speaking, stability in a system implies that small changes in the system input, in initial conditions or in system parameters, do not result in large changes in system output. In this sense, stability is a very salient characteristic of the transient performance of a system. Almost every practical system must possess the property that its zero-input response should decay to zero as time increases which guarantees, in turn, its well-damped behavior. Given the allowable region composed of boundaries of parameter variations permitted by stability considerations, we can then seek to improve the system performance.

A linear time-invariant (LTI) system is *stable* if the following two notions of system stability are satisfied:

1. When the system is excited by a *bounded input*, the result would be a *bounded output*.
2. In the absence of the input ($\mathbf{u}(t) = \mathbf{0}$), the output tends towards zero being the equilibrium state of the system, irrespective of initial

conditions. This stability concept is known as *asymptotic stability*, which clearly implies that the zero-input response decays to zero as time increases.

A simple equivalent expression would be that a LTI system having an $n \times n$ state-transition matrix $\Phi(t)$ with elements of $\phi_{ij}(t)$, $i, j = 1, \ldots, n$, is said to be *asymptotically stable* if

$$\lim_{t \to \infty} \phi_{ij}(t) = 0, \quad \text{for } i, j = 1, \ldots, n$$

We now develop several tests for indirectly determining when a system satisfies the foregoing notions of stability.

4.2 STABILITY CRITERIA

In this section, we are concerned with linear systems modeled by a single-input single-output transfer function:

$$\begin{aligned} G_f(s) &= \frac{C(s)}{R(s)} \\ &= \frac{\sum_{j=0}^{m} b_j s^j}{\sum_{j=0}^{n} a_j s^j}, \quad m < n \\ &= \frac{b_0 + b_1 s + b_2 s^2 + \cdots + b_m s^m}{a_0 + a_1 s + a_2 s^2 + \cdots + a_n s^n} \end{aligned} \quad (1)$$

We have learned from Chapter 2 that the output of (1), for zero initial condition, is given by

$$\begin{aligned} c(t) &= \mathcal{L}^{-1}[G_f(s) R(s)] \\ &= \int_0^\infty g(\tau) r(t - \tau) \, d\tau \end{aligned} \quad (2)$$

where $g(t) = \mathcal{L}^{-1}[G_f(s)]$ is the *impulse response* of the system. Taking the absolute value on both sides of (2) we obtain

$$|c(t)| = \left| \int_0^\infty g(\tau) r(t - \tau) \, d\tau \right| \quad (3)$$

But the absolute value of the integral is not greater than the integral of the absolute value of the integrand, thus (3) becomes

Stability and Frequency Response Method

$$|c(t)| \leq \int_0^\infty |g(\tau) r(t-\tau)| \, d\tau$$

$$\leq \int_0^\infty |g(\tau)| \, |r(t-\tau)| \, d\tau \qquad (4)$$

The first notion of stability (bounded-input bounded-output, BIBO) can be translated from (4) to read: if for every bounded input ($|r(t)| \leq B_i < \infty$), the output is bounded ($|c(t)| \leq B_0 < \infty$). From (4) we can write

$$|c(t)| \leq B_i \int_0^\infty |g(\tau)| \, d\tau \leq B_0 \qquad (5)$$

It is readily evident that *BIBO stability* condition entails from (5) that the impulse response $g(t)$ must be absolutely integrable; or equivalently the area under the absolute value curve of the impulse response $g(t)$ evaluated from $t=0$ to $t=\infty$ must be finite. Indeed, validating this condition is a tedious job. We therefore seek another route. Recall that the nature of $g(t)$ is dependent on the poles of the transfer function $G_f(s)$ which, in turn, are the roots of the characteristic equation. Discussion of this point follows.

4.2.1 Characteristics Equation

It is generally convenient for the case of LTI systems to test stability indirectly in the frequency domain rather than directly in the time domain. The connection between the two domains is established from

$$\zeta[\Phi(t)] = (sI - A)^{-1} = \Phi(s)$$

$$= \frac{\Psi(s)}{\Delta(s)}$$

$$= \frac{\mathrm{adj}(sI - A)}{\det(sI - A)} \qquad (6)$$

In matrix form, the partial fraction expansion of (6) is

$$\Phi(s0) = \sum_{i=1}^{r} \sum_{j=1}^{m_i} B_{ji} \left(\frac{1}{(s - s_i)^j} \right) \qquad (7)$$

where the s_i ($i = 1, 2, \ldots, r \leq n$, n the dimension of state space) are the r different roots of

$$\Delta(s) = \det(sI - A) = 0 \qquad (8)$$

and m_i are their respective multipliers. The matrix B_{ji} is the matrix of partial fraction coefficients. Inverse transforming (7) results in

$$\Phi(t) = \frac{\sum_{i=1}^{r} e^{s_i t} \sum_{j=1}^{m_i} B_{ji} t^{r-1}}{(r-1)!} \quad (9)$$

Before going further, let us illustrate the above procedure by a simple example.

Example 4.1. Consider a LTI system where

$$\Phi(s) = \begin{bmatrix} s+1 & 1 \\ -1 & s+1 \end{bmatrix} \left(\frac{1}{s^2 + 2s + 2} \right)$$

Note that $s^2 + 2s + 2 = (s + 1 + j)(s + 1 - j), j = \sqrt{-1}$. Partial fraction expansion yields

$$\Phi(s) = \begin{bmatrix} 1/2 & 1/2_j \\ -1/2_j & 1/2 \end{bmatrix} \left(\frac{1}{s+1+j} \right)$$

$$+ \begin{bmatrix} 1/2 & -1/2_j \\ 1/2_j & 1/2 \end{bmatrix} \left(\frac{1}{s+1-j} \right)$$

There are no repeated roots in this case, thus

$$s_1 = -1 - j, \qquad s_2 = -1 + j$$
$$m_1 = m_2 = 1, \qquad r = n = 2$$

In view of this information, we obtain

$$B_{11} = \begin{bmatrix} 1/2 & 1/2_j \\ -1/2_j & 1/2 \end{bmatrix}, \quad B_{12} = \begin{bmatrix} 1/2 & -1/2_j \\ 1/2_j & 1/2 \end{bmatrix}$$

and finally

$$\Phi(t) = e^{(-1-j)t} B_{11} + e^{(-1+j)t} B_{12}$$

We note from (9) that terms of the form $e^{s_i t}$ determine the general character of $\Phi(t)$. The roots of (9) thus have very special importance. These roots s_i of the *characteristic equation* $\Delta(s) = 0$ are known in matrix theory as the *eigenvalues* of the matrix A. Sometimes they are also called the *characteristic roots* of the system, or the system *eigenvalues*.

Stability of LTI systems is closely connected to the values of the characteristic roots. The situation is slightly complicated here by the

Stability and Frequency Response Method

fact that there may be cancellations between the numerator and the denominator polynomials in (6) or, equivalently

$$B_{qi} = B_{q+1,i} = \cdots = B_{m,i} = 0, \quad q \geq 1 \qquad (10)$$

for one or more of the characteristic roots s_i. This means that the system state-transition matrix, and hence the state vector, will appear to be of lower order than the characteristic equation. If all possible cancellations in (6) are performed, the resulting denominator polynomial is called the *minimal polynomal*. It is known in matrix theory that every eigenvalue of A appears as a zero of the minimal polynomial. That is, every simple eigenvalue will appear as a simple zero of the minimal polynomial and every multiple eigenvalue of A will appear as a zero (perhaps with reduced multiplicity) of the minimal polynomial.

A fundamental theorem revealing the relationship between the system characteristic roots and stability states that a free linear time-invariant system $\dot{x} = Ax$ is asymptotically stable if and only if ALL the system characteristic roots have negative real parts. This basic statement

Table 4.1. Response of Alternative Roots

	Types of Roots	Nature of Response	Figure
1.	Single root at $s = \pm \alpha$	$Ke^{\pm \alpha t}$	4.1 (1)
2.	h multiple roots at $s = \pm \alpha$	$(K_1 + K_2 t + \cdots + K_h t^{h-1})e^{\alpha t}$	4.1 (2)
3.	Complex conjugate pair at $s = \pm j\omega$	$Ke^{\pm \alpha t} \sin(\omega t + \beta)$	4.1 (3)
4.	Complex conjugate pairs of multiplicity h at $s = \pm \alpha \pm \omega$	$[K_1 \sin(\omega t + \beta_1) + K_2 t \sin(\omega t + \beta_2) + \cdots + K_h t^{h-1} \sin(\omega t + \beta_h)]e^{\pm \alpha t}$	4.1 (4)
5.	Simple complex conjugate pair on the $j\omega$-axis ($s = \pm j\omega$)	$K \sin(\omega t + \beta)$	4.1 (5)
6.	complex conjugate pairs of multiplicity h on the $j\omega$-axis	$K_1 \sin(\omega t + \beta_1) + K_2 t \sin(\omega t + \beta_2) + \cdots + K_h t^{h-t} \sin(\omega t + \beta_h)$	4.1 (6)
7.	Single root at origin ($s = 0$)	K	4.1 (7)
8.	h roots at origin	$K_1 + K_2 t + \cdots + K_h t^{h-1}$	4.1 (8)

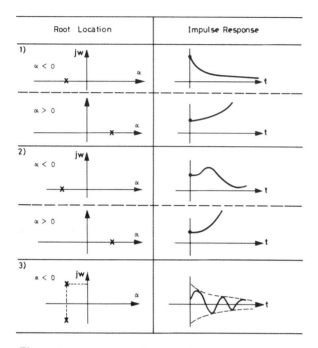

Figure 4.1 Response of alternative roots.

can be readily verified from (9). The characteristic roots may be real, complex conjugate, and may have multiplicity of various orders. The nature response terms contributed by all possible types of roots are given in Table 4.1 and have been depicted in Fig. 4.1.

Careful examination of the individual response emphasizes the foregoing theorem for the cases when $\alpha < 0$ (stable behavior) and $\alpha > 0$ (unstable behavior). When $\alpha = 0$ and single complex conjugate pair, sustained oscillations exist with constant amplitude. Roots on the $j\omega$-axis with multiplicity two or higher (cases 6 and 8) however contribute to unbounded behavior. A single root at the origin contributes a response term which is of constant magnitude. Multiple roots at the origin are, however producing a steady increase in response without bound as $t \to \infty$. These cases of unbounded response confirm the BIBO condition since the area under the response curve is infinite.

In view of the impact of parameter variations on the LTI system behavior, we have the following classification:

1. A linear system is *absolutely stable* with respect to a parameter of the system if it is *stable* for all values of this parameter.
2. A linear system is *conditionally stable* with respect to a parameter,

Stability and Frequency Response Method

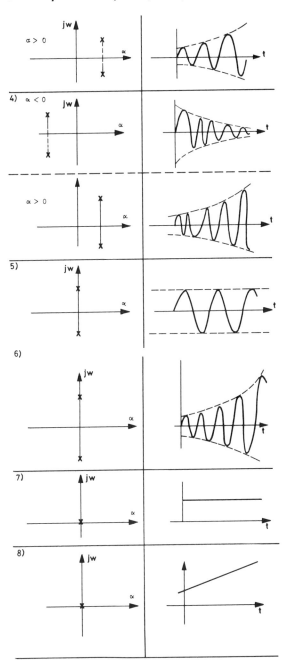

Figure 4.1 (Continued).

if it is *stable* for only certain bounded ranges of values of this parameter.

We should emphasize that the foregoing analysis hinges upon the determination of the locations of the characteristic roots. This direct solution is tedious and requires excessive effort. The following sections are devoted to a discussion of several stability techniques that do not require the direct calculation of the system characteristic roots.

4.2.2 Routh Criterion

We now describe the *Routh criterion* or *Routh stability test* for determining whether an equation such as the system characteristic equation has any right half-plane (RHP) roots. The test will be stated here without proof. For an $n \times n$ matrix A, the characteristic equation, which is of degree n, can be expressed as

$$\begin{aligned} \Delta(s) &= \det(sI - A) \\ &= s^n + a_{n-1}s^{n-1} + a_{n-2}s^{n-2} + \cdots + a_1 s + a_0 \\ &= 0 \end{aligned} \qquad (11)$$

The first step is to arrange the coefficients of (11) in two consecutive rows where the first row contains the odd coefficients (first, third, fifth, ...) and the second row contains the even coefficients (second, fourth, sixth, ...):

row 1: $\quad 1 \quad a_{n-2} \quad a_{n-4} \quad a_{n-6} \quad a_{n-8} \quad \cdots$

row 2: $\quad a_{n-1} \quad a_{n-3} \quad a_{n-5} \quad a_{n-7} \quad a_{n-9} \quad \cdots$

We then construct a third row

row 3: $\quad b_{n-1} \quad b_{n-3} \quad b_{n-5} \quad b_{n-7} \quad b_{n-9} \quad \cdots$

where

$$b_{n-1} = \frac{-1}{a_{n-1}} \begin{vmatrix} 1 & a_{n-2} \\ a_{n-1} & a_{n-3} \end{vmatrix} \qquad (12)$$

$$b_{n-3} = \frac{-1}{a_{n-1}} \begin{vmatrix} 1 & a_{n-4} \\ a_{n-1} & a_{n-5} \end{vmatrix} \qquad (13)$$

$$b_{n-5} = \frac{-1}{a_{n-1}} \begin{vmatrix} 1 & a_{n-6} \\ a_{n-1} & a_{n-7} \end{vmatrix} \qquad (14)$$

Stability and Frequency Response Method

and in general

$$b_{n-j} = \frac{-1}{a_{n-1}} \begin{vmatrix} 1 & a_{n-j-1} \\ a_{n-1} & a_{n-j-2} \end{vmatrix}, \quad j = 1, 3, 5, \ldots \quad (15)$$

Next, we form a fourth row

row 4: c_{n-1} c_{n-3} c_{n-5} c_{n-7} c_{n-9} $\cdots$

where in general

$$c_{n-j} = \frac{-1}{b_{n-1}} \begin{vmatrix} a_{n-1} & a_{n-j-2} \\ b_{n-1} & b_{n-j-2} \end{vmatrix}, \quad j = 1, 3, 5, \ldots \quad (16)$$

This procedure of forming a row from the preceding two rows is continued until only zero elements are obtained. The Routh test is as follows.

All the roots of the characteristic equation (11) have negative real parts if and only if the elements in the first column of the array

$$\begin{array}{cccccc} 1 & a_{n-2} & a_{n-4} & a_{n-6} & a_{n-8} & \cdots \\ a_{n-1} & a_{n-3} & a_{n-5} & a_{n-7} & a_{n-9} & \cdots \\ b_{n-1} & b_{n-3} & b_{n-5} & b_{n-7} & b_{n-9} & \cdots \\ c_{n-1} & c_{n-3} & c_{n-5} & c_{n-7} & c_{n-9} & \cdots \end{array}$$

have the same algebraic sign. The number of roots in the right half-plane is equal to the number of alternations of sign in the sequence $\{1, a_{n-1}, b_{n-1}, c_{n-1}, \ldots\}$.

Some illustrative examples are in order.

Example 4.2.   Let us determine whether the equation

$$s^4 + 8s^3 + 18s^2 + 16s + 5 = 0$$

has roots in the RHP or not.

First we form the two rows

row 1:   1   18   5
row 2:   8   16

Application of (15) gives

$$b_3 = \frac{-1}{8} \begin{vmatrix} 1 & 18 \\ 8 & 26 \end{vmatrix} = 16$$

$$b_1 = \frac{-1}{8} \begin{vmatrix} 1 & 5 \\ 8 & 0 \end{vmatrix} = 5$$

to form the third row

$$\text{row 3:} \quad b_3 \quad b_1$$

Next we form the fourth row

$$\text{row 4:} \quad c_3 \quad c_1$$

using (16)

$$c_3 = \frac{1}{16}\begin{vmatrix} 8 & 16 \\ 16 & 5 \end{vmatrix} = 13.5$$

$$c_1 = \frac{1}{16}\begin{vmatrix} 8 & 0 \\ 16 & 0 \end{vmatrix} = 0$$

Finally, the last row is

$$d_3 = \frac{1}{13.5}\begin{vmatrix} 16 & 5 \\ 13.5 & 0 \end{vmatrix} = 5$$

Thus the Routh array is

$$\begin{array}{ccc} 1 & 18 & 5 \\ 8 & 16 & \\ 16 & 5 & \\ 13.5 & & \\ 5 & & \end{array}$$

The elements of the first column are all positive, hence no RHP roots and the system is stable.

Example 4.3. Consider the following characteristic equation

$$s^4 + 3s^3 + 8s^2 + s + 9 = 0$$

The Routh array is given below:

row 1	1	8	9
row 2	3	1	
row 3	23/3	9	
row 4	$-58/2 \atop 3$		
row 5	9		

We observe two sign changes in the first column, so there must be exactly two RHP roots.

Note from the foregoing treatment that the Routh stability test yields

Stability and Frequency Response Method

only the number of RHP roots and says nothing about the locations of the system roots. There are two special cases which may occur in forming the Routh array.

Case I: The first term in a row is zero, but not all the other terms in that row are zero. The array may be computed by either of the following procedures:

1. Replace the zero by an arbitrarily small quantity ϵ. Complete the array as before then let $\epsilon \to 0$. The sign of ϵ is immaterial since it does not affect stability.
2. Replace s by z^{-1} in the characteristic equation. Then apply the Routh test to the new characteristic equation.

We illustrate Case I by an example.

Example 4.4. For the characteristic equation

$$s^5 + s^4 + 2s^3 + 2s^2 + 3s + 5 = 0$$

the Routh array is

row			
1	1	2	3
2	1	2	5
3	ϵ	-2	
4	$(2\epsilon + 2)/\epsilon$	5	
5	$(-4\epsilon - 4 - 5\epsilon^2)/(2\epsilon + 2)$		
6	5		

As $\epsilon \to 0$, the first element in row 4 has a positive sign (we are only interested in the signs and not in the magnitudes). Proceeding further, the first element in row 5 $(-4\epsilon - 4 - 5\epsilon^2)/(2\epsilon + 2)$ has a limiting value of -2 as $\epsilon \to 0$. By examining the terms in the first column of the Routh array, it is found that there are two changes in sign and hence the system is unstable having two poles in the right half s-plane.

Alternatively, the second procedure of replacing s by $(1/z)$ yields

$$5z^5 + 3z^4 + 2z^3 + 2z^2 + z + 1 = 0$$

for which the Routh array is

row			
1	5	2	1
2	3	2	1
3	$-4/3$	$-2/3$	
4	$1/2$	1	
5	2		
6	1		

There are two changes of sign in the first column of the Routh array as expected. This tells us that there are two z-roots in the right half z-plane or equivalently the number of s-roots in the RHP is two.

Case II: When all the elements in any one row are zero. A zero row indicates roots symmetric about the origin, including the $j\omega$-axis roots, real roots of equal magnitude but opposite sign, roots with quadrantal symmetry, etc. The polynomial whose coefficients are the elements of the row just above the row of zero is called an *auxiliary polynomial*. The polynomial gives the number and location of root pairs of the characteristic equation which are symmetrically located in the s-plane. Note that the order of the auxiliary polynomial is always even.

Because of a zero row in the array, the Routh's test breaks down. This situation is overcome by replacing the row of zeros in the Routh array by a row of coefficients of the polynomial generated by taking the first derivative of the auxiliary polynomial. The following example illustrates the procedure.

Example 4.5. Consider a sixth-order system with the characteristic equation

$$s^6 + 2s^5 + 8s^4 + 12s^3 + 20s^2 + 16s + 16 = 0$$

Beginning the array,

row 1	1	8	20	16
2	2	12	16	
3	2	12	16	
4	0	0		

We have encountered a row of zeros. We form an auxiliary equation using as coefficients the entries 2, 12, and 16 in the row just preceding the row of zeros. In general, the auxiliary equation contains even power of s and is of order $n - j + 2$, where n is the order of the characteristic equation, and j is the row of zeros. Here

$$n - j + 2 = 6 - 4 + 2 = 4$$

so the auxiliary equation is

$$2s^4 + 12s^2 + 16 = 0$$

The derivative of the polynomial with respect to s is

$$8s^3 + 24s = 0$$

Stability and Frequency Response Method

Now, the zeros in row 4 are replaced by the coefficients 8 and 24. The entire array thus becomes

row 1	1	8	20	16
2	2	12	16	
3	2	12	16	
4	8	24		
5	6	16		
6	8/3			
7	16			

We see that there is no change of sign in the first column of the new array. The roots of the auxiliary equation are

$$s = \pm\sqrt{2} \quad \text{and} \quad s = \pm j2$$

These two pairs of roots are also the roots of the original characteirstic equation. Since there is no sign change detected, we conclude that the sixth-order system has two LHP roots and four $j\omega$-axis roots. The system is conditionally stable.

4.2.3 Hurwitz Criterion

Reference is made to the characteristic equation (11). The Hurwitz stability test states that: it is necessary and sufficient for the stability of (11) that the n determinants formed from the coefficients $\{-a_0, a_1, \ldots, a_{n-1}\}$ be positive, where these determinants are taken as the principal minors of the following arrangement (called the Hurwitz determinant):

$$\begin{vmatrix} a_{n-1} & 1 & 0 & 0 & 0 & 0 & \cdots & 0 \\ a_{n-3} & a_{n-2} & a_{n-1} & 1 & 0 & 0 & \cdots & 0 \\ a_{n-5} & a_{n-4} & a_{n-3} & a_{n-2} & a_{n-1} & 1 & \cdots & 0 \\ \cdot & \cdot & \cdot & \cdot & \cdot & & & \cdot \\ \cdot & \cdot & \cdot & \cdot & \cdot & & & \cdot \\ \cdot & \cdot & \cdot & \cdot & \cdot & & & \cdot \\ a_{2n-1} & a_{2n-2} & \cdots & & \cdots & & \cdots & a_0 \end{vmatrix}$$

where the coefficients with indices larger than n or with negative indices are replaced by zeros.

Simply stated, the necessary and sufficient conditions for stability are

$$\Delta_1 = a_{n-1} > 0$$

$$\Delta_2 = \begin{vmatrix} a_{n-1} & 1 \\ a_{n-3} & a_{n-2} \end{vmatrix} > 0$$

$$\Delta_3 = \begin{vmatrix} a_{n-1} & 1 & 0 \\ a_{n-3} & a_{n-2} & a_{n-1} \\ a_{n-5} & a_{n-4} & a_{n-3} \end{vmatrix} > 0$$

.
.
.

$$\Delta_n = \text{entire arrangement of the Hurwitz array} > 0$$

When $\Delta_{n-1} = 0$, the system is conditionally stable. As an illustration, consider the following example.

Example 4.6. For the fourth-order system with characteristic equation

$$s^4 + 8s^3 + 18s^2 + 16s + 5 = 0$$

the Hurwitz arrangement is given by

$$\begin{vmatrix} 18 & 1 & 0 & 0 \\ 16 & 18 & 8 & 1 \\ 0 & 5 & 16 & 18 \\ 0 & 0 & 0 & 5 \end{vmatrix}$$

Hence,

$$\Delta_1 = 8 > 0$$

$$\Delta_2 = \begin{vmatrix} 8 & 1 \\ 16 & 18 \end{vmatrix} = 128 > 0$$

$$\Delta_3 = \begin{vmatrix} 8 & 1 & 0 \\ 16 & 18 & 8 \\ 0 & 5 & 16 \end{vmatrix} = 1728 > 0$$

$$\Delta_4 = \begin{vmatrix} 8 & 1 & 0 & 0 \\ 16 & 18 & 8 & 1 \\ 0 & 5 & 16 & 18 \\ 0 & 0 & 0 & 5 \end{vmatrix}$$

$$= 5\Delta_3 = 8640 > 0$$

It is thus evident that the system under consideration is stable.

Stability and Frequency Response Method

Despite the fact that Hurwitz and Routh stability criteria have been published independently, they are quite equivalent and in implementation the Routh array as presented here is more convenient and readily programmable. The title Routh–Hurwitz criterion will be given as a common designator to recognize the similarity of both criteria.

4.2.4 Relative Stability

The verification of stability by the Routh–Hurwitz criterion provides only a partial answer to the question of stability. The Routh–Hurwitz criterion ascertains the absolute stability of a system by determining if any of the roots of the characteristic equation lie in the RHP. However, if the system satisfies the Routh–Hurwitz criterion and is absolutely stable, we then proceed to determine its *relative stability* quantitatively by finding the settling time of the dominant roots of its characteristic equation. As we learned from Chapter 2, the settling time is inversely proportional to the real part of the dominant root. Thus, the relative stability can be specified by requiring that all the roots of the characteristic equation be more negative than a certain value. That is to say all the roots must lie to the left of the line $s = -\alpha_r (\alpha_r > 0)$. We now use the linear relation

$$s = w - \alpha_r$$

to shift the origin of the s-plane to $s = -\alpha_r$ and study the roots in the w-plane (see Figure 4.2). If the new characteristic equation in w satisfied the Routh–Hurwitz criterion, it implies that all the roots of the original characteristic equation are more negative than $-\alpha_r$.

Example 4.7. Consider the simple third-order equation

$$s^3 + 4s^2 + 6s + 4$$

Initially, we have the roots at -2 and $-1 \pm j$ in the s-plane. Setting $\alpha_r = 1$, we use $s = w - 1$ to obtain

$$(w - 1)^3 + 4(w - 1)^2 + 6(w - 1) + 4 = w^3 + w^2 + w + 1$$

for which the Routh array is established as

row		
1	1	1
2	1	1
3	0	0
4	1	0

Clearly there are roots on the shifted imaginary axis and the roots can be obtained from the auxiliary equation, which is

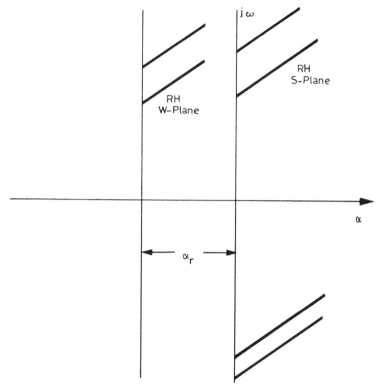

Figure 4.2 S-plane and W-plane.

$$w^2 + 1 = (w + j)(w - j)$$
$$= (s + 1 + j)(s + 1 - j)$$

This implies that the original stable system remains stable if shifting of axes took place up to $\alpha_r = 1$. Evidently for $\alpha_r > 1$. Some of the roots will be on the RHP.

4.2.5 Input-Output Stability

Thus far our focus has been on linear system stability through the zero-input response. It was clear that this is a form of internal stability. As we have mentioned earlier, another important notion concerns stability of system outputs or, loosely speaking, external stability. In the sequel, we wish to show that these two stability concepts are different in general. A linear system may have a nongrowing output and yet may

Stability and Frequency Response Method

be unstable in the sense of our earlier definition of stability. This discrepancy can be best illustrated by the following example.

Example 4.8 Consider the second-order system

$$\dot{\mathbf{x}} = \begin{bmatrix} \dot{x}_1 \\ \dot{x}_2 \end{bmatrix} = \begin{bmatrix} -2 & 1 \\ 6 & 1 \end{bmatrix} \begin{bmatrix} x_1 \\ x_2 \end{bmatrix} + \begin{bmatrix} 0 \\ 1 \end{bmatrix} u$$

$$y = \begin{bmatrix} -3 & 1 \end{bmatrix} \mathbf{x}$$

The characteristic equation $\Delta(s)$ is

$$\Delta(s) = \det[sI - A]$$

$$= \begin{vmatrix} s+2 & -1 \\ -6 & s+1 \end{vmatrix}$$

$$= (s+2)(s+1) - 6 = s^2 + 3s - 4$$

$$= (s+4)(s-1)$$

There is one RHP root at $+1$ and hence the system is unstable. However, if we examine the transfer function $Y(s)/U(s)$ we obtain

$$\frac{Y(s)}{U(s)} = \begin{bmatrix} -3 & 1 \end{bmatrix} \begin{bmatrix} s+2 & -1 \\ -6 & s-1 \end{bmatrix}^{-1} \begin{bmatrix} 0 \\ 1 \end{bmatrix}$$

$$= \begin{bmatrix} -3 & 1 \end{bmatrix} \begin{bmatrix} s-2 & 1 \\ 6 & s+1 \end{bmatrix} \begin{bmatrix} 0 \\ 1 \end{bmatrix} \frac{1}{s^2 + 3s - 4}$$

$$= \frac{s-1}{s^2 + 3s - 4}$$

$$= \frac{s-1}{(s-1)(s+4)} = \frac{1}{s+4}$$

It is thus obvious that the transfer function has one pole and this pole is a characteristic root. The other characteristic root, which happens to be in the RHP, is not a pole of the transfer function. Thus, if the input $u(t)$ is not growing with time, the output $y(t)$ will not grow with time. The system has a decaying output response for a nongrowing input, although the zero-input response is increasing. Hereafter, the definition of an input-output stability notion will be formally given and the relationship with zero-input stability will be presented. At this stage, the concept of norm is needed. It is simply a generalization of the notion of distance. We state it as follows: for a vector $\mathbf{z}$, the norm $\|\mathbf{z}\|$ must satisfy the following postulates:

1. $\|\mathbf{z}\| \geq 0$ for all $\mathbf{z}$
2. $\|\mathbf{z}\| = 0$ if and only if $\mathbf{z} = \mathbf{0}$
3. $\|a\mathbf{z}\| = |a|\,\|\mathbf{z}\|$ for all scalars a and all $\mathbf{z}$
4. $\|\mathbf{z} + \mathbf{m}\| \leq \|\mathbf{z}\| + \|\mathbf{m}\|$ (the triangle inequality)

For the state vector $\mathbf{x} = [x_1 \ x_2 \ \cdots \ x_n]^t$, an important norm called the *Euclidean norm* is defined by

$$\|\mathbf{x}\| = \sqrt{x_1^2 + x_2^2 + \cdots + x_n^2} \tag{17}$$

where the positive square root is taken. Notice that this is very similar to the distance from the origin to a point $x_1, \ldots, x_n$ in the n-dimensional space. Obviously, the Euclidean norm is always positive except at the origin ($\mathbf{x} = \mathbf{0}$) where it is zero. Multiplying the coordinates of $\mathbf{x}$ by any scale factor a, it is easy to see that the new norm is $|a|$ times the original norm. Verifying the triangle inequality is straightforward. We thus conclude that the Euclidean norm satisfies the foregoing postulates.

In terms of the norm, a time function $\mathbf{f}(t)$ is said to be *bounded* if there exists a finite constant F such that

$$\|\mathbf{f}\| < F < \infty \tag{18}$$

for all t. We note from (18) that a bounded function can never become infinite. Suppose that $f(t) = [K_1 \bar{e}^{\alpha_1 t} \ K_2 \bar{e}^{\alpha_2 t} \ \cdots \ K_n \bar{e}^{\alpha_n t}]$, $\alpha_j > 0$ for all j, then it is bounded by $\sqrt{K_1^2 + K_2^2 + \cdots + K_n^2}$, the norm at the initial time $t = 0$. Proceeding further, a system with input $\mathbf{u}(t)$ and output $\mathbf{y}(t)$ is said to be *zero-state bounded-input bounded-output* (*BIBO*) *stable* if, for zero initial condition ($\mathbf{x}(t_0) = \mathbf{0}$) and for all bounded inputs $\mathbf{u}(t)$, bounded on $t_0 < t < \infty$, the output $\mathbf{y}(t)$ is bounded on $t_0 \leq t < \infty$.

To relate the BIBO stability concept to transfer function, we use $Y(s)$, $U(s)$ as the Laplace transform of the zero-state output response $\mathbf{y}(t)$ of a LTI system and the applied input $\mathbf{u}(t)$, respectively. Thus

$$Y(s) = H(s)U(s) = [H_{ij}(s)]U(s) \tag{19}$$

where $H_{ij}(s)$ are the elements of the transfer function matrix. One of the basic results is that zero-state BIBO stability is guaranteed if and only if all poles of every transfer function $H_{ij}(s)$ lie strictly in the LHP and if the number of zeros of $H_{ij}(s)$ does not exceed the number of poles for all $H_{ij}(s)$.

As indicated by Example 4.8, zero-state BIBO stability does not imply the stability of the zero-input response. However, the stability of the zero-input response does guarantee BIBO stability. This is directly evident from the fact that when the transfer function $H(s) =$

Stability and Frequency Response Method

$C(sI - A)^{-1}B$ has all characteristic roots in the LHP, it will also be zero-state BIBO stable for *any* choice of input and output terminals.

4.2.6 Root-Locus Method

It becomes clear from the preceding sections that the characteristic equation $\Delta(s)$ of a system provides a valuable insight concerning the response of the system when the roots of the equation are determined. In order to locate the roots of the characteristic equation, one should seek methods other than analytic since analytic procedures are generally known to fail for systems higher than fourth-order. One of the most important orderly procedures that provides a simple correlation between the system parameters and the essential features of system response is the *root-locus*. It provides relevant information in a graphical manner on the *s*-plane which facilitates the rapid sketching of the root variation of locus.

Our discussion here deals with a general method for the closed-loop system transfer function (see Fig. 4.3). There is an adjustable gain K in the forward path, which we shall discuss in the sequel. The characteristic equation of the system is

$$\Delta(s) = 1 + KG(s)H(s) = 0 \qquad (20)$$

We note that (20) gives the characteristic roots (closed-loop poles) as a function of the parameter gain K. This functional dependence brings upon the title root-locus. From (20) it is quite clear that the characteristic roots satisfy the two conditions

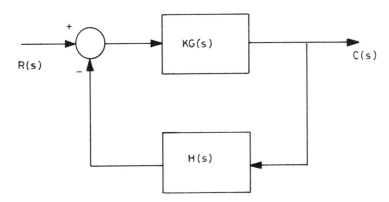

Figure 4.3 Closed-loop system.

$$K|G(s)H(s)| = 1 \quad \text{(magnitude)} \tag{21}$$

$$\arg|G(s)H(s)| = 180° \pm n(360°) \quad \text{(angle)} \tag{22}$$

For the case of subsequent development, we introduce

$$G_L(s) = G(s)H(s)$$

$$= \frac{\prod_{j=1}^{M}(s - z_j)}{\prod_{k=1}^{N}(s - p_k)} \tag{23}$$

where p_k and z_j are the poles and zeros of the loop transfer function $G_L(s)$. In view of (23), we can rewrite (20) in the form

$$\prod_{k=1}^{N}(s - p_k) + K \prod_{j=1}^{M}(s - z_j) = 0 \tag{24}$$

As K varies between 0 and ∞, (24) gives the pattern of closed-loop poles or root-loci. In particular, for $K = 0$ the roots of the characteristic equation occur for

$$\prod_{k=1}^{N}(s - p_k) = 0 \tag{25}$$

which means that the poles of $G_L(s)$ provide the starting points for the locus. On the other hand, as $K \to \infty$ the characteristic equation reduces to

$$\prod_{j=1}^{M}(s - z_j) = 0 \tag{26}$$

which implies that the zeros of $G_L(s)$ provide the receiving points of the locus. The first rule of root-locus construction is now ready.

Rule 1. The root-locus starts at the poles of the loop transfer function $G_L(s)$ and terminates at the zeros of $G_L(s)$.

Specializing the angle condition (22) to the real axis, it is readily seen that segments of the locus on the real axis (angle is multiple of 180°) exist only to the left of an odd number of poles and zeros of $G_L(s)$. Hence we have:

Rule 2. The locus segments on the real axis lie in a section to the left of an odd number of poles and zeros.

Since each locus starts at a pole of the loop function, we have

Rule 3. The number of separate segments (branches) of the locus is equal to the number of loop function poles.

Complex conjugate roots yield separate sections of the locus in the s-plane, thus:

Stability and Frequency Response Method

Rule 4. The root-locus must be symmetrical with respect to the real axis.

In view of Rule 1, the root-loci start at the poles and terminate at zeros. But in general the number of poles N exceeds the number of finite zeros M. Thus $N - M$ sections must terminate at infinity along asymptotic lines, whose center σ_A and angle θ_A are given below.

Rule 5. The number of locus sections terminating at infinity is equal to the excess of poles over zeros, with a center of asymptotes σ_A on the real axis given by

$$\sigma_A = \frac{\sum_{k=1}^{N} p_k - \sum_{j=1}^{M} z_j}{N - M} \tag{27}$$

and the angle of the asymptotes are

$$\theta_A = \frac{(2l + 1)180°}{N - M}, \quad l = 0, 1, 2, \ldots, N - M - 1 \tag{28}$$

For real line segments connecting two poles, the value of gain K starts with value zero at either end and varies some how to reach zero again at the other end. It is thus obvious that there exists a point between the two poles where K reaches a maximum. This point is called the *breakaway point* at which the locus departs from the real axis.

Rule 6. Breakaway points on the real axis where two or more branches of the root-locus depart form the real axis. Their location σ_b is obtained by maximizing

$$K(\sigma) = \frac{-1}{G_L(\sigma)} \tag{29}$$

or equivalently $dK(\sigma)/d\sigma = 0$ should yield σ_b.

To distinguish segment portions in the RHP from those in the LHP (representing instability), we need a guideline to identify these cases. We have the following rule.

Rule 7. The Rough–Hurwitz criterion provides the points at which the loci cross the imaginary axis.

When complex poles are encountered in the loop function $G_L(s)$, the locus will start at that complex pole in a direction determined by the following rule.

Rule 8. The departure angle from a complex poles is given by

$$\theta_D = 180° + \arg[G_L^+(s)] \tag{30}$$

where $\arg[G_L^+(s)]$ is the phase angle of $G_L(s)$ computed at the complex pole but dropping the contribution of that particular pole.

Several illustrative examples are now in order.

Example 4.9. A feedback system has the characteristic equation (20) in the form

$$1 + \frac{K}{s(s+1)(s+2)} = 0$$

It is simple to see that the loop function has three poles ($N = 3$) and no zeros ($M = 0$). The following information can be readily drawn:

1. There are three branches of the root-locus (Rule 3).
2. The branches of the root-locus originate with $K = 0$ from the open-loop poles $s = 0, -1$ and -2 (Rule 1).
3. The absence of finite zeros leads to (Rule 5) the situation that the three branches terminate on infinity, along the asymptotes whose angles with real axis are given by

$$\theta_A = \frac{(2l+1)180}{N - M}$$

$$= \frac{(2l+1)180}{3}, \quad l = 0, 1, 2$$

$$= 60, 180, 300$$

4. The center of the asymptotes is given by

$$\sigma_A = \frac{-1 - 2}{3} = -1$$

5. By Rule 2, the segments of the real axis between $(0, -1)$ and $(-2, \infty)$ lie on the root locus. Then, Rule 6 tells us that in the region $(0, -1)$ there will be a breakaway point at σ_b. Application of (29) gives

$$K(\sigma) = -\sigma(\sigma + 1)(\sigma + 2)$$

and thus $dK(\sigma)/d\sigma = -(3\sigma^2 + 6\sigma + 2) \to 0$ yielding

$$\sigma_b = -1 \pm \sqrt{2} = -0.423, \quad -1.577$$

Since the breakaway point must lie between $0, -1$, it is clear that $\sigma_b = -0.423$ is the actual point. The complete root-locus is schematically shown in Fig. 4.4. It remains to compute the intersection with the imaginary axis. From the characteristic equation

$$s(s+1)(s+2) + K = 0$$

we construct the Routh array as follows:

Stability and Frequency Response Method

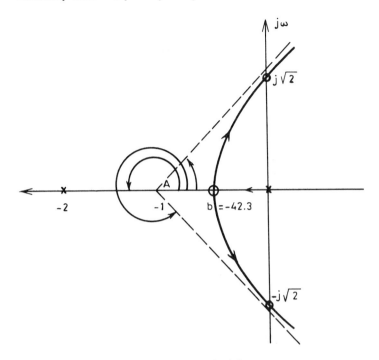

Figure 4.4 Root locus plot of example 4.8.

row 1	1	2
row 2	3	K
row 3	$(6-K)/3$	0
row 4	K	

and require that $K > 0$ and $6 - K > 0$. This means that $0 < K < 6$ keeps the system in the stable region. When $K = 6$, the auxiliary equation becomes

$$3s^2 + 6 = 0$$

On the imaginary axis $s = j\omega_c$ and thus $-\omega_c^2 + 2 = 0$ or $\omega_c = \pm\sqrt{2}$.

Example 4.10. A feedback control system has an open-loop transfer function

$$G_L(s) = \frac{1}{s(s^2 + 2s + 3)}$$

It is desired to find the root-locus as K is varied from 0 to ∞.

We have three poles ($N = 3$) at $s = 0$, $-1 + j\sqrt{2}$ and $-1 - 1j\sqrt{2}$ and no zeros ($M = 0$). Obviously, there is a real line segment (Rule 2) extending from the origin ($s = 0$) to infinity ($s = \infty$). Since there are no zeros, we conclude that the three segments (Rule 3) approach infinity along asymptotic lines that intersect at (Rule 5).

$$\sigma_A = -\frac{2}{3}$$

The angles are $\theta_A = 60°$, $180°$, and $300°$.

Two locus segments will start from the complex poles (Rule 4) and from (30) we have at $s = s_1 = -1 + \sqrt{2}j$

$$G_L^+(s_1) = \frac{1}{s_1[s_1 - (-1 - \sqrt{2}j)]}$$

$$= \frac{1}{(j2\sqrt{2})(-1 + \sqrt{2}j)}$$

Thus

$$\arg[G_L^+(s_1)] = -215.26$$

and

$$\theta_D = -180 - 215.26 = -35.26°$$

The intersection with the imaginary axis are found from the Routh array of the characteristic equation

$$s(s^2 + 2s + 3) + K = 0$$

as

row 1	1	3
row 2	2	K
row 3	(6 − K)/3	0
row 4	K	

For stability we require $K \leq 6$. At $K = 6$, the auxiliary equation $2s^2 + 6 = 0$ gives the frequency of oscillation $\omega_c = \pm\sqrt{3}$. The complete root plot is shown in Fig. 4.5.

Example 4.11. Consider the closed-loop system shown in Fig. 4.6. Sketch the root-locus as K varies.

Stability and Frequency Response Method

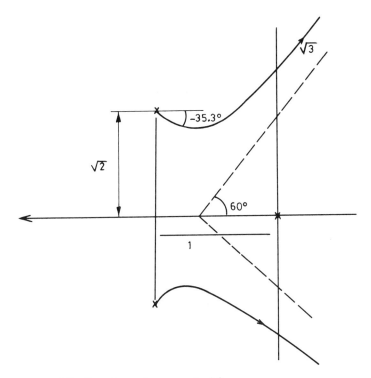

Figure 4.5 Root locus for example 4.9.

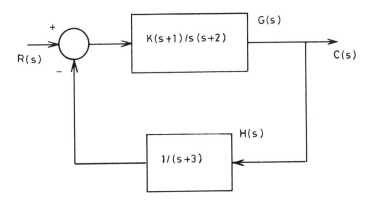

Figure 4.6 Example 4.6.

Initially, we have the loop-transfer function

$$G_L(s) = G(s)H(s)$$
$$= \frac{K(s+1)}{s(s+2)(s+3)}$$

for which the characteristic equation is

$$s(s+2)(s+3) + K(s+1) = 0$$
$$s^3 + 5s^2 + (6+K)s + K = 0$$

In this case $N = 3$, $M = 1$, and $(N - M) = 2$. The poles are at 0, -2, -3 and the zero is at -1. Thus we have three branches originating from 0, -2 and -3. Obviously, one will terminate at -1 and the other two will approach infinity along asymptotes centered at

$$\sigma_A = \frac{-2 - 3 + 1}{2} = -2$$

The angles are

$$\theta_A = 90°, 270°$$

Simple analysis shows that a breakaway point occurs between $s = -2$ and $s = -3$. Application of (29) gives

$$K(\sigma) = \frac{-\sigma(\sigma + 2)(\sigma + 3)}{\sigma + 1}$$

for which

$$\frac{dK(\sigma)}{d\sigma} = 2\sigma^3 + 8\sigma^2 + 10\sigma + 6$$

To obtain σ_b, some iterative procedure has to be employed in the region $(-2, -3)$ given the fact that $K(-2) = K(-3) = 0$. The result would be $\sigma_b = -2.45$. The complete plot is shown in Fig. 4.7.

Example 4.12. Sketch the root locus for the system whose open-loop transfer function is

$$G_L(s) = \frac{K}{(s+3)(s+1)(s^2 + 2s + 2)}$$

We will follow the rules to arrive at the full root-loci. There are four open-loop poles at -1, -3, $-1 \pm j$ and no open-loop zero. Thus $N = 4$, $M = 0$, and $N - M = 4$. The four branches start at -1, -3,

Stability and Frequency Response Method

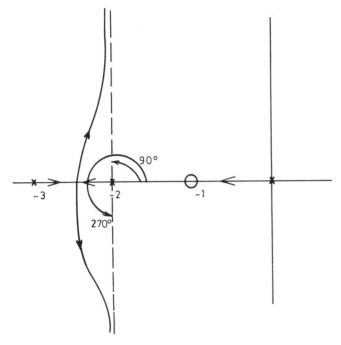

Figure 4.7 Root locus plot of example 4.10.

$-1+j$, and $-1-j$ and move away towards the outer regions as K increases. In the region $(-1, -3)$ there is a real line segment, and hence a breakaway point occurs. The center of asymptotes is given by

$$\sigma_A = \frac{-1-3-1-1}{4} = -1.5$$

with angles

$$\theta_A = \frac{(2l+1)180°}{4}, \quad l = 0, 1, 2, 3$$

$$= 45°, 135°, 225°, 315°$$

To determine the intersection with the imaginary axis, we form the characteristic equation

$$(s+1)(s+3)(s^2+2s+2) + K = 0$$

$$s^4 + 6s^3 + 13s^2 + 14s + (6+K) = 0$$

and then construct the array

row 1	1	13	$(6 + K)$
row 2	6	14	
row 3	64/6	$(6 + K)$	
row 4	$14 - 36(6 + K)/64$	0	
row 5	$(6 + K)$		

No sign changes in the first column requires that

$K > -6$ and $170 - 9K > 9K > 0$ or $-6 < K < 18.9$

At the imaginary axis, we have

$$\frac{-64\omega_c^2}{6} + \left(6 + \frac{170}{9}\right) = 0$$

which yields

$$\omega_c = \pm 1.53$$

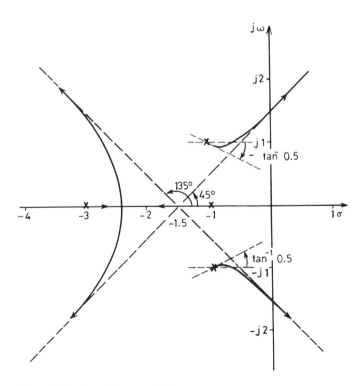

Figure 4.8 Complete root-loci.

Stability and Frequency Response Method

The breakaway point σ_b in the region $(-1, -3)$ is obtained by setting $dK(\sigma)/d\sigma = 0$ where $K(\sigma) = -1/(\sigma + 1)(\sigma + 3)(\sigma^2 + 2\sigma + 2)$. The result by iterative procedure is -2.4. At the pole $(-1 + j)$, the branches depart at angle

$$\theta_D = 180 - 90° - 90° - \tan^{-1}(0.5)$$
$$= -26.6°$$

and obviously by symmetry θ_D at $(-1 + j)$ would be $+26.6°$. The complete root-loci is displayed in Fig. 4.8.

4.2.7 Lyapunov Method

In this section, we discuss some kinds of stability exclusively in the time domain. We start with nonlinear systems and later specialize in linear systems. Consider the system

$$\dot{\mathbf{x}} = f(\mathbf{x}) \tag{31}$$

which is a time-invariant, nonlinear, and unforced dynamical system. Here, we focus on the stability of *equilibrium points*. System (30) is said to have an equilibrium point at $\mathbf{x} = \mathbf{x}_e$ if

$$\mathbf{f}(\mathbf{x}_e) = \mathbf{0} \tag{32}$$

Note that an equilibrium point is not unique and can always be shifted to the origin in state-space by a translation in coordinates. Initially, we introduce $\mathbf{x}_0$ as the initial state vector at $t = 0$ and $\Phi(t, \mathbf{x}_0)$ as solution (31). An equilibrium point $\mathbf{x}_e$ of (31) is said to be *stable* in the sense of Lyapunov if, for all $\epsilon > 0$, there exists a $\delta > 0$, so that if

$$\|\mathbf{x}_0 - \mathbf{x}_e\| < \delta \tag{33}$$

then

$$\|\Phi(t, \mathbf{x}_0) - \mathbf{x}_e\| < \epsilon \tag{34}$$

for all $t \geq 0$.

An interpretation of the above definition is that if the trajectory Φ remains within a specified distance from $\mathbf{x}_e$ provided the starting point $\mathbf{x}_0$ is close enough to $\mathbf{x}_e$. This fundamental idea is expressed geometrically for a second-order system in Fig. 4.9.

Frequently, it is desired that the trajectory actually approaches the equilibrium point as time progresses rather than just remaining in some specified neighborhood. To reflect this desire, we define an equilibrium point $\mathbf{x}_e$ of (30) to be *asymptotically stable* if it is stable and if for $\|\mathbf{x}_0 - \mathbf{x}_e\| < \delta$, $\lim_{t \to \infty} \Phi(t, \mathbf{x}_0) = \mathbf{x}_e$.

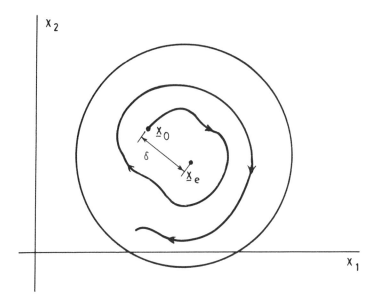

Figure 4.9 Illustration of a stable trajectory.

A generalization of this definition is that an equilibrium point x_e of system (30) is said to be *asymptotically stable in the large* if it is asymptotically stable for any initial condition x_0.

We take note that the foregoing stability definitions are expressed in terms of the time-domain solution of (30), for which an analytic solution is, unfortunately, not available in most cases. We thus need an indirect method of investigating stability and this is now available via the method of Lyapunov. The basis of this method is a scalar function $V(\mathbf{x})$ of the state vector having the following properties:

1. $V(\mathbf{x})$ is positive-definite for all $\mathbf{x}$, that is $V(\mathbf{x}) = 0$ for $\mathbf{x} = \mathbf{0}$ and $V(\mathbf{x}) > 0$ for $\mathbf{x} \neq \mathbf{0}$.
2. $-dV(\mathbf{x})/dt$ is positive-definite for all $\mathbf{x}$.
3. $V(\mathbf{x}) \to \infty$ as $\|\mathbf{x}\| \to \infty$.

A basic theorem of Lyapunov states that if for system (31) along with (32) there exists a continuous scalar function $V(\mathbf{x})$ possessing the above qualifications, then $\mathbf{x}_e = \mathbf{0}$ is asymptotically stable in the large. In the sequel we shall assume that the equilibrium point of interest is the origin.

Stability and Frequency Response Method

Example 4.13. For the system

$$\dot{x}_1 = -x_1 - x_2, \qquad \dot{x}_2 = x_1 - x_2$$

clearly, the origin **0** is an equilibrium point. Suppose that $V(\mathbf{x}) = V(x_1, x_2) = 1/2 x_1^2 + 1/2 x_2^2$. Since $V(0,0) = 0$ and $V(x_1, x_2) > 0$ for all $x_1, x_2 > 0$ then $V(\mathbf{x})$ is positive-definite. Moreover, $-\dot{V}(x_1, x_2) = x_1^2 + x_2^2$ which is quite easy to verify its positive-definiteness. We therefore conclude that $\mathbf{x}_e = \mathbf{0}$ is asymptotically stable in the large.

In system studies, $V(\mathbf{x})$ sometimes represents stored energy or distance to reflect practical situations. For higher-order systems, positive definite $V(\mathbf{x})$ can be expressed in the quadratic form $\mathbf{x}'P\mathbf{x}$:

$$\mathbf{x}'P\mathbf{x} = \begin{bmatrix} x_1 & x_2 & \cdots & x_n \end{bmatrix} \begin{bmatrix} p_{11} & \cdots & p_{1n} \\ \vdots & & \vdots \\ p_{n1} & \cdots & p_{nn} \end{bmatrix} \begin{bmatrix} x_1 \\ \vdots \\ x_n \end{bmatrix} \quad (35)$$

where $P = P^t$ is an $n \times n$ symmetric matrix. From matrix algebra, the quadratic form (35) is positive-definite if and only if all *principal minors* are positive

$$p_{11} > 0, \quad \begin{vmatrix} p_{11} & p_{12} \\ p_{21} & p_{22} \end{vmatrix} > 0,$$

$$\begin{vmatrix} p_{11} & p_{12} & p_{13} \\ p_{21} & p_{22} & p_{23} \\ p_{31} & p_{32} & p_{33} \end{vmatrix} > 0, \ldots, \det[P] > 0 \quad (36)$$

An interesting fundamental result is the version of the Lyapunov method for linear systems. It can be stated as the equilibrium point $\mathbf{x}_e = \mathbf{0}$ of the linear time-invariant system $\dot{\mathbf{x}} = A\mathbf{x}$ is asymptotically stable in the large if and only if, given any positive-definite symmetric matrix, Q, there exists a positive-definite symmetrix matrix, P, such that

$$A'P + PA = -Q \quad (37)$$

and $\mathbf{x}'P\mathbf{x}$ is a Lyapunov function for the LTI system.

The proof is straightforward and depends on satisfying the conditions of the method of Lyapunov. Note that the solution of Lyapunov equation (37) requires solving $n(n+1)/2$ simultaneous linear algebraic equations for the elements of P.

Example 4.14. Let

$$A = \begin{bmatrix} 0 & 1 \\ -5 & -6 \end{bmatrix}$$

and choose $Q = 1$. Then (37) becomes

$$\begin{bmatrix} 0 & -5 \\ 1 & -6 \end{bmatrix} \begin{bmatrix} p_1 & p_{12} \\ p_{12} & p_{22} \end{bmatrix} + \begin{bmatrix} p_{11} & p_{12} \\ p_{12} & p_{22} \end{bmatrix} \begin{bmatrix} 0 & 1 \\ -5 & -6 \end{bmatrix} = \begin{bmatrix} -1 & 0 \\ 0 & -1 \end{bmatrix}$$

where the symmetric character of P is used explicitly. Upon expansion, we obtain $2(2+1)/2 = 3$ equations as follows:

$$-10p_{12} = -1$$
$$p_{11} - 6p_{12} - 5p_{22} = 0$$
$$2(p_{12} - 6p_{22}) = -1$$

whose solution is given by

$$p_{11} = 11/10, \qquad p_{12} = 1/10, \qquad p_{22} = 1/10$$

Hence

$$P = \begin{bmatrix} 11/10 & 1/10 \\ 1/10 & 1/10 \end{bmatrix}$$

which in view of (36) can be easily checked to be positive-definite. Therefore the equilibrium point $x_e = 0$ is sympotically stable in the large.

It should be emphasized that LTI the origin $x_e = 0$ is the only equilibrium point. Solution algorithms of (37) are readily available in standard software packages including NAG, IMSL, and MATLAB.

This concludes the section on stability analysis and techniques. We next discuss methods of preserving stability properties.

4.3 FREQUENCY RESPONSE

This section is concerned with the study of frequency response analysis methods. The celebrated work of Nyquist and Bode has provided valuable tools in control engineering practice and has been widely accepted as a powerful technqiue in systems studies. Consider a linear system driven by a sinusoidal input of the form

$$r(t) = A\sin(\omega t) \tag{38}$$

Under steady-state, the system output as well as the signals at all other

Stability and Frequency Response Method

points in the system are sinusoidal. The steady-state output may be written as

$$c(t) = B \sin(wt + \phi) \tag{39}$$

The magnitude and phase relationship between the sinusoidal input and steady-state output of a system is termed the *frequency response*. In LTI systems, the frequency response is independent of the amplitude and phase of the input signal.

An interesting and revealing comparison of frequency and time-domain approaches is based on the relative stability studies of feedback systems. The Routh–Hurwitz criterion is a time-domain approach which establishes the system stability with relative ease, however its adoption in determining the relative stability is involved and requires repeated application of the criterion. The root-locus method is a very powerful time-domain approach as it reveals not only stability but also the actual time response of the system, though it is somewhat laborious, being a technique of determining the characteristic roots. We shall discuss here frequency-domain methods for extracting the information regarding stability as well as relative stability of linear systems.

4.3.1 Polar Plots

We consider that a LTI system is represented by its transfer function $G(s)$ which, as we repeatedly mentioned, relates the Laplace transform of the input $R(s)$ to the Laplace of the output $C(s)$. The sinusoidal transfer function $G(j\omega)$ is obtained simply by substituting $(j\omega)$ for s in $G(s)$.

$$G(j\omega) = |G(\omega)| \angle \phi(j\omega) \tag{40}$$

Note that the functional dependence of the magnitude $|G(\omega)|$ and phase $\phi(j\omega)$ on the radian frequency ω is emphasized. The sinusoidal transfer function (40) itself is a complex-valued function with ω as a parameter.

The polar plot is the locus of the phasor $|G(\omega)| \angle \phi(j\omega)$ in the complex plane as ω is varied from zero to infinity. We employ the standard notation of a positive phase angle measured counterclockwise from the positive real axis. For a particular ω, the end of phasor $G(j\omega)$ provides a point on the polar plot. In some occasions, we express $G(j\omega)$ in the rectangular form

$$G(j\omega) = X(\omega) + jY(\omega) \tag{41}$$

and attempt to derive functional relation in the cartesian plane.

To develop some properties of the polar plots, we put the sinusoidal transfer in quotient form as

$$G(j\omega) = \frac{K_m N(j\omega)}{(j\omega)^m D(jw)} \qquad (42)$$

Recall from Chapter 2 that m is the system type (excess pole-zero at origin). For convenience we let the numerator and denominator polynomials be in factored form as

$$N(j\omega) = (1 + j\omega\tau_a)(1 + j\omega\tau_b) \ldots (1 + j\omega\tau_M) \qquad (43)$$

$$D(j\omega) = (1 + j\omega\tau_1)(1 + j\omega\tau_z) \ldots (1 + j\omega\tau_P) \qquad (44)$$

We now analyze each component of (42) in terms of its phase:

1. The term $(j\omega)^m$ in the denominator contributes a constant angle ϕ_c

$$\phi_c = -m\pi/2 \qquad (45)$$

2. The numerator polynomial $N(j\omega)$ gives an angle ϕ_N which is the sum of the individual phases in (6)

$$\phi_N = \phi_a + \phi_b + \cdots + \phi_M \qquad (46)$$

$$\phi_a = \tan^{-1}(\omega\tau_a), \qquad \phi_b = \tan^{-1}(\omega\tau_b).$$

$$\vdots \qquad \qquad \vdots \qquad (47)$$

$$\phi_M = \tan^{-1}(\omega\tau_M)$$

3. The denominator polynomial $D(j\omega)$ provides an angle ϕ_D which can be seen from (44) to have the form

$$\phi_D = \phi_1 + \phi_2 + \cdots + \phi_P \qquad (48)$$

$$\phi_1 = -\tan^{-1}(\omega\tau_1), \qquad \phi_2 = -\tan^{-1}(\omega\tau_2).$$

$$\vdots \qquad \qquad \vdots \qquad (49)$$

$$\phi_P = -\tan^{-1}(\omega\tau_P)$$

The constant gain K_m provides nothing to phase. Thus from (42) through (49) we can write

$$\phi_{(j\omega)} = \phi_c + \phi_D + \phi_N \qquad (50)$$

Relating (40) to (42) we have

Stability and Frequency Response Method

$$|G(\omega)| = \frac{K_m |N(j\omega)|}{\omega^m |D(j\omega)|} \tag{51}$$

At low frequencies ($\omega \to 0$) we obtain from (51)

$$|G(0)| = \lim_{\omega \to 0} |G(\omega)|$$
$$= \infty, \quad m > 0$$
$$= K_0, \quad m = 0 \tag{52}$$

At higher frequencies ($\omega \to \infty$) is is obvious for proper systems that

$$|G(\infty)| = \lim_{\omega \to \infty} |G(\omega)| = 0 \tag{53}$$

From (47) and (49), it is readily seen that the angles ϕ_D and ϕ_N are zero for $\omega \to 0$, and thus

$$\phi_0 = \lim_{\omega \to 0} \phi_{(j\omega)} = \phi_c = -m\pi/2 \tag{54}$$

As ω increases progressively, we have the following limits:

$$\lim_{\omega \to \infty} \phi_D = -P\frac{\pi}{2}$$

$$\lim_{\omega \to \infty} \phi_N = +\frac{M}{\pi/2}$$

Thus

$$\phi(\infty) = \lim_{\omega \to \infty} \phi_{(j\omega)} = (M - P - m)\frac{\pi}{2} \tag{55}$$

At the points of intersection with real and imaginary axes, the corresponding frequencies are obtained from

$$Y(\omega) = 0, \quad X(\omega) = 0 \tag{56}$$

It is appropriate at this stage to consider an illustrative example.

Example 4.15. The transfer function of a simple RC filter is

$$G(s) = \frac{1}{1 + s\tau}, \quad \tau = RC \tag{57}$$

Therefore, the sinusoidal transfer function is

$$G(j\omega) = \frac{1}{1 + j\omega\tau}$$

$$= (1/\sqrt{1 + \omega^2\tau^2})|\tan^{-1}(-\omega\tau) \qquad (58)$$

In comparison with (3) we have

$$|G(\omega)| = \frac{1}{\sqrt{1 + \omega^2\tau^2}}$$

$$\phi(j\omega) = \tan^{-1}(-\omega\tau)$$

The cartesian compoents in (4) can be expressed as

$$X(\omega) = \frac{1}{1 + \omega^2\tau^2}, \qquad Y(\omega) = -\frac{\omega\tau}{1 + \omega\beta^2\tau^2}$$

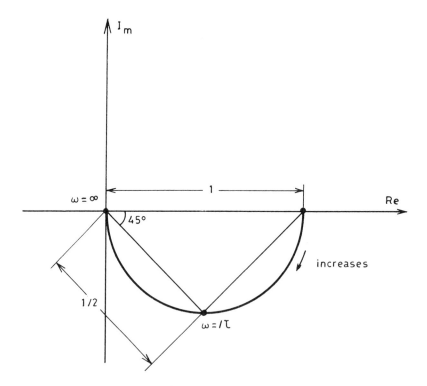

Figure 4.10 Polar plot of $1/(1 + j\omega\tau)$.

Stability and Frequency Response Method

The relevant points are:
1. At $\omega = 0$
$$|G(0)| = 1, \quad \phi(0) = 0$$
$$X(0) = 1, \quad Y(0) = 0$$

2. At $\omega = \infty$
$$|G(\infty)| = 0, \quad \phi(\infty) = -\pi/2$$
$$X(\infty) = 0, \quad Y(\infty) = 0$$

Note that for $\omega > 0$, $\phi(\omega)$ is always negative (phase lag) and $|G(\omega)|$ decreases monotonically. At $\omega = 1/\tau$, $|G(1/\tau)| = 1/\sqrt{2}$, $\phi = -\pi/4$, $X(1/\tau) = 1/2$ and $Y(1/\tau) = -1/2$. Eliminating ω, we can see that

$$X^2 + Y^2 = X \quad \text{or} \quad (X - 1/2)^2 + Y^2 = (1/2)^2$$

which is the locus of a circle with center at $(1/2, 0)$ and radius $= 1/2$. Since the phase is always negative, the polar plot is a semicircle in the fourth quadrant as displayed in Fig. 4.10.

4.3.2 Bode Logarithmic Plots

One of the most useful representations of a transfer function is a logarithmic plot which consists of two graphs, one giving the logarithm of the magnitude $|G(j\omega)|$ and the other phase angle of $G(j\omega)$ both plotted against frequency in logarithmic scale. These plots are termed *Bode logarithmic plots* after the pioneering work of H. W. Bode. An equivalent expression of (40) is

$$G(j\omega) = |G(j\omega)| e^{j\phi(\omega)} \tag{59}$$

Taking logarithms of both sides

$$\log G(j\omega) = \log|G(j\omega)| + j0.434\,\phi(\omega) \tag{60}$$

Thus the logarithm of the complex function $G(j\omega)$ resolves to the sum of a real part (given by the logarithm of the magnitude) and an imaginary part (represented by a scaled version of the phase angle). The real part is normally expressed in terms of the logarithmic gain in decibels (dB) denoted by A_{dB}, and defined as

$$A_{dB} = 20 \log|G(j\omega)| \tag{61}$$

which shows the dependence of the logarithmic gain on ω. Usually, the horizontal axis in Bode plots in log ω and on semilog paper the most frequent numbers are available. It is thus obvious that a wide range of frequencies can be easily displayed. More importantly, under logarithmic transformation, multiplication and division operations are converted to additions and subtractions, a basic feature which tremendously simplifies the work.

We now examine the logarithmic plot of the sinusoidal transfer function (42) along with (43) and (44), in terms of magnitude and phase. In general, we note that

$$20 \log |G(j\omega)| = 20 \log K + \{20 \log |1 + j\omega\tau_a| + \cdots$$
$$+ 20 \log |1 + j\omega\tau_M|\}$$
$$- 20m \log \omega - \{20 \log |1 + j\omega\tau_1| + \cdots$$
$$+ 20 \log |1 + j\omega\tau_P|\} \quad (62)$$

$$\phi(\omega) = \{\tan^{-1}(\omega\tau_a) + \cdots + \tan^{-1}(\omega\tau_M)\}$$
$$- m\pi/2 - \{\tan^{-1}(\omega\tau_1) + \cdots + \tan^{-1}(\omega\tau_P)\} \quad (63)$$

1. The constant gain K_m gives a magnitude in decibels, in light of (60), as

$$A_{dB} = 20 \log K \quad (64)$$

which is positive for $K > 1$, negative when $K < 1$ and in either case is independent of frequency. The associated phase plot is also constant at level 0; see Fig. 4.11. Note that the scale on horizontal axis is logarithmic and divided into equal bands, each of which is called a *decade*. To emphasize, the band between $\omega = 1$ and $\omega = 10$ is equal to that between $\omega = 100$ and $\omega = 1000$ and so on. The base value in the phase plot is $-180°$ which is (as we shall see later) in the border of stability.

2. The term $(j\omega)^m$ has the phase (45) and the log magnitude as a part of (60) is given by

$$A_{dB} = 20 \log \left| \frac{1}{(j\omega)^m} \right|$$
$$= 20 \log \frac{1}{\omega^m}$$
$$= -20m \log \omega \quad (65)$$

which describes a linear relationship between A_{dB} (on the vertical ordinate) and $\log \omega$ on the horizontal ordinate. The slope $(-20m)$ varies with m. A typical plot of the magnitude and phase of the term

Stability and Frequency Response Method

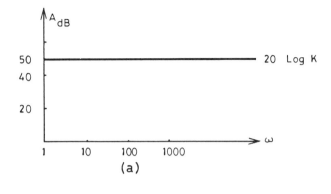

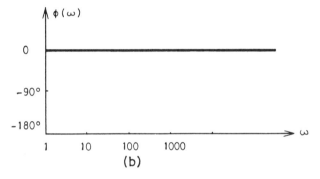

Figure 4.11 Magnitude and phase plots of a constant gain.

$(1/j\omega)^m$ appears in Fig. 4.12. Note that over a decade, the magnitude in decibels from (62) has a difference of $-20m$. For this reason, the slope in Fig. 4.12(a) is frequently labeled as $-20m$ dB/decade. So for type 1 system ($m = 1$) we have -20 dB/decade and for type 3 we obtain -60 dB/decade. Obviously, the inverse situation is obtained for term $(j\omega)^m$, that is the magnitude plot will be reflected about the vertical ordinate as a mirror (positive slope), and the phase plot will be reflected about the horizontal ordinate as a mirror (positive phase).

3. The term $1/(1 + j\omega\tau)$ from (43) is how analyzed. The log magnitude from (60) is

$$A_{dB} = 20 \log\left(\frac{1}{\sqrt{1 + \omega^2\tau^2}}\right)$$

$$= -20 \log \sqrt{1 + \omega^2\tau^2} \qquad (66)$$

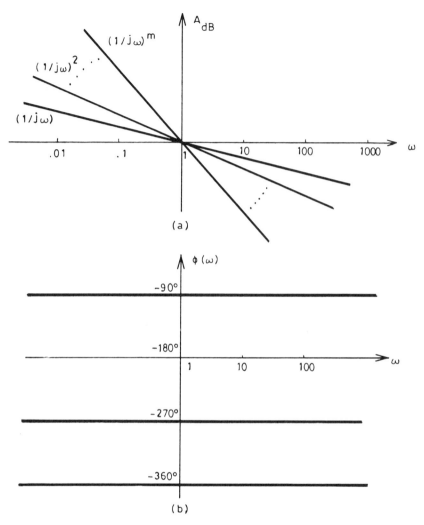

Figure 4.12 Magnitude and phase plots of $(1/j\omega)$.

a. At low frequency $(\omega\tau \ll 1)$
$$A_{dB} = 0 \text{ dB} \tag{67}$$
b. At higher frequency $(\omega\tau \gg 1)$
$$A_{dB} = -20 \log \omega\tau \tag{68}$$

Thus the magnitude function (64) can be replaced asymptotically by

Stability and Frequency Response Method 141

a straight line of 0 dB up to w_c, the difference between the actual and the asymptote is, from (64), equal to -3 dB. The phase is given by $\phi(\omega) = \tan^{-1}(-\omega\tau)$. The asymptotic plot has the characteristics:

Low frequency ($\omega\tau \ll 1$), $\phi \to 0$
High frequency ($\omega\tau \gg 1$), $\phi \to -\pi$
Corner frequency ($\omega\tau = 1$), $\phi = -\pi/2$

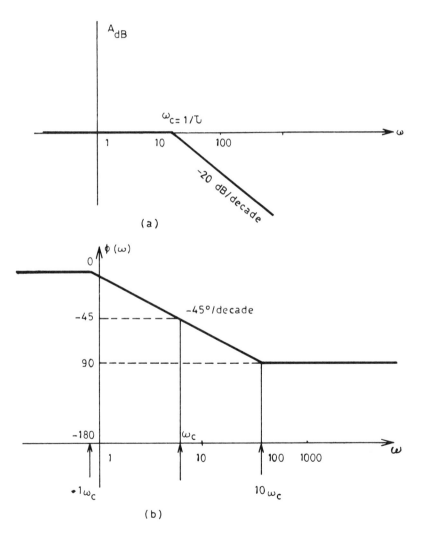

Figure 4.13 Asymptotic magnitude and phase plots of $1/(1 + j\omega\tau)$.

The exact curve can be plotted using some selected frequency. A reasonable approximation is to draw a straight line with a slope of $-45°$ per decade passing through $(-45°, \omega_c)$. The straight line extends to $(0, 0.1\,\omega_c)$ and $(-90°, 10\omega_c)$. The symptotic Bode plots are shown in Fig. 4.13.

Factors of the form $[1/(1 + j\omega\tau)]^p$ can be handled in a straightforward manner by extending the results of Fig. 4.13 such that

a. The slope after ω_c will be $(-20p)$ dB/decade
b. The slope in the phase plot will be $-45°$/decade.

Other information remains unchanged.

By the same token, terms like $(1 + j\omega\tau)^p$ will have logarithmic Bode plots which are the proper inverse of $[1/(1 + j\omega\tau)]^p$.

To sum up, the asymptotic Bode plot of any sinusoidal transfer function of the type (42) can be obtained by proper addition of the corresponding individual terms. The following example illustrates such a procedure.

Example 4.16. Construct the asymptotic Bode plot for $G(j\omega) = 1/j\omega(1 + j\omega)(1 + j0.2\omega)$. We have three terms: $1/j\omega$, $1/(1 + j\omega)$ and $1/(1 + j0.2\omega)$. The separate log magnitude plot is shown in Fig. 4.14(a) along with the corresponding asymptotic phase plot in Fig. 4.14(b).

The complete Bode plots (magnitude and phase) are displayed in Figs. 15(a) and 15(b).

When transfer functions have all of their poles and zeros in the left half of the s-plane, they are called *minimum phase*. The special class of transfer functions having a pole-zero pattern which is antisymmetric about the imaginary axis are called *all-pass systems*. Note that for every pole in the left half-plane, there is a zero in the mirror-image position. Common features of these systems are that their magnitude plot is unity at all frequencies and their phase plot vary from $0°$ to $-180°$ as ω is increased from 0 to ∞. Thus by making use of the foregoing cases, a transfer function which has one or more in the right half s-plane is known as *nonminimum phase transfer function*.

4.3.3 Stability Margins

We now introduce two concepts related to stability analysis: gain margin and phase margin. *Gain margin* is the factor by which the system gain can be increased to drive it to the verge of instability. More precisely, the gain margin (GM) may be defined as the reciprocal of the gain at the frequency at which the phase angle becomes $(180°)$, or phase crossover frequency, ω_p.

$$\text{GM} = 20\log|1/G(j\omega)|_{\omega=\omega_p}$$

Stability and Frequency Response Method

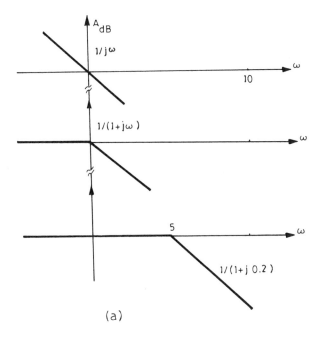

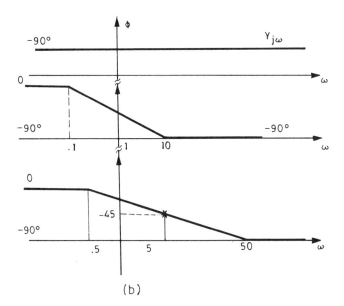

Figure 4.14 Individual asymptotic Bode plots for example 4.16.

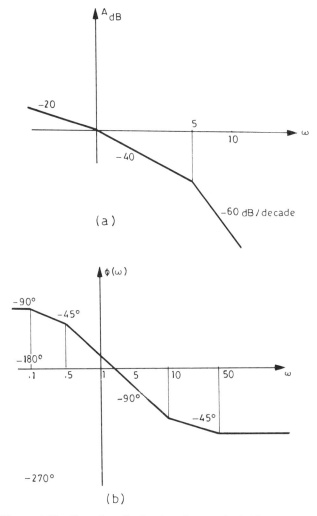

Figure 4.15 Complete Bode plot of example 4.16.

The frequency at which $|G(j\omega)H(j\omega)| = 1$ is called the *gain crossover frequency* ω_g. The *phase margin* (PM) may be defined as the amount of additional phase-lag at the gain crossover frequency required to bring the system to the verge of instability. The phase margin is always positive for stable feedback systems.

$$\text{PM} = \angle G(j\omega) H(j\omega)|_{\omega=\omega_g} + 180°$$

The stability margins are illustrated in Fig. 4.16.

Stability and Frequency Response Method

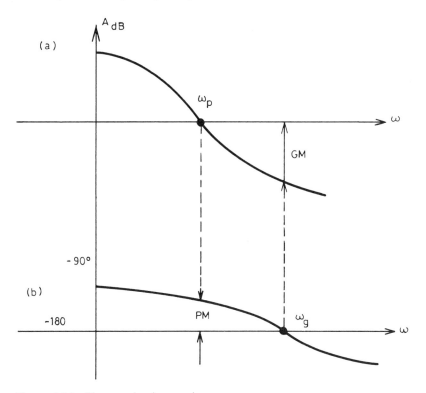

Figure 4.16 Phase and gain margins.

4.3.4 The Nyquist Criterion

The main concern here is with the mapping of contours in the complex s-plane by a function $F(s)$ in order to infer some stability properties in the frequency domain. The basis of the work comes from the theory of the function of a complex variable due to Cauchy. We illustrate his fundamental concept by an example. Define $s = \alpha + j\omega$ and consider a function $F(s) = 3s + 2$. The mapping of the s-plane square contour of side 2 to the $F(s)$-plane is accomplished through the relation $F(s)$ so that

$$F(s) = 3s + 2 = 3(\alpha + j\omega) + 2 = u + jv$$

Thus

$$u = 3\alpha + 2, \quad v = 3\omega$$

which shows that the contour has been mapped by $F(s)$ into a contour of an identical form, a square, with the center shifted by two units and the magnitude of a side multiplied by 3. This type of mapping which retains the angles of the s-plane contour on the $F(s)$-plane, is called a *conformal mapping*. We also note that a closed contour in the s-plane results in a closed contour in the $F(s)$-plane.

Cauchy theorem is concerned with the mapping of a function $F(s)$, which has a finite number of poles and zeros within the contour. It states that if a contour Γ_s in the s-plane encircled Z zeros and P poles of $F(s)$ and does not pass through any pole or zero of $F(s)$ as the transversal is in the clockwise direction along the contour, the correspoding contour Γ_F in the $F(s)$-plane encircles the origin of the $F(s)$ plane $N = Z - P$ times in the clockwise direction.

In order to utilize this concept in stability analysis we consider $F(s)$ as the characteristic equation

$$F(s) = 1 + P(s)$$
$$= \frac{K \Pi_{i=1}^{M} (s + z_i)}{\Pi_{j=1}^{N} (s + p_j)} \tag{69}$$

We have learned before that for a system to be stable, all the zeros of $F(s)$ must lie in the left half s-plane. Therefore we choose a contour Γ_s in the s-plane which encloses the entire right half s-plane, and we determine whether any zero of $F(s)$ lies within Γ_s by applying Cauchy theorem. That is, we plot Γ_F in the $F(s)$-plane and determine the number of encirclements of the origin N. The number of zeros of $F(s)$ with Γ_s (corresponding to unstable zeros of $F(s)$) is $Z = N + P$. When $P = 0$ in the usual case, we find that the number of unstable roots of the system is equal to N, the number of encirclements of the origin of the $F(s)$ plane. A typical Nyquist contour is shown in Fig. 4.17 which encloses the entire right half s-plane.

Instead of utilizing $F(s)$, we use $P(s) = F(s) - 1$ since it is typically available in factored form. We seek the mapping in the $P(s)$-plane. Accordingly, the number of clockwise encirclements of the origin in the $F(s)$-plane becomes the number of clockwise encirclements of the $(-1, 0j)$ point in the $P(s)$-plane. The *Nyquist stability criterion* may therefore be stated as follows. A feedback system is stable if and only if the contour Γ_p in the $P(s)$-plane does not encircle the point $(-1, 0)$ when the number of poles of $P(s)$ in the right half s-plane is zero ($P = 0$). Otherwise, when $P > 0$, the contour Γ_p has a number of counter-clockwise encirclements of the $(-1, 0)$ point equal to the number of poles of $P(s)$ with positive real parts.

Stability and Frequency Response Method

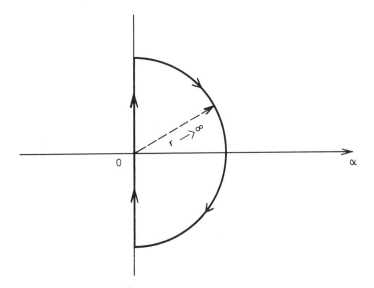

Figure 4.17 The Nyquist contour.

Example 4.17. A single-loop control system has

$$G(s)H(s) = \frac{K}{s(1 + \tau s)}$$

We select $P(s) = G(s) H(s)$ and seek to determine $\Gamma_p = \Gamma_{GH}$ in the $GH(s)$-plane. Initially the contour Γ_s in the s-plane is chosen as in Fig. 4.18(a) where an infinitesimal detour around the pole at the origin is effected by a small semicircle of radius $\gamma \to 0$. This detour is in compliance with Cauchy theorem to avoid passing through poles. We now consider the mapping of each portion:

1. The origin of the s-plane. We have $s = \gamma e^{j\phi}$ where ϕ varies from $-90°$ at $\omega = 0^-$ to $+90°$ at $\omega = 0^+$. But

$$\lim_{\gamma \to 0} G(s)H(s) = \lim_{\gamma \to 0}(K/\gamma e^{j\phi}) = \lim_{\gamma \to 0}(K/\gamma)\bar{e}^{j\phi}$$

   Therefore, the angle of the contour in the $GH(s)$-plane changes from $90°$ at $\omega = 0^-$ to $-90°$ at $\omega = 0^+$, passing through $0°$ at $\omega = 0$. The radius of the contour is infinite as shown in Fig. 4.18(b).

2. The portion from $\omega = 0^+$ to $\omega = +\infty$. Since $s = j\omega$

$$G(s)H(s)\,|_{s=j\omega} = GH(j\omega) = \frac{K}{j\omega(j\omega\tau + 1)}$$

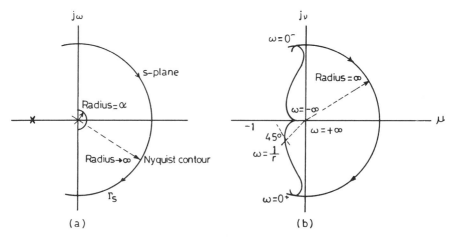

Figure 4.18 Nyquist contour and mapping of example 4.17.

But

$$\lim_{\omega \to +\infty} GH(j\omega) = \lim_{\omega \to +\infty} \left| \frac{K}{\tau \omega^2} \right| \angle \left\{ -\left(\frac{\pi}{2}\right) - \tan^{-1}(\omega \tau) \right\}$$

Therefore the magnitude approaches zero at an angle of $-180°$.

3. The portion from $\omega = +\infty$ to $\omega = -\infty$. In the s-plane, Γ_s is represented by $re^{j\phi}$, thus in the $GS(s)$-plane we obtain

$$\lim_{r \to \infty} GH(s)|_{s=re^{j\phi}} = \lim_{r \to \infty} \left| \frac{k}{\gamma^2} \right| \bar{e}^{2j\phi}$$

as ϕ changes from $+90°$ (at $\omega = +\infty$) to $-90°$ at $\omega = -\infty$. Thus the contour moves from an angle of $-180°$ (at $\omega = +\infty$) to an angle of $180°$ at $\omega = -\infty$. The magnitude of $GH(s)$-contour (as $r \to \infty$) is always zero or a constant.

4. The portion from $\omega = -\infty$ to $\omega = 0^-$. In this case the portion of the contour Γ_s from $\omega = -\infty$ to $\omega = 0^-$ is mapped by the function $GH(s)$ as

$$GH(s)|_{s=-j\omega} = GH(-j\omega)$$

which is symmetrical to the portion in item 2 above.

Now, to examine the stability, we note that there are no poles within the right half s-plane ($P = 0$). Therefore for this system to be stable, we require $N = Z = 0$ and the contour Γ_{GH} must not encircle the $(-1, 0)$

Stability and Frequency Response Method

point in the GH-plane. On examining Figure 4.18(b), we find that no matter what K and τ are, the contour remains safely far from the $(-1, 0)$ point and thus the system is always stable.

From this example, we can draw the following rules:

1. The polar plot of $GH(s)$ will be symmetric in the $GH(s)$-plane about the u-axis. It is therefore sufficient to construct the contour Γ_{GH} for the frequency range $0^+ < \omega < +\infty$ in order to investigate the stability.
2. The magnitude of $GH(s)$ as $s = \gamma e^{j\phi}$ and $\gamma \to \infty$ will normally approach zero or a constant.

4.4 SENSITIVITY ANALYSIS

Earlier in Chapter 2 the sensitivity of a transfer function $T(s)$ to variation of a parameter h has been defined as

$$S_h^T = \frac{d(\ln T)}{d(\ln h)}$$

$$= \frac{\partial T/T}{\partial h/h} \qquad (70)$$

We are going to discuss here some special forms of (70) pertinent to frequency response methods. Of particular interest is the sensitivity measure in terms of the positions of the roots of the characteristic equation. This stems from the fact that these roots represent the dominant modes of transient response. The *root sensitivity* of a transfer function $T(s)$ can be defined as

$$S_h^{p_j} = \frac{\partial p_j}{\partial (\ln h)} = \frac{\partial p_j}{(\partial h/h)} \qquad (71)$$

where p_j equals the jth root of the function

$$T(s) = K_r \frac{\prod_{k=1}^{M}(s + z_k)}{\prod_{j=1}^{N}(s + p_j)} \qquad (72)$$

The root sensitivity relates the change in the location of the root in the s-plane to the change in the parameter. Relating the root sensitivity (71) to the logarithmic sensitivity (70), we can easily find out by differentiating (72) with respect to h that

$$S_h^T = \frac{\partial \ln K_r}{\partial h} - \sum_{j=1}^{n} \frac{\partial r_j/\partial \ln h}{s + r_j}$$

where we have used $\partial z_k/\partial \ln h = 0$ to imply the independence of the zeros of $T(s)$ on the parameter h. When the gain of the system is independent of the parameter h, the first term in the last equation goes to zero and with the aid of (70) it results in

$$S_h^T = -\sum_{j=1}^{n} \frac{S_j^{p_j}}{s + r_j} \qquad (73)$$

which provides a direct link between the two sensitivity measures.

The importance of the root sensitivity in systems control comes from its wide use in the root-locus method. An example will illustrate the process of evlauting the root sensitivity.

Example 4.18. A unity feedback system has $G(s) = K/s(s + \sigma)$. The characteristic equation is

$$1 + \frac{K}{s(s + \sigma)} = 0 \quad \text{or} \quad s^2 + \sigma s + K = 0$$

Consider $K = K_0$ and $\sigma = \sigma_0$ are the nominal values and let

$$K = K_0 \pm \Delta K, \qquad \sigma = \sigma_0 \pm \Delta \sigma$$

For simplicity let $K_0 = 0.5$ and $\sigma_0 = 1$ and focus on the effect of K. Then the characteristic equation becomes

$$s^2 + s + K_0 \pm \Delta K = s^2 + s + 0.5 \pm \Delta K \qquad (74)$$

or, alternatively

$$1 + \left(\pm \frac{\Delta K}{s^2 + s + 0.5}\right) = 1 + \left(\pm \frac{\Delta K}{(s + p_1)(s + \hat{p}_1)}\right) = 0 \qquad (75)$$

where $p_1 = -0.5 + j0.5$. Therefore, the effects of changes in the gain can be evaluated from the root-locus of Fig. 4.19. For a 20% change in K, we have $\pm \Delta K = \pm 0.1$. The root locations for a gain $K = 0.4$ and $K = 0.6$ are determined by root locus methods. The results are displayed in Fig. 4.19.

For $K = 0.6$, one of the roots of (74) is $p_1 + \Delta p_1 = -0.5 + j0.592$ resulting in $\Delta p_1 = j0.052$. On the other hand, for $K = 0.4$, the root becomes $p_1 + \Delta p_1 = -0.5 \pm j0.387$ yielding $\Delta p_1 = -j0.113$. Hence the root sensitivity for p_1 is

Stability and Frequency Response Method

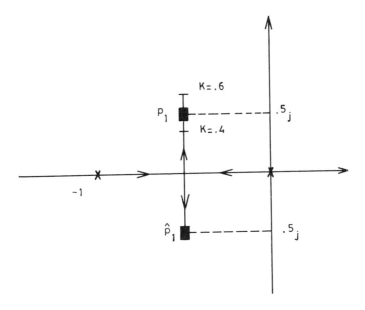

Figure 4.19 Root locus of example 4.16.

$$S_{\Delta K}^{p_1} = \frac{\Delta p_1}{\Delta K/K}$$

$$= \frac{j0.09}{0.2}$$

$$= j0.45 = 0.45 \angle 90° \tag{76}$$

When the change in gain is positive. For negative changes, we have

$$S_{\Delta K}^{p_1} = \frac{\Delta p_1}{\Delta K/K}$$

$$= -\frac{j0.113}{0.2}$$

$$= -j0.565 = 0.565 \angle -90° \tag{77}$$

It should be emphasized that the angle of the root sensitivity indicates the direction the root would move as the parameter K changes. Had we considered the variation in σ instead of K, the corresponding characteristic equation would have been of the form (for $\sigma_0 = 1$)

$$s^2 + s + \Delta\sigma s + K = 0 \tag{78}$$

or, in root-locus form, it becomes

$$1 + \frac{\Delta \sigma s}{s^2 + s + K} = 0 \tag{79}$$

It is ascertained that the denominator in the second term is the unaltered characteristic equation with $\Delta \sigma = 0$. For $\sigma_0 = 1$ and $\Delta \sigma = 20\%$, it can be easily shown that

$$S_{\sigma+}^{p_1} = \frac{\Delta p_1}{\Delta \sigma / \sigma}$$

$$= 0.16 \angle -131°/0.2$$

$$= 0.8 \angle -131° \tag{80}$$

$$S_{\sigma-}^{p_1} = \frac{\Delta p_1}{\Delta \sigma / \sigma}$$

$$= 0.125 \angle 38°/0.2$$

$$= 0.625 \angle 38°$$

Again as $\Delta \sigma / \sigma$ decreases, the sensitivity measures will approach equality in magnitude and a difference in angle of 180°.

4.5 SOME NOTES

We now present some notes concerning the frequency response methods.

1. In the root-locus method, the open-loop gain can be determined for a specified damping of the dominant roots. A damping line (making an angle $\theta = \cos^{-1} \alpha$ with the negative real axis) is drawn for the specified damping. The intersection with the root-locus gives the desired root, at which the open-loop gain is computed. Indeed, some iterative procedure will be employed if accurate results are needed to determine the intersection points.

2. All the rules that have been discussed for constructing the root-locus of single loop systems can be easily extended to multiple loop systems. The main idea is to write down the overall transfer function which will eventually contain all the gains. Different root contours will subsequently be drawn by varying one gain factor (single or composite) at a time.

Example 4.19. Consider the multiloop system shown in Fig. 4.20. It

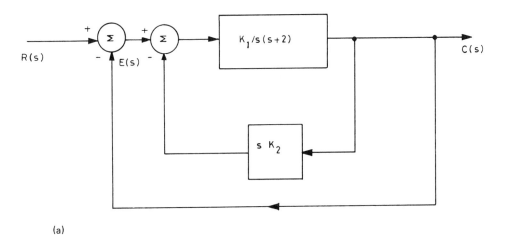

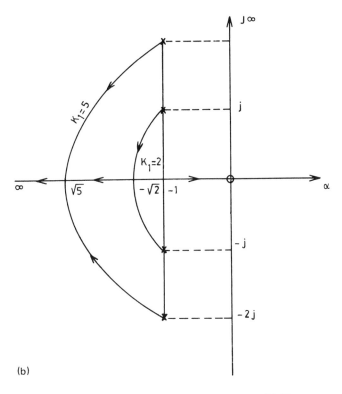

Figure 4.20 (a) A feedback multiloop system. (b) Root contours.

is easy to write

$$G(s) = \frac{C(s)}{E(s)}$$

$$= \frac{K_1/s(s+2)}{1 + sK_1K_2/s(s+2)}$$

$$= \frac{K_1}{s(s+2) + sK_1K_2}$$

and therefore the characteristic equation $1 + G(s) = 0$ takes the form

$$s(s+2) + sK_1K_2 + K_1 = 0$$

or alternatively,

$$1 = \frac{K_1K_2s}{s(s+2) + K_1} = 0$$

which has the roots

$$p_1 = -1 + j\sqrt{K_1 - 1}, \quad p_2 = -1 - j\sqrt{K_1 - 1}$$

The root contours plotted for various values of K_1 with $K = K_1K_2$ varying from zero to infinity are shown in Fig. 4.20(b).

3. In the case where experimental frequency response data are available, the Bode plots will be of great value in determining the transfer function. This is done by fitting an asymptotic log-magnitude plot to the desired accuracy. The slopes of the symptotes and the corner frequencies are then used to construct the transfer function according to the previous rules of Bode plots.

This concludes the section on system analysis by frequency response methods. In the next chapter, we shall use these methods in the system design, that is selecting appropriate compensators to meet some desired specifications.

4.6 PROBLEMS

1. Utilizing the Routh–Hurwitz criterion, determine the stability of the following polynomials:
   a. $s^5 + s^3 + 2s^2 + 2s + K$
   b. $s^4 + s^3 + 2s^2 + 6s + 8$
   c. $s^6 + 3s^5 + 5s^4 + 9s^3 + 8s^2 + 6s + 4$
   d. $s^4 + 20Ks^3 + 5s^2 + 10s + 15$

Stability and Frequency Response Method

For all cases, determine the number of roots, if any, in the right half-plane. Determine the range of K, cases (a) and (d), which results in a stable system.

2. A feedback control system has

$$G(s) = \frac{K(s + 40)}{s(s + 10)}, \quad H(s) = \frac{1}{(s + 20)}$$

   a. Determine the limiting value of gain K for a stable system.
   b. For the gain which results in conditional stability, determine the magnitude of the imaginary roots.
   c. Reduce the gain to one-half the magnitude of the conditinal value and hence determine the relative stability of the system.

3. Determine the range of values of K ($K > 0$) such that the characteristic equation

$$s^3 + 3(K + 1)s^2 + (7K + 5)s + (4K + 7) = 0$$

   has roots more negative than -1.

4. A unity feedback control system has

$$G(s) = \frac{2(s + 4)(s + 8)}{s(s + 5)^2}$$

   Determine the relative stability of the system by examining the location of the roots in the s-plane.

5. Draw the root locus for the following open-loop transfer functions when $0 < K < \infty$:
   a. $GH(s) = K/s(s + 1)^2$
   b. $GH(s) = K(s + 1)/(s + 2)(s + 3)$
   c. $GH(s) = K/(s + 1)(s^2 + s + 1)$

6. The open-loop tranfer function of standard feedback control system has two elements in cascade

$$G(s) = \frac{25}{s(s + 4)}, \quad G_c(s) = \frac{as + 1}{bs + 1}$$

   together with $H(s)\% = 1\%$. By drawing root contours, discuss the effect of varying a and b on system dynamics. Determine suitable values of a and b so that settling time is less than 1 s and the dominant roots have damping ratio greater 0.5.

7. A unity feedback system has

$$G(s) = \frac{K}{s^2(s + 2)}$$

a. By sketching the root locus, show that the system is unstable for all values of K.
b. Add a zero at $s = -a$, $0 \leq a < 2$ and show that this results in a stable system.
c. Determine and compare the root sensitivity of the dominant roots to variations in
   Gain $K = 5$
   Open-loop pole at $s = -2$
   Open-loop zero at $s = -1$

8. Sketch the root-locus for a unity feedback system with a forward transfer function given by

$$G(s) = \frac{K(s+4)(s+40)}{s(s+0.1)(s+0.2)(s+100)(s+200)}$$

9. A single-axis attitude control system is modeled in a standard feedback configuration with

$$G(s) = \left(\frac{1}{Js^2}\right)\left(\frac{K_m s}{1 + \tau_m s}\right)$$

$$H(s) = K_1\left(\frac{s}{a+1}\right)$$

Sketch root-loci for the parameter $K = K_1 K_m/Ja\tau_m$ as a variable with a and τ_m fixed for the following cases:
a. $A > 1/\tau_m$
b. $a < 1/\tau_m$
c. $a \to \infty$ (no rate feedback)

10. A model of a hydroelectric alternator, turbine, and penstock has a forward transfer function

$$G(s) = \frac{1}{(Ms+D)} G_1(s)\left[1 + \frac{\tau_m s}{1 + (\tau_m s/2)}\right]$$

$$G_1(s) = \frac{25(1+5s)}{(1+s/0.0205)(1+s/2.518)(1+s/7.6615)}$$

$$H(s) = \frac{1}{R}$$

show that the system is unstable for $\tau_m = 1$, $R = 0.05$, $M = 10$ and $D = 1.0$. By changing the loop gain using the Bode plots find the maximum value of loop gain to retain stability.

11. Sketch the polar plot of the frequency response for the following

Stability and Frequency Response Method

transfer functions:
a. $G(s) = (1 + \tau_1 s)/(1 + \tau_2 s)$
b. $G(s) = (1 + s/2)/s^2$
c. $G(s) = 1/(1 + s)(1 + 2s)$
d. $G(s) = K/s(1 + 0.25s)(1 + s/16)$
for $K = 10$ and $K = 100$.

12. An excitation control system is represented in the standard feedback configuration by

$$G_1(s) = \frac{K_g}{1 + T_g s}$$

$$G_2(s) = \frac{1}{K_e + T_e s}$$

$$G_3(s) = \frac{K_a}{1 + T_a s}$$

$$G(s) = G_1(s)G_2(s)G_3(s)$$

$$H(s) = \frac{K_r}{1 + T_r s}$$

with $K_a = 40$, $T_a = 0.1$, $K_e = -0.05$, $T_e = .1$, $K_g = 1$, $T_g = 1$, $K_r = 1$ and $T_r = 0.05$. Use Bode logarithmic plots to check the stability of the system.

NOTES AND REFERENCES

In this chapter we have considered the concept of the stability of a feedback control system and have developed different frequency-response approaches to examine the stability conditions. Initially, definition of a stable system in terms of both bounded system response and location of the poles of the system transfer function in the s-plane were outlined and analyzed. Subsequently, the Routh–Hurwitz stability criterion was introduced. The notion of relative stability of a feedback control system was then established.

The movement of the closed-loop roots of the characteristic equation is investigated on the s-plane as the system parameters are varied by utilizing the root-locus method. Being a graphical method, the root-locus can be deployed to obtain a tentative schematic that can be used for the analysis of the initial control design. Sensitivity of the root-locus to undesired parameter variations was assessed through a root sensitivity measure.

Frequency-response methods including polar plots, Bode algorithmic plots and Nyquist stability criterion were presented and analyzed. Several examples were considered throughout the chapter to illustrate the concepts presented. The fundamental notions of relative stability measures: gain margin and phase margin are introduced as indices of the transient performance.

The topics covered so far should set the scene for the next chapter on system control design. Additional material and detailed illustrative examples can be found in Ogata (1979), Perkins and Cruz (1969), D'Azzo and Houpis (1981), Dorf (1980), El-Hawary (1984), and Kuo (1982). An excellent survey of the art is documented in MacFarlane (1979). Applications to process control and chemical engineering can be found in Weber (1973) and Stephanapoulous (1984).

5
Introduction to Design

5.1 THE DESIGN PROBLEM

The performance of a feedback control system is of primary importance. Hence, the design of a feedback control system is considered the fundamental function that the control engineer performs. In chapter 3, some quantitative measures of performance were developed. We have found that a suitable control system is generally stable and that it results in an acceptable response to input commands. Systems analysis, presented at length in the preceding chapters, has provided a framework to design and construction of practical control systems. While some of the available design techniques can be drawn from analysis, in almost all cases the design procedure is ad hoc wherein analysis methods are implemented iteratively.

Every control system designed for a specific application has to meet some desired performance measures. However, a feedback control system that provides an optimum performance without any necessary adjustments is rare indeed. Usually one finds it necessary to compromise among the many conflicting and demanding specifications and to adjust the system parameters to provide a suitable and acceptable performance when it is not possible to obtain all the desired optimum requirements.

We have discussed two basic approaches of characterizing the performance of a control system:

Approach 1. By a set of specifications in either the time domain and/or in the frequency domain such as peak overshoot, setting time, gain margin, phase margin, steady-state error, and the like.

Approach 2. By minimizing (or maximizing) a certain function of some variables and/or parameters of the control system.

In some situations, other constraints are imposed on the control system design which will limit the selection of components and devices. These constraints are frequently represented by availability of component size, weight, power supply, and limits of economic cost.

The choice of the controller and the associated elements is dictated initially by the performance specifications and later by the physical and practical constraints. This gives the direction that to finalize the control design we have to pass through two eventual stages, selection of controller configuration and implementation of control system. The first stage is referred to as *compensation* and the resulting control element is called a *compensator*. By allowing further gain adjustment, one can strive to satisfy the other constraints. The design problem may therefore be stated as follows: Given a controlled system and a set of performance requirements, design appropriate compensators so that the overall control system will meet the given requirements.

5.2 SOME GUIDELINES

As we knew from the previous chapters, the performance of a control system may be characterized either in the time domain or in the frequency domain. Some measures of performance include requirements of a certain peak time, maximum overshoot and settling time for a step input. In addition, it is frequently desirable to specify the maximum allowable steady-state error for several test signal inputs and disturbance inputs. These performance specifications may be defined through selecting the closed-loop poles and zeros of a transfer function at desirable positions. In this regard, the root locus method can be readily used and by further adding a suitable compensating network one can achieve appropriate root configuraton.

Alternatively, the performance measures of a feedback control system may be described in the frequency domain. These measures include phase margin of the system, peak of the closed-loop response, resonant frequency, and bandwidth. Again, by adding a suitable network with

Introduction to Design

adjustable parameters, the system specifications may be satisfied. Design of these compensation networks is developed in terms of the frequency response resulting from the polar plot, Bode asymptotic diagrams, and the like. In practice, Bode plots are preferred in conjunction with cascade networks.

The foregoing discussions emphasize that the compensation of a system is concerned with the alteration (reshaping) of the frequency response or the root locus of the system in order to obtain a suitable system performance. Looked at in this light, compensation by root-locus methods is accomplished by altering the root-loci, compensation by frequency-response methods implies changing the frequency characteristics and both aim at satisfying the cited specifications.

In the time domain, the objective is to control a system, expressed by a state-variable model, with an input signal which is a function of several measurable state variables. Essentially, one develops a state-variable controller which operates on the information supplied through measurements. The selection of controller parameters may be attained by several compensation methods including system optimization. Next, we look into the various design approaches which, for convenience and ease in exposition, are categorized into three sections: compensation methods (mainly frequency domain), state variable design (based on time domain) and parameter optimization (yielding optimal schemes).

5.3 COMPENSATION METHODS

In the sequel, we focus on cascade compensation in which a compensating network $G_c(s)$ is placed in front of the plant $G_p(s)$ within a unity feedback configuration as shown in Fig. 5.1. Let the compensator $G_c(s)$ be expressed in the form

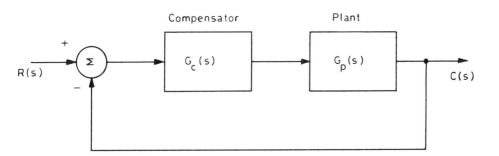

Figure 5.1 Cascade compensation scheme.

$$G_c(s) = K_c \frac{\Pi_{j=1}^M (s + z_j)}{\Pi_{k=1}^N (s + p_k)} \tag{1}$$

where K_c is the gain; z_j and p_k as the zeros and poles of the compensating network. The design problem is to determine M, the number of zeros; N, the number of poles; the values of poles and zeros; and the gain K_c in order to meet the performance specifications.

Before we proceed, some special and relevant cases of $G_c(s)$ are analyzed first:

1. *Gain adjustment.* In this case, $G_c(s)$ becomes

$$G_c(s) = K_c \tag{2}$$

which is obviously frequency independent. Thus it will be of limited use except for cases where pure amplification ($K_c > 1$) or pure attenuation ($K_c < 1$) of signals is required.

2. *Proportional-derivative (PD) compensator.* It has the transfer function

$$G_c(s) = K_p + sK_d = K_d\left(s + \frac{K_p}{K_d}\right) = K_d(s + z_1) \tag{3}$$

In comparison with (1) we see that $K_c = K_d$ and $z_1 = K_p/K_d$. Thus the inclusion of a PD compensator is to add a zero at an adjustable position. When dealing with the root-locus, such an addition causes a left shift and if the zero is located near the origin it will generally have a stabilizing effect. To shed some light on this fact, it is sufficient to consider the root-locus of the transfer functions

$$G_p(s) = \frac{K_1}{s[s + (1/\tau_1)][s + (1/\tau_2)]}$$

$$G_c(s)G_p(s) = \frac{K_2[s + (1/\tau)]}{s[s + (1/\tau_1)][s + (1/\tau_2)]}$$

for two cases: (1) $\tau_2 < \tau < \tau_1$; (2) $\tau_1, \tau_2, < \tau$. The result is shown in Fig. 5.2. It is easy to see from (3) that

$$G_c(j\omega) = K_p \sqrt{\omega + \left(\frac{k_p}{k_d}\right)^2} \angle \tan^{-1}\left(\frac{\omega K_p}{K_d}\right) \tag{4}$$

and thus $\phi(\omega)$ varies from 0° to 90° continuously; that is positive. This means that the output always leads the input and a PD compensator is referred to as a *lead compensator*. The corresponding phase and gain margins are increased as can be seen from Fig. 5.3. Despite these

Introduction to Design

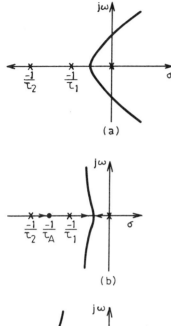

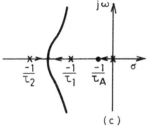

Figure 5.2 Root locus of some transfer functions.

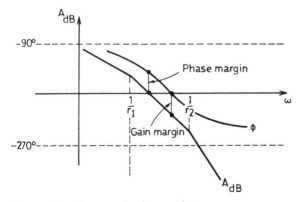

Figure 5.3 Phase and gain margins.

nice properties, a PD compensator cannot be physically realized using passive circuit elements. Note that for small values of s (low frequency), PD compensator (3) reduces to (2). We conclude that a PD compensator acts as a high pass filter (see Fig. 5.4), allows amplification of noise, and therefore yields undesirable behavior.

3. *Proportional-integral (PI) compensator.* The compensator has the form

$$G_c(s) = K_p + \frac{K_i}{s}$$

$$= \frac{K_p[s + (K_i/K_p)]}{s} \qquad (5)$$

which when compared with (1) yields $K_c = K_p$, $z_1 = K_i/K_p$ and $p_k = 0$. Therefore, the addition of a cascade PI compensator results in increasing the order and type of the system by one (has pole at origin) and inserting one more zero at K_i/K_p. Consequently, if the steady-state error of the system is constant for a test signal, then the PI compensator will cancel it out for the same test signal. The increase in system order can cause instability.

The PI sinusoidal transfer function is

$$G_c(j\omega) = K_p \frac{\sqrt{[\omega^2 + (K_i/K_p)^2]}}{\omega} \angle \phi(\omega) \qquad (6)$$

$$\phi(\omega) = \tan^{-1}\left(\frac{\omega K_p}{K_i}\right) - 90° \qquad (7)$$

The phase angle $\phi(\omega)$ varies from $-90°$ (at $\omega = 0$) to $0°$ at $\omega \to \infty$ (high frequency); that is, it is always negative. The output always lags the input and a PI compensator is referred to as a *lag compensator*.

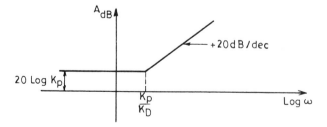

Figure 5.4 Magnitude plot of a PD compensator.

Introduction to Design

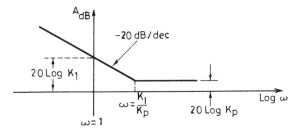

Figure 5.5 Magnitude plot of a PI compensator.

By similarity to the PD case, the PI has the effect of pulling the root-locus to the right. Note that at $\omega \to \infty$; PI reduces to (3) and thus it cuts off high frequency factors. It is essentially a low pass filter as shown in Fig. 5.5.

We have so far seen the two extremes of compensator networks. In the subsequent sections, we shall examine compensator networks that are easily realizable using passive elements and yield particular behavior.

5.3.1 Lead Compensator

The general form of the lead compensator is

$$G_c(s) = \frac{s + z_1}{s + p_1}$$

$$= \frac{(s + 1/\tau)}{(s + 1/\alpha\tau)}, \quad \alpha = \frac{z_1}{p_1} < 1 \quad \tau > 0$$

$$= \frac{\alpha(1 + s\tau)}{1 + \alpha\tau s} \tag{8}$$

Thus it has a zero at $s = -1/\tau$ and a pole at $s = -1/\alpha\tau$ with the zero closer to the origin than the pole; see Figure 5.6. Transfer function (8) can be realized by an electric network shown in Fig. 5.7(a).

The polar plot of the frequency response of the lead compensator can be obtained from

$$G_c(j\omega) = \alpha\left(\frac{1 + j\omega\tau}{1 + j\omega\alpha\tau}\right), \quad \alpha < 1$$

At $\omega = 0$, the compensator has a gain of $\alpha < 1$. In frequency domain

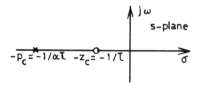

Figure 5.6 Pole and zero of a lead compensator.

compensation technique, it is convenient to cancel this dc attention with an amplification $A = 1/\alpha$, as shown in Fig. 5.7(b). The sinusoidal transfer function of the unity-gain lead compensator is then given by

$$G_c(j\omega) = \frac{1 + j\omega\tau}{1 + j\omega\alpha\tau}, \quad \alpha < 1 \tag{9}$$

Multiplying and dividing by the complex conjugate of the denominator, we obtain

$$G_c(j\omega) = \frac{(1 + \alpha\omega^2\tau^2) + j\omega\tau(1 - \alpha)}{1 + \omega^2\alpha^2\tau^2}$$

$$= X(\omega) + jY(\omega) \tag{10}$$

where

$$X(\omega) = \frac{1 + \alpha\omega^2\tau^2}{1 + \omega^2\alpha^2\tau^2} \tag{11}$$

$$Y(\omega) = \frac{\omega\tau(1 - \alpha)}{1 + \omega^2\alpha^2\tau^2} \tag{12}$$

To drive the locus of the polar plot we note from $X(\omega)$, $Y(\omega)$ that

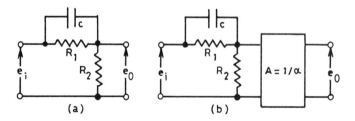

Figure 5.7 Network realization of a lead compensator. (a) Normal circuit. (b) Adding an amplifier to cancel d.c. attenuation.

Introduction to Design

$$\frac{X(\omega) - 1}{Y(\omega)} = \alpha\omega\tau$$

which when used in the real part $X(\omega)$ and manipulated, results after simplification in

$$Y^2(\omega) + \left[X(\omega) - \frac{1+\alpha}{2\alpha}\right]^2 = \left(\frac{1-\alpha}{2\alpha}\right)^2 \qquad (13)$$

It is obvious that (13) or (11) and (12) represents a circle (or part of circle) with its center on the real axis X_c and radius R_c

$$X_c = \frac{1+\alpha}{2\alpha}, \qquad R_c = \frac{1-\alpha}{2\alpha}$$

which yields

$$X_c - R_c = 1, \qquad X_c + R_c = \frac{1}{\alpha}$$

The geometry of the polar plot is displayed in Fig. 5.8 in the form of a semicircle in the first quadrant. From (10) we have

$$G_c(0) = 1, \qquad \lim_{\omega \to \infty} G(j\omega) = \frac{1}{\alpha}$$

and from (11) and (12)

$$X(0) = 1, \qquad Y(0) = 0$$

$$X(\infty) = \frac{1}{\alpha}, \qquad Y(\infty) = 0$$

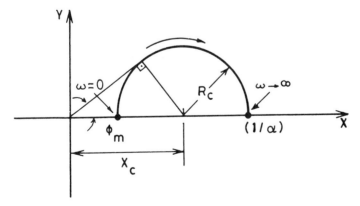

Figure 5.8 Polar plot of a lead compensator.

The phase angle at any radian frequency is

$$\phi(\omega) = \tan^{-1}(\omega\tau) - \tan^{-1}(\alpha\omega\tau) \tag{14}$$

But since $\alpha < 1$, then we conclude that $\phi(\omega) \geq 0$. Using the condition $d\phi(\omega)/d\omega = 0$ in (14), we find that the maximum phase-lead ϕ_m occurs at the frequency

$$\omega_m = \frac{1}{\tau\sqrt{\alpha}} = \sqrt{\frac{1}{\tau}\frac{1}{\alpha\tau}} \tag{15}$$

Alternatively from Fig. 5.8 we have

$$\sin \phi_m = \frac{R_c}{X_c} = \frac{1-\alpha}{1+\alpha} \tag{16}$$

This gives α in terms of ϕ_m as

$$\alpha = (1 - \sin \phi_m)/(1 + \sin \phi_m) \tag{17}$$

As seen from (15), ω_m is the geometric mean of the two corner frequencies of the compensator. From (9), or equivalently from Fig. 5.8, we obtain

$$|G(j\omega_m)|^2 = X_c^2 - R_c^2 = \frac{1}{\alpha}$$

$$= \frac{|1 + j\omega_m \tau|}{|1 + j\omega_m \alpha\tau|} \tag{18}$$

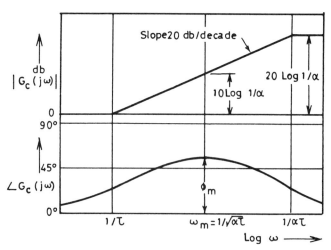

Figure 5.9 Bode plot of a lead compensator.

Introduction to Design

Close examination of (16) indicate that for $\phi_m \geq 60°$, $\alpha \leq 0.08$. Therefore, it is advisable to use two cascaded lead networks when the desired phase is greater than 60°. The Bode diagram of the lead compensator (with gain $A = 1/\alpha$) is given in Fig. 5.9. It is easily seen that the high frequency signals are amplified by a factor of $1/\alpha > 1$, while the low frequency signals undergo unit amplification. Thus the signal/noise ratio at the output of the lead compensator is poorer than at its input. In practice, we chose $\alpha \geq 0.1$ to prevent further deterioration in signal/noise ratio. To sum up, the synthesis of lead compensator requires the determination of two parameters: the constant gain (zero-to-pole ratio) α and the zero time constant τ.

5.3.2 Lag Compensator

In this section we discuss the properties of a lag compensator, which has the s-plane representation of Fig. 5.10(a). It has a pole at $(-1/\sigma\tau)$ and a zero at $(-1/\tau)$ with the zero located to the left of the pole on the negative real axis. The general form of the tag transfer function

$$G_c(s) = \frac{s + z_2}{s + p_2}$$

$$= \frac{s + (1/\tau)}{s + (1/\sigma\tau)}, \quad \sigma = \frac{z_2}{p_2} > 1 \quad \tau > 0 \tag{19}$$

Realization of (19) is achieved with an electric network of the form in Fig. 5.10(b). In the sequel we consider

$$G_c(s) = \frac{1 + \tau s}{1 + \sigma\tau s}, \quad \sigma > 1 \tag{20}$$

as a unity-gain lag compensator leaving an additional degree of freedom in the form of a multiplicative factor $(1/\sigma)$ for impedance matching. Since $\sigma > 1$ the steady-state output has a lagging phase with respect to

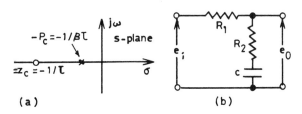

Figure 5.10 (a) Pole-zero pattern. (b) Network realization.

the sinusoidal input and hence the name lag network. The corresponding Bode diagram is drawn in Fig. 5.11.

The polar plot of the frequency response is derived using

$$G_c(j\omega) = \frac{1 + j\omega\tau}{1 + j\sigma\omega\tau}$$
$$= X(\omega) + jY(\omega) \tag{21}$$

with

$$X(\omega) = \frac{1 + \sigma\omega^2\tau_2}{1 + \sigma^2\omega^2\tau^2}$$

$$Y(\omega) = \frac{\omega\tau(1 - \sigma)}{1 + \sigma^2\omega^2\tau_2}$$

It is easy to see that

$$\frac{Y(\omega)}{(1 - \sigma)X(\omega)} = \omega\tau \tag{22}$$

On eliminating $\omega\tau$, we obtain

$$\left[X(\omega) - \frac{(\sigma + 1)}{2\sigma}\right]^2 + Y^2(\omega) = \left(\frac{\sigma - 1}{2\sigma}\right)^2 \tag{23}$$

This is the equation of a circle with center on the real axis X_c and

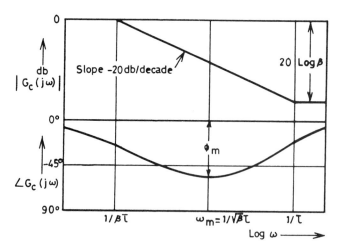

Figure 5.11 Bode plot of a lag compensator.

Introduction to Design

radius R_c given by

$$X_c = \frac{\sigma + 1}{2\sigma}, \quad R_c = \frac{\sigma - 1}{2\sigma}$$

such that

$$X_c + R_c = 1, \quad X_c - R_c = \frac{1}{\sigma}$$

The polar plot of the lag compensator is shown in Fig. 5.12. Observe the analogy with the case of the lead compensator.

From (21), the phase angle is given by

$$\phi(\omega) = \tan^{-1}(\omega\tau) - \tan^{-1}(\sigma\omega\tau) \quad (24)$$

Since $\sigma > 1$, $\phi(\omega) < 0$ with a maximum phase ϕ_m occurring at

$$\omega_m = \frac{1}{\tau\sqrt{\sigma}} = \sqrt{\frac{1}{\tau}\frac{1}{\sigma\tau}} \quad (25)$$

with

$$\phi_m = \sin^{-1}\left(\frac{1-\sigma}{1+\sigma}\right)$$

From Figure 5.11, it is observed that the lag network has a dc gain of unity while it offers a high frequency gain of $1/\sigma$. Since $\sigma > 1$, it means that the high frequency noise is attenuated in passing through the network whereby the signal/noise ratio is improved, in contrast to the

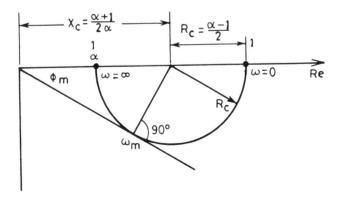

Figure 5.12 Polar plot of a lag compensator.

lead network. A typical choice of σ is 10. In brief, the synthesis of the lag compensator amounts to the determination of two parameters: the zero-to-pole ratio σ and the zero time constant τ.

5.3.3 Lead-Lag Compensator

The lead-lag compensator is a combination of a lag compensator and a lead compensator. As we knew before, the lag section has one real pole and one real zero with the pole to the right of zero. The lead section also has one real pole and one real zero but the zero is to the right of the pole. It is therefore natural to express the lead-lag compensator in the form

$$G_c(s) = \frac{s + (1/\tau_2)}{s + (1/\sigma\tau_2)} \frac{s + (1/\tau_1)}{s + (1/\alpha\tau_1)}, \quad \alpha < 1, \quad \sigma > 1 \quad (27)$$

In Fig. 5.13, a single electric network that realizes (27) is given. It is customary to use $\alpha\sigma = 1$ which constraints the choice of α and σ. Under this condition (27) can be cast into the form

$$G_c(s) = \frac{s + (1/\tau_2)}{s + (1/\sigma\tau_2)} \frac{s + (1/\tau_1)}{s + (\sigma/\tau_1)}$$

$$= \frac{s + z_2}{(s + p_2)} \frac{s + z_1}{s + p_1}, \quad \sigma > 1, \quad \sigma = \frac{z_2}{p_2} = \frac{p_1}{z_1} \quad (28)$$

The s-plane representation of (28) is shown in Fig. 5.14. By simple construction, the polar plot of the lead-lag transfer function (28) can be obtained as shown in Fig. 5.15. It is easy to see that the compensator

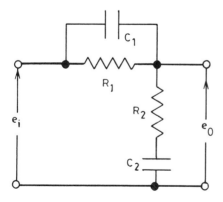

Figure 5.13 Electric lead-lag network.

Introduction to Design

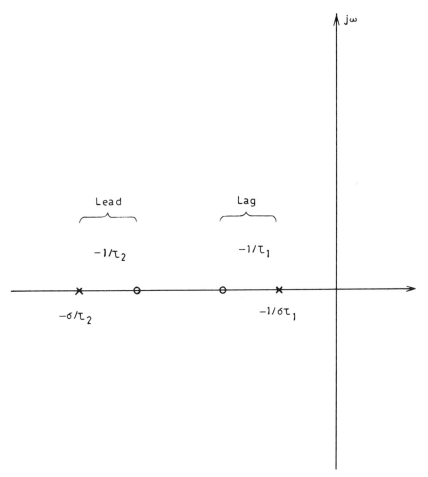

Figure 5.14 The s-plane representation of lead-lag compensator.

acts as lag network for $0 < \omega_1$ and as a lead network $\omega_1 < \omega < \infty$ where ω_1 (the frequency at which the phase angle is zero) is

$$\omega_1 = \frac{1}{\sqrt{\tau_1 \tau_2}} \tag{29}$$

The corresponding Bode plot is presented in Fig. 5.16.

To conclude, the synthesis of the lead-lag compensator requires the determination of three parameters: the lead-zero time constant τ_1, the lag-zero time constant τ_2, and the zero-to-pole ratio σ.

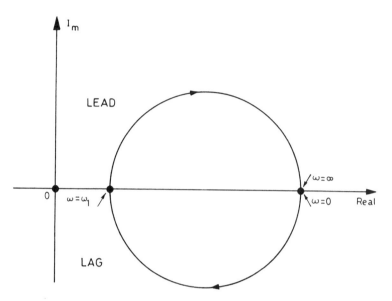

Figure 5.15 Polar plot of a lead-lag compensator.

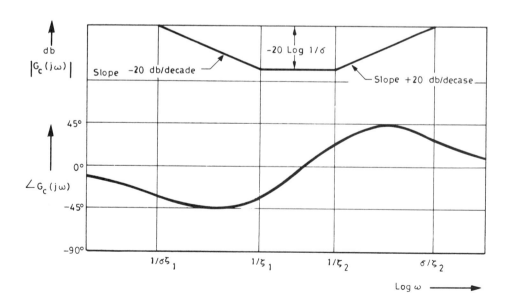

Figure 5.16 Bode plot of lead-lag network.

Introduction to Design

5.3.4 Technical Considerations

It has been mentioned already that when the system behavior is not desirable, it is often required to add cascade and/or feedback compensators such that the resulting compensated system would have certain desirable characteristics. There are two sets of "performance specifications", one concerns time domain and the other, frequency domain. The elements of time domain and frequency domain specifications are summed up in Table 5.1, while their detailed definitions were given earlier in Chapter 3. Recall that two parameters are needed to mechanize either phase-lead or phase-lag compensators while three parameters are required when using phase lead-lag compensators. This in turn limits the performance measures to be satisfied. In Table 5.2, we present a comparison of the technical properties of lead and lag compensators in order to guide their selection to suit desirable performance.

The immediate question is how to select a compensator? In general, there are two situations in which compensation is required. In the first case, the system is absolutely unstable and the compensation is required to stabilize as well as to achieve a specified performance. In the second case, the system is stable but the compensation is required to obtain the desired performance. The systems which are type-2 or higher, are usually absolutely unstable. For these types of systems, clearly a lead compensator is needed to increase the margin of stability.

In type-1 and type-0 systems, stable operation is always possible if the gain is sufficiently reduced. In such cases any of the three compen-

Table 5.1. Performance Specifications

Time Domain		Frequency Domain	
Symbol	Meaning	Symbol	Meaning
C_p	Maximum overshoot	ω_m	Frequency of peak magnitude
M_p	Percent overshoot	M_m	Peak value of magnitude
τ_s	Settling time		
τ_p	Peak time	PM	Phase margin
τ_r	Rise time	GM	Gain margin
ξ	Damping ratio		
ω_n	Undamping natural frequency		
ω_d	Damped natural frequency		
e_{ss}	Steady-state error		

Table 5.2. Properties of Compensators

Lead Compensator	Lag Compensator
• Speeds up the transient response and boots the gain at high frequency	• Improve steady-state error while nearly preserving its transient response
• Increases the margin of stability	• Provides high gain at low frequency
• Causes shift of the gain crossover frequency to higher values	• Yields attention that will shift the gain crossover frequency to a lower frequency point
• Increases system bandwidth and gives faster dynamic response	• Decreases system bandwidth but suppresses high frequency noise

sators (lead, lag, or lead-lag) may be used to obtain the desired performance.

5.3.5 A Synthesis Procedure

We shall explain below one way of picking up the parameters of a lead compensator. Suppose that the design specifications require a given $|G(j\omega_s)|$ and $\phi(\omega_s)$ at a given radian frequency ω_s. The problem then is to find α and τ of the lead compensator to meet these requirements. A straightforward method would be by calculating the rectangular components $X(\omega_s)$ and $Y(\omega_s)$ using the given information as

$$X(\omega_s) = |G(j\omega_s)| \cos \phi_s, \qquad Y(\omega_s) = |G(j\omega_s)| \sin \phi_s$$

But we know that

$$\frac{X(\omega_s) - 1}{Y(\omega_s)} = \frac{\omega_s \tau}{\alpha}$$

and this yields

$$\tau = \alpha \left[\frac{Y(\omega_s)}{\omega_s (X_s(\omega_s) - 1)} \right] \tag{30}$$

From (16), $X(\omega_s)$ satisfies

$$X(\omega_s) = \frac{1 + \alpha \omega_s^2 \tau^2}{1 + \alpha^2 \omega_s^2 \tau^2}$$

Thus substituting for τ in the above equation and rearranging, we

Introduction to Design

obtain

$$\alpha = \frac{X_s^2(\omega_s) - 1}{X_s^2(\omega_s) + Y(\omega_s)}$$

Indeed there are numerous ways to solve this problem; further details are found in the references cited at the end of the chapter. Next examine some examples to illustrate the design of compensators.

5.3.6 Examples

Example 5.1. Design a lead compensator for the unity gain closed-loop system with open-loop tranfer function

$$G_p(s) = \frac{K}{s^2}$$

Such that

$$M_p \leq 20, \qquad \tau_s \leq 4 \text{ s}$$

The second specification implies that

$$\tau_s = 4/\xi\omega_n \leq 4 \text{ s} \quad \text{or} \quad \xi\omega_n \geq 1$$

since $M_p = \exp[-(\xi/\sqrt{1-\xi^2})\pi]$ thus

$$\xi = \sqrt{\frac{\delta^2}{1+\delta^2}}, \qquad \delta = \frac{1}{\pi}\ln\left(\frac{1}{M_p}\right)$$

As a result, the requirement that $M_p \leq 20$ entails $\xi \geq 0.456$.

It can be shown that the phase margin for a second-order system is related to ξ by

$$\frac{1}{\tan^2\phi_m} = \frac{1}{2}\left(\sqrt{1 + \frac{1}{4}\xi^4} - 1\right)$$

Given $\xi = 0.455$, we obtain $\phi_m = 48.1477°$. The phase angle of the uncompensated system is $-180°$ for all frequencies; thus the phase margin will be provided by the compensating network. The best we can do is to let the phase margin occur at maximum phase lead; thus

$$\sin \phi_m = \sin 48.1477°$$

$$= \frac{1-\alpha}{1+\alpha}$$

results in $\alpha = 0.1462$. With a lead compensator $G_c(s) = (1 + \tau s)/(1 + \tau s \alpha)$, the corresponding open-loop transfer function ($s = j\omega$) has the form

$$G_c(j\omega)G_p(j\omega) = -\frac{K(1 + j\omega\tau)}{\omega^2(1 + j\omega\alpha\tau)}$$

which has the components:

$$X(\omega) = -\frac{K(1 + \omega^2\tau^2\alpha)}{\omega^2(1 + \omega^2\alpha^2\tau^2)}$$

$$Y(\omega) = -\frac{K\omega\tau(1 - \alpha)}{\omega^2(1 + \omega^2\alpha^2\tau^2)}$$

The phase angle of the compensated system is

$$\tan\phi = \frac{\omega\tau(1 - \alpha)}{(1 + \omega^2\tau^2\alpha)}$$

At ω_ϕ, the phase angle is $\phi_m = 458.1477$ with $\alpha = 0.1462$. We have

$$\tan 48.1477 = \frac{\omega_\phi\tau(1 - 0.1462)}{1 + 0.1462\omega_\phi^2\tau^2}$$

$$1.1164 = \frac{0.8538\omega_\phi\tau}{1 + 0.1462\omega_\phi^2\tau^2}$$

which when simplified and solved for ω_ϕ, gives

$$\omega_\phi\tau = 0.4388 \quad \text{or} \quad 0.42$$

But at ω_ϕ we have

$$|G_c(jw_\phi)G_p(\omega_\phi)| = 1$$

thus

$$1 = \frac{K^2(1 + \omega_\phi^2\tau^2)}{\omega_p^4(1 + \omega_\phi^2\tau^2\alpha^2)}$$

with $K = 10$, $\omega_\phi\tau = 0.4338$ or 0.42 we obtain $\omega_\phi = 3.298$ or 3.229. Thus $\tau = 0.1315$ or 0.1277 and we have two possible designs.

1. $G_c(s) = (1 + 0.1315s)/(1 + 0.0192s)$
2. $G_c(s) = (1 + 0.1277s)/(1 + 0.0187s)$

Note that in view of the simplicity of $G_p(s)$, the treatment was almost analytic.

Introduction to Design

Example 5.2. Consider the system of Fig. 5.17 which has the open-loop transfer function

$$G_p(s) = \frac{K}{s(1+s)(4+s)}$$

It is desired to compensate the system to meet the following specifications:

$$\text{damping ratio } \xi = 0.5$$
$$\text{undamped natural frequency } \omega_n = 2$$

Given the above specifications, the dominant closed-loop poles are found to be

$$s_r = -\xi\omega_n \pm j\omega_n\sqrt{1-\xi^2}$$
$$= -1 \pm j1.73$$

Locating these poles along with the open-loop one in Fig. 5.18, we observe that the angle contribution required from a lead compensator pole-zero pair is

$$\phi = -180 - \underline{/G(s_r)}$$
$$= -180 - (-120 - 30 - 90)$$
$$= 60°$$

We place the compensator zero to the left of the desired pole (say $s = -1.2$) to ensure the dominance condition. Connecting this zero to s_r and placing the compensator pole by making an angle of $\phi = 60°$, we obtain the dotted portion on Fig. 5.18.

The position of the pole is at -4.95. From the magnitude criterion,

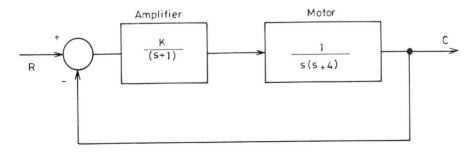

Figure 5.17 System of example 5.2.

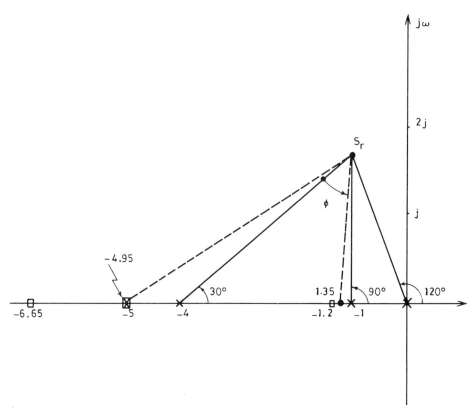

Figure 5.18 A lead compensator for example 5.2.

we see that $K = 30.4$. The open-loop transfer function of the compensated system becomes

$$G(s) = G_p(s)G_c(s)$$

$$= \frac{30.4(s + 1.2)}{s(s + 1)(s + 4)(s + 4.95)}$$

for which $K_v = \lim_{s \to 0} sG(s) = 30.4 \times 1.2/1 \times 4 \times 4.95 = 1.84$.

From the rules of the root-locus of the compensated system, it is readily seen that the closed-loop transfer function can be written as

$$\frac{C(s)}{R(s)} = \frac{30.4(s + 1.2)}{(s + 1.35)(S + 6.65)(s + 1 + j1.73)(s + 1 - j1.73)}$$

Introduction to Design

whose impulse response is given by

$$c(t) = 2.94\bar{e}^t \sin(1.73t - \tan^{-1}0.216) - 0.27\bar{e}^{1.35t} + 0.89\bar{e}^{6.65t}$$

By examining the above relation, we see that the desired poles have large residue (first term) in comparison to the other two terms which ensures the dominance condition.

5.4 STATE VARIABLE DESIGN

5.4.1 Concept of Controllability and Observability

In modern control engineering, two basic design issues have to be resolved. The first issue is related to the ability of transferring the control system from arbitrary initial condition to another desired condition in finite time upon applying an external control function. The second issue somehow bears the inverse meaning of the first; it concerns the ability of deducing the initial system state from finite records of the measurable (output) variables.

Credit is lent to the pioneering work of Kalman who conceptualized the fundamental concepts of *controllability* and *observability* of a system. We now define these concepts starting with controllability. A system is said to be *completely state controllable* if, for any initial time t_0, it is possible to generate an unconstrained control vector $\mathbf{u}(t)$ that will take any given initial state $\mathbf{x}(t_0)$ to any final state $\mathbf{x}(t_f)$ in a finite time interval $t_p \leq t \leq t_f$.

If this is true for all initial times t_0 and all initial states $\mathbf{x}(t_0)$, the system is *completely controllable*. For a linear system, the final state may be taken to be the origin without loss of generality. We next define observability. A system is said to be *completely observable*, if every arbitrary initial state $\mathbf{x}(t_0)$ can be completely determined from measurements of the output vector $\mathbf{y}(t)$ over a finite time interval.

It must be emphasized that controllability is a property of the coupling between the input and the state, and this involves the matrices A and B. On the other hand, observability is a property of the coupling between the state and the output, and thus involves the matrices A and C.

A general r-input, m-output linear time-invariant system

$$\dot{\mathbf{x}} = A\mathbf{x} + B\mathbf{u} \tag{31}$$

$$\mathbf{y} = C\mathbf{x} \tag{32}$$

is completely controllable if and only if the rank of the composite

matrix
$$Q_c = [B \quad AB \quad A^2B \quad \cdots \quad A^{n-1}B] \tag{33}$$
is n, the dimension of system (31). A concise statement of controllability is that the pair (A, B) is controllable if and only if the rank of Q_c is n. On the other hand, system (31) and (32) is said to be completely observable if and only if the rank of the composite matrix
$$Q_0 = [C' \quad A'C' \quad \cdots \quad (A')^{n-1}C'] \tag{34}$$
is n. By the same token, this condition is referred to as the pair (A, C) being observable.

It can be easily verified that when the triplet (A, B, C) is in the controllable canonical form, the composite matrix Q_c is the identity matrix. Similarly, if the triplet (A, B, C) is in the observable canonical form then the composite matrix Q_0 is the identity matrix.

Example 5.4. A linear system of the type (31) and (32) with
$$A = \begin{bmatrix} -0.75 & -0.25 \\ -0.50 & -0.50 \end{bmatrix}, \quad B = \begin{bmatrix} 1 \\ 1 \end{bmatrix}, \quad C = [4 \quad 2]$$

Is the system completely controllable and completely observable?
Using (33) with $n = 2$ we obtain
$$Q_c = [B \quad AB]$$
$$= \begin{bmatrix} 1 & \cdot & -1 \\ 1 & \cdot & -1 \end{bmatrix}$$

which obviously has rank = 1. Turning to (34) we obtain
$$Q_0 = [C' \quad A'C']$$
$$= \begin{bmatrix} 4 & -4 \\ 2 & -2 \end{bmatrix}$$

and again it has rank 1. Therefore the system is neither completely controllable nor completely observable. The subsequent sections are devoted to the design of freedback compensators for linear, constant coefficient multivariable systems using state-variable models.

5.4.2 State Feedback

In this case, the system is given by (31) and (32) and we use the state feedback control law

Introduction to Design

$$\mathbf{u} = F\mathbf{v}(t) - K\mathbf{x}(t) \tag{35}$$

The result is a state variable feedback system shown in Fig. 5.19. In (35), the $r \times n$ matrix K is the unknown feedback gain. Combining (31) and (35) gives

$$\dot{\mathbf{x}} = [A - BK]\mathbf{x} + [BF]\mathbf{v} \tag{36}$$

which describes the closed-loop system and indicates that the closed-loop eigenvalues are roots of

$$\Delta_s(\lambda) \triangleq \det[\lambda I - A + BK] = 0 \tag{37}$$

One of the basic results is that if (and only if) the open-loop system (A, B) is completely controllable, then any set of desired closed-loop eigenvalues $\Omega = \{\lambda_1, \lambda_2, \ldots, \lambda_n\}$ can be achieved using a constant state feedback matrix K.

Note that if Ω contains complex conjugate pairs of eigenvalues, then real feedback gains are guaranteed. The problem at hand is to find a matrix K which yields n specified set of eigenvalues $\lambda_j \in \Omega$. From (37) and the properties of determinants we have

$$\begin{aligned}
\Delta_s(\lambda) &= \det\{[\lambda I - A][I + (\lambda I - A)^{-1}BK]\} \\
&= \det[\lambda I - A] \det[I + (\lambda I - A)^{-1}BK] \\
&= \Delta(\lambda) \det[I + \Phi(\lambda)BK] \\
&= \Delta(\lambda) \det[I + BK\Phi(\lambda)] \\
&= \Delta(\lambda) \det[I + K\Phi(\lambda)B] \tag{38}
\end{aligned}$$

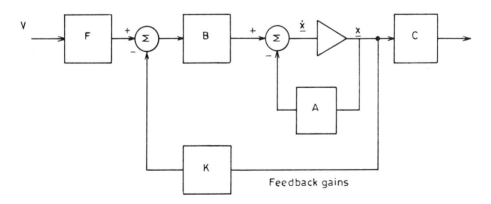

Figure 5.19 A state variable feedback system.

The identity matrix I in (38) has an appropriate dimension compatible with the matrix product involved. The condition $\Delta_s(\lambda) = 0$ for each $\lambda_j \in \Omega$ can be accomplished by forcing the $r \times r$ determinant in (38) to vanish. A sufficient condition for $\det[I + K\Phi(\lambda_j)B]$ to be zero is when any column or row is zero. We focus on column, define the jth column of I as $\mathbf{e}_j$ and the jth column of $N(\lambda_k) = \Phi(\lambda_k)B$ as $\mathbf{n}_j$. Thus λ_j is a root of $\Delta_s(\lambda_k)$ if K is selected to satisfy

$$\mathbf{e}_j + K\mathbf{n}_j(\lambda_k) = 0$$

or alternatively

$$K\mathbf{n}_j(\lambda_k) = -\mathbf{e}_j \tag{39}$$

Iterating on (39) by finding n linearly independent columns $\mathbf{n}_a(\lambda_1), \mathbf{n}_b(\lambda_2), \ldots, \mathbf{n}_s(\lambda_n)$ from the extended matrix

$$N(\lambda) \triangleq [N(\lambda_1) \quad N(\lambda_2) \quad \cdots \quad N(\lambda_n)]$$

and rearranging we obtain

$$K = -[\mathbf{e}_a \mathbf{e}_b \quad \cdots \quad \mathbf{e}_s][\mathbf{n}_a(\lambda_1) \quad \cdots \quad \mathbf{n}_s(\lambda_n)]^{-1} \tag{40}$$

Note that the subscripts $a, b, \ldots, s$ in (40) are arbitrarily coded. The above algorithm is due to Brogan [49] and has been extended recently by Fahmy and O'Reilly [50, 51]. Controllability of the pair (A, B) is sufficient that $\text{rank}[N(\lambda_k)] = r$ for each λ_k. We next illustrate the implementation of the above algorithm.

Example 5.5. Consider a linear system with

$$A = \begin{bmatrix} 0 & 4 \\ 0 & 5 \end{bmatrix}, \quad B = \begin{bmatrix} 1 & 0 \\ 0 & 1 \end{bmatrix}$$

The system is controllable but open-loop unstable. Then

$$\Phi(\lambda) = \begin{bmatrix} \lambda - 5 & 4 \\ 0 & \lambda \end{bmatrix} \begin{bmatrix} 1 \\ \lambda(\lambda - 5) \end{bmatrix}$$

$$N(\lambda) = \Phi(\lambda)B$$

$$= \begin{bmatrix} 1/\lambda & 4/\lambda(\lambda - 5) \\ 0 & 1/(\lambda - 5) \end{bmatrix}$$

If the desired eigenvalues are $\lambda_1 = -3$ and $\lambda_2 = -6$. Then we obtain

$$\mathbf{n}_1(\lambda_1) = \begin{bmatrix} -1/3 \\ 0 \end{bmatrix}, \quad \mathbf{n}_2(\lambda_1) = \begin{bmatrix} 1/6 \\ -1/8 \end{bmatrix}$$

$$\mathbf{n}_1(\lambda_2) = \begin{bmatrix} -1/6 \\ 0 \end{bmatrix}, \quad \mathbf{n}_2(\lambda_2) = \begin{bmatrix} 2/33 \\ -1/11 \end{bmatrix}$$

Application of (40) with $\mathbf{n}_1(\lambda_1)$ and $\mathbf{n}_2(\lambda_2)$ gives

$$K = -\begin{bmatrix} 1 & 0 \\ 0 & 1 \end{bmatrix}\begin{bmatrix} -1/3 & 2/33 \\ 0 & -1/11 \end{bmatrix}^{-1}$$

$$= \begin{bmatrix} 3 & 2 \\ 0 & 11 \end{bmatrix}$$

Alternatively, using $\mathbf{n}_2(\lambda_1)$ and $\mathbf{n}_1(\lambda_2)$ results in

$$K = -\begin{bmatrix} 0 & 1 \\ 1 & 0 \end{bmatrix}\begin{bmatrix} 1/6 & -1/6 \\ -1/8 & 0 \end{bmatrix}^{-1}$$

$$= \begin{bmatrix} 6 & 8 \\ 0 & 8 \end{bmatrix}$$

It is easily verified that both of these solutions yield $\lambda_1 = -3$ and $\lambda_2 = -6$. The gain matrix is not unique and the remaining freedom can be exploited to satisfy other requirements.

5.4.3 Output Feedback

When some of the states are not directly measurable, the feedback control law cannot be used. Rather we use (32) for the measurable states or outputs together with the output feedback law.

$$\mathbf{u}(t) = L\mathbf{v}(t) - K_0\mathbf{y}(t) \tag{41}$$

Combining (31), (32), and (41) we arrive at

$$\dot{\mathbf{x}} = [A - BK_0C]\mathbf{x} + [BL]\mathbf{v} \tag{42}$$

which describes the closed-loop system as shown in Fig. 5.20 and has the eigenvalues as roots of

$$\begin{aligned}
\Delta_0(\lambda) &= \det[\lambda I - A + BK_0C] \\
&= \det\{[\lambda I - A][I + (\lambda I - A)^{-1}BK_0C]\} \\
&= \det[\lambda I - A]\det[I + (\lambda I - A)^{-1}BK_0C] \\
&= \Delta(\lambda)\det[I + \Phi(\lambda)BK_0C] \\
&= \Delta(\lambda)\det[I + K_0C\Phi(\lambda)B] \tag{43}
\end{aligned}$$

where the cyclic properties of matrix product under determinant oper-

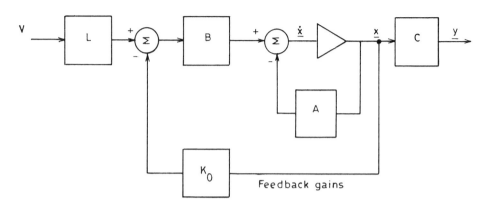

Figure 5.20 An output feedback system.

ations have been used. Define $R(\lambda) = C\Phi(\lambda)B$ and note that it has dimension of $m \times nr$. It can be shown that the number of linearly independent columns $\mathbf{r}_j(\lambda_k)$ will never exceed rank $|C|$. Davison [52] has shown that when (A, B) is completely controllable and C has full rank $m \leq n$, then m of the n eigenvalues of the closed-loop system can be arbitrarily specified to some degree of accuracy. We let the m independent columns $\mathbf{r}_j(\lambda_k)$ form an $m \times m$ matrix S_0. Then, following the same development as the state feedback case we obtain

$$K_0 S_0 = -[\mathbf{e}_a(\lambda_1) \quad \cdots \quad \mathbf{e}_n(\lambda_m)] \tag{44}$$

Note that $\mathbf{e}_f(.)$ is an r-vector. Inverting (44) we obtain

$$K_0 = -[\mathbf{e}_a(\lambda_1) \quad \cdots \quad \mathbf{e}_h(\lambda_m)] S_0^{-1} \tag{45}$$

We illustrate the above procedure by an example.

Example 5.6. An unstable but completely controllable system is described by

$$A = \begin{bmatrix} -1 & 1 & 0 \\ 0 & -2 & 0 \\ 0 & 0 & 3 \end{bmatrix}, \quad B = \begin{bmatrix} 0 & 0 \\ 0 & 1 \\ 1 & 0 \end{bmatrix}$$

$$C = \begin{bmatrix} 0 & 0 & 1 \\ 1 & 0 & 0 \end{bmatrix}$$

Here $n = 3$, $r = 2$, and $m = 2$. It is required to find an output feedback

Introduction to Design

matrix which gives the closed-loop eigenvalues at $\lambda_1 = -5$ and $\lambda_2 = -6$.

A straightforward application of the foregoing algorithm results in

$$R(\lambda) = C[\lambda I - A]^{-1} A$$

$$= \begin{bmatrix} 1/(\lambda - 3) & 0 \\ 0 & 1/(\lambda + 1)(\lambda + 2) \end{bmatrix}$$

Using $\mathbf{r}_1(\lambda_1)$ and $\mathbf{r}_2(\lambda_2)$, we obtain

$$K_0 \begin{bmatrix} -1/8 & 0 \\ 0 & 1/20 \end{bmatrix} = -\begin{bmatrix} 1 & 0 \\ 0 & 1 \end{bmatrix}$$

$$K_0 = -\begin{bmatrix} 1 & 0 \\ 0 & 1 \end{bmatrix} \begin{bmatrix} -1/8 & 0 \\ 0 & 1/20 \end{bmatrix}^{-1}$$

$$= \begin{bmatrix} 8 & 0 \\ 0 & -20 \end{bmatrix}$$

Using this result in (43) leads to $\Delta_0(\lambda) = (\lambda + 6)(\lambda - 3)(\lambda + 5)$ which shows that the two desired roots have been achieved, but the system is still unstable due to the root $\lambda_3 = 3$.

The algorithms developed thus far are frequently called pole placement or eigenvalue assignment techniques.

5.4.4 Decoupling Control

We shall consider a particular class of (31) and (32) in which the inputs and outputs are equal to number m. In this case the transfer function $H(s)$ becomes square. The problem under consideration is that of reducing a system with m inputs and m outputs to decoupled (diagonal or block-diagonal) form using a state feedback control

$$\mathbf{u}_d(t) = F_d \mathbf{v}(t) - K_d \mathbf{x}(t) \tag{46}$$

By Laplace transforming (31), (32), and (46) and manipulating, we obtain the transfer function $H(s)$ as

$$H(s) = C[sI - A + BK_d]^{-1} BF_d \tag{47}$$

it is required to select $m \times m$ matrix F_d and $m \times M$ matrix K_d so that $H(s)$ is diagonal and nonsingular. Simple analysis on the inverse transform of (47) using the Cayley–Hamilton theorem would show that the family of matrices

$$C[A - BK_d]^j BF_d, \quad j = 0, 1, \ldots, n - 1 \tag{48}$$

must all be diagonal if the system is decoupled. This is in line with the work of Falb and Wolovich [53] and Morse and Wonham [54, 55]. It turns out that system (31) and (32) can be decoupled using state feedback law (46) if and only if the following $m \times m$ matrix is nonsingular

$$W = \begin{bmatrix} \mathbf{c}_1 A^{\theta_1} B \\ \mathbf{c}_2 A^{\theta_2} B \\ \cdot \quad \cdot \quad \cdot \\ \cdot \quad \cdot \quad \cdot \\ \cdot \quad \cdot \quad \cdot \\ \mathbf{c}_m A^{\theta_m} B \end{bmatrix}$$

where $\mathbf{c}_k$ is the kth row of C and

$$\theta_j = \min_j \{j \mid \mathbf{c}_j A^j B \neq \mathbf{0}; j = 0, 1, \ldots, N - 1\}$$

or

$$\theta_j = n - 1, \quad \text{if } \mathbf{c}_j A^j B = \mathbf{0} \text{ for all } j \qquad (49)$$

the end result is that one set of decoupling matrices is

$$F_d = W^{-1} \quad \text{and} \quad K_d = W^{-1} \begin{bmatrix} \mathbf{c}_1 A^{\theta_1+1} \\ \cdot \\ \cdot \\ \cdot \\ \mathbf{c}_m A^{\theta_m+1} \end{bmatrix} \qquad (50)$$

which produces $H(s)$ of the form

$$H(s) = \text{diag}[s^{\theta_1-1} \quad s^{\theta_2-1} \quad \cdots \quad s^{\theta_m-1}] \qquad (51)$$

This is often called *integrator decoupled form*.

Example 5.7. Applying the decoupling control procedure to

$$A = \begin{bmatrix} 2 & 1 & 0 \\ 0 & -2 & 0 \\ 0 & 0 & 4 \end{bmatrix}, \quad B = \begin{bmatrix} 0 & 0 \\ 0 & 1 \\ 1 & 0 \end{bmatrix}$$

$$C = \begin{bmatrix} 0 & 0 & 1 \\ 0 & 0 & 0 \end{bmatrix}$$

since $\mathbf{c}_1 A^0 B = [1 \quad] \neq \mathbf{0}$ we set $\theta_1 = 0$. Next $\mathbf{c}_2 A^0 B = \mathbf{0}$ but $\mathbf{c}_2 AB = [0 \quad 1] \neq \mathbf{0}$ so that $\theta_2 = 1$. Therefore (48) now reads

$$W = \begin{bmatrix} \mathbf{c}_1 B \\ \mathbf{c}_2 AB \end{bmatrix} = \begin{bmatrix} 1 & 0 \\ 0 & 1 \end{bmatrix}$$

Introduction to Design

which is nonsingular and hence the decoupling is possible. From (50) we obtain

$$F_d = \begin{bmatrix} 1 & 0 \\ 0 & 1 \end{bmatrix}$$

$$K_d = \begin{bmatrix} 1 & 0 \\ 0 & 1 \end{bmatrix} \begin{bmatrix} \mathbf{c}_1 A \\ \mathbf{c}_2 A^2 \end{bmatrix}$$

$$= \begin{bmatrix} 0 & 4 & 4 \\ 4 & -4 & 0 \end{bmatrix}$$

which when substituted in (47) gives

$$H(s) = \begin{bmatrix} 1/s & 0 \\ 0 & 1/s^2 \end{bmatrix}$$

This complies with (51). In general, we can further add state feedback around the decoupled system to improve the performance by shifting the eigenvalues to desired locations.

Before proceeding further, we need to evaluate the three feedback control schemes. State feedback though efficient and flexible requires the availability of measuring all state variables. The result is always a stable closed-loop system. However, in several systems applications some of the states are not directly measurable. Output feedback schemes operate on the external variables but do not guarantee the stability of the closed-loop system. Decoupling schemes are generally limited and in any case they give intermediate results that should be supplemented by other pole placement techniques.

To improve upon the state feedback control scheme, we next present an observer-based control scheme.

5.4.5 Observer-Based Feedback Control

Luenberger [56] has shown that the state of the linear system can be observed (estimated or reconstructed) from the records of inputs and outputs. This process of reconstruction is accomplished by a linear dynamic system called an observer. It has the form

$$\dot{\mathbf{x}}_s = A_s \mathbf{x}_s + B_s \mathbf{u} + C_s \mathbf{y} \tag{52}$$

where $\mathbf{x}_s$ the $n \times 1$ vector approximation (estimate) to $\mathbf{x}$, the original state vector. The observer matrices A_s, B_s, C_s have dimensions $n \times n$, $n \times r$, $n \times m$, respectively. In terms of (31) and (32), we define the observation (estimation) error as

$$\mathbf{e}(t) = \mathbf{x}(t) - \mathbf{x}_s(t) \tag{53}$$

and by making use of (31), (32), and (52) we obtain

$$\dot{\mathbf{e}} = A\mathbf{x} - A_s\mathbf{x}_s + B\mathbf{u} - B_s\mathbf{u} - C_sC\mathbf{x}$$
$$= [A - C_sC]\mathbf{x} - A_s\mathbf{x}_s + [B - B_s]\mathbf{u} \tag{54}$$

On selecting

$$B_s = B \tag{55}$$
$$A_s = A - C_sC \tag{56}$$

then (54) via (53) reduces to

$$\dot{\mathbf{e}} = A_s\mathbf{e} \tag{57}$$

which describes a free (unforced) dynamic system. If the eigenvalues of A_c all have negative real parts, then an asymptotically stable error equation results. This entails that $\mathbf{e}(t) \to \mathbf{0}$ or equivalently $\mathbf{e}_s(t) \to \mathbf{x}(t)$ as $t \to \infty$. Recall that the matrix C_s is still undetermined. From the preceding sections, we can say that when the pair (A, C) is completely observable then C_s can be found by pole placement techniques. The values of the desired poles will eventually determine the estimation rate. From (57) the characteristic polynomial is given by

$$\Delta_e(\lambda) = \det[\lambda I - A_s]$$
$$= \det[\lambda I - A + C_sC]$$
$$= \det[\lambda I - A]\det[I + \Phi(\lambda)C_sC]$$
$$= \Delta(\lambda)\det[I + C\Phi(\lambda)C_s] \tag{58}$$

Complete observability of the pair (A, C) guarantees that n linearly independent rows can be selected from the rows of $C\Phi(\lambda)$. It is interesting to note that

$$\det[\lambda I - A + C_sC] = \det[\lambda I - A^t + C^tC_s^t] \tag{59}$$

which is exactly the same form as (37) but with (A, B, K) replaced by (A^t, C^t, C_s^t). In this sense we say that the observer design problem is the dual of the state feedback design problem. We can therefore use the state feedback algorithm with the foregoing changes, or alternatively work on (58) and define $\pi_j^t(\lambda_k)$ and the jth row of $C\Phi(\lambda_k)$ at λ_k. Thus

$$\pi_j^t(\lambda_k)C_s = -\mathbf{e}_j^t \tag{60}$$

where $\mathbf{e}_j^t$ is the jth row of I. Iterating on (60) for each eigenvalue we

Introduction to Design

arrive at

$$C_s = \begin{bmatrix} \pi_a^t(\lambda_1) \\ \pi_b^t(\lambda_2) \\ \cdot \\ \cdot \\ \cdot \\ \pi_h^t(\lambda_n) \end{bmatrix}^{-1} \begin{bmatrix} e_a^t \\ e_b^t \\ \cdot \\ \cdot \\ \cdot \\ e_h^t \end{bmatrix} \qquad (61)$$

An example will illustrate the above steps.

Example 5.8. A third-order system is described by the matrices

$$A = \begin{bmatrix} -2 & -2 & 0 \\ 0 & 0 & 1 \\ 0 & -3 & -4 \end{bmatrix}, \qquad B = \begin{bmatrix} 1 & 0 \\ 0 & 0 \\ 0 & 1 \end{bmatrix}$$

$$C = \begin{bmatrix} 1 & 0 & 1 \\ 0 & 1 & 0 \end{bmatrix}$$

It is required to design an observer with poles at $\lambda_1 = \lambda_2 = -5$ and $\lambda_3 = -6$. We need to find A_s, B_s, and C_s. From (55), we obtain

$$B_s = \begin{bmatrix} 1 & 0 \\ 0 & 0 \\ 0 & 1 \end{bmatrix}$$

To utilize (58), we first calculate its components:

$\Phi(\lambda) = [\lambda I - A]^{-1} =$

$$\begin{bmatrix} 1/\lambda + 2 & -2(\lambda+4)/(\lambda+1)(\lambda+2)(\lambda+3) & -2/(\lambda+1)(\lambda+2)(\lambda+3) \\ 0 & (\lambda+4)/(\lambda+1)(\lambda+3) & 1/(\lambda+1)(\lambda+3) \\ 0 & -3/(\lambda+1)(\lambda+3) & \lambda/(\lambda+1)(\lambda+3) \end{bmatrix}$$

$$C\Phi(\lambda) = \begin{bmatrix} 1/(\lambda+2) & (-5\lambda-14)/(\lambda+1)(\lambda+2)(\lambda+3) \\ 0 & (\lambda+4)/(\lambda+1)(\lambda+3) \end{bmatrix}$$

$$\begin{matrix} (\lambda^2+2\lambda-2)/(\lambda+1)(\lambda+2)(\lambda+3) \\ 1/(\lambda+1)(\lambda+3) \end{matrix}$$

Then, to implement (61) we use

$$\pi_1^t(-5) = [-1/3 \quad -11/24 \quad 13/24]$$

$$\pi_2^t(-5) = [0 \quad -1/8 \quad 1/8]$$
$$\pi_2^t(-6) = [0 \quad -2/15 \quad 1/15]$$
$$\mathbf{e}_1^t = [1 \quad 0]; \qquad \mathbf{e}_2^t = [0 \quad 1]$$

$$C_s = \begin{bmatrix} -1/3 & -11/24 & -13/24 \\ 0 & -1/8 & -1/8 \\ 0 & -2/15 & -1/15 \end{bmatrix}^{-1} \begin{bmatrix} 1 & 0 \\ 0 & 1 \\ 0 & 1 \end{bmatrix}$$

$$= \begin{bmatrix} 3 & -8 \\ 0 & 7 \\ 0 & -1 \end{bmatrix}$$

and finally from (56) we have

$$A_s = \begin{bmatrix} -5 & 6 & -3 \\ 0 & -7 & 1 \\ 0 & -2 & -4 \end{bmatrix}$$

The observers described in the foregoing section are used to produce $\mathbf{x}_s$; an estimate of $\mathbf{x}$. It is therefore tempting to look for a feedback scheme to yield some design objectives. The most appealing one would be a state feedback based on $\mathbf{x}_s$. The result is called observer-based feedback scheme as shown in Fig. 5.21. The composite system (original plus observer) is of order $2n$ and is represented by

$$\dot{\mathbf{x}} = A\mathbf{x} + B\mathbf{u} \tag{62}$$

$$\mathbf{y} = C\mathbf{x} \tag{63}$$

$$\dot{\mathbf{x}}_s = A_s\mathbf{x}_s + B_s\mathbf{u} + C_s\mathbf{y} \tag{64}$$

$$\mathbf{u} = -K_s\mathbf{x}s + F_s\mathbf{v} \tag{65}$$

Combining (62) to (65) using (55) we obtain

$$\begin{bmatrix} \dot{\mathbf{x}} \\ \dot{\mathbf{x}}_s \end{bmatrix} = \begin{bmatrix} A & -BK_s \\ C_sC & A_s - BK_s \end{bmatrix} \begin{bmatrix} \mathbf{x} \\ \mathbf{x}_s \end{bmatrix} + \begin{bmatrix} BF_s \\ BF_s \end{bmatrix} \mathbf{v} \tag{66}$$

The $2n$ closed-loop eigenvalues are roots of

$$\Delta_{to}(\lambda) = \det \begin{bmatrix} \lambda I - A & BK \\ -C_sC & \lambda I - A_s + BK \end{bmatrix} = 0 \tag{67}$$

using similarity transformation with the aid of (56), we obtain

Introduction to Design

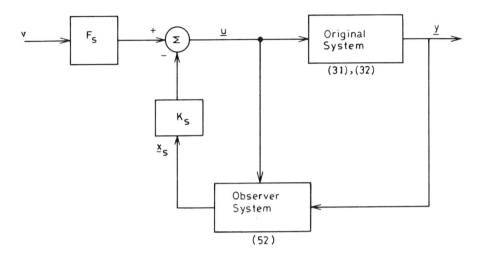

Figure 5.21 An observer-based feedback control scheme.

$$\Delta_{to}(\lambda) = \det\left\{\begin{bmatrix} I & 0 \\ -I & I \end{bmatrix}\begin{bmatrix} \lambda I - A & BK_s \\ -C_s C & \lambda I - A_s + BK_s \end{bmatrix}\begin{bmatrix} I & 0 \\ I & I \end{bmatrix}\right\}$$

$$= \det\begin{bmatrix} \lambda I - A + BK_s & BK_s \\ 0 & \lambda I - A_s \end{bmatrix}$$

$$= \det[\lambda I - A + BK_s]\det[\lambda I - A_s]$$

$$= \Delta_s(\lambda)\Delta_e(\lambda) \tag{68}$$

which identifies a separation in design: by proper selection of K_s, n of the closed-loop eigenvalues can be specified and by proper selection of C_s, the remaining n eigenvalues of the observer can be specified. Indeed, this design pattern is guaranteed provided the triple (A, B, C) is completely controllable and completely observable. There is no restriction on the choice of $2n$ eigenvalues, however good designs are frequently obtained when the observer poles are located deeper in the s-plane (large and negative magnitude).

5.4.6 Discussions

The preceding sections have considered the design of the feedback compensators of linear multivariable systems using either

1. Complete information about system state variables (state feedback)

2. Partial information about system state variables (output feedback), or
3. Estimation of the state variables plus an appropriate state feedback (observer-based feedback)

A common objective is the achievement of suitable pole locations in order to ensure satisfactory transient response. An additional design objective which cannot arise in SISO systems is the achievement of a decoupled or noninteracting system. This means that each input component affects just one output component, or possibly some prescribed subset of output components. The main results have been exposed in each case and subsequent extensions can be found in the cited references.

5.5 PARAMETER OPTIMIZATION

In the foregoing sections, our approach to the control design problem has been to select the configuration of the overall system by introducing compensators and then choose the parameters of the compensators to meet the given specifications on performance. Both time domain and frequency domain specifications are considered. Cascade and feedback compensation schemes are developed based on different methods. The parameters of these compensators are selected to give as closely as possible the desired system performance. In general, it may not be possible to satisfy all the desired specifications. However, an acceptable one can be attained that is not unique. The situation is particularly pronounced in multi-input, multi-output systems.

A more tractable methodology is that of parameter optimization in which the performance specifications are lumped into a single index. As has been discussed in Chapter 3, several performance indices are described. In general, we adopt the notion of minimizing the performance index to yield the "best" parameters. We introduce in the next section a simple motivating example to pave the way for further elaboration.

5.5.1 Quadratic Performance Criteria

There are many criteria by which system performance might be judged. Steady-state or transient response, reliability, cost, energy, consumption, and weight are all possible criteria. For analytic design, we must be able to relate the performance criterion mathematically to the design parameters to be selected. A popular performance index involves the

Introduction to Design

integral of the sum of squares of several system variables. Such indices are called quadratic performance criteria. One attractive feature of these criteria is that they are mathematically tractable. In addition, most of the time, they have a reasonable engineering interpretation in real problems.

Let us consider the following example. A unity feedback position control system has a forward path transfer function

$$G(s) = \frac{K}{s} \tag{69}$$

For a unit-step input, we define $e(t)$ as the error signal and $c(t)$ as the output signal. Thus

$$e(t) = 1 - c(t) \tag{70}$$

$$\dot{e}(t) = -\dot{c}(t), \qquad \ddot{e}(t) = -\ddot{c}(t) \tag{71}$$

It is easy to write in the s-domain

$$\frac{E(s)}{R(s)} = \frac{1}{1 + G(s)}$$

$$= \frac{s}{s + K}$$

Since $R(s) = 1/s$, thus $E(s) = 1/(s + K)$ and $e(t) = e^{-Kt}$. We wish to compute the value of K that minimizes

$$I = \int_0^\infty \{e^2(t) + \lambda[\ddot{e}(t)]^2\}\, dt, \quad \lambda > 0 \tag{72}$$

In view of (71), the output acceleration could be limited in the manner of (72). The parameter λ represents a weighting of the relative importance of the integral square error and integral square output acceleration. To proceed further, we first note that

$$\ddot{e} = K^2 e^{-Kt}$$

which when substituted in (72) produces

$$I = \int_0^\infty [e^{-2Kt} + \lambda K^4 e^{-2Kt}]\, dt$$

$$= \frac{1 + \lambda K^4}{2K} \tag{73}$$

The minimum of I is obtained when

$$\frac{\partial I}{\partial K} = -\frac{1}{2K^2} + \frac{3\lambda K^2}{2} = 0 \tag{74}$$

which yields

$$K^* = (3\lambda)^{-1/4}$$

As a check we note that $\partial^2 I/\partial K^2 = 3\lambda K + (1/K^3) > 0$. From (74) in (73) we obtain

$$I^* = \frac{2}{3(s\lambda)^{1/4}} \tag{75}$$

It is interesting to note that when $\lambda = 0$, no constraint is imposed on the output acceleration but from (74), infinite gain is obtained. For $\lambda \ll 1$, errors will be penalized more than output acceleration, leading to large values of $\ddot{e}(t)$ producing small values of $e(t)$. Conversely, if $\lambda \gg 1$ the acceleration is penalized more heavily than the system error. In practice, the designer chooses some compromise so that the relative importance of the system performance is contrasted with the importance of the limit output acceleration.

5.5.2 Optimization Using Parseval's Theorem

In the preceding section we introduced the quadratic performance index, which has the general form

$$I = \int_0^\infty \sum_{j=1}^h f_j^2(t)\, dt \tag{76}$$

where the $f_j(t)$ are various signals of importance. In the example discussed above, $f_1 = e$ and $f_2 = \ddot{e}$. In what follows, we present a method of finding I in terms of the system design parameters. The method involves Parseval's theorem and a table. Consider a term in the quadratic index (76)

$$I_k = \int_0^\infty f_k^2(t)\, dt \tag{77}$$

Under the assumption that $f_k(t)$ is Laplace transformable, we may write

$$\begin{aligned} I_k &= \int_0^\infty f_k(t) f_k(t)\, dt \\ &= \int_{-0}^\infty f_k(t) \left[\frac{1}{2\pi j} \int_{c-j\infty}^{c+j\infty} F_k(s) e^{st}\, ds \right] dt \end{aligned} \tag{78}$$

Introduction to Design

Interchanging the order of performing the two integrations in (78) we obtain

$$I_k = \frac{1}{2\pi j} \int_{c-j\infty}^{c+j\infty} F_k(s) \left[\int_0^\infty f_k(t) e^{st} dt \right] ds$$

$$= \frac{1}{2\pi j} \int_{c-j\infty}^{c+j\infty} F_k(s) F_k(-s) ds$$

Hence

$$I_k = \int_0^\infty f_k^2(t) dt$$

$$= \frac{1}{2\pi j} \int_{c-j\infty}^{c+j\infty} F_k(s) F_k(-s) ds \tag{79}$$

This result is known as Parseval's theorem. The value of the right-hand integral in (78) can easily be found from published tables [35] provided the $F_k(s)$ can be written in the form $B(s)/A(s)$

$$B(s) \triangleq b_0 + b_1 s + \cdots + b_{n-1} s^{n-1}$$
$$A(s) \triangleq a_0 + a_1 s + \cdots + a_n s^n \tag{80}$$

where the roots of $A(s)$ are in the left half of the complex plane. Results up to fourth-order are given in Table 5.3.

Example 5.9. Consider the system in Fig. 5.22 with $G(s) = 100/s^2$ and $R(-s) + (1/s)$. Determine the optimal values of the parameters K_1 and K_2 such that (i) $I_e = \int_0^\infty e^2(t) dt$ is minimized, and (ii) $I_u = \int_0^\infty u^2(t) dt = 0.1$.

From Figure 5.22 we obtain

$$E(s) = \frac{s + 100K_1 K_2}{s^2 + 100K_1 K_2 s + 100K_1}$$

$$C(s) = \frac{100K_1}{s(s^2 + 100K_1 K_2 s + 100K_1)}$$

$$= \frac{100 U(s)}{s^2}$$

Therefore

$$U(s) = \frac{sK_1}{s^2 + 100K_1 K_2 s + 100K_1}$$

Table 5.3. A Table of Definite Integrals

$$I = \frac{1}{2\pi_j} \int_{c-j\infty}^{c+j\infty} \left[\frac{B(s)B(-s)}{A(s)A(-s)}\right] ds$$

$$B(s) = \sum_{k=0}^{n-1} b_k s^k, \qquad A(s) = \sum_{k=0}^{n} a_k s^k$$

$$I_1 = \frac{b_0^2}{2a_0 a_1}$$

$$I_2 = \frac{b_1^2 a_0 + b_0^2 a_2}{2a_0 a_1 a_2}$$

$$I_3 = \frac{b_2^2 a_0 a_1 + (b_1^2 - 2b_0 b_2) a_0 a_3 + b_0^2 a_2 a_3}{\Delta_1}$$

$$\Delta_1 = 2a_0 a_3(-a_0 a_3 + a_1 a_2)$$

$$I_4 = [b_3^2(-a_0^2 a_3 + a_0 a_1 a_2) + (b_2^2 - 2b_1 b_2) a_0 a_1 a_4$$

$$+ (b_1^2 - 2b_0 b_2) a_0 a_3 a_4 + b_0^2(-a_1 a_4^2 + a_2 a_3 a s_4)]/\Delta_2$$

$$\Delta_2 = 2a_0 a_4(-a_0 a_3^2 - a_1^2 a_4 + a_1 a_2 a_3)$$

For part (i) using Table 5.3, we have

$$b_1 = 1, \qquad b_0 = 100 K_1 K_2$$
$$a_2 = 1, \qquad a_1 = 100 K_1 K_2, \qquad a_0 = 100 K_1$$
$$n = 2$$

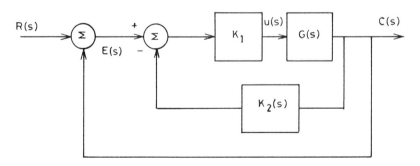

Figure 5.22 Control system of example 5.7.

Introduction to Design

which yields
$$I_e = \frac{b_1^2 a_0 + b_0^2 a_2}{2 a_0 a_1 a_2}$$
$$= \frac{1 + 100 K_1 K_2^2}{200 K_1 K_2}$$

Turning to part (ii) we have
$$b_0 = 0, \quad b_1 = k_1, \quad n = 2$$
$$a_2 = 1, \quad a_1 = 100 K_1 K_2, \quad a_0 = 100 K_1$$

which produces
$$I_u = \frac{K_1}{200 K_2} = 0.1 \tag{81}$$

One way to deal with this problem is to form a new index I
$$I = I_e + \lambda I_u \tag{82}$$
$$= \frac{1 + 100 K_1 K_2^2}{200 K_1 K_2} + \frac{\lambda K_1}{200 K_2}$$

Conditions of optimality imply
$$\frac{\partial I}{\partial K_1} = 0, \quad \frac{\partial I}{\partial K_2} = 0$$

which results in
$$\lambda K_1^2 = 1 \tag{83}$$
$$100 K_1 K_2^2 - 1 - \lambda K_1^2 = 0 \tag{84}$$

Combining (81), (83), and (84) we arrive at
$$\lambda = 0.25, \quad K_1 = 2, \quad K_2 = 0.1$$

As we have seen, the use of Table 5.3 is helpful in expressing any quadratic performance index in terms of the design parameters. Such a procedure quickly becomes unwieldy as the order of the system and the number of design parameters is increased. Thus, we naturally seek computation-oriented procedures for accomplishing parameter optimization. One possibility is discussed in the next section.

5.5.3 The Optimum State Regulator

Here we consider a special case of the general quadratic performance index of (73) in which the $f_k(t)$ are either

- Weighted plant state variables $x_k(t)$, or
- Weighted plant input $u_j(t)$

$$I = \int_0^\infty \left[\sum_{k=1}^n q_k x_k^2(t) + \sum_{j=1}^m r_j u_j^2(t) \right] dt \qquad (85)$$

Using the matrix representation for state variables and inputs

$$\mathbf{x} = \begin{bmatrix} x_1(t) \\ x_2(t) \\ \cdot \\ \cdot \\ \cdot \\ x_n(t) \end{bmatrix}, \quad \mathbf{u} = \begin{bmatrix} u_1(t) \\ u_2(t) \\ \cdot \\ \cdot \\ \cdot \\ u_m(t) \end{bmatrix} \qquad (86)$$

equation (85) can be cast into the form

$$I = \int_0^\infty [\mathbf{x}'Q\mathbf{x} + \mathbf{u}'R\mathbf{u}]\, dt \qquad (87)$$

where

$$Q = \operatorname{diag}[q_1 \; q_2 \; \cdots \; q_n], \quad R = \operatorname{diag}[r_1 \; r_2 \; \cdots \; r_m] \qquad (88)$$

A more general index can be considered where Q and R have off-diagonal entries. However, the diagonal case is usually easier to understand physically, and thus is most often used in practice. For the index (87) to be physically meaningful, the contribution of any trajectory to $\mathbf{x}'Q\mathbf{x}$ must be nonnegative. Therefore we require $q_k \geq 0$ for all k. Furthermore, the cost of any control should be strictly positive, and thus we require $r_j > 0$ for all j.

With $q_k \geq 0$, there is a possibility that the optimum system could be unstable. This can happen if an unstable mode does not affect the performance index (87). In view of (88) the form $\mathbf{x}'Q\mathbf{x}$ is a sum of squares and thus an unstable mode will not affect the performance index I if and only if that mode is not observable from the state variables appearing in $\mathbf{x}'Q\mathbf{x}$, considering those state variables as system outputs. To avoid these difficulties we shall require that all state variables are observable from those x_ks in $q_k x_k^2$ corresponding to $q_k \neq 0$. Furthermore, we shall require the system to be completely state controllable. If this was not true, unstable uncontrollable modes may appear in the resulting optimum system. Clearly, this is undesirable physically. It can be shown that when the plant is completely state controllable and completely observable, the resulting optimal system is asymptotically stable.

Introduction to Design

The plant is described in state form by

$$\dot{x} = Ax + Bu, \quad x(0) = x_0 \tag{89}$$

We shall further restrict the problem by requiring **u**, the control input to the plant, to be constructed from the state variables by linear, time-invariant feedback

$$u = Kx \tag{90}$$

where K is an $m \times n$ matrix of constant gains. The problem then is one of parameter optimizations. Choose K to minimize the quadratic performance index (87) subject to the linear dynamics (89).

It must be emphasized that the performance index (87) puts penalty for both large deviations in the state variables from zero and for the use of large control inputs. A system design in this manner is called an *optimum state regulator*.

The derivation of the optimal K is straightforward, although it involves some matrix calculus. The details are deferred to Appendix C. The results are the following:

1. The optimal matrix K minimizing (87) is given by

$$K = -R^{-1}B'P \tag{91}$$

where the matrix P is the positive-definite, symmetric solution ($P = P'$) to the matrix Riccati equation

$$PA + A'P - PBR^{-1}B'P + Q = 0 \tag{92}$$

2. The optimal vlaue of the performance index is

$$I^* = x'(0)Px(0) \tag{93}$$

Since P is an $n \times n$ matrix, (92) represents n^2 simultaneous, nonlinear, algebraic equations for the n^2 elements of P. In view of the symmetric feature of P, only $n + (n^2 - n)/2 = n(n+1)/2$ elements are unknowns. Note that in the light of (91) through (93) the gain K is independent of the initial conditions x_0. Also, the optimum state regulator, which is known to be asymptotically stable, forces the eigenvalues of the matrix $[A - BR^{-1}B'P]$ to lie always in the left half s-plane. A block diagram representation of the optimal state regulator is given in Fig. 5.23.

Example 5.10. Determine the control law which minimizes the performance index

$$J = \int_0^\infty (2x_1^2 + 2u^2)\,dt$$

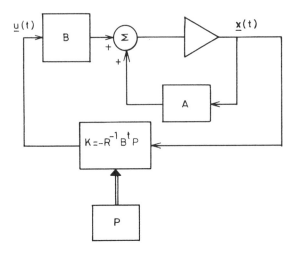

Figure 5.23 Block diagram of optimum state-regulator system.

for the system

$$\begin{bmatrix} \dot{x}_1 \\ \dot{x}_2 \end{bmatrix} = \begin{bmatrix} 0 & 1 \\ 0 & 0 \end{bmatrix} \begin{bmatrix} x_1 \\ x_2 \end{bmatrix} + \begin{bmatrix} 0 \\ 1 \end{bmatrix} u$$

We have for this problem

$$A = \begin{bmatrix} 0 & 0 \\ 0 & 1 \end{bmatrix}, \quad B = \begin{bmatrix} 0 \\ 1 \end{bmatrix}, \quad Q = \begin{bmatrix} 2 & 0 \\ 0 & 0 \end{bmatrix}, \quad R = [2]$$

Thus (92) can be expressed as

$$\begin{bmatrix} p_{11} & p_{12} \\ p_{12} & p_{22} \end{bmatrix} \begin{bmatrix} 0 & 1 \\ 0 & 0 \end{bmatrix} + \begin{bmatrix} 0 & 0 \\ 1 & 0 \end{bmatrix} \begin{bmatrix} p_{11} & p_{12} \\ p_{12} & p_{22} \end{bmatrix}$$
$$- \begin{bmatrix} p_{11} & p_{12} \\ p_{12} & p_{22} \end{bmatrix} \begin{bmatrix} 0 \\ 1 \end{bmatrix} [1/2][0 \quad 1] \begin{bmatrix} p_{11} & p_{12} \\ p_{12} & p_{22} \end{bmatrix} + \begin{bmatrix} 2 & 0 \\ 0 & 0 \end{bmatrix} = \begin{bmatrix} 0 & 0 \\ 0 & 0 \end{bmatrix}$$

Upon simplification we obtain

$$-\frac{1}{2}p_{12}^2 + 2 = 0, \quad \frac{p_{11} - p_{12}p_{22}}{2} = 0$$

$$-\frac{p_{22}^2}{2} + 2p_{12} = 0$$

the solution of which yields

Introduction to Design

$$P = \begin{bmatrix} 2\sqrt{2} & 2 \\ 2 & 2\sqrt{2} \end{bmatrix}$$

which is positive-definite. From (91) we obtain

$$K = -\begin{bmatrix} R \\ 2 \end{bmatrix} [0 \ 1] \begin{bmatrix} 2\sqrt{2} & 2 \\ 2 & 2\sqrt{2} \end{bmatrix} \begin{bmatrix} x_1 \\ x_2 \end{bmatrix}$$

$$= -x_1 - \sqrt{2} x_2$$

As a check, the closed-loop system matrix is given by

$$A_c = A - BR^{-1}B'P$$

$$= \begin{bmatrix} 0 & 1 \\ 0 & 0 \end{bmatrix} - \begin{bmatrix} 0 \\ 1 \end{bmatrix} [1/2][0 \ 1] \begin{bmatrix} 2\sqrt{2} & 2 \\ 2 & 2\sqrt{2} \end{bmatrix}$$

$$= \begin{bmatrix} 0 & 1 \\ 1 & -\sqrt{2} \end{bmatrix}$$

whose eigenvalues are $(-\sqrt{2} \pm \sqrt{2}_j)/2$ which are stable as expected.

5.5.4 The Optimum Output Regulator

Consider the plant (89) along with

$$\mathbf{y}(t) = C\mathbf{x}(t) \tag{94}$$

where $\mathbf{y}(t)$ is a $p \times 1$ output vector. The use of (94) is to reflect the limited number of measurable states. Now the problem becomes that of finding the control law $\mathbf{u}(t) = K_0\mathbf{x}(t)$ so that the following quadratic performance index is minimized.

$$I = \int_0^\partial [\mathbf{y}'Q_0\mathbf{y} + \mathbf{u}'R\mathbf{u}] \, dt \tag{95}$$

where $Q_0 \geq 0$ and $R > 0$ as usual. This problem can be reduced in form to the optimum state regulator if we substitute (94) into (95) and obtain

$$I = \int_0^\infty [\mathbf{x}'C'Q_0C\mathbf{x} + \mathbf{u}'R\mathbf{u}] \, dt \tag{96}$$

Comparing (87) with (96), we observe that the two indices are identical in form by replacing $C'Q_0C$ in (96) to obtain Q in (87). Therefore the results of the optimal output regulator are

1. The optimal matrix K_0 minimizing (96) is given by

$$K_0 = -R^{-1}B'P_0 \tag{97}$$

where the matrix P_0 is the positive-definite symmetric solution ($P_0 = P_0'$) to the matrix Riccati equation

$$P_0 A + A'P_0 - P_0 B R^{-1} B' P_0 + C' Q_0 C = 0 \tag{98}$$

2. The optimal value of the performance index is

$$I = \mathbf{x}'(0) P_0 \mathbf{x}(0) \tag{99}$$

Recall that the objective of (95) is to bring and keep the output $\mathbf{y}(t)$ near zero without using an excessive amount of control energy. In some cases, it is desired to steer the output $\mathbf{y}(t)$ very close to a reference $\mathbf{r}(t)$, which is frequently called the *tracking problem*.

This can be accomplished by minimizing

$$I = \int_0^\infty [\mathbf{e}' Q_e \mathbf{e} + \mathbf{u}' R \mathbf{u}]\, dt \tag{100}$$

where $Q_e \geq 0$, $R > 0$, and $\mathbf{e}(t) = \mathbf{y}(t) - \mathbf{r}(t)$. To reduce this problem to the output regulator form, we consider only those $\mathbf{r}(t)$ that can be generated by arbitrary initial conditions $\mathbf{z}(0)$ in the system

$$\dot{\mathbf{z}} = A\mathbf{z}(t), \quad \mathbf{r}(t) = C\mathbf{z}(t) \tag{101}$$

We now define a new vector $\mathbf{w}(t) = \mathbf{x}(t) - \mathbf{z}(t)$; it is easy to show that

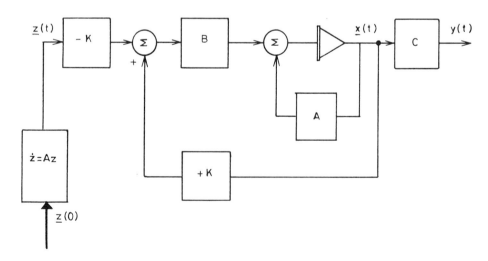

Figure 5.24 Block diagram of optimum output tracking system.

Introduction to Design

it satisfies
$$\dot{\mathbf{w}} = A\mathbf{w} + B\mathbf{u}, \quad \mathbf{e} = C\mathbf{w} \tag{102}$$

Applying results of the output regulator problem immediately gives that the optimal control law for the tracking problem is

$$\mathbf{u} = -R^{-1}B'P_e\mathbf{w}$$
$$= -R^{-1}B'P_e[\mathbf{x} - \mathbf{z}] \tag{103}$$

where P_e is the solution of

$$P_e A + A' P_e - P_e B R^{-1} B' P_e + C' Q_e = 0 \tag{104}$$

In Fig. 5.24, the system for output tracking is presented. We next turn attention to another approach to the parameter optimization.

5.5.5 Spectral Factorization

In what follows, we develop a transfer function approach to optimization design of linear plants. Given a plant described by a proper $G(s) = C(s)/U(s)$, it is required to find the transfer function $T^*(s)$ of the overall system (plant plus a cascade compensator) that minimizes the quadratic performance index

$$I = \int_0^\infty \{[r(t) - c(t)]^2 + \lambda u^2(t)\} \, dt$$

where $\lambda > 0$, $u(t)$ is the actuating signal, $c(t)$ is the actual output and $r(t)$ is the desired output (reference input) of the system. In the frequency domain, the performance index is expressed as

$$I = \frac{1}{2}\pi_j \int_{-j\infty}^{j\infty} \{[R(s) - C(s)][R(-s) - C(-s)] + \lambda U(s)U(-s)\} \, ds \tag{105}$$

We have the following relations

$$T(s) = \frac{C(s)}{R(s)}, \quad U(s) = \frac{T(s)R(s)}{G(s)} \tag{106}$$

From (106) in (105), it becomes

$$I = \frac{1}{2}\pi_j \int_{-j\infty}^{j\infty} \left\{ [T(s) - 1][T(-s) - 1] + \frac{\lambda T(s)T(-s)}{G(s)G(-s)} \right\} r(s)R(-s) \, ds \tag{107}$$

The optimization problem is formulated as: find a stable $T^*(s)$ that minimizes the performance index (107). By definition $I[T^*(s)] \leq I[T(s)] \leq I_\epsilon[T^*(s) + \epsilon \Delta T(s)]$ where $\Delta T(s)$ is some arbitrary stable transfer function. Conditions of optimality entail that $d/d\epsilon(I_\epsilon) = 0$ and $d^2/d\epsilon^2(I_\epsilon) > 0$. To obtain I_ϵ, we use $T(s) = T^*(s) + \epsilon \Delta T(s)$ in (107) and rearrange the result in the form

$$I_\epsilon = I_a + \epsilon(I_b + I_c) + \epsilon^2 I_d \tag{108}$$

where

$$I_a = \frac{1}{2\pi j} \int_{-j\infty}^{j\infty} \left\{ [T^*(s) - 1][T^*(-s) - 1] + \frac{\lambda T(s) T^*(-s)}{G(s)G(-s)} \right\} R(s) R(-s) \, ds$$

$$I_b = \frac{1}{2\pi j} \int_{-j\infty}^{j\infty} \left\{ [T^*(-s) - 1] + \frac{\lambda T^*(-s)}{G(s)G(-s)} \right\} R(s) R(-s) \Delta T(s) \, ds$$

$$I_c = \frac{1}{2\pi j} \int_{-j\infty}^{j\infty} \left\{ [T^*(s) - 1] + \frac{\lambda T^*(s)}{G(s)G(-s)} \right\} R(s) R(-s) \Delta T(-s) \, ds$$

$$I_d = \frac{1}{2\pi j} \int_{-j\infty}^{j\infty} \left\{ 1 + \frac{\lambda}{G(s)G(-s)} \right\} R(s) R(-s) \Delta T(s) \Delta T(-s) \, ds$$

Expressing $G(s)$ as $N(s)/D(s)$, we then rewrite the integrand of I_d as $\{[N(s)N(-s) + \lambda D(s)D(-s)]/N(s)N(-s)\} R(s) R(-s) \Delta T(s) \Delta T(-s)$. But the term $[N(s)N(-s) + \lambda D(s)D(-s)]$ is symmetric about the imaginary axis; that is, if α is a root, so is $-\alpha$. It then follows that there are no roots on the imaginary axis. Moreover, $\lambda > 0$ and the polynomials $N(s)$ and $D(s)$ have no common factor. Thus, for all ω, the polynomial $[N(j\omega)N(-j\omega) + \lambda D(j\omega)D(-j\omega)] = |N(j\omega)|^2 + \lambda |D(j\omega)|^2 \neq 0$. In view of these features we may write

$$N(s)N(-s) + \lambda D(s)D(-s) = P^*(s) P^*(-s) \tag{109}$$

where $P^*(s)$ consists of all left half s-plane roots of the composite polynomial in (109). Clearly, $P^*(-s)$ consists of all right half s-plane and this suggests the title *spectral factorization*. It is now easy to infer that $I_d > 0$ and by further examination, $I_b = I_c$. The foregoing analysis leads to the conditions of optimality as $I_b = 0$ and $I_d > 0$. Hence

$$\frac{1}{2\pi j} \int_{-j\infty}^{j\infty} \left\{ [T^*(s) - 1] + \frac{\lambda T^*(s)}{G(s)G(-s)} \right\} R(s) R(-s) \Delta T(-s) \, ds = 0 \tag{110}$$

Obviously, stable design requires that all the poles of $T(s)$, $T^*(s)$, and $\Delta T(s)$ be in the left half s-plane. Conversely $T(-s)$, $T^*(-s)$, and $\Delta T(-s)$ should have all poles located in the right half s-plane. In view of these conditions, (110) is satisfied if all the poles of the function

Introduction to Design

$$F = \left[T^*(s) - 1 + \frac{\lambda T^*(s)}{G(s)G(-s)} \right] R(s)R(-s)$$

lie in the right half s-plane. Manipulating F,

$$F = \left[1 + \frac{\lambda}{G(s)G(-s)} \right] T^*(s)R(s)R(-s) - R(s)R(-s)$$

$$= V(s)V(-s)R(s)R(-s)T^*(s) - R(s)R(-s)$$

where $V(s)$ and $R(s)$ have no poles or zeros in the right half s-plane. Proceeding further, we obtain

$$\frac{F}{V(-s)R(-s)} = V(s)R(s)T^*(s) - \frac{R(s)}{V(-s)}$$

$$= V(s)R(s)T^*(s) - W_+(s) - W_-(s)$$

where $W_+(s)$ is composed of the part involving all the poles in the left half s-plane. Rearranging the above relation, we have

$$V(s)R(s)T^*(s) - W_+(s) = \frac{F}{V(-s)R(-s)} + W_-(s) \quad (111)$$

It must be emphasized in (111) that all the poles of the left-hand side are in the left half s-plane and all the poles of the right-hand side are in the right half s-plane. In view of their independence, each must be equal to zero. Thus

$$T^*(s) = \frac{W_+(s)}{V(s)R(s)}$$

$$= \frac{W_+(s)N(s)}{P^*(s)R(s)} \quad (112)$$

where $P^*(s)$ is given in (109). Once (112) is determined, we move on to derive the compensator transfer function $G_c(s)$ since

$$T^*(s) = \frac{G(s)G_c(s)}{1 + G_c(s)G(s)}$$

and we conclude that

$$G_c(s) = \frac{T^*(s)}{G(s)[1 - T^*(s)]} \quad (113)$$

Example 5.11. To illustrate the foregoing design procedure, we consider

$$G(s) = \frac{100}{s^2}, \quad \lambda = 0.25$$

and note that

$$N(s) = N(-s) = 100, \quad D(s) = D(-s) = s^2$$
$$P^*(s)P^*(-s) = 10000 + 0.25s^4$$

We define $P^*(s) = p_0 + p_1 s + p_2 s^2$ and therefore

$$[p_0 + p_1 s + p_2 s^2][p_0 - p_1 s + p_2 s^2] = 10000 + 0.25s^4$$

Equating similar terms, we obtain

$$p_0 = 100, \quad p_1 = 10, \quad p_2 = 0.5$$

Thus

$$P^*(s) = 100 + 10s + 0.5s^2$$
$$V(-s) = \frac{P^*(-s)}{N(-s)}$$
$$= \frac{100 - 10s + 0.5s^2}{100}$$

For $R(s) = 1/s$ we analyze the term $R(s)/V(-s)$

$$\frac{R(s)}{V(-s)} = \frac{100}{s(0.5s^2 - 10s + 100)}$$
$$= W_+(s) + W_-(s)$$

which gives $W_+(s) = 1/s$ and in conclusion, $T^*(s)$ from (112) reads

$$T^*(s) = \frac{100}{100 + 10s + 0.5s^2}$$

and the required compensator from (113) is $G_c(s) = 20s/(s + 20)$.

5.5.6 Design Considerations

We now have some experience in the optimization of linear systems with quadratic indices. Our study has included the optimal state and output regulators and the transfer function approach. The optimum design involves two major difficulties:

1. How are the weighting terms Q and R (or equivalently λ) selected?
2. How is the optimum control computed for complex, high-order systems?

Introduction to Design

On the first problem, the physical meaning of the variables in the index, and hence their relative importance, can be discussed only within the context of the particular application. The diagonal entries in Q, q_j, indicate the relative importance to be assigned to the "energies" represented by the integrals of the squares of the state variables. The diagonal entries of R indicate the relative importance of control energies to be supplied by the various inputs. Note that the ratio Q/R expresses a trade-off between the internal and external energies. Sometimes we choose $Q = I$ and adjust R according to some other criterion. This is the same role of the factor λ.

On the second problem, computationally efficient algorithms have now been developed for computing matrix equations and software support using high-level languages are readily and commercially available.

A final word is that the spectral factorization becomes quite complex for multi-input multi-output systems. However, the state-variable approach (state and output regulators) yields proportional gains which do not provide satisfactory performance in most cases. In short, the optimum design problem has several solution approaches; yet each has its own limitations.

5.6 PROBLEMS

1. A plant is described by the transfer function

$$G(s) = \frac{K}{s(s + 10)(s + 1000)}$$

   It is desired that the overshoot be approximately 7.5% for a step input and the settling time of the system be 400 ms. Design a phase-lead compensator by using root-locus methods. Select the compensator zero at $s = -20$ and determine the compensator pole. Hence find the resulting system K_r.

2. A unity feedback control system has the open-loop transfer function

$$G(s) = \frac{K}{s(s + 2)(s + 6)}$$

   It is desired to have a velocity error constant of $K_r = 20$, a phase margin around 45°, and a closed-loop bandwidth > 4 rad/s. (Hint: use two identical cascaded phase-lead networks).

3. Consider the open-loop transfer function

$$G(s) = \frac{K}{s(1+s)(1+0.2s)}$$

Determine K so that the dominant closed-loop poles have a damping ratio of 0.5 for a unity feedback. Find the undamped natural frequency ω_n.

4. Design a phase-lag compensator for the open-loop transfer function

$$G(s) = \frac{K}{s(s+4)(s+5)}$$

to achieve a damping ratio of 0.7 and velocity error constant of 30.

5. Consider the third-order system

$$G(s) = \frac{80}{s(1+0.02s)(1+0.05s)}$$

   a. Find the gain and phase margins and the gain crossover frequency. Is the system stable?
   b. Design a phase-lag cascade compensator to achieve the following performance specifications: (1) phase margin of approximately 30°; (2) gain margin of about 7 dB.
   c. Design a phase-lead compensator to achieve the specifications of part (b) above.

6. A second-error system has

$$G(s) = \frac{50}{s(s+10)}$$

Design a lead compensator to achieve the following specifications:
   a. The velocity error constant $K_r = 100$
   b. Phase margin $= 45°$
   c. An error of less than 2% is achieved for sinusoidal input functions of frequency up to 1 rad/s

7. A linear system is described by

$$A = \begin{bmatrix} -2 & 1 & 0 \\ 0 & -2 & 0 \\ 0 & 0 & 4 \end{bmatrix}, \quad B = \begin{bmatrix} 0 & 0 \\ 0 & 1 \\ 1 & 0 \end{bmatrix}$$

Find a constant state feedback matrix K which yields closed-loop poles at $(-2, -3, -4)$.

Introduction to Design

8. Repeat Problem 7 if the desired set of eigenvalues is
   a. $(-1, -1, -1)$
   b. $(-2, -2, -15)$
   c. $(-3, -2, \pm j)$
9. Specify a constant output feedback matrix K_0 so that the system

$$\dot{x} = \begin{bmatrix} -1 & 0 \\ 0 & -4 \end{bmatrix} x + \begin{bmatrix} 0 & 1 \\ 1 & 0 \end{bmatrix} u$$

$$y = \begin{bmatrix} 0 & 1 \end{bmatrix} x$$

   will have a closed-loop eigenvalue at -10.
10. Repeat Problem 9 when

$$y = \begin{bmatrix} 1 & 1 \\ 0 & 1 \end{bmatrix} x$$

   and the desired closed-loop eigenvalues are -10 and -10.
11. Can the system described by

$$A = \begin{bmatrix} 1 & 0 & 0 \\ 0 & 0 & 1 \\ 0 & 1 & 0 \end{bmatrix}, \quad B = \begin{bmatrix} 1 & 0 \\ 1 & 0 \\ 1 & 1 \end{bmatrix}, \quad C = \begin{bmatrix} 1 & 0 & 0 \\ 0 & 0 & 0 \end{bmatrix}$$

   be decoupled?
12. Find the matrices F_d and K_d such that the state feedback reduces the system

$$A = \begin{bmatrix} -2 & -2 & 0 \\ 0 & 0 & 1 \\ 0 & -3 & -4 \end{bmatrix}, \quad B = \begin{bmatrix} 1 & 0 \\ 0 & 0 \\ 0 & 1 \end{bmatrix}$$

$$C = \begin{bmatrix} 1 & 0 & 0 \\ 0 & 1 & 1 \end{bmatrix}$$

   to integrator decoupled form. Find the decoupled transfer matrix also.
13. A unity feedback control system has an open-loop transfer function $G(s) = 4/s(s + \alpha)$. The output is required to track unit step. Find the value of α that minimizes integral square error.
14. The standard unity feedback configuration has $G(s) = 100/s^2$, $G_c(s) = sK_1/(s + K_2)$, and $R(s) = 1/s$. Determine the optimal

value of parameters K_1 and K_2 such that

$$I = \int_0^\infty [e^2(t) + 0.25u^2(t)]\, dt \to \text{minimum}$$

15. Repeat Problem 14 when

$$G_c(s) = K_1\left(\frac{1 + K_2 s}{1 + \alpha K_2 s}\right)$$

It is required to find K_1, K_2, and α.

16. Some performance indices other than squared error indices can be optimized, using Parseval's theorem and Table 5.3. Consider the following:
    a. Show that $\zeta[te(t)] = -dE(s)/ds$ and thus

    $$\int_0^\infty t^2 e^2\, dt = \frac{1}{2\pi_j}\int_{-j\infty}^{j\infty}\left[-\frac{dE(s)}{ds}\right]\left[-\frac{dE(s)}{ds}\right]ds$$

    b. Find a proportional gain K in cascade with a $G(s) = 1/s$ in a unity feedback such that $\int_0^\infty t^2 e^2\, dt$ is minimized subject to

    $$\int_0^\infty u^2\, dt = 1/2$$

17. Consider the plant

    $$\dot{\mathbf{x}} = \begin{bmatrix} 1 & 0 \\ -1 & 2 \end{bmatrix}\mathbf{x} + \begin{bmatrix} 1 \\ 0 \end{bmatrix}u$$

    Check the system stability and controllability. Find the control law and the optimal index that minimizes

    $$I = \int_0^\infty [\mathbf{x}'Q\mathbf{x} + \mathbf{u}'R\mathbf{u}]\, dt$$

    for the following cases:
    a. $Q = I$, $R = 1$
    b. $Q = I$, $R = 100$
    c. $Q = I$, $R = 0.01$

    and compare the obtained results.

NOTES AND REFERENCES

Several alternative approaches to the design and compensation of feedback control systems have been considered. We started by introducing

Introduction to Design

cascade compensators within a unity feedback configuration. This included lead, lag, and lead-lag networks and were found useful for alternating the shape of the frequency response of a system. For more elaboration and further studies, the readers are referred to D'Azzo and Houpis (1981), Anand (1984), Chen (1970), Eveleigh (1972), Shinners (1978), Coffey (1970), Dorf (1980), and El-Hawary (1984).

We next moved to state-variable design through state, output, observer-based, and decoupling feedback schemes. The main technique was eigenvalue assignment whose objective is to position the desired eigenvalues (derived from the performance specifications) at particular locations in the left half of the s-plane. For related topics and computational algorithms, the readers are advised to consult Luenberger (1971, 1979), Morse and Wonham (1970), Wonham (1967), Davison (1970), Brogan (1985), De Russo et al. (1965), and Armstrong (1980).

Finally, we discussed the important topic of parameter optimization through the use of quadratic performance indices. Here we emphasized both the time domain and frequency domain performance measures and demonstrated the interlink through Parseval's theorem. Matrix Riccati equation was shown to be a basic computational load in the development of optimal control laws. State and output regulators were examined in detail. For further reading, consult MacFarlane (1972), Sage and White (1977), Anderson and Moore (1971), Kwakernaak and Sivan (1972), Bryson and Ho (1975), Lapidus and Luus (1967), Elgerd (1967), Kalman (1964), Kriendler and Jameson (1972), Perkins and Cruz (1971), and the references cited therein.

6
Methods of Analysis

6.1 INTRODUCTION

We have described in Part I of this book, control systems where the flow of information and processing of data were continuous in time. Almost all the devices used were analog using physically realizable elements. The relevance of Part I was to cover the fundamental concepts of control analysis and methods which were in common use by practitioners about three decades ago. In this part, we study control sytems in which signals and information will appear only at certain selected instants; perhaps at regular intervals. This discontinuous (discrete) form of signal brings about the use of data converters (analog to digital (A/D) or digital to analog (D/A)) and the introduciton of digital electronics and components. We will generally refer to this class of systems as discrete time, discrete or digital. The advantages of Laplace transform methods for analyzing continuous-time linear systems have been demonstrated in Part I and Appendix A. In the simplest sense, perhaps, they convert differential equations into algebraic equations that are usually easy to manipulate and visualize. We can think of the transform technique as being auxiliary in the process of linear-system analysis. That is, it is an aid rather than a necessity, although frequently a problem solution that is easy to obtain with the transform

Methods of Analysis

approach may be extremely difficult to achieve with a strictly time domain approach.

We should hope that the same general benefits are available for discrete-time linear systems. This is in fact the case and the transformation that provides this is called the Z transform. It will be found that many operations can be better understood or performed in the Z domain than in the discrete-time domain.

6.2 THE Z TRANSFORM

Rather than introducing the Z transform as an abstract mathematical formula we will attempt to relate it to past experience. Consider the three related signals shown in Fig. 6.1. The first, $f(t)$, is a continuous-time signal and is the only one that is a function in the strict mathema-

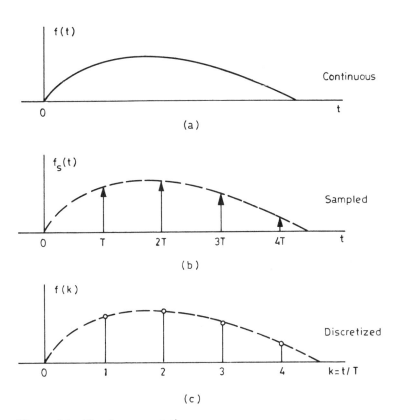

Figure 6.1 Signal representations.

tical sense. The sequence of impulses can be obtained in theory with a balanced modulator that multiplies a periodic unit-impulse train with period T by $f(t)$

$$f_s(t) = f(t) \sum_{k=0}^{\infty} \delta(t - kT)$$

$$= \sum_{k=0}^{\infty} f(kT)\delta(t - kT) \qquad (1)$$

The representation in Fig. 6.1(c) is that of a sequence, $f(k)$, and nothing more.

6.2.1 Definition and Properties

Recall from Appendix A that the unilateral Laplace transform of $f_s(t)$ is $F_s(s)$

$$F_s(s) = \sum_{k=0}^{\infty} f(kT)e^{-kTs} \qquad (2)$$

which is a periodic function; that is, $F_s(s) = F_s(s + 2_{kj}\pi)$, $k = 0, 1, 2, \ldots$. This periodicity complicates the process of performing the inverse Laplace transformation, taking away some of the attractive features that occur when periodic functions of s appear. On the other hand, if we let $s = (1/T) \ln z$, or $z = e^{sT}$, then we obtain a function of z called the unilateral Z transform of $f(kT)$

$$F(z) = \sum_{k=0}^{\infty} f(kT)z^{-k} \qquad (3)$$

Since $|z| = |e^{sT}| = |e^{\sigma T}e^{j\omega T}| = |e^{\sigma T}|$, it is easy to see that when $\sigma < 0$ (i.e., when in the left half of the s-plane), then $|z| < 1$, and thus the left half of the s-plane corresponds to the interior of the unit circle in the complex Z-plane. In the same way the right half of the s-plane corresponds to the exterior of the unit circle in the Z-plane. The $j\omega$ axis in the s-plane and the unit circle in the Z-plane correspond. The results should not be surprising since we already know that a stable continuous system has a characteristic equation with roots in the left half of the s-plane. In a similar fashion, a stable discrete system has a characteristic equation with roots inside the unit circle in the Z-plane. From this point onwards we leave the Laplace transform and concern ourselves only with the Z transform. Sometimes, it is said that (3) is the fundamental defining relation for the Z transform of a time function and it must be memorized. It is the preferred starting point for proving,

Methods of Analysis

"from first principles," properties of the Z transform. It also furnishes a numerical method for the calculation of Z transforms of particular time functions.

In the following, we present some of the properties of the Z transform that constitute pairs themselves and also enable us to deduce other pairs.

Linearity

The transform of a linear combination of functions is the sum of the transforms of the individual functions. Given that

$$Z[af(kT)] = aF(z), \quad Z[bg(kT)] = bG(z)$$

then

$$Z[af(kT) + bg(kT)] = aF(z) + bG(z)$$

In the sequel we shall drop T for notational simplicity.

Shifting (Delay)

The transform of a delayed sequence $f(k - m)$ is found by substitution from (3)

$$Z[f(k - m)] = \sum_{k=0}^{\infty} f(k - m)z^{-k} \qquad (4)$$

With $n = k - m$ we obtain

$$\begin{aligned}
Z[f(k - m)] &= \sum_{n=-m}^{\infty} f(n)z^{-n-m} \\
&= z^{-m} \sum_{n=-m}^{\infty} f(n)z^{-n} \\
&= z^{-m}\left[\sum_{n=0}^{\infty} f(n)z^{-n}\right] \\
&\quad + z^{-m}\left[\sum_{n=-m}^{-1} f(n)z^{-n}\right] \\
&= z^{-m}F(z) + z^{-m}\sum_{n=-m}^{-1} f(n)z^{-n} \qquad (5)
\end{aligned}$$

That is, a delay of m samples leads to a transform that is z^{-m} multiplied by the transform of the undelayed samples plus terms due to $f(k)$ for $k<0$. If $f(k)=0$ for $k<0$, the last term in (5) vanishes and we obtain

$$Z[f(k-m)] = z^{-m}F(z), \quad f(k)=0, \quad k<0 \tag{6}$$

In terms of the unit sequence

$$U(k-m) = \begin{cases} 1, & k \geq m \\ 0, & k \leq m \end{cases} \tag{7}$$

We can rewrite (6) in a more meaningful way

$$Z[f(k-m)U(k-m)] = z^{-m}F(z) \tag{8}$$

Shifting (Advance)

In a parallel development, it can be shown that

$$Z[f(k+m)] = \sum_{k=0}^{\infty} f(k+m)z^{-k}$$

$$= z^m f(z) - z^m \sum_{n=0}^{m-1} f(n)z^{-n} \tag{9}$$

In the same way

$$[f(k+m)U(k+m)] = z^m F(z) - z^m \sum_{n=0}^{m-1} f(n)z^{-n} \tag{10}$$

Convolution-Summation

Given the three sequences linked together by the expression

$$y(k) = \sum_{m=0}^{k} x(k)h(k-m) \tag{11}$$

Applying (3) to (11) and making use of (6) with some manipulation, we arrive at

$$Y(z) = X(z)H(z) \tag{12}$$

where $\{y(k), Y(z)\}$, $\{x(k), X(z)\}$, and $\{h(k), H(z)\}$ are three Z transform pairs. This result is not unexpected. It simply states that the Z transform of the zero-state response of a linear, casual, shift-invariant discrete system is given by the product of the transform of the input

Methods of Analysis

Table 6.1 Z Transform Pairs

Time Function $f(k)$	Z Transform $F(z)$
$kf(k)$	$-z\dfrac{dF(z)}{dz}$
$a^n f(k)$	$F\left(\dfrac{z}{a}\right)$
$f(0)$	$\lim_{z\to\infty} F(z)$
$f(1)$ (initial value)	$\lim_{z\to\infty} zF(z)$
$\lim_{n\to\infty} f(k)$ (final value)	$\lim_{z\to 1}\left(1 - \dfrac{1}{z}\right)F(z)$
$\delta(k)$	1
$\delta(k - m)$	z^{-m}
$U(k)$	$\dfrac{z}{z - 1}$
$U(k - m)$	$z^{-m}\left(\dfrac{z}{z-1}\right)$
$kU(k)$	$\dfrac{z}{(z-1)^2}$
a^k	$\dfrac{z}{z - a}$
$k^2 U(k)$	$\dfrac{z^2 + z}{(z-1)^3}$

$X(z)$ and a discrete transfer function $H(z)$. In Table 6.1, a summary of some of the important Z transform pairs is presented.

Example 6.1. Find the Z transform of the sequence $\{1, 2, 4, 8, 16, 32, \ldots\}$. This sequence has the general term 2^k, $k = 0, 1, \ldots$. From (3) with $T = 1$ we obtain

$$F(z) = \sum_{k=0}^{\infty} 2^k z^{-k}$$
$$= 1 + 2z^{-1} + 4z^{-2} + 8z^{-3} + 16z^{-4} + \cdots$$

which is an infinite geometric progression with radix $2z^{-1}$ and initial term = 1. The sum is $1/(1 - 2z^{-1}) = z/(z - 2)$.

Example 6.2. Use the final-value theorem to find $f(\infty)$ if $F(z) = z/(z - a)$, $|a| < 1$. We have

$$\lim_{k \to \infty} f(k) = \lim_{z \to 1} \frac{(1 - 1/z)z}{z - a}$$

$$= \lim_{z \to 1} \frac{z - 1}{z - a}$$

$$= 0, \quad |a| < 1$$

This is certainly an expected result since $f(k) = a^k U(k)$ which has $\lim_{k \to \infty} a^k U(k) =$ for $|a| < 1$.

It may be helpful to consider z as a shift operator that moves a time function one discrete step forward. This is obvious from the advance shifting property. An alternative useful concrete visualization of z is to consider z^{-1} as representing unit delay that can be easily realized in practice.

6.2.2 The Pulse Transfer Function

Given a dynamic element, the pulse transfer function $G(z)$ is defined by the relation

$$G(z) = \frac{Y(z)}{U(z)} \tag{13}$$

where $U(.)$ and $Y(.)$ are the input and output variables, respectively. Note that the function $G(z)$ relates the input and output signals only at discrete times separated by intervals T. The relationship between the continuous, sampled, and discrete signals is displayed in Fig. 6.2. Observe that $G(z) = Y_s(s)/U_s(s)$. In this regard, the absence or presence of a sampler in a signal line affects the Z transform of the system

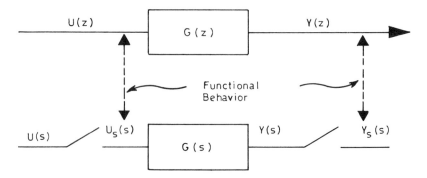

Figure 6.2 Transfer functions: continuous and discrete.

Methods of Analysis

very significantly. Some helpful rules that guide the manipulation in the z domain are:

1. It is always assumed in taking the Z transform of a time function that a sampler exists. If this is not the case, a fictitious sampler is introduced.
2. If $Y(s) = G(s)U(s)$, then it follows that

$$Y(z) = Y_s(s) = [G(s)U(s)]_s$$
$$\neq G_s(s)U_s(s) = G(z)U(z) \qquad (14)$$

3. Let $Y(s) = G(s)U_s(s)$, then

$$Y_s(s) = [G(s)U_s(s)]_s$$
$$= G_s(s)U_s(s)$$
$$= G(z)U(z) \qquad (15)$$

The difference between (14) and (15) should be always borne in mind. Intuitively, expression (14) illuminates a simple fact. Once two continuous functions have been combined by convolution, they must thenceforth be treated as an entity for purposes of Z transformation.

We shall demonstrate how one can obtain (15) for emphasis. Since

$$Y_s(s) = \sum_{k=-\infty}^{+\infty} Y(s + jk\omega_s)$$
$$= \sum_{k=-\infty}^{\infty} G(s + jk\omega_s)U_s(s + jk\omega_s)$$
$$= U_s(s) \sum_{k=-\infty}^{+\infty} G(s + jk\omega_s)$$
$$= U_s(s)G_s(s) \qquad (16)$$

which holds since a sampled function is periodic. The above analysis shows that if at least one of the continuous functions has been sampled before continuous convolution, then Z transformation of the product is equal to the product of the Z transformations. Since sampling an already sampled signal has no further effect, we have $[U_s(s)]_s = U_s(s)$.

As a consequence of the foregoing discussions, we derive below the pulse transfer function of distinct cases:

1. *Z transfer functions in cascade.* With reference to Fig. 6.3(b), we have

$$Y_s(s) = G_{2s}(s)X_s = G_{2s}(s)X_s(s)$$

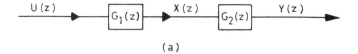

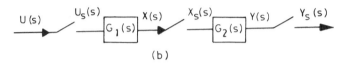

Figure 6.3 Z-transfer functions in cascade.

thus

$$\frac{Y_s(s)}{U_s} = \frac{Y(z)}{U(z)}$$

$$= G_2(z)G_1(z) \quad (17)$$

which is equivalent to Fig. 6.3(a).

2. *S transfer functions in cascade plus an input sampler.* With reference to Fig. 6.4, we have

$$\frac{Y_s(s)}{U_s(s)} = \frac{Y(z)}{U(z)}$$

$$= \{G_1(s)G_2(s)\}$$

$$\triangleq G_1G_2(z) \quad (18)$$

Again comparison of (17) and (18) is needed to emphasize the role played by the sampler.

3. *Z transfer functions in a feedback loop.* We consider Fig. 6.5(a) and write

$$\frac{Y(z)}{U(z)} = \frac{G(z)}{1 + G(z)H(z)} \quad (19)$$

Figure 6.4 S-transfer functions in cascade plus a sample at the input.

Methods of Analysis

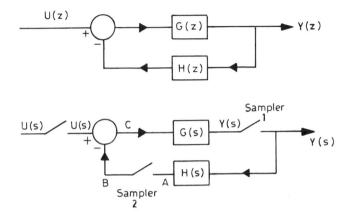

Figure 6.5 Z-transfer functions in a feedback loop.

which gives the input-output relationship of the closed-loop in a straightforward way. Now we consider Fig. 6.5(b) and analyze it step-by-step. The signal at point A is $H(s)Y_s(s)$ and due to the effect of sampler 2, at point B it becomes $[H(s)Y_s(s)]_s = H_s(s)Y_s(s)$. At point C, we have the signal $[U_s(s) - H_s(s)Y_s(s)]$ which when applied to $G(s)$ gives

$$Y(s) = G(s)[U_s(s) - H_s(s)Y_s(s)]$$

After sampler 1, we obtain

$$Y_s = [G(s)[U_s(s) - H_s(s)Y_s(s)]]_s$$
$$= G_s(s)U_s(s) - G_s(s)H_s(s)Y_s(s)$$

Thus

$$Y(z) = G(z)U(z) - G(z)H(z)Y(z)$$

and finally with simple rearrangement

$$\frac{Y(z)}{U(z)} = \frac{G(z)}{1 + G(z)H(z)} \tag{20}$$

which is identical with (19).

6.2.3 Inverse Transform

The formula for obtaining the time response $f(kT)$ from $F(z)$, called the *inverse Z transform*, is

$$f(kT) = \frac{1}{2\pi j} \int_c F(z) z^{k-1} \, dz \qquad (21)$$

where c is a circle in the complex plane that encircles all the singularities of $[F(z)z^{k-1}]$ or equivalently the loci of points where $F(z)z^{k-1} \to \infty$. For physical applications the function $F(z)$ in (21) is rational and causal. Under these restricted conditions, the expression is easily evaluated using the theory of residues in complex analysis. The *residue* of a function $F(s)$ at a particular pole p_j, is defined in terms of a Laurent expansion of the function in the region of p_j. The Laurent series can be written as

$$F(z) = a_{-k}(z - p_j)^{-k} + \cdots + a_{-1}(z - p_j)^{-1} + a_0 \\ + a_1(z - p_j) + \cdots \qquad (22)$$

then the residue of $f(z)$ at the pole p_j is defined to equal to a_{-j}. Complex variable theory states that when the poles of $F(z)$ are simple, then

$$\frac{1}{2\pi j} \int F(z) z^{k-1} \, dz = \sum_{m=1}^{n} r_m \qquad (23)$$

where r_m is the residue of the integrant at the mth pole and determined by

$$r_m = \lim_{z \to p_m} (z - p_m) F(z) z^{k-1} \qquad (24)$$

To illustrate (21)–(24) we consider the following example.

Example 6.3. Use $F(z) = z/(z-1)(z-4)(z-6)$ which has simple poles at $p_1 = 1$, $p_2 = 4$, $p_3 = 6$.

$$f(kt) = \sum_{m=1}^{3} \lim_{z \to p_j} \left[\frac{(z - p_j)z}{(z-1)(z-4)(z-6)} \right] z^{k-1}$$

$$= \frac{1}{15} - \frac{4^k}{6} + \frac{3}{5} 6^k$$

In the case of repeated roots, another formula is needed. Let the pole z_0 be of order m, then the residue at that pole is given by

$$r_0 \vert_{z_0} = \frac{1}{(m-1)!} \lim_{z \to z_0} \frac{d^{m-1}}{z^{m-1}} [(z - z_0)^m F(z) z^{k-1}] \qquad (25)$$

Alternative methods for Z-inverse transform include

Methods of Analysis

1. Partial fraction expansion of $F(z)/z$ followed by use of Z transform pairs.
2. Long division and then recognition of inverse transforms from tables.

These methods are illustrated in the following examples.

Example 6.4. Find the inverse transform of

$$F(z) = \frac{z}{z^2 - 6z + 8}$$

Applying the partial fraction method by writing

$$\frac{F(z)}{z} = \frac{1}{z^2 - 6z + 8}$$

$$= \frac{1}{(z-2)(z-4)}$$

$$= \frac{A}{z-2} + \frac{B}{z-4}$$

Thus

$$A = \frac{(z-2)F(z)}{z}\bigg|_{z=2} = \frac{1}{(z-4)}\bigg|_{z=2} = -\frac{1}{2}$$

$$B = \frac{(z-4)F(z)}{z}\bigg|_{z=4} = \frac{1}{(z-2)}\bigg|_{z=4} = \frac{1}{2}$$

Therefore

$$F(z) = -\frac{1}{2}\left(\frac{z}{z-2} - \frac{z}{z-4}\right)$$

From Table 6.1, we have

$$F(k) = -\frac{1}{2}(2^k - 4^k)U(k)$$

Example 6.5. Find the inverse transform of

$$F(z) = \frac{3z^2 - z}{(z-1)((z-z)^2}$$

This function has one simple pole at $p_1 = 1$ and double ($m = 2$) pole

at $p_2 = 2$. From (24), we obtain
$$r_1 = \lim_{z \to 1}(z-1)F(z)z^{k-1}$$
$$= \lim_{z \to 1} \frac{3z^2 - z}{(z-2)^2} = 2$$

Application of (25) with $m = 2$ gives
$$r_2 = \lim_{z \to 2} \frac{d}{dz}\left[\frac{(3z^2 - z)z^{k-1}}{z-1}\right]$$
$$= [5(k-1)2^{k-1} + 11(2^{k-1}) + 5(2^k)]U(k)$$
$$= [5k(2^{k-1}) - 2^{k+1}]U(k)$$

Thus
$$f(k) = r_1 + r_2$$
$$= [2 + 5k(2^{k-1}) - 2^{k+1}]U(k)$$

Example 6.6. It is desired to verify the results of Example 6.5 using long division. We have

$$\begin{array}{r} (3/z) + (14/z^2) + (46/z^3) + \cdots \\ z^3 - 5z^2 + 8z - 4 \overline{\big)\, 3z^2 - z } \\ 3z^2 - 15z + 24 + (12/z) \\ \hline 14z - 24 + (12/z) \\ 14z - 70 + (112/z) - (56/z^2) \\ \hline 46 - (100/z) + (56/z^2) \\ 46 \ldots \end{array}$$

Thus
$$F(z) = \frac{3}{z} + \frac{14}{z^2} + \frac{46}{z^3} + \cdots$$

Hence
$$f(k) = 3\delta(k-1) + 14\delta(k-2) + 46\delta(k-3) + \cdots$$

Comparing this result with the others in Example 6.5 for $k = 0, 1, 2, 3$ shows that they are exactly the same; that is $f(0) = 0$, $f(1) = 3$, $f(2) = 14$, $f(93) = 46$.

6.2.4 Poles and Zeros

It has been recognized from Part I that the utility of the s-plane in continuous-time control analysis and design is well established. We aim in this section to describe the basic material in the Z-plane for discrete-time systems. By presenting parallel development, we will establish useful correspondences between s-plane and z-plane representations.

Simply stated, we deal with the Z-plane as the standard complex number plane taking values $z = a + jb$. Define $G(z) = N(z)/D(z)$; then the poles and zeros are the complex-valued roots of the equations $D(z) = 0$ and $N(z) = 0$, respectively.

In view of the fact that $G(z) = G_s(s)$, and the pole-zero pattern of $G_s(s)$ can be displayed in the s-plane, then we can reach a solid relationship between the two planes. It is convenient to replace T by $2\pi/\omega_s$, where ω_s is the sampling frequency and, as usual we substitute $s = \sigma + j\omega$. As the poles and zeros of a transfer function $G(z)$ move in the Z-plane, so the corresponding poles and zeros of $G_s(s)$ move in the s-plane.

By construction, the relation $z = e^{e^T}$ provides a mapping from s to z, which we denote by R. Given any value of s, $R(s) = e^{e^T}$ accounts for the mapped value of z. Obviously, R^{-1} represents the inverse mapping from z to s. For $s = \sigma \pm j\omega$, $T = 2\pi/\omega_s$ we have

$$R(s) = R(\sigma \pm j\omega)$$
$$= e^{(\sigma \pm j\omega)(2\pi/\omega_s)}$$
$$= e^{2\pi\sigma/\omega_s} e^{\pm j(2\pi\omega/\omega_s)} \quad (26)$$

So that $R(s)$ denotes a vector in the z-plane with $|R| = e^{2\pi\sigma/\omega_s}$ and $LR = e^{\pm j2\pi\omega/\omega_s}$. Some specific relations are described below:

1. When $\sigma = 0$; then for any ω (represent the imaginary axis in the s-plane), $M(j\omega)$ accounts for the circumference of a unit circle. Thus the imaginary axis (as a demarcation line between stability and instability) is mapped into the unit circle (separates the stable modes from the unstable ones).
2. Let σ be fixed at $\sigma = \alpha > 0$ and let ω move from $\omega = 0$ to $\omega = \pm\omega_s/2$. The resulting values of $z = e^{2\pi\alpha/\omega_s} e^{j2\pi\omega/\omega_s}$ a circle of radius $r = e^{2\pi\alpha/\omega_s}$.
3. Let L be the lines of constant damping factor in the s-plane (see Fig. 6.6(a)). The damping factor $\xi = \cos\alpha$. The locus in the z-plane is generated by

$$z = \exp\left[-2\pi - \frac{\omega}{\omega_s Z}\tan\left(\frac{\pi}{2} - \cos^{-1}\xi\right)\right]\exp(\pm 2\pi\omega/\omega_s)$$

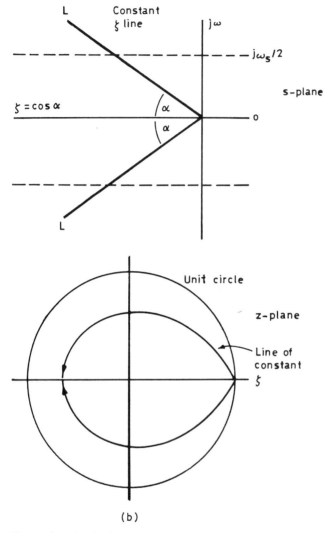

Figure 6.6 Loci of constant damping factor in the s- and z-planes.

as ω varies from $\omega = 0$ to $\omega = \pm\omega_s/2$. The resulting curve is a logarithmic spiral and shown in Fig. 6.6(b).

From the foregoing discussion, it is now possible to make sketches of the unforced (free) time response of a discrete system corresponding to particular pole locations. Figure 5.7 includes some of the relevant cases.

Methods of Analysis

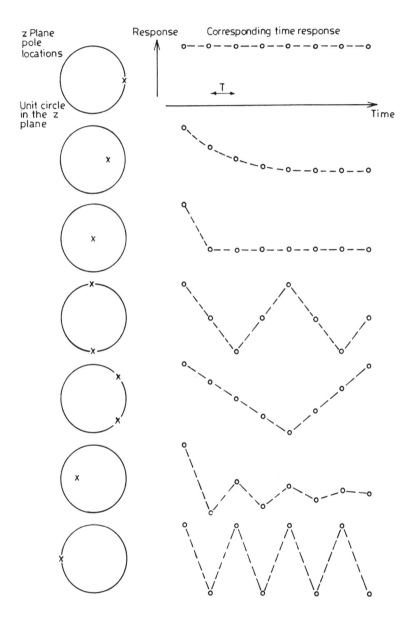

Figure 6.7 Some pole locations and the corresponding sequence.

6.3 DIFFERENCE EQUATIONS

Difference equations arise in their own right. In electrical, mechanical, and other systems, when a structure repeats itself one obtains difference equations. A difference equation generally takes the form

$$g[k, y(k), y(k + 1), \ldots, y(k + n)] = 0 \qquad (28)$$

A linear, shift-invariant form of (28) can be cast into

$$\sum_{m=0}^{N} a_m y(k + m) = x_f(k) = \sum_{m=0}^{M} b_m x(k + m) \qquad (29)$$

where $\{x(k)\}$ is an input sequence, $\{y(k)\}$ is an output sequence and $x_f(k)$ is the forcing function sequence. As we learned from the continuous case, the *complete solution* is obtained as the sum of the *complementary function* (homogeneous solution or zero-input response) and the *particular solution* (foerced solution or zero-state response). Specifically, the solution of

$$\sum_{m=0}^{N} a_m y_{zi}(k + m) = 0 \qquad (30)$$

gives $y_{zi}(k)$ as the zero-input response. By analogy to the continuous case, the zero-point response is found by assuming the form $y_{zi}(k) = z^k$ and substituting in (30) to obtain

$$\sum_{m=0}^{N} a_m z^{k+m} z^k \sum_{m=0}^{k} a_m z^m = 0$$

which specifies the *characteristic equation* that has N roots. For the class of shift-invariant, causal systems starting at $k = 0$, the zero-state response is

$$y_{zs}(k) = \sum_{m=0}^{N} x_f(m) g(k - m) \qquad (31)$$

where $g(.)$ is the unit function response (similar to the unit impulse response in the continuous case). We shall not elaborate on this method further since it is somehow classical, however, we shall use the Z transform as the basic tool.

6.3.1 Solution by the Z Transform

The Z transform method basically does for discrete systems what the Laplace (Fourier) transform does for continuous systems. The procedure involves:

Methods of Analysis

1. Z transformation of the difference equation using the transform properties
2. Manipulation of the transformed expression
3. Inverse Z transformation to yield the time solution

To link the three steps above with the components of the complete solution, we consider (29) again and apply the Z transform to obtain

$$\sum_{m=0}^{N} a_m z^m \left[Y(z) - \sum_{j=0}^{m-1} y(j) z^{-j} \right] = \sum_{m=0}^{M} b_m z^m \left[X(z) - \sum_{j=0}^{m-1} x(j) z^{-j} \right] \quad (32)$$

Notice that, as in the case of the Laplace transform, the initial conditions appear explicitly. Solving algebraically for $Y(z)$, we obtain

$$Y(z) = \frac{\sum_{m=0}^{M} b_m z^m X(z)}{\sum_{m=0}^{N} a_m z^m} - \frac{\sum_{m=0}^{M} b_m z^m \sum_{j=0}^{m-1} x(j) z^{-j}}{\sum_{m=0}^{N} a_m z^m}$$

$$+ \frac{\sum_{m=0}^{N} a_m z^m \sum_{j=0}^{m-1} y(j) z^{-j}}{\sum_{m=0}^{N} a_m z^m}$$

$$= Y_{zs}(z) + Y_{zi}(z) \quad (33)$$

Recall that the pulsed transfer function is defined by $Y_{zs}(z) = G(z)X(z)$ (when all initial conditions are zero), then from (33) we obtain

$$G(z) = \frac{\sum_{m=0}^{M} b_m z^m}{\sum_{m=0}^{N} a_m z^m} \quad (34)$$

We now illustrate the Z transform method by examples.

Example 6.7. Find the closed-form unit function response for the system that is modeled by

$$3y(k+2) + 9y(k+1) + 6y(k) = x(k+2)$$

Applying the Z transform with zero initial conditions we obtain

$$3z^2 Y(z) + 9z Y(z) + 6Y(z) = z^2 X(z)$$

from which the pulsed transfer function is deduced as

$$G(z) = \frac{Y(z)}{X(z)}$$

$$= \frac{z^2}{3z^2 + 9z + 6}$$

Since the degree of the numerator polynomial is the same as that of the denominator polynomial, a unit function $\delta(k)$ is present in the unit function response. That is, $G(z)$ may be written

$$G(z) = \frac{1}{3} - \frac{1}{3}\frac{3z+2}{z^2+3z+2}$$

$$= \frac{1}{3} - \frac{1}{3}\left[\frac{A}{z+1} + \frac{B}{z+2}\right]$$

$$= \frac{1}{3} - \frac{1}{3}\left[\frac{1}{z+1} - \frac{4/3}{z+2}\right]$$

Consequently, by inverse Z transformation we obtain the unit function response

$$g(k) = \frac{1}{3}\delta(k) + \frac{1}{3}[(-1)^{k-1} - 4(-2)^{k-1}]U(k-1)$$

Example 6.8. Find the response of the compound savings account system $[y(-1) = 0]$

$$y(k) = \left[1 + \frac{i}{n}\right]y(k-1) + x(k)$$

where i is the yearly interest rate, n is the number of compounding periods per year and $x(k)$ is the total deposits during the kth period. Take $i = 10\%$, $n = 2$ and consider the cases (1) $x(k) = \delta(k)$, (2) $x(k) = U(k)$, (3) $x(k) = kU(k)$.

Applying Z transform, we have

$$Y(z) - 1.05z^{-1}Y(z) - 1.05y(-1) = X(z)$$

Since $y(-1) = 0$, thus

$$Y_{zs}(z) = \frac{zX(z)}{z - 1.05}$$

1. $X(z) = 1$, $Y_{zs}(z) = z/(z - 1.05)$ which gives

$$y_{zs}(k) = (1.05)^k U(k)$$

2. $$X(z) = \frac{z}{(z-1)}, \quad Y(z) = \frac{z}{z-1}\frac{z}{z-1.05}$$

$$= 1 + \frac{20(1.05)^2}{(z-1.05)} - \frac{20}{z-1}$$

$$= \delta(k) + 20[(1.05)^{k+1} - 1]U(k-1)$$

Methods of Analysis

3.
$$X(z) = \frac{z}{(z-1)^2}$$

$$Y_{zs}(z) = \frac{z}{z - 1.05} \frac{z}{(z-1)^2}$$

$$= \frac{400(1.05)^2}{z - 1.05} - \frac{20}{(z-1)^2}$$

$$= \frac{400(1.1)}{z - 1}$$

which yields

$$y_z = 20[20(1.05)^{k+1} - k + 2]U(k-1)$$

Example 6.9. Find the zero-input response for the system

$$y(k) = y(k-1) - \frac{1}{4}y(k-2) + x(k)$$

This system has the standard form

$$y(k+2) - y(k+1) + \frac{1}{4}y(k) = x(k+2)$$

The characteristic equation is easily obtained

$$z^2 - z + \frac{1}{4} = 0 \quad \text{or} \quad \left(z - \frac{1}{2}\right)\left(z - \frac{1}{2}\right) = 0$$

We have a pair of roots at $z = 1/2$. The zero-input response is

$$y_{zi}(k) = d_1\left(\frac{1}{2}\right)^k + d_2\left(\frac{1}{2}\right)^k$$

at $k = 0$, $y_{zi}(0) = 0 = d_1 - 0$ give $d_1 = 0$ and at $k = 1$, $y_{zi}(1) = 1 = (1/2)d_1 + (1/2)d_2$ gives $d_2 = 2$. Thus

$$g(k) = 2(k-1)\left(\frac{1}{2}\right)^{k-1} U(k-1)$$

6.3.2 Numerical Solution

Recall the difference equation (29) and given the input sequence $x(k)$ for all k, we can rewrite it in the expanded form

$$a_N y(k+N) + a_{N-1} y(k+N-1) + \cdots + y(k)$$
$$= b_M x(k+M) + b_{M-1} x(k+M-1) + \cdots + x(k) \quad (35)$$

Define $\alpha_j = a_{N-j}/a_N$, $\beta_k = b_k/a_N$, and solve for $y(k+N)$

$$y(k+N) = -\alpha_1 y(k+N-1) - \alpha_2 y(k+N-2) - \cdots$$
$$-\alpha_N y(k) + \beta_M x(k+M) + \cdots$$
$$+ \beta_0 x(k) \quad (36)$$

Through direct substitution in (36) we can iteratively calculate $y(k)$ at all k. Such calculations are ideally suited for computer implementation. In general we will have the following array

k	$x(k)$	$x(k+1)$	$\cdots$	$x(k+M)$	$y(k)$	$x(yk+1)$	$\cdots$	$y(k+N)$
0								
1								
2								
3								
$\vdots$								

which calls for an illustration

Example 6.10. Obtain a numerical solution of

$$y(k+2) = x(k) + 5y(k+1) - 6y(k)$$

when $x(k) = 1$ for all k and zero initial conditions. The corresponding table is

k	$x(k)$	$y(k)$	$y(k-1)$	$y(k+2)$
0	1	0	0	1
1	1	0	1	6
2	1	1	6	25
2	1	1	6	25
3	1	6	25	90

Observe the upward shift between the columns of $y(k)$, $y(k+1)$, $y(k+2)$; a fact which simplifies the computer processing for high-order systems.

6.3.3 Computer Realization

Based on the fact that the quantity $y(k-1)$ is obtained from $y(k)$ by one-step delay, we can mechanize the implementation of (29) using

Methods of Analysis

three basic elements: multipliers (for the different coefficients), unit delay (or shift registers) and summers (to continue multi-inputs and produce one output). It will be convenient to start from the expression (36). There is a close resemblance between computer realization schemes of discrete-time systems and nonrecursive digital filters. The following example will illustrate the above concepts.

Example 6.11. Develop computer realization schemes for the discrete systems.

1. $y(k) = x(k) - 3x(k-1) + 2x(k-2) + x(k-3)$
2. $y(k) + 3y(k-1) + 2y(k-2) = -x(k) + 3x(k-2)$

The first system is a typical nonrecursive digital filter in which the input $x(k)$ and its delayed factors are combined to produce the output. The block diagram is given in Fig. 6.8. Of interest is to compute the unit function response.

$$g(k) = \delta(k) - 3\delta(k-1) + 2\delta(k-2) + \delta(k-3)$$

$k = 0 \quad g(0) = 1$
$k = 1 \quad g(1) = -3$
$k = 2 \quad g(2) = 2$
$k = 3 \quad g(3) = 1$
$k > 3 \quad g(k) = 0$

For the second system, delayed output sequences contribute to the present output. The result is presented in Fig. 6.9 and is obtained from

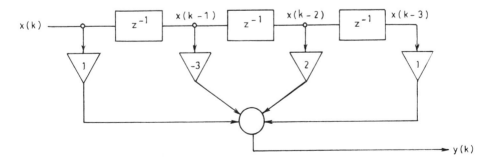

Figure 6.8 Block diagram of a nonrecursive filter.

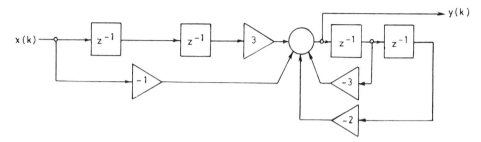

Figure 6.9 Realization of discrete system.

processing

$$y(k) = -x(k) + 3x(k-2) - 3y(k-1) - 2y(k-2)$$

More will be said on the realization aspects in the next sections.

6.4 DISCRETIZATION

The purpose of this section is to cover the basic material necessary for pursuing the design of single-input single-output (SISO) control loops of discrete (digital) systems. We shall draw heavily on the use of Z transform. The process of discretization is of fundamental importance both from a practical and a theoretical viewpoint, and consequently, several schemes will be developed to allow them to be evaluated in context.

With the introduction of digital devices in control system configurations, interests were shifted to discontinuous control of cotinuous-time systems. In such configurations, continuous signals are usually available at the input port. These include proper plant signals or particular variables. Most of the time, the output port delivers discrete-time commands (in response to plant signals) or discrete-time estimates of process performance (for a particular variable). Notice that in passing the signal from the input port to the output port, data-type converts and/or samplers are encountered.

6.4.1 Principles and Issues

From now onwards, we shall treat digital algorithms and digital simulations as one entity. This is generally valid except that digital algorithms are most likely implemented in real time whereas digital simulations operate on recorded data. In both schemes, time discretization is util-

Methods of Analysis

ized to mean the mechanism of converting a differential equation or continuous-time transfer function $G(s)$ to a difference equation or a discrete-time transfer function $G(z)$. Discretization turns out to be more interesting than might have been expected. We know that a continuous system is characterized either by its time response, frequency response or by pole-zero locations. None of the discretization methods to be presented preserves all three characteristics exactly; that is, there will always be some information lost.

The basis of discretization arises from the following consideration. Let $f(t)$ be a continuous function which is only known at equally spaced time instants, $f(1), f(2), f(3), \ldots$. suppose we wish to integrate $f(t)$ over a given interval T. Usually, we fit a polynomial $p(t)$ through a number of points $f(j), j \geq 1$, and then integrate $p(t)$. This corresponds to

$$\int_0^T f(t)\,dt \simeq \int_0^T p(t)\,dt \tag{37}$$

which implies that $p(t)$ is expressed in terms of $f(kT)$, $k = 0, 1, \ldots$ since $f(t)$ is only known at these points. In implementation, $p(t)$ in (37) is specified a priori. A zero-order polynomial would seem to be the simplest case. Define

$$v(k) = \int_{(k-1)T}^{KT} f(t)\,dt \tag{38}$$

Two versions are readily available:

2. $V(k) - V(k-1) = Tf(k-1)$, in which the approximation is pursued by a forward rectangle. By Z transformation we have

$$V(z) - z^{-1}V(z) = Tz^{-1}F(z)$$

Thus

$$\frac{V(z)}{F(z)} = \frac{Tz^{-1}}{1 - z^{-1}}$$

$$D_f(z) = \frac{T}{z - 1} \tag{39}$$

2. $V(k) - V(k-1) = Tf(k)$, in which the approximation is pursued by a backward rectangle. By Z transformation we have

$$V(z) - z^{-1}V(z) = TF(z)$$

Thus

$$\frac{V(z)}{F(z)} = \frac{T}{1-z^{-1}}$$

$$D_b = \frac{Tz}{z-1} \qquad (40)$$

Interpretation of (39) or (40) is the discretization of $f(t)$ by zero-order polynomial corresponds to processing $F(z)$ through the function $D(z)$ or $D_b(z)$. Observe the difference between both functions, the next section is exclusively devoted to the description of different methods of discretization.

6.4.2 Different Schemes

In the sequel, we will develop different forms of the function $D_j(z)$ relating $F(z)$ to $V(z)$ where the subscript j corresponds to the given scheme.

Trapezoidal Integration

In this scheme, $p(t)$ is approximated by a first-order polynomial connecting the ordinate $f(k-1)$ to the ordinate $f(k)$. This leads to the rule

$$V(k) - V(k-1) = \frac{T}{2}[f(k) + f(k-1)]$$

which implies that the area under integration is replaced by the product of the mid-ordinate and the time interval. By Z transformation we obtain

$$V(z) - z^{-1}V(z) = \frac{T}{2}[F(z) + z^{-1}F(z)]$$

and manipulating

$$\frac{V(z)}{F(z)} = D_t(z)$$

$$= \frac{T}{2}\left[\frac{1+z^{-1}}{1-z^{-1}}\right]$$

$$= \frac{T}{2}\left[\frac{z+1}{z-1}\right] \qquad (41)$$

The three approximation methods expressed by (39), (40), (41) will

Methods of Analysis

now be utilized in conjunction with other rules to accomplish the time discretization requirements.

Differential Mapping Methods

The idea of this method is to map a continuous-time system (described by a differential equation, or $G(s)$) into a discrete-time (expressed by a difference equation or $G(z)$) via one of the functions $D_f(z)$, $D_b(z)$, or $D_t(z)$. One of the obvious strategies is to use the forward difference approximation

$$\left.\frac{dy}{dt}\right|_{t=kT} = \frac{y(k+1) - y(k)}{T} \tag{42}$$

Based on (42), the discrete approximation

$$G_f(z) = G(s)|_{s=(z-1)/T} \tag{43}$$

is obtained. Another strategy would be the backward difference approximation

$$\left.\frac{dy}{dt}\right|_{t=kT} = \frac{y(k) - y(k-1)}{T} \tag{44}$$

which leads to the discrete approximation

$$G_b(z) = G(s)|_{s=(1-z^{-1})/T} \tag{45}$$

By the same token the trapezoidal approximation can be considered as a third strategy yielding

$$G_t(z) = G(s)|_{s=(2/T)[(z-1)/(z+1)]} \tag{46}$$

It is interesting to observe that (43), (45), and (46) are easily seen to be exactly equivalent to the three integration methods (39), (40), and (41). As mentioned earlier, the dynamic features of the resultant difference equations vary significantly and this will be explained in the solved examples.

Pole-Zero Mapping

We know that the solid relation between the continuous s-plane and the discrete Z-plane is given by $Z = e^{sT}$. A straightforward method would be to directly substitute this relation into $G(s)$ to hopefully produce $G(z)$. This mehtod suffers from the complicated mathematics involved. However, we can use the relation $Z = e^{sT}$ to map the poles and zeros of $G(s)$ into the Z-plane. The next step is to seek ways to synthesize a pulsed-transfer function $G(z)$ having the mapped pole-zero pattern and to provide a constant factor corresponding to the correct steady-state gain. Difficulties usually arise when dealing with the zeros. It is left as an open issue to look at in specific cases, as we shall see in some examples.

Truncated Series

Since $s = (1/T) \ln z$, we can expand it into series as

$$s = \frac{2}{T}\left[h + \frac{h^3}{3} + \frac{h^5}{5} + \cdots\right], \quad |h| < 1 \tag{47}$$

$$h = \frac{1 - z^{-1}}{(1 + z^{-1})} \tag{48}$$

A little algebra on (47) yields

$$s^{-1} = \frac{T}{2}\left(\frac{1}{h} - \frac{h}{3} - \frac{4h^3}{45} \cdots\right) \tag{49}$$

The series (49) can be truncated arbitrarily to meet practical constraints. Higher-order powers of $1/s$ are frequently needed and can be obtained from (49) with suitable approximation. It is suggested to use formulae

$$s^{-1} \approx \frac{T}{2h}$$

$$= \frac{T}{2}\frac{1 + z^{-1}}{1 - z^{-1}} \tag{50}$$

$$s^{-2} \approx \frac{T^2}{4}\left[\frac{1}{h^2} - \frac{2}{3}\right]$$

$$= \frac{T^2}{12}\left[\frac{1 + 10z^{-1} + z^{-1}}{(1 - z^{-1})^3}\right] \tag{51}$$

$$s^{-3} \approx \frac{T^3}{2}\left[\frac{z^{-1} + z^{-2}}{(1 - z^{-1})^3}\right] \tag{52}$$

Impulse-Invariant Method

The idea behind this mehtod is the fact that a perfect discretization scheme of $G(s)$ would be $G(z) = Z[GD(s)]$ where $D(s)$ is the dynamic element that helps in reconstructing the continuous signals from a finite number of samples. A practically useful form of $D(s)$ is the zero-order

Methods of Analysis

hold (ZOH) whose transfer function is

$$D(s) = \frac{1 - e^{-Ts}}{s} \qquad (53)$$

Therefore, we use

$$G(z) = Z\frac{(1 - e^{-Ts})G(s)}{s} \qquad (54)$$

as a candidate discretization scheme. Note that by this method the impulse response of both continuous and discrete systems is the same, hence the title of the method.

Step Response Matching

Given the sampled values $\{h(0), h(t), h(2T), \ldots\}$ of the continuous unit step response, then the Z transform gives $H(z) = \{h(0), h(T)z^{-1}, h(2T)z^{-2}, h(3T)z^{-3}, \ldots\}$. The Z transform $U(z)$ of a unit step applied at time kT, $k = 0$ is given by $U(z) = z/(z - 1)$. Therefore for matching the step response, the required discretized transfer function can be readily obtained from

$$\begin{aligned} G(z) &= \frac{H(z)}{U(z)} \\ &= \frac{z-1}{z}[0 + h(T)z^{-1} + h(2T)z^{-2} + \cdots] \end{aligned} \qquad (55)$$

which is a long sequence to be rearranged to yield a closed-form of $G(z)$. Note that the knowledge of $G(s)$ was not needed but rather the input-output data is required.

Trajectory Approximation

Instead of approximating the time derivatives by either of the formulae (43), (45), (46), we first obtain the solution of the differential equation. Then we approximate the resulting trajectory over each time step. The approach in principle is applicable to lower- or higher-order systems. However, it is only suitable for computer calculations.

This completes our discussions on the discretization schemes. Some illustrative examples follow.

Example 6.12. It is required to examine the discretization methods for the continuous system

$$G(s) = \frac{1}{s^2 + 0.4s + 0.68}$$

The system has poles $-0.2 \pm j0.8$ which is stable. Using (43) with $T = 1$ results in

$$G_f(z) = \frac{1}{(z-1)^2 + 0.4(z-1) + 0.68}$$

$$= \frac{1}{z^2 - 1.6z + 1.28}$$

which has poles at $0.8 \pm j0.8$. From (45), we obtain

$$G_b(z) = \frac{1}{(1-z^{-1})^2 + 0.4(1-z^{-1}) + 0.68}$$

$$= \frac{1}{z^{-2} - 2.4z^{-1} + 2.08}$$

which yields the poles at $0.576 \pm j0.3384$. Observe particularly that $G(s)$, $G_b(z)$ are stable transfer functions whereas $G_f(z)$ is not stable. Using the rule

$$s = \frac{2(z-1)}{z+1}$$

from (46), we obtain the pulsed transfer function $G_t(z)$

$$G_t(z) = \frac{1}{4(z-1)/(z+1)^2 + 0.4[(z-1)/(z+1)] + 0.68}$$

$$= \frac{z^2 + 2z + 1}{5.48z^2 - 6.64z + 3.88}$$

The values of the poles are found to be $0.606 \pm j0.584$ which means that $G_t(z)$ is a stable transfer function. It is easy to see that $G_t(z)$ yields quite accurate results in terms of pole positions.

Example 6.13. Given $G(s) = (s+1)/(s+2)(s+3)$ which has zero at -1 and two poles at -2 and -3. According to pole-zero mapping, we express $G(z)$ in the form

$$G(z) = \frac{(z+1)(z - e^{-T})}{(z - e^{-2T})(z - e^{-3T})}$$

where the term $z+1$ has been used to replace the zero at infinity.

Methods of Analysis

This transfer function implies that any change in input is transmitted instntaneously to the output since the orders of the numerator and denominator are equal.

Example 6.14. Solve the first-order differential equation

$$a_1(dy/dt) + a_0 y = x(t), \quad y(0) = y_0$$

numerically. Using the approximation (44) yields

$$\frac{a_1}{T}[y(k) - y(k-1)] + a_0 y(k) = x(k)$$

Alternatively we obtain

$$(a_0 T + a_1) y(k) - a_1 y(k-1) = Tx(k)$$

choosing $a_1 = 1$, $a_0 = 2$, $y(0) = 0$, $x(t) = $ unit step and $T = 0.5$ or 0.25, the results of simulating the continuous systems and the approximate discrete are plotted in Fig. 6.10. As expected, the smaller T gives the better approximation.

Example 6.15. Find the difference equation for the continuous system

$$\frac{d^2 y}{dt^2} + 6\frac{dy}{dt} + 5y = x(t)$$

using $T = 0.02$. For simplicity we employ

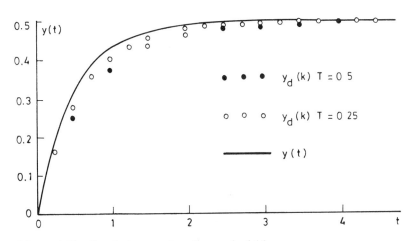

Figure 6.10 Simulation results of example 6.14.

$$\frac{dy}{dt} = \frac{y(k) - y(k-1)}{T}$$

$$\frac{d^2y}{dt^2} = \frac{y(k) - 2y(k-1) + y(k-2)}{T^2}$$

as the approximation scheme. Thus

$$2500[y(k) - 2y(k-1) + y(k-2)] + 300[y(k) - y(k-1)]$$
$$+ 5y(k) = x(k)$$

Rearranging

$$y(k) = \frac{1}{2805}[x(k) + 5300y(k-1) - 2500y(k-2)]$$

6.5 PROBLEMS

1. Derive Z transforms corresponding to the following Laplace functions
   a. $1/s^2$
   b. $1/(s+\alpha)(s+\beta)$
   c. $1/s(s+\alpha)^2$
   d. $ab/s(s+a)(s+b)$
2. Find the Z transform of
   a. $k^3 U(k)$
   b. $(a^k/k!)U(k)$
   c. $ka^k U(k)$
   d. $k^2 a^k U(k)$
   e. $U(k) - U(k-2)$
   f. $\sum_{m=0}^{n} \delta(m)$
3. A process of transfer function

$$G(s) = \frac{10}{s(s+\alpha)}$$

is preceded by a zero-order hold with sampling interval T. Determine the transfer function $G_d(z)$ of the combination. Investigate the effect of changing the sampling interval. Take $\alpha = 2$ and $T = 1$ as base values.

4. Show that for a periodic sequence $f(k)U(k) = f(k+N)U(k)$, the Z transform is $[z^N/(z^N - 1)]F_0(z)$ where $F_0(z) = \sum_{k=0}^{N-1} f(k)z^{-k}$ is the Z transform of the first cycle of $f(k)U(k)$.

Methods of Analysis

5. Find $f(k)$ for the following $F(z)$
   a. $z^2/(z - a)^2$
   b. $z(z - 2)/(z^2 - 4z + 16)$
   c. z^{-3}
   d. $Tz/(z - 1)^2$
   e. $z^2/(z^2 - 4z + 3)$
   f. $1/(z - 2)(z - 1)^2$
6. Use the final-value theorem to find $f(\infty)$ for
   a. 1
   b. $z/(z - 1)$
   c. $z/(z - a)$
   d. $1/(z - a)^m$
   e. $(z + 2)/(z - 2)^2$
7. Two numerical integrators of the type (46) are cascaded. What is the overall transfer function? What is the response to $x(k) = U(k)$?
8. A discrete system is described by $y(k) - 2y(k - 1) = 3x(k - 1)$
   a. Is it stable?
   b. A feedback element, $H(z)$, whose input-output is characterized by $y(k) - \alpha y(k - 1) = \beta x(k - 1)$, is added to stabilize the system. Find α and β so that the new system has a double pole at $z = 0.5$.

NOTES AND REFERENCES

The purpose of this chapter has been to provide an introduction to the fundamental tools and concepts of discrete-time systems. It covers the standard material on Z transform and difference equations. Then it goes on to stress the relationship between elements and operations in the s-plane and their counterpart in the Z-plane. In particular, the chapter addresses the important topic of discretization and related issues.

For further reading, it is suggested to consult Astrom and Wittenmark (1987), Iserman (1981), Astrom (1983), Leigh (1985) for basic and related topics. The classical work on sampled-data systems and their wide applications can be found in Jury (1958) and Ragazzini and Franklin (1958). A thorough treatment of small- and large-scale discrete dynamical systems has been covered by Mahmoud and Singh (1984). Digital control systems are discussed in depth by many authors including Kuo (1984), Franklin and Powell (1981). Reference should always be made to the pioneering work of Kalman and Bertram (1958, 1959) and Kalman and Kopcke (1958). The next chapter will deal with the design algorithms using frequency domain (Z domain) and the time domain (state-space) methods.

7
Design Algorithms

7.1 STABILITY TESTS

In the studies of linear feedback control systems with continuous plants, we have seen that the most important performance criterion of all is overall system stability. If a system is unstable, then to speak of its transient or steady-state response would be insignificant, since the system is virtually useless unless it is first stabilized by some means of compensation. This section is concerned with stability tests for linear discrete-time systems. We focus on methods that are unique to discrete systems.

Recall that a discrete-time system is stable if all the roots of the characteristics equation

$$F(z) = a_n z^n + a_{n-1} z^{n-1} + \cdots + a_2 z^2 + a_1 z + a_0 = 0 \quad (1)$$

lie inside the unit circle in the Z-plane. We present the Jury test for examining the stability of discrete-time systems of the type (1).

7.1.1 Jury Test

Referring to the polynomial of (1) with $a_n > 0$, the first step in the Jury test is the formulation of the following table using the coefficients of the polynomial:

Design Algorithms

Row	z^0	z^1	z^2	z^{n-k}	z^{n-1}	z^n
1	a_0	a_1	a_2	a_{n-k}	a_{n-1}	a_n
2	a_n	a_{n-1}	a_{n-2}	a_k	a_1	a_0
3	b_0	b_1	b_2	b_{n-k}	b_{n-1}	
4	b_{n-1}	b_{n-2}	b_{n-3}	b_k	b_0	
5	c_0	c_1	c_2	c_{n-2}		
6	c_{n-2}	c_{n-3}	c_{n-4}	c_0		
$2n-5$	p_0	p_1	p_2	p_3		
$2n-4$	p_3	p_2	p_1	p_0		
$2n-3$	q_0	q_1	q_2			

Note that the elements of row $2k + 2$ consist of the coefficients of the row $2k + 1$ written in reverse order. The elements in the table are defined as

$$b_k = \begin{vmatrix} a_0 & a_{n-k} \\ a_n & a_k \end{vmatrix}, \quad c_k = \begin{vmatrix} b_0 & b_{n-1-k} \\ b_{n-1} & b_k \end{vmatrix}$$

$$d_k = \begin{vmatrix} c_0 & c_{n-2-k} \\ c_{n-2} & c_k \end{vmatrix} \cdots$$

$$\cdots q_0 = \begin{vmatrix} p_0 & p_3 \\ p_3 & p_0 \end{vmatrix}, \quad q_2 = \begin{vmatrix} p_0 & p_1 \\ p_3 & p_2 \end{vmatrix}$$

The necessary and sufficient conditions for the polynomial $F(z)$ to have no roots on and outside the unit circle in the Z-plane are

1. $$F(1) > 0$$

2. $$F(-1) > 0, \quad n \text{ even}$$
$$< 0, \quad n \text{ odd}$$

3. $$\left. \begin{array}{c} |a_0| < a_n \\ |b_0| > |b_{n-1}| \\ |c_0| > |c_{n-2}| \\ |d_0| > |d_{n-3}| \\ \cdot \\ \cdot \\ \cdot \\ |q_0| > |q_2| \end{array} \right\} n - 1 \text{ constraints}$$

We now illustrate the Jury test by examples.

Example 7.1. The polynomial

$$z^3 + 3.5z^2 + 3.5z + 1 = (z + 1)(z + 0.5)(z + 2) = 0$$

is to be tested by Jury's method. We have $F(1) = 9 > 0$, but $F(-1) = 0$ which violates condition 2 that for $n = 3$, $F(-1)$ must be less than zero. There is no need to proceed further and we conclude that the polynomial has at least one root lying on or outside the unit circle. This is verified from the position of roots $(-0.5, -1, -2)$.

Example 7.2. Consider the polynomial

$$F(z) = z^4 - 1.368z^3 + 0.4z^2 + 0.08z + 0.002$$
$$= 0$$

Applying conditions 1 and 2 gives

$$F(1) = 0.114 > 0, \qquad F(-1) = 2.69 > 0$$

Since $n = 4$ is even, these conditions satisfy the first two requirements. Next, we tabulate the coefficients in the following manner.

Row	z^0	z^1	z^2	z^3	z^4
1	0.002	0.08	0.4	−1.368	1
2	1	−1.368	0.4	0.08	0.002
3	−1	1.368	−0.399	−0.083	
4	−0.083	−0.399	1.368	−1	
5	0.993	−1.385	0.512		

From the above tabulation, we have

$$|a_0| = 0.002 < a_4 = 1$$
$$|b_0| = 1 > |b_3| = 0.083$$
$$|c_0| = 0.993 > |c_2| = 0.512$$

Since all these constraints are satisfied, the system represented by the polynomial as its characteristic equation is stable.

It should be remarked that the Hurwitz or Routh tests can be used but the results would be incomplete since they would fail to indicate whether any root of (1) satisfies $|z| \leq 1$. Indeed direct methods include

1. Factorizing (1) and obtaining specific locations for the poles
2. Determining the positions of the roots using computer-implemented numerical schemes

The above methods do not need to be discussed further. We now turn to another method which transforms the problem into the S-plane (or an analog of the S-plane) and then applies the Hurwitz or Routh criterion.

7.1.2 The W-Plane Method

We have learned before that transformation of a polynomial in z into a polynomial in s could be achieved in theory at least by replacing z by e^{sT}. In practice it is much easier to use an alternative transformation. The mapping $P: z \to w$ described by

$$P(z) = \frac{1+w}{1-w} \qquad (2)$$

maps the exterior of the unit circle into the right half W-plane and maps the interior of the unit circle in the Z-plane onto the left half of the W-plane. Thus $|Z| > 1$ in the Z-plane implies that $w > 0$ in the w-plane and a stability test can be carried out on the transformed problem in the W-plane by methods usually used in the S-plane. Let us illustrate this by an example.

Example 7.3. To test the polynomial equation

$$z^4 - 2z^3 + 1.75z^2 - 0.75z + 1.125 = 0$$

for roots on or outside the unit circle. Set $z = (1+w)/(1-w)$ to produce the transformed polynomial

$$5.625w^4 + 6w^3 + 3.25w^2 + w + 0.125 = 0$$

The Routh-Hurwitz table can be put as

row 4	5.625	1	0.125
row 3	6	0.125	0
row 2	2.3125	0	
row 1	0.6757		
row 0	0.125		

Since there is no sign change in the first column, then no roots of the equation in w are in the right half-plane. This confirms that the original polynomial in z has all roots inside the unit circle. It helps to note that the roots of the original polynomial in z are at $z = 0.5, 0.5, 0.5 \pm j0.5$. As a closing remark to this section, we can say that (2) is a special case of the bilinear transformation

$$P(z) = \frac{aw+b}{cw+d} \qquad (3)$$

with $a = b = d = 1$ and $c = -1$. Sometimes (2) is termed the "w-transformation". Had we set $a = b = c = 1$ and $d = -1$ and replace w by r we would obtain

$$P(z) = \frac{r+1}{r-1} \tag{4}$$

which is called the "*r*-transformation". There is no real advantage gained by choosing one transformation over the other. We now move to the development of design algorithms for digital control.

7.2 ALGORITHMS OF DIGITAL CONTROLLER DESIGN

In this section, we are concerned with the following question. Given a continuous process (plant) equipped with adequate measurement and actuating devices and a digital computer closing the control loop (see Fig. 7.1), how is the computer to be programmed to achieve a particular control objective? We provide hereafter design algorithms to answer the above question. Implicit in the statement of the question is the fact that a sufficiently accurate mathematical model of the process is readily available. Otherwise, a modeling procedure should somehow be included.

7.2.1 Algorithm 1

Reference is made to Fig. 7.2 which shows a process modeled by a transfer function $G_p(s)$ and preceded by a hold device $G_h(s)$. This is the practice that we follow throughout. For simplicity, we employ a zero-order hold (ZOH) and define

$$\begin{aligned} G(s) &= G_h(s)G_p(s) \\ &= \frac{1 - e^{-sT}}{s} G_p(s) \end{aligned} \tag{5}$$

As the continuous-time transfer function of the combined process plus ZOH. Let $G_t(z)$ be the desired transfer function of the closed-loop system; thus

$$G_t(z) = \frac{D(z)G(z)}{1 + D(z)G(z)} \tag{6}$$

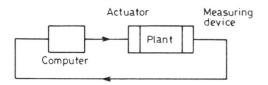

Figure 7.1 Closed-loop diagram control of continuous plant.

Design Algorithms

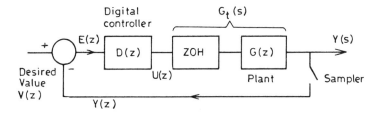

Figure 7.2 A continuous process with ZOH and digital controller $D(z)$.

where $D(z)$ is the transfer function of the digital controller which can be determined from (6) as

$$D(z) = \frac{G_t(z)}{1 - G_t(z)} \frac{1}{G(z)} \qquad (7)$$

For the important class of processes described by

$$G_p(s) = \frac{Ke^{-sT_1}}{1 + sT_2} \qquad (8)$$

and noting that

$$G(z) = Z\left[1 - e^{-sT} \frac{G_p(s)}{s}\right]$$

$$= (1 - z^{-1})Z\left[\frac{G_p(s)}{s}\right] \qquad (9)$$

with some manipulations we can rewrite (7) in the form

$$D(z) = \frac{G_t(z)}{1 - G_t(z)} \frac{z - e^{-T/T_2}}{Kz^{-T_1/T}(1 - e^{-T/T_2})} \qquad (10)$$

Some useful comments stand out:

1. Typical industrial processes are represented by (8) in which the model parameters are defined by:
   a. T_1 is the delay time.
   b. T_2 is the rise time.
   c. K is the steady-state response level for a unit step input.
2. The transfer function $G_t(z)$ must be chosen so that the controller transfer function $D(z)$ is realizable and stable. In practice, this involves avoiding attempts to cancel out either process dead-time or process zeros that are outside the unit circle in the Z-plane.

Any error in cancellation, however small, can lead to an unstable system.
3. A further restriction on $G_r(z)$ results from the fact that the controller $D(z)$ must not have a numerator whose order exceeds that of the denominator. To illustrate this, let $G(z) = 1/(z-1)(z-0.5)$ and $G_r(z) = 1/z$ then it follows from (7) that

$$D(z) = \frac{(z-1)(z-0.5)z^{-1}}{1-z^{-1}}$$

$$= z - 0.5 \qquad (11)$$

which is not realizable since it will require a knowledge of future values of the input. However, let $G_r(z) = 1/z^2$ and $G(z) = 1/(z-1)(z-0.5)$ then (7) gives

$$D(z) = \frac{(z-1)(z-0.5)z^{-2}}{1-z^{-2}}$$

$$= \frac{z-0.5}{z+1} \qquad (12)$$

which is a physically realizable controller.
4. In general, typical industrial control requirements have to be converted into a desired closed-loop transfer function $G_r(z)$. One should make use of the interrelationships between performance-related system parameters (such as rise time, settling time, overshoot, bandwidth) and design coefficients.

7.2.2 Algorithm 2

This algorithm is derived from a design approach based on digital approximation to the process open-loop step response. With reference to Fig. 7.3, define the sequence g_j, $j = 1, \ldots$ as the increments in the output response due to unit step input. It is clear that the sequence g_j converges so long as the response is monotonic and nonoscillatory. The response to a unit step input is given by

$$Y(z) = \frac{G_p(z)}{1-z^{-1}} \qquad (13)$$

where $G_p(z)$ is the process transfer function. But from the diagram, we have

$$Y(z) = g_1 z^{-1} + (g_1 + g_2) z^{-2} + \cdots$$

$$= \sum_{k=1}^{\infty} \sum_{j=1}^{k} g_j z^{-k} \qquad (14)$$

Design Algorithms

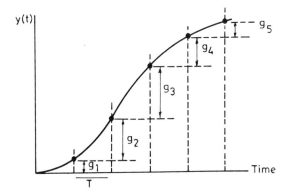

Figure 7.3 Open-loop step response.

Hence, from (13) and (14) we obtain

$$G_p(z) = (1 - z^{-1}) \sum_{k=1}^{\infty} \sum_{j=1}^{k} g_j z^{-k}$$

$$= \sum_{i=1}^{\infty} g_i z^{-i} \quad (15)$$

Seeking simplicity in the design we truncate (15) to yield

$$G(z) = \sum_{i=1}^{n-1} \frac{g_i z^{-i} + g_n z^{-n}}{1 - pz^{-1}} \quad (16)$$

where p is a parameter governing the rate of decay of the sequence $\{g_i\}$, $i = n, \ldots$ and is chosen so that the correct steady-state value is obtained for the output y. In terms of y_{ss}, the steady-state value of y for a unit step, the parameter p takes the form

$$p = 1 - \frac{g_n}{y_{ss} - \sum_{i=1}^{n-1} g_i} \quad (17)$$

A finite-time settling controller can be designed provided that the system response can be expressed as a finite polynomial in the operator z. The control input $u(k)$ can be put in the form

$$u(k) = K_I \sum_{j=0}^{k} [v(j) - y(j)] - \sum_{m=1}^{n} K_m x_m(k) \quad (18)$$

where x_m represents the mth element in the state vector of the process and

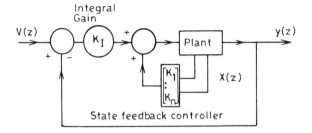

Figure 7.4 Implementation of finite settling time controller.

$$K_I = \frac{1}{\sum_{j=1}^{n} b_j} \tag{19}$$

$$\left.\begin{array}{l} b_j = g_j - pg_{j-1}, \quad j = 2, \ldots, n \\ b_1 = g_1, \quad K_1 = 0 \\ K_2 = K_3 = \cdots = K_{n-1} = K_I \\ K_n = \dfrac{1 + p - (\sum_{i=1}^{n-1} g_i / K_I)}{g_n} \end{array}\right\} \tag{20}$$

The implementation of the above algorithm is shown in Fig. 7.4.

7.2.3 Algorithm 3

Starting from the fundamental idea of Ziegler and Nichols (1942) that an open-loop stable process may be approximated by a first-order model in series with a dead-time element, we can write the continuous transfer function

$$G_p(s) = \frac{Ke^{-sT_1}}{1 + sT_2} \tag{21}$$

A three-term continuous controller (PID) is suggested in the form

$$D(s) = C\left(1 + \frac{1}{T_I s} + T_D s\right) \tag{22}$$

where

$$\left.\begin{array}{l} C = \dfrac{1.2 T_2}{K T_1}, \\ T_I = 2T_1 \\ T_D = 0.5 T_1 \end{array}\right\} \tag{23}$$

Design Algorithms

Using a two-term (PI) controller, the rule is

$$D(s) = C\left[1 + \frac{1}{T_I s}\right] \quad (24)$$

with

$$C = \frac{0.9 T_2}{K T_1} \quad (25)$$

In the proportional case (one-term controller)

$$D(s) = C$$

$$= \frac{T_2}{K T_1} \quad (26)$$

In any of the three cases (22), (24) or (26), we can use simple forward rectangular approximation to both differentiation and integration to obtain the discrete-time form. The three-term digital controller can be written as

$$u(k) = C\left\{e(k) + \frac{T}{T_I}\sum_{i=0}^{k} e(i) + \frac{T_D}{T}[e(k) - e(k-1)]\right\} + u(0) \quad (27)$$

Incrementing (27), we obtain the following algorithm

$$\Delta u(k) = u(k) - u(k-1)$$
$$= C\left\{[e(k) - e(k-1)]\right.$$
$$\left. + \frac{T}{T_I} e(k) + \frac{T_D}{T}[e(k) - 2e(k-1) + e(k-2)]\right\} \quad (28)$$

We make use of (27) when calculating the magnitude and of (28) when evaluating the gradient. Bearing in mind the great simplicity, (27) and (28) can be utilized interchangeably.

7.2.4 Algorithm 4

According to this algorithm, we first convert the Z-plane by the transformation

$$w = \frac{2}{T}\frac{z-1}{z+1} \quad (29)$$

Let ω_w be the imaginary part of w and ω be the actual frequency.

Hence we have the relation

$$\omega_n = \frac{2}{T} \tan \frac{wT}{2} \tag{30}$$

It is easy to see that for T in the range $0.1 \to 1$, ω_w is a good estimate of ω. We can now use (30) to transform a transfer function $G_p(z)$ into $G_p(s)$ and then design a cascade compensator using the Bode asymptotic plots.

To illustrate the above procedure, we consider a process described by $G(s) = 1/s^2$ preceded by ZOH such that

$$G_h G_p(s) = \frac{1 - e^{-sT}}{s} \frac{1}{s^2}$$

which leads to

$$G_h G_p(z) = \frac{T^2}{2} \frac{z + 1}{(z - 1)^2}$$

Using (29) in the form $z = (2 + Tw)/(2 - Tw)$ we arrive after simplification at

$$G_h G_p(w) = \frac{1 - (w/2T)}{w^2}$$

which is the desired transfer function before using the Bode plot plus lead compensator $(1 + wT_c)/(1 + \alpha w T_c)$, $\alpha < 1$, lag compensator $(1 + \alpha w T_c)/(1 + w T_c)$, $\alpha < 1$, or lag-lead compensator having the transfer function $[1 + (w/a)][1 + (w/c)]/[1 + (w/b)][1 + (w/d)]$, $b/a = c/d > 1$. Sometimes, it is required to enforce pole-zero cancellation to redistribute the pole-zero pair having close pole and zero positions (dipole) which would improve the steady-state gain without affecting the dynamic behavior.

7.2.5 Some Comments

By looking into the four algorithms, we can easily identify the following features:

1. Algorithm 1 is basically a continuous-time modeling to derive $G_p(s)$ followed by transformation to obtain $G_p(z)$ and then applying digital control design.
2. Digital modeling coupled with digital design is the main route of Algorithm 2.
3. A different procedure is pursued in Algorithm 3 in which continu-

Design Algorithms

ous-time modeling is followed by continuous controller design and then a discretization of the resulting system is performed.
4. Algorithm 4 is characterized by an entire design in the scaled w-plane.

It should be borne in mind that at the outset we determine $D(z)$ or equivalently the control signal $u(\ )$ is specified in terms of the error $e(\ )$. All the algorithms apply to unity feedback configuration.

7.2.6 On-Off Control

This is an algorithm which specifies the input control $u(k)$ in the form

$$u(k) = U_c, \quad e(k) > 0$$
$$u(k) = -U_c, \quad e(k) \leq 0 \tag{31}$$

where U_c is a preselected constant level and $e(k)$ is the error signal. Algorithm (31) is of switching type and can be entirely suitable for applications where long time constants smooth the effects of the rectangular waveforms that will be applied to the plant. It is very simple and most importantly it requires only an on-off actuator. In practice, the cost savings on the actuator are usually the biggest incentive to use the algorithm. Note that by definition, algorithm (31) is independent of the process dynamics. Some applications require a special version of (31) in which the controller is excited only for positive amplitudes of the error; that is $u(k) = U_c$ for $e(k) > 0$ and zero otherwise.

In the on-off control, the only parameter to consider is the magnitude U_c which is usually selected to avoid deriving the plant or other system elements into saturation or into the nonlinear region. Keeping U_c at relatively small values is recommended in practice.

Next, we present another design tool based on the root-locus.

7.3 THE ROOT-LOCUS DIAGRAM

It has been shown in Chapter 4 that the root-locus technique is quite useful and popular in the analysis and design of linear control systems with continuous data. The root-locus diagram of continuous systems is essentially a plot of the roots of the characteristic equation as a function of the loop gain K (K varies from 0 to ∞). Since the closed-loop Z transfer function $C(z)/R(z)$ is a rational function of z and the characteristic equation is an algebraic equation in z, therefore we can state that the construction of the Z-plane root-loci is a straightforward replica of the conventional root-locus diagram and all the rules are carried over in a similar fashion.

7.3.1 Transient Response and Relative Stability

Once the root-locus diagram of a given discrete system is constructed in the Z-plane, the absolute stability as well as the relative stability of the system can be observed from the root loci. *Absolute stability* requires that all roots of the characteristic equation lie inside the unit circle in the Z-plane. *Relative stability* is deduced from the behavior of the root-loci with respect to the constant damping factor, the constant frequency loci and the constant damping ratio loci.

The constant damping loci are a family of concentric circles centered at the origin of the Z-plane; the radius of the circle corresponding to a given damping factor σ_1 is $e^{\sigma_1 T}$, where T is the sampling period. The constant frequency loci are straight lines emanating from the origin of the Z-plane at angles of $\theta = \omega T$ radius measured from the positive real axis. These loci and their s-plane counterparts are shown in Figs. 7.5 and 7.6, respectively. For a constant damping ratio ξ, the constant ξ loci in the Z-plane are shown to be a family of logarithmic spirals, except for $\xi = 0$ and $\xi = 1$. Typical constant ξ loci for $\xi = 0.5$ are shown in Fig. 7.7 for the s-plane and the Z-plane.

7.3.2 Illustrative Examples

We present three examples to demonstrate the potential of the root-locus diagram.

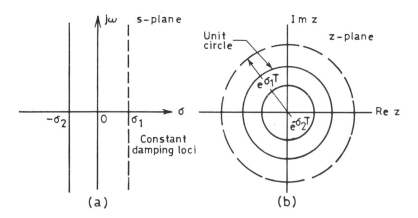

Figure 7.5 Constant damping loci.

Design Algorithms

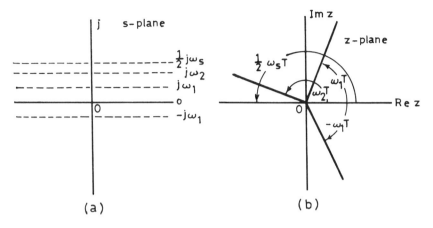

Figure 7.6 Constant frequency loci.

Example 7.4. Consider the closed-loop system with sampled-data shown in Fig. 7.8. The open-loop Z-transfer function of the system is given by

$$G_p = Z\left[\frac{K}{s(s+1)}\right]$$

$$= K\frac{z(1-e^{-T})}{(z-1)(z-e^{-T})} \quad (32)$$

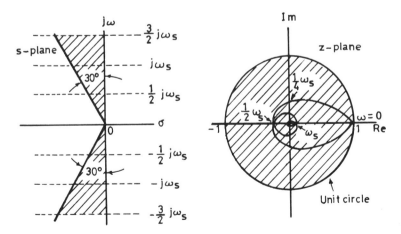

Figure 7.7 Constant damping ratio for $E = 0.5$.

Chapter 7

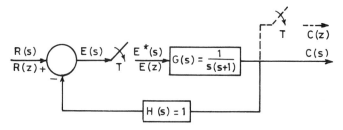

Figure 7.8 Closed-loop system of example 7.4.

Since the system has unity-feedback, $H(s) = 1$, the characteristic equation is

$$1 + G_p(z) = 0 \tag{33}$$

As we know, the root-loci of the system are the plots of the roots of (33) when K is varied from 0 to ∞. We shall consider the sampling period $T = 1$, then (32) reads

$$G_p(z) = 0.632K \frac{z}{(z-1)(z-0.368)} \tag{34}$$

$G_p(z)$ has two poles at $z = 1$, $z = 0.368$, and one zero at $z = 0$. We can proceed to construct the root-loci in the z-plane in the light of the rules presented earlier in Chapter 4. In summary we have the following:

1. These are two loci starting from $z = 0.368$ and $z = 1$. One of the loci terminates at $z = 0$, and the other locus goes to $-\infty$.
2. The parts of the real axis between (0.368 and 1) and between ($-\infty$ and the origin) constitute sections of the loci.
3. There are two breakaway points that can be easily shown to be located at $z = -0.62$ and $z = 0.62$.
4. Let $z = x + jy$, then from (34) we have

$$G_p(z) = 0.632K \frac{x + jy}{[(x-1) + jy][(x-0.368) + jy]}$$

From the angle criterion

$$\angle G_p(z) = \tan^{-1}\frac{y}{x} - \tan^{-1}\frac{y(2x - 1.368)}{(x-1)(x-0.368) - y^2}$$

$$= (2k + 1)\pi \tag{35}$$

On taking the tangent of both sides of (35) we obtain

Design Algorithms

$$\tan \angle G_p(z) = \tan(2k+1)\pi = 0$$

$$= \frac{W_1}{W_2} \tag{36}$$

where

$$W_1 = \frac{y}{x} - \frac{y(2x - 1.368)}{(x-1)(x-0.368) - y^2} \tag{37}$$

$$W_2 = 1 + \frac{y}{x} W_1 \tag{38}$$

Simplifying (36)–(38) we reach

$$x^2 + y^2 = 0.368 \tag{39}$$

which is an equation of a circle with the center at the origin and radius 0.62.

The root-loci plot is shown in Fig. 7.9. Note that by the magnitude criterion we obtain $K = 4.32$ at $z = -1$.

Had we changed the sampling period such that $T = 2$, the same

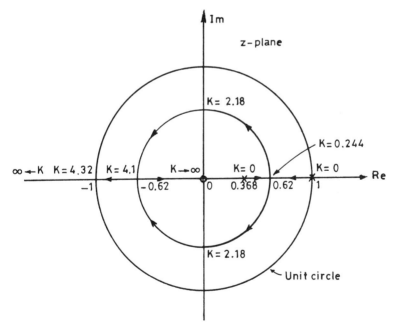

Figure 7.9 Root locus plot of example 7.4.

pattern of root-loci is obtained and (39) becomes $x^2 + y^2 = 0.135$. In general, the equation of a circle takes the form $x^2 + y^2 = e^{-T}$ for arbitrary T.

Example 7.5. This example examines the effect of including a ZOH unit in the previous system. The new system configuration is shown in Fig. 7.10. It is easy to see that the open-loop transfer function is given by

$$G_h(z)G_p(z) = Z\left[\frac{(1-e^{-Ts})K}{s^2(s+1)}\right]$$

$$= K(1-z^{-1})\left[\frac{Tz}{(z-1)^2} - \frac{z}{z-1} + \frac{z}{z-e^{-T}}\right]$$

$$= \frac{K[(T-1+e^{-T})z - (Te^{-T} - 1 + e^{-T})]}{(z-1)(z-e^{-T})}$$

$$= \frac{K(T-1+e^{-T})(z+a)}{(z-1)(z-e^{-T})} \quad (40)$$

where

$$a = \frac{1 - e^{-T} - Te^{-T}}{T - 1 + e^{-T}} \quad (41)$$

The open-loop Z transfer function $G_h(z)G_p(z)$ has two poles at $z = 1$ and $z = e^{-T}$ and one zero at $z = -a$. It is readily seen that the root-loci in the Z-plane must start from $z = 1$ and $z = e^{-T}$ and one of the loci terminates at $z = -a$, whereas the other locus terminates at $-\infty$. Proceeding further, the real part of the root-loci lies in the section between $z = e^{-T}$ and $z = 1$ and between $-\infty$ and $z = -a$ on the real

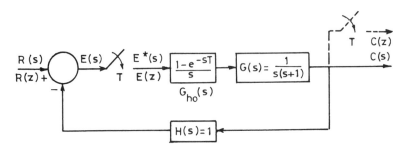

Figure 7.10 Closed-loop system of example 7.5.

Design Algorithms

axis. The complex conjugate section of the loci is obtained by substituting $z = x + jy$ and (40) into the phase angle requirement and simplifying to yield

$$(x + a)^2 + y^2 = a^2 + a(a + 1)e^{-T} \qquad (42)$$

which represents a circle of center at $(-a, 0)$ and radius of $\sqrt{a^2 + a + (a + 1)e^{-T}}$. For $T = 2$, $a = 0.524$ and the radius is 1.0022. The complete root-loci is depicted in Fig. 7.11. We take note that for the system with ZOH, the root-loci may intersect the unit circle at complex conjugate points rather than on the real axis as in the case when the ZOH is absent. The critical value of K for stability is 1.45.

Example 7.6. We consider the system shown in Fig. 7.10 and analyzed in the previous example. For $T = 1$ the root-locus plot is constructed and displayed in Fig. 7.12, along with some constant loci. Inspection of the root plot shows that the locus intersects the $\xi = 0.5$ locus at the point that corresponds to $K = 0.49$. Therefore, if the forward gain of the system is 0.49, the damping ratio of the second-order system is equal to 50%. This result is significant, since in this particular system the forward transfer function between the sampler and the output

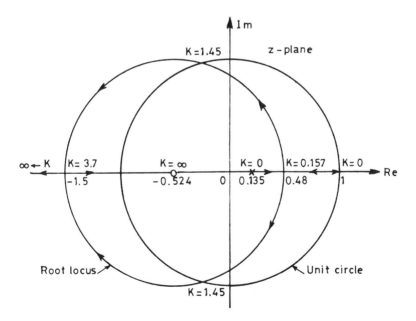

Figure 7.11 Root locus plot example 7.5.

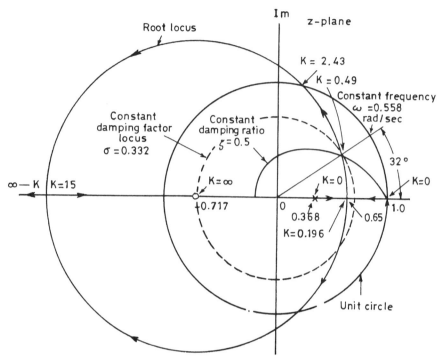

Figure 7.12 Root locus plot with different loci.

provides sufficient attention at frequencies higher than $\omega_s/2 = \pi$, so that only the constant ξ locus in the primary strip is necessary. By use of the constant damping and the constant frequency loci, the damping factor and ω of the system when $K = 0.49$ are found to be 0.332 and 0.558, respectively.

This example provides some knowledge about the relevance of using the root-locus as a control design method. The constant loci determine desirable regions in which selected poles can be positioned to satisfy design requirements. Modification of the plot can be made if necessary to meet all conditions on system performance parameters.

Now going back to the digital controller $D(z)$, which can generally be put in the form

$$D(z) = \frac{E_0(z)}{E_i(z)}$$

$$= \frac{\sum_{k=0}^{n} a_k z^{-k}}{\sum_{j=0}^{m} b_j z^{-j}} \tag{43}$$

Design Algorithms

where $b_0 \neq 0$, m and n are positive integers and $E_0(z)$ and $E_i(z)$ represent the Z transforms of the output and the input signal, respectively. In the previous sections, it has been implicitly assumed that once $D(z)$ is obtained it can be constructed by a combination of ZOH unit and passive networks. This is particularly useful when conversion onto the w-plane is employed. In the next section, we look into methods to realize (43) using digital components.

7.4 DESIGN REALIZATION

A control algorithm represented by $D(z)$ must be realized in the digital computer in terms of unit delays (shift registers) and constant multipliers. In principle, a particular transfer function has a number of alternative realizations. From a mathematical standpoint, the alternative realizations are completely equivalent but, considered as alternative programming approaches, they differ in general in terms of ease of programming, computational efficiency, noise transmission, and sensitivity to parameters errors.

7.4.1 Direct Programming

We start from (43), cross-multiply and take the inverse Z transform to yield

$$b_0 e_2(t) - \sum_{k=1}^{m} b_k e_2(t - kT) = \sum_{j=0}^{n} a_j e_1(t - jT) \tag{44}$$

where T is the sampling period and $\{e_2(t), E_0(z)\}$, $\{e_1(t), E_i(z)\}$ are Z transform pairs. Solving (44) for $e_2(t)$ gives

$$e_2(t) = \frac{1}{b_0} \sum_{j=0}^{n} a_j e_1(t - jT) - \sum_{k=1}^{m} b_k e_2(t - kT) \tag{45}$$

It is evident from (45) that the present value of the output $e_2(t)$ depends on the present and the past information of the input and also on the past information of the output signal.

To implement (45) by a digital computer program, two basic mathematical operations are required. The first operation is the *data shift* in which the sample at output is a one-sampling period delay of the input.

The second operation involves arithmetic manipulation of either the input or output through multiplication by constants, addition, and subtraction. The block diagram representation of the direct programming to realize (45) is shown in Fig. 7.13 in which it is clear that

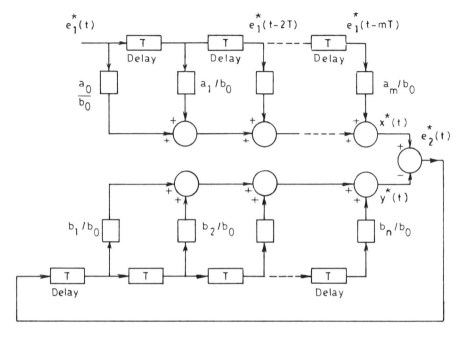

Figure 7.13 Block diagram for direct programming realization.

$$e_2(t) = x(t) - y(t) \tag{46}$$

defines the output signal as the difference between two channels where

$$x(t) = \frac{1}{b_0} \sum_{j=0}^{n} a_j e_1(t - jT) \tag{47}$$

$$y(t) = \frac{1}{b_0} \sum_{k=1}^{m} b_k e_2(t - kT) \tag{48}$$

represent the input channel and output channel, respectively. Note the use of a delay element (data shifter) and different weighting factors are instrumental in generating the output samples.

7.4.2 Cascade Programming

A complex pulse transfer function $D(z)$ may be conveniently factorized and written as the product of a number of simple pulse transfer functions, each realizable by a simple digital program. In this case, realization of $D(z)$ may then be represented by the series of cascaded digital

Design Algorithms

programs of the simple pulse transfer functions. Writing (43) in factored form, we have

$$D(z) = \frac{E_2(z)}{E_1(z)}$$

$$= \frac{a_0 \prod_{k=1}^{m}(1 + c_k z^{-1})}{b_0 \prod_{j=1}^{n}(1 + d_j z^{-1})} \quad (49)$$

where $-c_k$ and $-d_j$ are the zeros and the poles of $D(z)$, respectively. For proper transfer functions $n > m$ and $D(z)$ may be written as the product of n simple pulse transfer functions

$$D(z) = \prod_{k=1}^{n} D_k(z) \quad (50)$$

where

$$D_1(z) = \frac{a_0(1 + c_1 z^{-1})}{b_0(1 + d_1 z^{-1})} \quad (51)$$

$$D_k(z) = \frac{1 + c_k z^{-1}}{1 + d_k z^{-1}}, \quad k = 2, 3, \ldots, m \quad (52)$$

$$D_k(z) = \frac{1}{1 + d_k z^{-1}}, \quad k = m+1, \ldots, n \quad (53)$$

The digital program of $D(z)$ is shown in block form in Fig. 7.14(a) as the series connection of n simple digital programs. The schematic diagram of the digital program of $D_j(z)$, $j = 2, 3, \ldots, m$, is illustrated in Fig. 7.14(b).

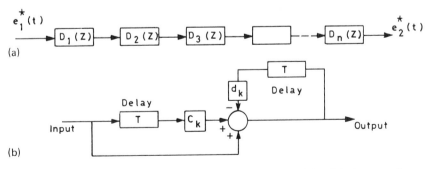

Figure 7.14 (a) Realization by cascade programming. (b) Digital processing of $D_k(z)$.

7.4.3 Parallel Programming

An alternative programming method to realize $D(z)$ is through parallel combination of a number of simple digital programs. This type of programming involves taking the partial fraction expansion of $D(z)$. We rewrite (43) in the form

$$D(z) = \frac{1}{b_0} \sum_{k=1}^{n} \frac{B_k}{1 - d_k z^{-1}} \qquad (54)$$

where, for simplicity, it is assumed that $D(z)$ has only simple poles. Like cascade programming, we express (54) as

$$D(z) = \sum_{j=1}^{n} D_j(z) \qquad (55)$$

$$D_j(z) = \frac{1}{b_0} \frac{B_j}{1 - d_j z^{-1}} \qquad (56)$$

Therefore, $D(z)$ can be realized by the n simple digital programs for $D_1(z), D_2(z), \ldots, D_n(z)$ operating in parallel. The resulting interconnection is shown in Fig. 7.15(a) and the individual digital program is depicted in Fig. 7.15(b).

7.4.4 Examples

In what follows, we present some examples to illustrate the realization methods.

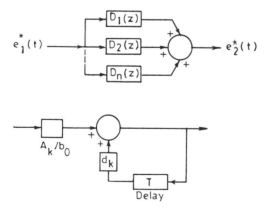

Figure 7.15 (a) Realization by parallel programming. (b) Digital processing of $D_k = (1/b_0) A_k / (1 - d_k z^{-1})$.

Design Algorithms

Example 7.7. Consider the digital controller

$$D(z) = \frac{1 - 0.6z^{-1}}{1 - 0.34z^{-1}}$$

$$= \frac{E_2(z)}{E_1(z)}$$

which is of the type (43). Cross-multiplying and taking the Z transform yields

$$e_2(t) = e_1(t) - 0.6e_1(t - T) + 0.34e_2(t - T)$$

which is similar to (45) with $n = m = 1$. The block diagram of digital computer programming is shown in Fig. 7.16, which is of the direct form.

Example 7.8. Let

$$D(z) = \frac{z + 2}{z^2 + 6z + 5}$$

This can be expressed in line of parallel programming as

$$D(z) = \frac{z + 2}{(z + 1)(z + 5)}$$

In terms of partial fractions expansion

$$D(z) = \frac{0.25}{z + 1} + \frac{0.75}{z + 5}$$

$$= \frac{E_2(z)}{E_1(z)}$$

The digital realization Fig. 7.17, follows immediately.

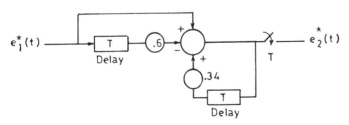

Figure 7.16 Realization of pulse-transfer function.

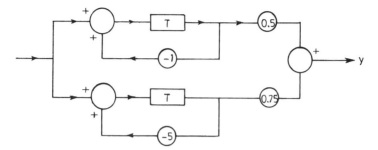

Figure 7.17 Realization by parallel programming.

Example 7.9. To illustrate cascade programming, consider

$$D(z) = \frac{1}{z^3 + 0.8z^2 + 0.16z + 0.01}$$

$$= \frac{1}{(z + 0.1)(z + 0.2)(z + 0.5)}$$

which corresponds to (49) with $m = 0$, $n = 3$, $d_1 = 10$, $d_2 = 5$, $d_3 = 2$, and $b_0 = 0.01$. $D(z)$ can then be represented as a sequence of elements as in Fig. 7.18.

Example 7.10. A discrete-time, unity feedback control system with digital compensation is shown in Fig. 7.19. The controlled process of the system has the transfer function

$$G_p(s) = \frac{K}{s(s + 1)}$$

Let the sampling period $T = 1$. Simple calculation shows that the open-loop transfer function of the system without compensation is

$$G_h G_p(z) = 0.368K \frac{z + 0.717}{(z - 1)(z - 0.368)}$$

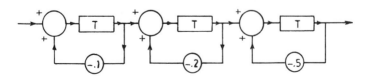

Figure 7.18 Realization by the factorization method.

Design Algorithms

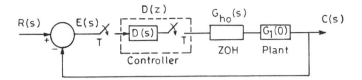

Figure 7.19 A unity feedback control system of example 7.10.

The root-locus diagram of the uncompensated system is plotted in Fig. 7.20 from which we note that the complex conjugate part of the root loci is a circle with center at $z = -0.717$ and a radius of 1.37. For all values of $k > 2.43$, the closed-loop system becomes unstable. Let us assume that K is set at this marginal value so that the two characteristic roots are on the unit circle as shown in Fig. 7.20.

Suppose that the digital controller $D(z)$ is of the phase-lead type

$$D(z) = \frac{\alpha\tau + 1}{\tau + 1} \frac{z + (1 - \alpha\tau)/(1 + \alpha\tau)}{z + (1 - \tau)/(1 + \tau)}$$

where $\alpha > 1$, $\tau > 0$. The constant factor $(1 + \alpha\tau)/(1 + \tau)$ in $D(z)$ is necessary, since the insertion of the digital controller should not affect the velocity error constant K_v while improving the stability of the system. In other words, $D(z)$ must satisfy the condition $\lim_{z \to 1} D(z) = 1$.

Our objective is to determine α and τ so that the system in Fig. 7.19 is stable. We knew before that the addition of an open-loop zero has the effect of pulling the root-loci toward it, whereas an additional open-loop pole has a tendency to push the loci away. A reasonable

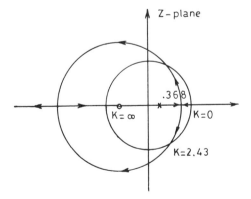

Figure 7.20 Root locus diagram of example 7.10.

compromise is $\alpha = 2.0$ and $\tau = 0.4$ so that the digital controller reads

$$D(z) = 1.286 \, \frac{z + 0.111}{z + 0.429}$$

and the open-loop transfer function of the compensated system is

$$D(z)G_hG_p(z) = 0.473K \, \frac{(z + 0.111)(z + 0.717)}{(z - 1)(z - 0.368)}$$

which is stable. Similar result is obtained when $\alpha = 3$ and $\tau = 0.1$.

We observe that the damping ratios are less than 10% which would yield 70% overshoot. The system is slightly improved but changing τ would result in an unstable system. Since the original system is on the verge of instability, the phase-lead compensation is generally ineffective. We would then try phase-lag compensation with the transfer function

$$D(z) = \frac{1 + \alpha\tau}{1 + \tau} \, \frac{z + (1 - \alpha\tau)(1 + \alpha\tau)}{z + (1 - \tau)/(1 + \tau)}$$

with $\alpha < 1$.

From experience, phase-lag compensation is effective when τ is large. Selecting $\tau = 100$ and $\alpha = 0.2$ gives

$$D(z) = 0.191 \, \frac{z - 0.905}{z - 0.98}$$

from which the compensated system is now described by

$$D(z)G_hG_p(z) = 0.07K \, \frac{(z - 0.905)(z + 0.717)}{(z - 1)(z - 0.368)(z - 0.98)} \tag{57}$$

which is well-damped and stable.

Example 7.11. In Section 7.24, it was shown that the process $G(s) = 1/s^2$ cascaded by a ZOH unit and both transformed in the scaled w-plane has the combined transfer function

$$G_hG_p(w) = \frac{[1 - (w/2T)]}{w^2}$$

with $T = 10$.

We set $w = j\omega$ to obtain the complex transfer function

$$G_hG_p(\omega) = \frac{j\omega}{2T} - \frac{1}{\omega^2}$$

Design Algorithms

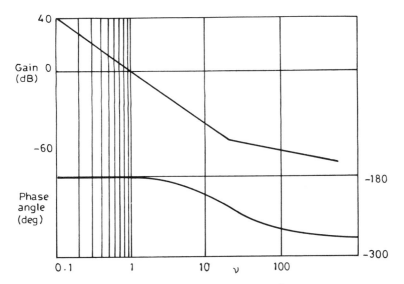

Figure 7.21 Bode plots for $G(jv) = [jv120 - 1]1v^2$.

the Bode plot of which is given in Fig. 7.21. The plot indicates that severe stability problems can be expected unless a controller is included. It is proposed to use a lead-compensator to help in increasing the phase angle in the region where $\omega < 2$. Thus the proposed compensator is expressed as

$$D(\omega) = 0.316 \frac{1 - (j\omega/0.3)}{1 + (j\omega/3)}$$

and the compensated system has the Bode plot in Fig. 7.22. Note that the factor 0.316 which is equivalent to -10 dB would yield sufficient phase margin. Transforming back to the Z-plane yields with $T = 10$

$$D(z) = 0.49 \frac{z + 0.2}{z + 0.875}$$

which is the desired digital controller.

Example 7.12. A feedback control system is shown in Fig. 7.23 where

$$G_p(s) = \frac{K}{s(1 + 0.1s)(1 + 0.5s)}$$

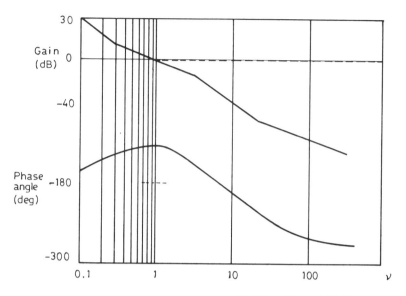

Figure 7.22 Bode plots for $G(jv)D(jv) = \dfrac{[jv/20 - 1][1 + jv10.3]}{3.16v[1 + jv/3]}$

The sampling period $T = 0.5$. It is desired to design a digital controller for the unity feedback system to meet the following specifications

$K_v \geq 1.5$, Phase margin $\geq 50°$, Resonant peak $M_p \leq 1.3$.

The open-loop transfer function of the uncompensated system is given by

$$G_h(s)G_p(s) = \left[\frac{1 - e^{-Ts}}{s}\right]\frac{K}{s(1 + 1.0s)(1 + 0.5s)}$$

The Z transform of $G_h(s)G_p(s)$ with $T = 0.5$ is

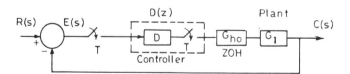

Figure 7.23 A feedback control system of example 7.12.

Design Algorithms

$$G_h G_p(z) = 0.13K \frac{(z + 1.31)(z + 0.54)}{z(z - 1)(z - 0.368)}$$

Applying the final-value theorem of Z transform, we found that for $K_v = 1.5$, the gain $K = 3$. On substituting $z = (1 + w)/(1 - w)$ and $K = 3$ yields

$$G_h G_p(w) = 0.75 \frac{(1 - w)(1 + 0.9w)(1 - 0.134w)}{w(1 + w)(1 + 2.17w)}$$

In Fig. 7.24, the Bode plot is presented for the uncompensated system, from which it is clear that the system is critically stable with zero phase margin. To obtain a phase margin of 50° while maintaining the velocity

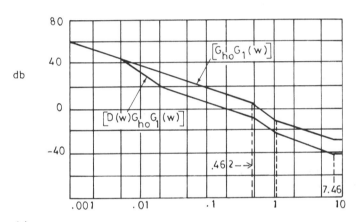

(a)

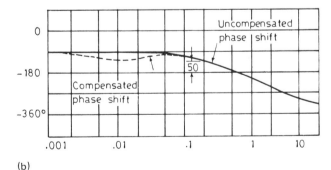

(b)

Figure 7.24 Bode diagram of example 7.12 (a) Magnitude plot. (b) Phase plot.

constant K_v at 1.5, it is suggested to use a phase-lag compensator. The transfer function of the compensator has the form

$$D(w) = \frac{1 + \alpha \tau w}{1 + \tau w}$$

To realize the required phase margin, the gain crossover of the system should be shifted downward from $\omega_\phi = 0.6$ to $\omega_\phi = 0.2$. The Bode plot of the uncompensated system at $\omega_\phi = 0.2$ has a gain of 12 dB. Therefore, the phase-lag transfer function $D(w)$ must provide some attention such that

$$\alpha = 10^{-12/20} = 0.25$$

Choosing the upper corner frequency of $D(w)$ at one decade below $\omega_\phi = 0.2$, we have $1/\alpha\tau = 0.02$ and thus $1/\tau = 0.005$. The digital controller is then described by

$$D(w) = \frac{1 + 50w}{1 + 100w}$$

which leads to the open-loop of the compensated system

$$D(w)G_h G_p(w) = 0.75 \frac{(1 + 50w)(1 - w)(1 + 0.9w)(1 - 0.314w)}{w(1 + 100w)(1 + w)(1 + 2.17w)}$$

The Bode plot of the compensated system is now plotted in Fig. 7.24, from which it is clear that the system performance is satisfactory, where

$$M_p = 1.2$$
$$\text{Phase margin} = 50°$$
$$\text{Gain margin} = 12 \text{ dB}$$
$$\text{Bandwidth} = 1.5 \text{rad/s}$$

Substituting $w = (z - 1)/(z + 1)$ we finally reach

$$D(z) = 0.25 \frac{(z - 0.96)}{(z - 0.99)}$$

which is the desired transfer function of the digital controller.

7.5 PROBLEMS

1. A process G is modeled by

$$y(k + 1) + 0.4y(k) = u(k)$$

Design Algorithms

and is preceded by a compensator $D(z)$. It is required that the open-loop poles of $G(z)D(z)$ be at $z = \pm 0.5j$. Find a $D(z)$ to achieve this. Sketch the step response of both $G(z)$ and $G(z)D(z)$.

2. The forward path of a unity feedback configuration has $G(z) = 0.2(z + 0.8)/(z - 1)(z - 0.6)$ in cascade with a compensator $D(z)$. Design $D(z)$ such that the closed-loop system is stable and has zero steady-state position error to a ramp input signal.

3. The transfer function of a unity feedback system

$$G(s) = \frac{K}{s(10 + s)(2 + s)}$$

is preceded by ZOH unit and digital controller $D(z)$ with $T = 0.5$. Develop an expression of the compensated system when $D(z)$ is of the phase-lead form. Sketch the root-locus diagram, with the following cases:
   a. $\alpha = 0$, $0 < \tau < \infty$
   b. $0 < \alpha < \infty$, $0 < \tau < \infty$

4. The fixed parts of a discrete-time system with ZOH is described by

$$G_h G_p(s) = \frac{1 - e^{-Ts}}{s^2(s + 1)}$$

The sampling period is 0.1. Design a digital controller meet the following specifications:
   a. $\xi = 0.707$
   b. $T_{max} \leq 0.3$
   c. Overshoot 10%
   d. $K_v \geq 5$

5. Repeat Problem 4 but with the requirement to achieve dead-beat response when the input is a unit step function.

6. A process has the transfer function $G(z) = Ke^{-sT}/s$. When put into closed-loop with unity feedback, the system oscillates at 0.01 Hz. Determine T. Suggest a procedure for a three-term controller to control the process.

7. Derive an expression of a digital controller $D(z)$ on series with a ZOH unit and a process having

$$G_p(s) = \frac{Ke^{-sT_1}}{1 + sT_2}$$

and a closed-loop system $H(z)$ within a unity feedback. Find $D(z)$ such that the resulting response is dead-beat to a step input.

8. A unity feedback system has open-loop transfer function

$$G_p(z) = 0.736 \frac{z + 0.716}{(z - 1)(z - 0.368)}$$

Assume that the sampling period $T = 1$. Sketch the Bode plots in the W-plane and hence determine the gain and phase margins of the system.

9. Sketch realizations of the algorithm

$$D(z) + \frac{z + 1}{(z - 1)(z^2 - 1.1z + 0.3)}$$

using the three approaches, direct, cascade, and parallel programming.

10. A process described by

$$G_p(s) = \frac{2}{(s + 0.5)(s^2 + 2s + 4)}$$

is put in closed-loop using either a proportional controller or a proportional-integral controller. Use the Ziegler–Nichols settings to determine the coefficients of the controllers.

NOTES AND REFERENCES

Given the fact that the principle of compensation of discrete-time systems is quite similar to that of systems with continuous data, this chapter has outlined methods for control design using continuous-time (analog) networks. However, due to the sampling operation, the design does present more problems and difficulties than the continuous case. Hence, we examined the use of pulsed-data network or digital controllers and further discussed realization schemes. We have given particular attention to the use of the Bode plot in the scaled w-plane and root-locus in the Z-plane as basic design tools. Several examples were given to illustrate the various concepts introduced. For further details one can consult standard works on sampled-data systems: Ziegler and Nichols (1942), Takahashi et al. (1970), and Mori (1957). Updated treatment of related subjects can be found in Goff (1966), Jury (1958), Astrom and Wittenmark (1984), Kuo (1982), and Kalman and Bertam (1959). On the application of digital control in typical case studies, the reader is referred to Ragazzani and Franklin (1958), Andrews (1982), Woolvet (1979), Uronen and Yliniemi (1977), Unbehauen et al. (1976), and Ortegar (1982).

8
Optimal Design Methods

8.1 INTRODUCTION

In the previous chapter the control synthesis problem was solved using different design algorithms. The main design parameters have been the compensator coefficients, and the presentations have been limited to single-input single-output systems. In this chapter a more general control problem is discussed. The plant is still assumed to be linear, but it may be time-varying and have several inputs and outputs. The control problem is solved using an alternative technique which is based on optimal control theory.

The problem of optimal control has received a great deal of attention during the past decade owing to increasing demand for systems of high performance and to the ready availability of digital computers. Optimal control methods are generally attractive because they handle multivariable systems easily and allow the designer to quickly determine many good candidate values for the feedback control matrix.

The synthesis problem is formulated hereafter to minimize a criterion, which is a quadratic function of the status and the control signals. It will be shown that the resulting optimal controller is linear. Properties of such a controller will be revealed.

8.2 LINEAR QUADRATIC CONTROL

We shall study the dynamic optimization problem in which it is required to select a sequence of control variables (or controller) which produce, via a set of difference equations, the sequence of state variables and at the same time to minimize a prescribed cost functional.

8.2.1 Problem Formulation

The design problem is specified in terms of the plant model, the cost functional, and the admissible control signals.

The Plant Model

It is assumed that the process to be controlled is described by the state model

$$\frac{d\mathbf{x}}{dt} = A\mathbf{x} + B\mathbf{u} \tag{1}$$

where A and B may be time-varying matrices. The model (1) can be sampled with the input $\mathbf{u}(t)$ being constant over the sampling period. Thus, the solution of (1) can be put in the form

$$\mathbf{x}(t) = \Phi(t, kh)\,\mathbf{x}(kh) + \Psi(t, kh)\,\mathbf{u}(kh) \tag{2}$$

where $\Phi(t, kh)$ is the fundamental matrix of (1) satisfying

$$\frac{d\Phi(t, kh)}{dt} = A(t)\,\Phi(t, kh), \qquad \Phi(t, t) = I \tag{3}$$

and

$$\Psi(t, kh) = \int_{kh}^{t} \Phi(t, T)\, B(T)\, dT \tag{4}$$

For simplicity the time arguments are dropped to yield

$$\mathbf{x}(k + 1) = \Phi\mathbf{x}(k) + \Psi\mathbf{u}(k) \tag{5}$$

which is the desired state-space model. We take note that (5) could be obtained directly in some typical applications.

The Cost Functional

The control objective is to minimize the cost functional

$$J = \int_{0}^{Nh} [\mathbf{x}^t(t)\, Q_1\, \mathbf{x}(t) + 2\mathbf{x}^t(t)\, S_1 \mathbf{u}(t) + \mathbf{u}^t\, R_1 \mathbf{u}(t)]\, dt \tag{6}$$

Optimal Design Methods

where the matrices Q_1, R_1, and S_1 are symmetric and positive-definite. The cost functional (6) is expressed in continuous time. It is now transformed into a discrete-time cost functional to suite the model (5) for digital control applications. Integrating (6) over intervals of length h gives

$$J = \sum_{k=0}^{N-1} J(k) \tag{7}$$

where

$$J(k) = \int_{kh}^{kh+h} [\mathbf{x}^t(t) Q_1 \mathbf{x}(t) + 2\mathbf{x}^t(t) S_1 \mathbf{u}(t) + \mathbf{u}^t(t) R_1 \mathbf{u}(t)] \, dt \tag{8}$$

Bearing in mind that $\mathbf{u}(t)$ is constant over the sampling period and using (2) in (8), we arrive at

$$J(k) = \mathbf{x}^t(kh) Q\mathbf{x}(kh) + 2\mathbf{x}^t(kh) S\mathbf{u}(kh) + \mathbf{u}^t(kh) R\mathbf{u}(kh) \tag{9}$$

where

$$Q = \int_{kh}^{kh+h} \Phi^t(r, kh) Q_1 \Phi(r, kh) \, dr \tag{10}$$

$$S = \int_{kh}^{kh+h} \Phi^t(r, kh) [Q_1 \Psi(r, kh) + S_1] \, dr \tag{11}$$

$$R = \int_{kh}^{kh+h} [\Psi^t(r, kh) Q_1 \Psi(r, kh) + 2\Psi^t(r, kh) S + R_1] \, dr \tag{12}$$

Minimizing the cost functional (6) when the control $\mathbf{u}(kh)$ is constant over the sampling period is thus minimizing the discrete-time cost functional

$$J = \sum_{k=0}^{N-1} [\mathbf{x}^t(k) Q\mathbf{x}(k) + 2\mathbf{x}^t(k) S\mathbf{u}(k) + \mathbf{u}^t(k) R\mathbf{u}(k)] \tag{13}$$

where the matrices Q, S, R are given by (10)–(12), respectively. In the sequel, we consider that $Q \geq 0$, $R > 0$, and $S = 0$. If $S \neq 0$, we can still eliminate it by introducing a new control

$$\mathbf{v}(k) = \mathbf{u}(k) + [SR^{-1}]^t \mathbf{x}(k) \tag{14}$$

so that the system (5) will become

$$\mathbf{x}(k+1) = \{\Phi - \Psi[SR^{-1}]^t\} \mathbf{x}(k) + \Psi \mathbf{v}(k) \tag{15}$$

and the cost functional (13) in this case will be

$$J = \sum_{k=0}^{N-1} [\mathbf{x}^t(k)\{Q - SR^{-1}S^t\}\mathbf{x}(k) + \mathbf{v}^t(k) R\mathbf{v}(k)] \qquad (16)$$

That the cross product between states and control is eliminated in (14) and (15) is now obvious. Therefore the use of (5) and (13) with $S = 0$ corresponds to the use of (15) and (16) for a specified nonzero value of S.

Admissible Control Laws

We consider the sampling process to be periodic with the control signal constant over the sampling periods. When the state variables are all measurable, we will have the case of *complete state information*. Otherwise, we will have the case of *incomplete state information* in which some of the state variables are not known exactly. In the former, we will have the control law based upon the states whereas in the latter, it will be based on the outputs.

The Problem

The optimal control problem is now defined to be determining the admissible control signal that minimizes the cost functional (12) when the plant model is characterized by (5). The design parameters are the matrices in the cost functional and the sampling period. This is frequently called *the linear-quadratic (LQ)-control problem*.

8.2.2 Derivation of the Optimal Sequence

Given the initial state $\mathbf{x}(0) = \mathbf{x}_0$, we solve the LQ control problem by the method of Lagrange multipliers. In this method, there will be a Lagrange multiplier vector, $\boldsymbol{\lambda}(k + 1)$, for each value of k. The procedure is to adjoin (5) to J in (12) with the multiplier sequences as

$$J_a = \sum_{k=0}^{N-1} \{\mathbf{x}^t(k) Q\mathbf{x}(k) + \mathbf{u}^t(k) R\mathbf{u}(k) + \boldsymbol{\lambda}^t(k+1)$$
$$\times [\Phi \mathbf{x}(k) + \Psi \mathbf{u}(k) - \mathbf{x}(k+1)]\} \qquad (17)$$

and attempt to find the minimum of J_a which will correspond to that of J, with respect to $\mathbf{x}(k)$, $\mathbf{u}(k)$, and $\boldsymbol{\lambda}(k)$. Note that the subscript of $\boldsymbol{\lambda}$ is normally arbitrary, but we will take it to be $k + 1$ because this choice will yield a particularly easy form of the equations later on. For

Optimal Design Methods

convenience, we define a scalar sequence, often called the Hamiltonian, $H(k)$

$$H(k) = \mathbf{x}'(k)\, Q\mathbf{x}(k) + \mathbf{u}'(k)\, R\mathbf{u}(k) + \boldsymbol{\lambda}'(k+1)[\Phi \mathbf{x}(k) + \Psi \mathbf{u}(k)] \quad (18)$$

The use of (18) in (17) and changing the indices of summation yields

$$J_a = -\boldsymbol{\lambda}'(N)\mathbf{x}(N) + \sum_{k=1}^{N-1}[H(k) - \boldsymbol{\lambda}'(k)\mathbf{x}(k)] + H(0) \quad (19)$$

The necessary conditions of optimality could be obtained by setting the total differential change in J_a to zero. The result is, for $k = 0, 1, \ldots, N-1$

$$\mathbf{x}(k+1) = \Phi \mathbf{x}(k) + \Psi \mathbf{u}(k), \qquad \mathbf{x}(0) = \mathbf{x}_0 \quad (20)$$

$$\boldsymbol{\lambda}(k) = \Phi'\boldsymbol{\lambda}(k+1) + Q\mathbf{x}(k), \qquad \boldsymbol{\lambda}(N) = \mathbf{0} \quad (21)$$

$$R\mathbf{u}(k) + \Psi'\boldsymbol{\lambda}(k+1) = \mathbf{0} \quad (22)$$

From (21) we obtain

$$\mathbf{u}(k) = -R^{-1}\Psi'\boldsymbol{\lambda}(k+1) \quad (23)$$

Seeking to derive the optimal control law, we consider (20), (21), (23) and introduce the transformation

$$\boldsymbol{\lambda}(k) = P(k)\,\mathbf{x}(k) \quad (24)$$

and substitute it into (22) with the aid of (20) to yield

$$\mathbf{u}(k) = R^{-1}\Psi' P(k+1)\,\mathbf{x}(k+1)$$

$$= -R^{-1}\Psi' P(k+1)[\Phi \mathbf{x}(k) + \Psi \mathbf{u}(k)]$$

Solving for $\mathbf{u}(k)$ we obtain

$$\mathbf{u}(k) = -L^{-1}(k)\Psi' P(k+1)\Phi \mathbf{x}(k)$$

$$= -G(k)\,\mathbf{x}(k) \quad (25)$$

where

$$L(k) = [R + \Psi' P(k+1)\,\Psi] \quad (26)$$

$$G(k) = L^{-1}(k)\,\Psi' P(k+1)\,\Phi \quad (27)$$

Proceeding further, we substitute (24) into (21) to arrive at

$$P(k)\mathbf{x}(k) = \Phi' P(k+1)\,\mathbf{x}(k+1) + Q\mathbf{x}(k) \quad (28)$$

which, upon using (20), reduces to

$$P(k)\,\mathbf{x}(k) = \Phi' P(k+1)[\Phi\,\mathbf{x}(k) + \Psi \mathbf{u}(k)] + Q\mathbf{x}(k) \quad (29)$$

Next we use (25) for $\mathbf{u}(k)$ in (29) to obtain

$$P(k)\mathbf{x}(k) = \Phi' P(k+1)\Phi \mathbf{x}(k) + Q\mathbf{x}(k)$$
$$- \Phi' P(k+1)\Psi L^{-1}(k)\Psi' P(k+1)\Phi \mathbf{x}(k) \qquad (30)$$

Since (30) must hold for arbitrary $\mathbf{x}(k)$, it follows that

$$P(k) = \Phi' P(k+1)[I - \Psi L^{-1}(k)\Psi' P(k+1)]\Phi + Q \qquad (31)$$

which describes a backward difference equation in $P(k)$. From (21) and (24), the end condition on the sequence $\{p(k)\}$ is obtained as

$$P(N) = 0 \qquad (32)$$

This completes the derviation of the optimal sequences for the LQ-control problem. Let us summarize the entire procedure:

1. Set k (iteration index) $= N$, $P(k) = 0$, and $G(k) = 0$
2. Compute $G(k-1)$ from (26), (27) and store it.
3. Solve (31) for $P(k-1)$. If $k = 0$, stop. Otherise set $k-1 \leftarrow k$ and go back to Step 2.

When the iteration stops, we use the stored gains $\{G(k)\}$ together with (25) in (20) to determine the optimal state sequence. $P(k)$ is often called *the discrete Riccati matrix*, the basic properties of which are examined in the next section.

8.2.3 Some Properties

One should initially observe that the optimal feedback gains $\{G(k)\}$ are time-varying; however, they are independent of the initial state $\mathbf{x}_0$. Perhaps it would be informative at this stage to evaluate the optimal value of the cost function J. To do this, we consider (16) under the validity of (20), (21), (22), and $S = 0$. By letting

$$\boldsymbol{\lambda}'(K+1)\Phi = \boldsymbol{\lambda}'(k) - \mathbf{x}'(k)Q\mathbf{x}(k)$$

from (21) and

$$\boldsymbol{\lambda}'(k+1)\Psi = -\mathbf{u}'(k)R$$

from (22), then (17) reduces to

$$J_a^* = \sum_{k=0}^{N-1} [\boldsymbol{\lambda}'(k)\mathbf{x}(k) - \boldsymbol{\lambda}'(k+1)\mathbf{x}(k+1)]$$

which upon iteration gives

$$J_a^* = \mathbf{x}_0' P(0) \mathbf{x}_0 \qquad (33)$$

This shows that the optimal cost functional depends only on the initial state and initial Riccati matrix.

Optimal Design Methods

By transposing (31), it is easily verified that $P(k) = P^t(k)$, which implies that the Riccati matrix is symmetric. This fact is usually exploited in the numerical computation to reduce the number of unknowns from n^2 (a square matrix of order n) to $1/2n(n+1)$ (a symmetric square matrix of order n).

It should be emphasized that the state feedback controller (25) is linear. This control law is applicable to multivariable and time-varying systems. By changing the relative magnitude between the elements in the weighting matrices, it is easy to compromise between the speed of the recovery and the magnitude of the control signals. With the assumptions on the system properties (reachable and detectable) and the cost functional (symmetric and positive-definite), the resulting LQ controller will always yield a stable closed-loop system. This means that the closed-loop system governed by

$$\mathbf{x}(k+1) = [\Phi - \Psi G(k)] \mathbf{x}(k) \tag{34}$$

has eigenvalues located within the unit circle in the complex plane.

In the optimal control of linear plants with infinite-time horizon, steady-state solution of (31) exists and a constant Riccati matrix P is obtained ($N \to \infty$). From (25) we obtain

$$\begin{aligned} \mathbf{u}(k) &= -G\mathbf{x}(k) \\ &= -[R + \Psi^t P \Psi]^{-1} \Psi^t P \Phi \, \mathbf{x}(k) \end{aligned} \tag{35}$$

where P is the solution of the algebraic Riccati equation

$$\begin{aligned} P &= \Phi^t P[I + \Psi R^{-1} \Psi^t P]^{-1} \Phi + Q \\ &= \Phi^t [P^{-1} + \Psi R^{-1} \Psi^t]^{-1} \Phi + Q \end{aligned} \tag{36}$$

The solution of (36) can be achieved by eigenvalue-eigenvector methods. On taking the Z transform of (20)–(23) we obtain

$$zX(z) - z\mathbf{x}_0 = \Phi X(z) - \Psi R^{-1} \Psi^t [z\Omega(z) - z\mathbf{p}_0],$$
$$\Omega(z) = \Phi^t [z\Omega(z) - z\mathbf{p}_0] + QX(z)$$

where $X(z)$ and $\Omega(z)$ are the Z transforms of $\mathbf{x}$ and $\boldsymbol{\lambda}$, respectively. Solving for $X(z)$ and $\Omega(z)$, we obtain

$$\begin{bmatrix} X(z) \\ \Omega(z) \end{bmatrix} = \begin{bmatrix} zI - \Phi & z\Psi R^{-1}\Psi^t \\ -z^{-1}Q & z^{-1}I - \Phi^t \end{bmatrix}^{-1} \begin{bmatrix} z\mathbf{x}_0 - z\Psi R^{-1}\Psi^t \mathbf{p}_0 \\ -\Phi^t \mathbf{p}_0 \end{bmatrix} \tag{37}$$

Careful examination of (37) indicates that the components of $X(z)$ and $\Omega(z)$ are rational functions in z except at singular points given by

$$\det\begin{bmatrix} zI - \Phi' & z\Psi R^{-1}\Psi' \\ -z^{-1}Q & z^{-1}I - \Phi' \end{bmatrix} = 0 \tag{38}$$

which can equivalently be stated as

$$\det[zI - \Phi]\det[z^{-1}I - \Phi' + Q(zI - \Phi)^{-1}\Psi R^{-1}\Psi'] = 0 \tag{39}$$

It is readily evident from (39) that it is the product of two polynomials, one in z and the other in z^{-1}. This implies that the eigenvalues of (39) are formed by the pairs $(z_j, 1/z_j)$, $j = 1, \ldots, n$. In view of the fact that the closed-loop system (34) is asymptotically stable, then n eigenvalues of (39) having module strictly less than one are characteristic value of (34) and the remaining n characteristic values are greater than one. Next we present illustrative examples.

8.2.4 Examples

Two examples will be worked out: the first is second-order and the second is fifth-order.

Example 8.1. A system of the type (5) and (12) with

$$\Phi = \begin{bmatrix} 0 & 1 \\ 0 & 0 \end{bmatrix}, \quad \Psi = \begin{bmatrix} 0 \\ 1.4142 \end{bmatrix}$$

$$Q = \begin{bmatrix} 1 & -1 \\ -1 & 1 \end{bmatrix}, \quad R = 1$$

$$S = 0, \quad N \to \infty$$

Direct solution of (36) gives the steady-state Riccati matrix

$$P = \begin{bmatrix} 1 & -1 \\ -1 & 1.5 \end{bmatrix} \tag{40}$$

for which the closed-loop system matrix (34) has the value

$$[\Phi - \Psi G] = \begin{bmatrix} 0 & 1 \\ 0 & 0.5 \end{bmatrix}$$

whose eigenvalues $(0, 0.5)$ are within the unit circle as expected. The gain matrix G is constant in this case and is given by

$$G = [0 \quad 0.3536]$$

Example 8.2. A discrete model of a steam power system can be reduced to (5) with

Optimal Design Methods

$$\Phi = \begin{bmatrix} 0.915 & 0.051 & 0.038 & 0.015 & 0.038 \\ -0.03 & 0.889 & -0.0005 & 0.046 & 0.111 \\ -0.006 & 0.468 & 0.247 & 0.014 & 0.048 \\ -0.715 & -0.022 & -0.021 & 0.24 & -0.024 \\ -0.148 & -0.003 & -0.004 & 0.09 & 0.026 \end{bmatrix}$$

$$\Psi = \begin{bmatrix} 0.0098 \\ 0.122 \\ 0.036 \\ 0.562 \\ 0.115 \end{bmatrix}$$

The open-loop eigenvalues of this model are $\{0.8928 \pm j0.0937, 0.2506 \pm j0.0252, 0.0295\}$. We solve the problem of determining the optimal control sequence with

$$R = 1, \quad Q = 0.1 I_5$$

using the eigenvalue-eigenvector method. The result is

$$P = \begin{bmatrix} 0.95 & 0.066 & 0.044 & -0.005 & 0.037 \\ 0.066 & 0.665 & 0.021 & 0.043 & 0.075 \\ 0.044 & 0.021 & 0.109 & 0.001 & 0.004 \\ -0.005 & 0.043 & 0.001 & 0.11 & 0.006 \\ 0.037 & 0.075 & 0.004 & 0.006 & 0.111 \end{bmatrix}$$

which is symmetric and positive-definite. From (25), the gain matrix has the form

$$G = [-0.035 \quad 0.1 \quad 0.001 \quad 0.023 \quad 0.012]$$

which yields the closed-loop eigenvalues as $\{0.0298, 0.2465 \pm j0.0246, 0.88839 \pm j0.0877\}$, which are stable. To avoid variation in the initial state, the expected value of the optimal cost is taken as the trace of the Riccati matrix which gives

$$J_a^* = \text{Tr}[P] = 1.945$$

Simulation of the problem using other values of Q, R were undertaken and the output is summarized in Table 8.1.

8.3 FILTERING AND PREDICTION

The development of an LQ controller has been based on the assumption that all state variables are available for direct measurements. In practice, only finite states (outputs) are allowable for recording externally.

Table 8.1 Simulation Results

	$R = 1 \quad Q = I_5$	$R = 0.1 \quad Q = I_5$
Closed-loop eigenvalues	{0.0301, 0.1936, 0.2432, 0.8463 ± j0.0375}	{0.0415 ± j0.0053, 0.252, 0.7262, 0.893}
j^{*a}	16.85	15.04
G	[−0.1378 0.555 0.0121 0.1468 0.0708]	[−0.1556 1.1636 0.041 0.3632 0.155]

Optimal Design Methods

The problem of estimating the state variables of linear dynamical systems which are subject to stochastic disturbances is considered in this section. For related background material, Appendix D contains a brief account of random variables and Gauss–Markov process. We shall focus attention on the filtering problem and derive the Kalman filter.

8.3.1 The Estimation Problem

Consider a dynamic system whose state as a function of time is an n-dimensional discrete-time stochastic process $\{\mathbf{x}(k)\}$; $k \in I_t = \{0, 1, 2, \ldots\}$. Suppose that we have made a sequence of measurements $\mathbf{y}(0), \mathbf{y}(1), \ldots, \mathbf{y}(m)$, at consecutive discrete instants, which are related to $\mathbf{x}(k)$ by means of an appropriate measurement system. We wish to utilize the measurement data in some way to infer the value of $\mathbf{x}(k)$. Let us assume that the sequence $\{\mathbf{y}(j), j = 0, \ldots, m\}$ is a discrete-time stochastic process.

Given the measurement records $\{\mathbf{y}(0), \ldots, \mathbf{y}(m)\}$, we denote (at least for the time being) an estimate of $\mathbf{x}(k)$ based on these measurements by $\hat{\mathbf{x}}(k|m)$. As a function of the measurements, define the estimate of the state to be

$$\hat{\mathbf{x}}(k|m) = \mathbf{g}_k[\mathbf{y}(j), \quad j = 0, \ldots, m] \tag{41}$$

We can then state the following. *The estimation problem is one of determining $\mathbf{g}_k[.]$ in an appropriate way.*

The Filtering Problem

The filtering problem is obtained by setting $k = m$ in (41). It therefore means the recovery at time k of some information about $\mathbf{x}(k)$, which corresponds to $\hat{\mathbf{x}}(k|k)$, using measurement data up till time k. One should note the following points:

1. We wish to obtain the approximate value of $\mathbf{x}(k)$ at time k.
2. The measurement records are available at time k and not at a later time.
3. All the measurement records up to time k are used in estimating the state.

The reason for stressing these points is to distinguish the filtering problem from the prediction and smoothing problems to be defined below.

The Smoothing Problem

This problem results from the estimation problem stated previously when $k < m$. The smoothing problem thus differs from the filtering

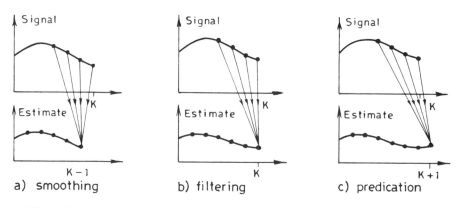

Figure 8.1 Estimation pattern.

problem in that the information about $\mathbf{x}(k)$, in the form of measurement data, need not become available at time k and measurement records derived later than time k can be utilized in obtaining information about $\mathbf{x}(k)$. It should be noted that:

1. There is a delay in producing the estimate of the state.
2. More data records are used then in the filtering problem.

The Prediction Problem

Here we use $k > m$ in (41). The aim of the prediction problem is to obtain at time k, information about $\mathbf{x}(k + s)$ for some $s > 0$. It therefore represents a forecast, that is, we wish to guess how $\mathbf{x}(.)$ will behave after a specified period of time.

In the sequel we shall consider only the filtering problem because of its wide use in control and systems applications. Figure 8.1 illustrates the different problems.

8.3.2 Principal Methods of Obtaining Estimates

From now onwards, we shall limit out discussions to linear, discrete-time dynamical systems of the type discussed in Appendix D, where $\{\mathbf{x}(k)\}$ and $\{\mathbf{y}(k)\}$ are Gaussian random processes. Our purpose is to indicate how knowledge of the value taken by one random variable, the measurement $\mathbf{y}(k)$ in our case, can provide information about the value taken by a second random variable the state $\mathbf{x}(k)$. In particular, we wish to solve the following problem (filtering problem).

Find at time k some information about $\mathbf{x}(k)$ from $[\mathbf{y}(0), \mathbf{y}(1), \ldots, \mathbf{y}(k)]$

Optimal Design Methods

Let the information that we are seeking be summarized by the vector $\hat{\mathbf{x}}(k)$, the *estimate* of $\mathbf{x}(k)$. Since both $\{\mathbf{x}(k)\}$ and $\{\mathbf{y}(k)\}$ are random processes, so is $\hat{\mathbf{x}}(k)\}$. In general, $\hat{\mathbf{x}}(k)$ will not be equal to $\mathbf{x}(k)$. It is thus reasonable to define

$$\tilde{\mathbf{x}}(k) = \mathbf{x}(k) - \hat{\mathbf{x}}(k) \tag{42}$$

as the *estimation error*. Since the estimate $\hat{\mathbf{x}}(k)$ can be derived in several ways, the problem of interest would then be to find an estimate, which is a function of $\mathbf{y}(k)$, such that it is *optimal* with respect to some criterion. In addition, it is necessary to ensure that the estimates possess certain convergence properties with respect to the real values of the state.

Minimum Variance Estimate

Define $\mathbf{Y}(k) = \{\mathbf{y}(0), \ldots, \mathbf{y}(k)\}$; then an average measure of the estimation error in (38) is provided by

$$E[\{\mathbf{X} - \hat{\mathbf{x}}\}^t C\{\mathbf{X} - \hat{\mathbf{x}}\} \mid \mathbf{Y}(k)] \tag{43}$$

where C is a nonnegative definite symmetrical weighting matrix. We note in (43) that

1. $\hat{\mathbf{x}}$ is a fixed vector which needs to be determined from a knowledge of $\mathbf{Y}(k)$.
2. The average measure is a scalar quantity which is convenient for comparison.
3. It has a zero value when the estimate is exact.

The *minimum variance estimate* $\hat{\mathbf{x}}$ is defined as one for which

$$E[\|\mathbf{X} - \hat{\mathbf{x}}\|_C^2 \mid \mathbf{Y}(k)] \leq E[\|\mathbf{X} - \mathbf{r}\|_C^2 \mid \mathbf{Y}(k)] \tag{44}$$

for all vectors $\mathbf{r}$, determined in some way from $\mathbf{y}(k)$ and where $\|\mathbf{h}\|_C^2 = \mathbf{h}^t C \mathbf{h}$. It should be observed in (44) that $\mathbf{r}$ in general depends on $\mathbf{y}$ but it is independent of $\mathbf{x}$. The right-hand side of (44) can be written as

$$E[\|\mathbf{X} - \mathbf{r}\|_C^2 \mid \mathbf{Y}(k)] = \int_{-\infty}^{+\infty} (\mathbf{x} - \mathbf{r})^t C(\mathbf{x} - \mathbf{r}) p(\mathbf{x} \mid \mathbf{y}) \, d\mathbf{x} \tag{45}$$

which, by adding and subtracting appropriate terms, can be put in the form

$$E[\|\mathbf{X} - \mathbf{r}\|_C^2 \mid \mathbf{Y}(k)] = \left[\mathbf{r}^t - \int_{-\infty}^{+\infty} \mathbf{x}^t p(\mathbf{x} \mid \mathbf{y}) \, d\mathbf{x}\right] C \left[\mathbf{r} - \int_{-\infty}^{+\infty} \mathbf{x} p(\mathbf{x} \mid \mathbf{y}) \, d\mathbf{x}\right]$$
$$+ \int_{-\infty}^{+\infty} \mathbf{x}^t C \mathbf{x} p(\mathbf{x} \mid \mathbf{y}) \, d\mathbf{x} - \left\|\int_{-\infty}^{+\infty} \mathbf{x} p(\mathbf{x} \mid \mathbf{y}) \, d\mathbf{x}\right\|_C^2 \tag{46}$$

It is evident that the right-hand side of (46) has a unique minimum when $\mathbf{r} = E[\mathbf{X} | \mathbf{Y}(k)]$, which implies that

$$\hat{\mathbf{x}} = E[\mathbf{X} | \mathbf{Y}(k)] = \int_{-\infty}^{+\infty} \mathbf{x} p(\mathbf{x} | \mathbf{y}) \, d\mathbf{x} \qquad (47)$$

We now conclude that the minimum variance estimate $\hat{\mathbf{x}}$ is the conditional mean estimate; that is, the conditional expectation of $\mathbf{X}$ given $\mathbf{Y}(k)$.

The value of the average mean square error associated with the estimate of $\hat{\mathbf{x}}$ can be obtained from (46) by substituting $\mathbf{y} = \hat{\mathbf{x}}$ to yield

$$E[\|\mathbf{X} - \hat{\mathbf{x}}\|_C^2 | \mathbf{Y}(k)] = E[\|\mathbf{X}\|_C^2 | \mathbf{Y}(k)] - \|\hat{\mathbf{x}}\|_C^2 \qquad (48)$$

The estimate $\hat{\mathbf{x}}$ is often called the least-squares estimate or the minimum mean-square estimate. It is interesting to observe that the above analysis is carried out for arbitrary probability densities. For a given configuration of stochastic processes, all that is needed is to evaluate the conditional probability density function.

In the light of our discussions in Appendix D, we wish to emphasize that what we have established is a procedure by which one can compute an estimate (known vector) of a random process given a particular set of measurements (value of another random process). Such a procedure is essentially a rule of association between the measurement values and the value of the estimate. We therefore define the random variable which has $\hat{\mathbf{x}}$ as a particular value as $\hat{\mathbf{X}}$. One should think of $\hat{\mathbf{X}}$ as a function which depends on $\mathbf{y}$ or $\mathbf{Y}(k)$ and generates $\hat{\mathbf{x}}$. It is frequently called an *estimator* of $\mathbf{X}$, and from (47) we obtain

$$\hat{\mathbf{X}} = E[\mathbf{X} | \mathbf{Y}] \qquad (49)$$

as the *minimum variance estimator*.

A major property of the conditional mean estimate (minimum variance estimate) $\hat{\mathbf{x}}$ is that it is an *unbiased estimate*, that is,

$$E[\mathbf{X} - \hat{\mathbf{x}} | \mathbf{Y}(k)] = E[\mathbf{X} | \mathbf{Y}(k)] - \hat{x}$$
$$= \hat{\mathbf{x}} - \hat{\mathbf{x}} \qquad (50)$$
$$= 0$$

The above expression shows that the conditional expected error in using $\hat{\mathbf{x}}$ as an estimate of $\mathbf{x}$, given $\mathbf{Y}(k)$, is zero.

Maximum Likelihood Estimate

We can define the conditional probability density function $p\{\mathbf{Y}(k) | \mathbf{x}(k)\}$ as the *likelihood function*. This is a function of $\mathbf{x}(k)$

Optimal Design Methods

whose maximum indicates the most likely value of the sequence $Y(k) = \{y(0), \ldots, y(k)\}$ that we obtain using the parameters $x(k)$.

In many cases we will maximize the logarithm of this function. If the logarithm has a continuous first derivative, then a necessary condition for a *maximum likelihood estimate* $\hat{x}_m$ can be obtained by differentiating

$$\log[p\{Y(k)|x(k)\}]$$

with respect to $x(k)$, so that we have

$$\frac{\partial}{\partial x(k)} \log[p\{Y(k)|x(k)\}] = 0$$

$$x(k) = \hat{x}_m \quad (51)$$

From Bayes' rule, it is easy to see that $p\{Y(k)|x(k)\} = p\{Y(k), x(k)\}/p\{x(k)\}$ so that the maximum likelihood estimate requires the prior data $p\{x(k)\}$.

Maximum a Posteriori Estimate

This estimate is obtained by considering the a posteriori probability density function $p\{x(k)|Y(k)\}$. It deals with the inverse problem of the maximum likelihood. The *maximum a posteriori estimate* $\hat{x}_a$ is the value of $\hat{x}$ which maximizes the distribution $p\{x(k)|Y(k)\}$. Similarly to (51) it is given by

$$\frac{\partial}{\partial x} \log[p\{x(k)|Y(k)\}] = 0$$

$$x(k) = \hat{x}_m \quad (52)$$

The maximum likelihood and maximum a posteriori estimates can be related through Bayes' rule

$$p\{x(k)|Y(k)\} = p\{x(k), Y(k)\}/p\{Y(k)\}$$
$$= p\{Y(k)|x(k)\} p\{x(k)\}/p\{Y(k)\}$$

When the a priori distribution is uniform, both estimates are identical.

We note that the three estimates described thus far are presented for arbitrary probability distributions. It has been verified that when the random vectors are normally distributed (Gaussian) the maximum a posteriori estimates are precisely the conditional mean estimates. For this reason, we will adopt the minimum variance as our criterion for determining the optimal estimator (filter).

8.3.3 Development of the Kalman Filter Equations

Our objective here is to develop the mechanism by which the best estimate of $x(k)$, given the measurement sequence $Y(k) = \{y(0), \ldots, y(k)\}$, can be obtained. For linear discrete models with Gaussian random processes, this mechanism is known as the *Kalman filter*. In his original derivation of the discrete filter, Kalman (1960) used the concept of orthogonal projections. Subsequent to this work, several methods have been developed to derive the discrete Kalman filter. First, we present the optimal filtering problem and state the associated assumptions.

The Optimal Filtering Problem

The problem of interest is to estimate, at each discrete instant, in an optimal way, the state of a linear dynamical system using noisy measurements of the output records. The general model of the system is of the form

$$x(k+1) = A(k)x(k) + G(k)w(k)$$
$$y(k) = H(k)x(k) + v(k) \qquad (53)$$

The assumptions concerning the model and disturbances are summed up below.

Assumption 1. $\{w(k)\}$ and $\{v(k)\}$ are Gaussian white noise sequences such that

$$E[w(k)] = 0, \quad E[v(k)] = \mathbf{0}$$
$$E[w(k)w^t(j)] = Q\delta_{kj}$$
$$E[v(k)v^t(j)] = R\delta_{kj}$$

Assumption 2. The random processes $\{w(k)\}$ and $v(k)\}$ are uncorrelated, that is

$$E[w(k)v^t(j)] = 0, \quad \text{for all } k \text{ and } j$$

Assumption 3. The initial state $x(0)$ is a Gaussian random vector with mean $E[x(0)] = m(0)$ and covariance

$$E[\{x(0) - m(0)\}\{x(0) - m(0)\}^t] = W(0)$$

Assumption 4. The initial state $x(0)$ and the noise processes $\{w(k)\}$ and $\{v(k)\}$ are uncorrelated, that is

Optimal Design Methods

$$E[\mathbf{x}(0)\,\mathbf{w}^t(k)] = 0, \quad \text{for all } k$$
$$E[\mathbf{x}(0)\,\mathbf{v}^t(k)] = 0, \quad \text{for all } k$$

Assumption 5. The elements of the system matrices $A(k)$, $G(k)$, and $H(k)$ are known.

It should be noted that Assumption 2 is not strictly necessary. It is convenient though, since the final expression for the filter would be much simpler in this case. We adopt the minimization of the conditional variance (43) as our criterion for determining the best (optimal) estimate.

We saw in Section 8.3.2 that the optimal estimate which minimizes the conditional variance is the conditional mean estimate

$$\begin{aligned}\hat{\mathbf{x}}(k\,|\,k) &= E[\mathbf{x}(k)\,|\,\mathbf{y}(0),\dots,\mathbf{y}(k)]\\ &= E[\mathbf{x}(k)\,|\,\mathbf{Y}(k)]\end{aligned} \quad (54)$$

Note that the argument k is used twice in defining the optimal estimate. The k to the left of the conditioning bar denotes the discrete instant at which the estimate is required whereas the other k denotes the discrete instant up to which the output records are available. Therefore, the estimate $\hat{\mathbf{x}}(k\,|\,k-1)$ is the estimate of the state $\mathbf{x}(k)$ at the instant k, given the sequence of measurements up to $k-1$, that is

$$\begin{aligned}\hat{\mathbf{x}}(k\,|\,k-1) &= E[\mathbf{x}(k)\,|\,\mathbf{y}(0),\dots,\mathbf{y}(k-1)]\\ &= E[\mathbf{x}(k)\,|\,\mathbf{Y}(k-1)]\end{aligned} \quad (55)$$

which is actually a one-step predictor (see Section 8.3.1). By convention, we will define $\hat{\mathbf{x}}(k-1\,|\,k-1)$ for $k=0$ (that is $\hat{\mathbf{x}}(-1\,|\,-1)$) to be $\mathbf{m}(0)$, that is, the expected value of $\mathbf{x}(0)$ given no measurements. For the same reason, the initial value of the associated error covariance matrix $P(-1,-1)$ is taken to be $W(0)$. We can now state the basic optimal filtering problem. For the linear, discrete-time system of (53) under Assumptions (1)–(5), determine the estimates

$$\hat{\mathbf{x}}(k\,|\,k-1) \text{ and } \hat{\mathbf{x}}(k\,|\,k)$$

defined by (54) and (55) and the associated error covariance matrices $p(k\,|\,k-1)$ and $p(k\,|\,k)$.

Solution Procedure

The approach to the development of the Kalman filter equations can be divided into a number of distinct steps.

Step 1: Transition of the State $x(k-1)$ **to** $x(k)$. We assume that the estimate $\hat{x}(k-1\,|\,k-1)$ is known and we wish to determine the one-step predictor $\hat{x}(k\,|\,k-1)$. Rewrite the dynamic model in the form

$$x(k) = A(k-1)\,x(k-1) + G(k-1)\,w(k-1) \tag{56}$$

On taking the conditional mean of (56), we obtain

$$E[x(k)\,|\,Y(k-1)] = A(k-1)\,E[x(k-1)\,|\,Y(k-1)]$$
$$+ G(k-1)\,E[w(k-1)\,|\,Y(k-1)] \tag{57}$$

In view of Assumptions (1), (2), and (4), the random vector $w(k-1)$ is independent of $Y(k-1)$ so that

$$E[w(k-1)\,|\,Y(k-1)] = E[w(k-1)] = 0$$

and hence (57) reduces to

$$\hat{x}(k\,|\,k-1) = A(k-1)\,\hat{x}(k-1\,|\,k-1) \tag{58}$$

Next, we determine the error covariance matrix associated with the one-step predictor, that

$$P(k\,|\,k-1) = E[\{x(k) - \hat{x}(k\,|\,k-1)\}$$
$$\times \{x(k) - \hat{x}(k\,|\,k-1)\}'\,|\,Y(k-1)] \tag{59}$$

We note that the vector $[x(k) - \hat{x}(k\,|\,k-1)]$ is independent of the sequence $Y(k-1)$. This simplifies (59) to

$$P(k\,|\,k-1) = E[\{x(k) - \hat{x}(k\,|\,k-1)\}\{x(k) - \hat{x}(k\,|\,k-1)\}'] \tag{60}$$

Using (56) and (58), the one-step prediction error can be written as

$$x(k) - \hat{x}(k\,|\,k-1) = A(k-1)[x(k-1) - \hat{x}(k-1\,|\,k-1)]$$
$$+ G(k-1)\,w(k-1) \tag{61}$$

On expanding (60) with the aid of (61), we obtain

$$P(k\,|\,k-1) = A(k-1)\,P(k-1\,|\,k-1)\,A'(k-1) + A(k-1)$$
$$\times E[\{x(k-1) - \hat{x}(k-1\,|\,k-1)\,w'(k-1)\,|\,Y(k-1)]$$
$$+ G(k-1)\,QG'(k-1) + G(k-1)\,E[w(k-1)$$
$$\times \{x(k-1) - \hat{x}(k-1\,|\,k-1)\}'\,|\,Y(k-1)]\,G'(k-1)$$

Since $w(k-1)$ has zero mean and $x(k-1)$ is a function of $x(0)$ and $\{w(0), \ldots, w(k-1)\}$ but not of $w(k-1)$, then

Optimal Design Methods

$$E[\mathbf{x}(k-1)\mathbf{w}^t(k-1)] = 0$$

Also

$$E[\hat{\mathbf{x}}(k-1|k-1)\mathbf{w}^t(k-1)|\mathbf{Y}(k-1) = \hat{\mathbf{x}}(k-1)E[\mathbf{w}^t(k-1)] = 0$$

so that the error covariance matrix could be determined from the expression

$$P(k|k-1) = A(k-1)P(k-1|k-1)A^t(k-1)$$
$$+ G(k-1)QG^t(k-1) \quad (62)$$

Step 2: One-Step Prediction of the Filtered Estimate. Next, we wish to express the estimate of $x(k)$ given the measurement sequence $\mathbf{Y}(k)$. To accomplish this, we use the maximum a posteriori. First, the conditional density function $p\{\mathbf{x}(k)|\mathbf{Y}(k)\}$ can be written as

$$p\{\mathbf{x}(k)|\mathbf{Y}(k)\} = p\{\mathbf{x}(k)|\mathbf{Y}(k-1), \mathbf{y}(k)\}$$

where we separate out the last measurement from the previous measurement records $\mathbf{Y}(k-1)$. Then applying Bayes' theorem to the above expression leads to

$$p\{\mathbf{x}(k)|\mathbf{Y}(k)\} = p\{\mathbf{x}(k)|\mathbf{Y}(k-1), \mathbf{y}(k)\}$$
$$= p\{\mathbf{x}(k), \mathbf{Y}(k-1), \mathbf{y}(k)\}$$
$$/ p\{\mathbf{Y}(k-1), \mathbf{y}(k)\}$$
$$= p\{\mathbf{x}(k), \mathbf{Y}(k-1)\}p\{\mathbf{y}(k)|\mathbf{x}(k), \mathbf{Y}(k-1)\}$$
$$/ p\{\mathbf{Y}(k-1), \mathbf{y}(k)\}$$
$$= p\{\mathbf{y}(k)|\mathbf{x}(k), \mathbf{Y}(k-1)\}p\{\mathbf{x}(k)|\mathbf{Y}(k-1)\}$$
$$/ p\{\mathbf{y}(k)|\mathbf{Y}(k-1)\} \quad (63)$$

To compute (63) we consider the observation equation

$$\mathbf{y}(k) = H(k)\mathbf{x}(k) + \mathbf{v}(k) \quad (64)$$

We see that knowledge of $\mathbf{x}(k)$ implies that the only random vector left is $\mathbf{v}(k)$, which is independent of $\mathbf{Y}(k)$ in view of assumptions (1), (2), and (4). We can thus write

$$p\{\mathbf{y}(k)|\mathbf{x}(k), \mathbf{Y}(k-1)\} = p\{\mathbf{y}(k)|\mathbf{Y}(k-1)\}$$

which, when substituted into (63), gives

$$p\{\mathbf{x}(k)|\mathbf{Y}(k)\} = p\{\mathbf{y}(k)|\mathbf{x}(k)\}p\{\mathbf{x}(k)|\mathbf{Y}(k-1)\}$$
$$/ p\{\mathbf{y}(k)|\mathbf{Y}(k-1)\} \quad (65)$$

In order to determine the maximum a posteriori estimate using (65), it will only be necessary to evalute the probability densities of the numerator since the denominator is not an explicit function of $\mathbf{x}(k)$.

For a given $\mathbf{x}(k)$, $\mathbf{y}(k)$ is a Gaussian random vector with mean

$$E[\mathbf{y}(k)\,|\,\mathbf{x}(k)] = H(k)\mathbf{x}(k)$$

since $E[\mathbf{v}(k)\,|\,\mathbf{x}(k)] = E[\mathbf{v}(k)] = \mathbf{0}$. The covariance matrix is given by

$$E[\{\mathbf{y}(k) - H(k)\mathbf{x}(k)\}\{\mathbf{y}(k) - H(k)\mathbf{x}(k)\}^t] = E[\mathbf{v}(k)\mathbf{v}^t(k)]$$
$$= R(k)$$

Thus, $p\{\mathbf{y}(k)\,|\,\mathbf{x}(k)\}$ is $N(H(k)\mathbf{x}(k), R(k))$.

Turning to the a priori density function $p\{\mathbf{x}(k)\,|\,\mathbf{Y}(k-1)\}$, it is easy to show that this function is actually a Gaussian of mean $\hat{\mathbf{x}}(k\,|\,k-1)$ and covariance $p(k\,|\,k-1)$. Hence, $p\{\mathbf{x}(k)\,|\,\mathbf{Y}(k-1)\}$ is $N(\hat{\mathbf{x}}(k\,|\,k-1), p(k\,|\,k-1))$. To sum up, we can now write the a posteriori density function $p\{\mathbf{x}(k)\,|\,\mathbf{Y}(k)\}$ in (65) as

$$p\{\mathbf{x}(k)\,|\,\mathbf{Y}(k)\} = K \exp[-\tfrac{1}{2}\{\mathbf{y}(k) - H(k)\mathbf{x}(k)\}^t R^{-1}(k)$$
$$\times \{\mathbf{y}(k) - H(k)\mathbf{x}(k)\} + \{\mathbf{x}(k) - \hat{\mathbf{x}}(k\,|\,k-1)\}^t$$
$$\times P^{-1}(k\,|\,k-1)\{\mathbf{x}(k) - \hat{\mathbf{x}}(k\,|\,k-1)\}] \quad (66)$$

where the factor K takes into account the denominator $p\{\mathbf{y}(k)\,|\,\mathbf{Y}(k-1)\}$ in (65).

In order to develop the maximum a posteriori estimate (which is identical with the conditional mean estimate in this case), we can differentiate the logarithm of (65) with respect to $\mathbf{x}(k)$ and set the result to zero to obtain the estimate $\hat{\mathbf{x}}(k\,|\,k)$. If we do this, we obtain

$$H^t(k)R^{-1}(k)\{\mathbf{y}(k) - H(k)\mathbf{x}(k)\} - P^{-1}(k\,|\,k-1)$$
$$\times \{\mathbf{x}(k) - \hat{\mathbf{x}}(k\,|\,k-1)\} = 0$$

For $\mathbf{x}(k) = \hat{\mathbf{x}}(k\,|\,k)$ and rearranging, we obtain

$$\hat{\mathbf{x}}(k\,|\,k) = \hat{\mathbf{x}}(k\,|\,k-1) + [H^t(k)R^{-1}(k)H(k)$$
$$\times H^t(k)R^{-1}(k)\{\mathbf{y}(k) - H(k)\hat{\mathbf{x}}(k\,|\,k-1)\} \quad (67)$$

which is the new value of the state estimate given a new observation.

Finally, to calculate the variance of the estimation error we use the matrix identity

$$[I - (M + N)^{-1}M] = (M + N)^{-1}N$$

with $M = H^t(k)R^{-1}H(k)$, $N = P^{-1}(k\,|\,k-1)$ in (67) to arrive at

Optimal Design Methods

$$\hat{x}(k\,|\,k) = [H^t(k)R^{-1}(k)H(k) + P^{-1}(k\,|\,k-1)]^{-1}$$
$$\times [H^t(k)\,R^{-1}(k)\,\mathbf{y}(k) + P^{-1}(k\,|\,k-1)\,\hat{x}(k\,|\,k-1)] \quad (68)$$

Using (64) in (67) and after some algebraic manipulation, the result is

$$\mathbf{x}(k) - \hat{x}(k\,|\,k) = -[H^t(k)\,R^{-1}(k)\,H(k) + P^{-1}(k\,|\,k-1)]$$
$$\times [H^t(k)R^{-1}(k)\mathbf{v}(k) - P^{-1}(k\,|\,k-1)\{\mathbf{x}(k)$$
$$- \hat{x}(k\,|\,k-1)\}]$$

In the above expression we note that $\mathbf{v}(k)$ and $\mathbf{x}(k)$ are independent, $\mathbf{v}(k)$ and $\hat{x}(k\,|\,k-1)$ are also independent and $\mathbf{v}(k)$ has zero mean. By virtue of these facts, it can be readily shown that

$$P(k\,|\,k) = E[\{\mathbf{x}(k) - \hat{x}(k\,|\,k)\}\{\mathbf{x}(k) - \hat{x}(k\,|\,k)\}^t$$
$$= [H^t(k)R^{-1}(k)H(k) + P^{-1}(k\,|\,k-1)]^{-1} \quad (69)$$

which can be alternatively written, using the well-known matrix inversion lemma

$$[F_1 + F_2 F_3 F_2^{t-1}] = F_1^{-1} - F_1^{-1} F_2 [F_3^{-1} + F_2^t F_1^{-1} F_2]^{-1} F_2^t F_1^{-1} \quad (70)$$

with $F_1 = P^{-1}(k\,|\,k-1)$, $F_2 = H^t(k)$, $F_3 = R^{-1}(k)$ and the indicated inverses exist

$$P(k\,|\,k) = P(k\,|\,k-1) - P(k\,|\,k-1)\,H^t(k)[H(k)P(k\,|\,k-1)$$
$$\times H^t(k) + R(k)]^{-1}\,H(k)P(k\,|\,k-1) \quad (71)$$

and subsequently we write (68) as

$$\hat{x}(k\,|\,k) = \hat{x}(k\,|\,k-1) + P(k\,|\,k-1)\,H^t(k)[H(k)P(k\,|\,k-1)$$
$$\times H^t(k) + R(k)]^{-1}\{\mathbf{y}(k) - H(k)\,\hat{x}(k\,|\,k-1)\} \quad (72)$$

In summary, (58), (62), (71), and (72) constitute the equations of the optimal minimum variance filter.

Close examination of (72) will reveal that the optimal estimate $\hat{x}(k\,|\,k)$ is the sum of the one-step predictor $\hat{x}(k\,|\,k-1)$ and the difference between the *actual output* $\mathbf{y}(k)$ and the *predicted output* $H(k)\hat{x}(k\,|\,k-1)$, weighted by the term

$$K(k) = P(k\,|\,k-1)H^t(k)[R(k) + H(k)P(k\,|\,k-1)\,H^t(k)]^{-1} \quad (73)$$

which is often called the *filter gain*. The quantity $\tilde{\mathbf{y}}(k\,|\,k-1) = \{\mathbf{y}(k) - H(k)\,\hat{x}(k\,|\,k-1)\}$ under assumptions (1)–(5) and the minimum error variance criterion turns out to be a white noise stochastic process frequently called the *innovations process*. The reason for this is that it contains all of the new information in the measurement **y**(k). The procedure of computing the Kalman filter is carried out recursively in the following order. Given $P(0|-1) = W(0)$ and $\mathbf{x}(0|-1) = m(0)$

1. Compute the filter gain using

$$K(k) = P(k|k-1)H'(k)[R(k) + H(k)P(k)|(k-1)H'(k)]^{-1}$$

2. Compute the state estimate vector

$$\hat{\mathbf{x}}(k|k) = \hat{\mathbf{x}}(k|k-1) + K(k)\{\mathbf{y}(k) - H(k)\hat{\mathbf{x}}(k|k-1)\} \quad (74)$$

3. Compute the error covariance matrix

$$P(k|k) = [I - K(k)H(k)]P(k|k-1) \quad (75)$$

These equations can be represented as shown in Figure 8.2. It is interesting to note that (71), which enables us to compute the error covariance, is a matrix equation of the Riccati type.

Occasionally, it is required to determine the function $\hat{\mathbf{x}}(k+1|k)$ in the mean square error sense. This can be obtained directly from (58), (72), and (73) as

$$\hat{\mathbf{x}}(k+1)|k) = [A(k) - K^+(k)H(k)]\hat{\mathbf{x}}(k|k-1) + K^+(k)\mathbf{y}(k) \quad (76)$$

where $K^+(k)$ is defined by

$$K^+(k) = A(k)P(k|k-1)H'(k)[R(k) + H(k)P(k|k-1)H'(k)]^{-1}$$
$$= A(k)K(k) \quad (77)$$

and is sometimes called the *Kalman filter gain*.

It is remarked that the estimate $\hat{\mathbf{x}}(k+1|k)$ is actually the one-stage predictor at the discrete instant k given the measurement records $\mathbf{Y}(k)$.

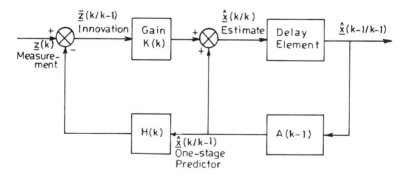

Figure 8.2 Block diagram of the Kalman predictor.

Optimal Design Methods 301

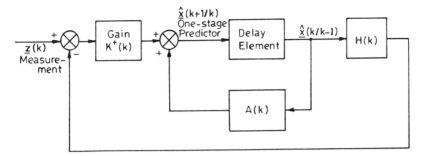

Figure 8.3 Structure of the one-stage Kalman predictor.

The reader should distinguish between this quantity and $\hat{x}(k\,|\,k)$ which is the filtered estimate at k. For this reason, we have used different symbols for the corresponding gains. Figure 8.3 gives the structure of the one-step predictor.

Next, we present a number of properties of the Kalman filter (or Kalman one-stage predictor).

Some Important Properties

With reference to the development of the celebrated Kalman filter we now list some of its important properties:

1. Consideration of (74) or (77) shows that the Kalman filter has the same structure as the process. In fact, the Kalman filter is a linear, discrete-time, finite-dimensional system. Sometimes the estimate $\hat{x}(k\,|\,k)$ is called the linear, unbiased, minimum variance estimate. Note that all the computations are performed recursively.

2. The input to the filter, or the one-step predictor is the noise process $\{y(k)\}$. The output is $\{\hat{x}(k\,|\,k)\}$ for the filter and $\{\hat{x}(k\,|\,(k-1)\}$ for the predictor. Obviously, the output sequence (estimate) depends on the input sequence (measurement); but the interesting thing is that the conditional error covariance matrix is actually independent of $Y(k)$, see (71), and of $Y(k-1)$ as in (62). More importantly, the gain $K(k)$, or $K^+(k)$, is also independent of $Y(k)$. Because of these, both the error covariance matrices $p(k\,|\,k-1)$, $P(k\,|\,k)$ and the gains $K(k)$, $K^+(k)$ can be precomputed.

3. As discussed earlier, the processes $\{x(k)\}$ and $\{y(k)\}$ are jointly Gaussian, which in turn implies that $\{x(k)\,|\,Y(k-1)\}$ is Gaussian. We also saw that the conditional density $P\{x(k)\,|\,Y(k-1)\}$ has mean of $\{\hat{x}(k\,|\,k-1)$ and error covariance of $P(k\,|\,k-1)$. It follows that the

Kalman filter equations provide an updating procedure for the entire conditional probability density function of $\mathbf{x}(k)$.

4. Consider the case of linear, shift-invariant discrete systems of the form

$$\mathbf{x}(k+1) = A\mathbf{x}(k) + G\mathbf{w}(k)$$
$$\mathbf{y}(k) = H\mathbf{x}(k) + \mathbf{v}(k) \tag{78}$$

where the matrices A, G, H are constant and, in addition, the noise processes are white and stationary. In general, $P(k|k)$ and therefore $K(k)$ will not be constant. This means that the Kalman filter will normally be time-varying despite time invariance and stationary in the process.

In fact, time invariance of a linear process driven by stationary white noise is necessary but not sufficient for stationarity of the state and output processes (Meditch, 1969). Normally, asymptotic stability of the noise-free system is also required. This corresponds to the condition that all of the eigenvalues of the system matrix be strictly less than one in absolute value.

We can now state the following. For a linear process of the form (78), which is time-invariant and driven by stationary white noise, time invariant filters (constant error covariance and gain) are obtained when either the process is asymptotically stable ($|\lambda_j(A)| < 1$) or the $[A, H]$ is completely detectable and the pair $[A, GD]$ is completely stabilizable where $DD^t = Q$.

8.3.4 Examples

Example 8.3. Let us consider discrete state estimation for the scalar process

$$x(k+1) = ax(k) + w(k)$$
$$y(k) = x(k) + v(k)$$

where w and v are $N(0, r)$ and $N(0, q)$, respectively. Assume that $W(0) = \alpha$ and $m(0) - \beta$. We wish to show one cycle of computing the Kalman filter equations. Given that $\hat{x}(-1|-1) = \alpha$, then (58) yields at $k = 0$, $\hat{x}(0|-1) = \beta$. From (62), with $P(-1|-1) = \alpha$, we obtain

$$P(0|-1) = \sigma + q$$

Then, using (73)–(75), we obtain at $k = 0$

$$K(0) = P(0|-1)[r + P(0|-1)]^{-1}$$
$$= \frac{\alpha + q}{r + \alpha + q}$$

Optimal Design Methods

$$\hat{x}(0|0) = \hat{x}(0|-1) + K(0)\{z(0) - \hat{x}(0|-1)\}$$
$$= [1 - K(0)]\hat{x}(0|-1) + K(0)z(0)$$
$$= \frac{\beta r}{r + \alpha + q} + \frac{(\alpha + q)z(0)}{r + \alpha + q}$$

$$P(0/0) = [1 - K(0)]P(0|-1)$$
$$= \frac{r(\alpha + q)}{r + \alpha + q}$$

Example 8.4. We consider a constant x of which we record n successive measurements $y(k)$, each one of which is $N(0, r)$. Let the initial estimate $x(0)$ be $N(m, W)$. We want to show that for arbitrary s

$$\hat{x}(s|s) = \frac{rm + w \sum_{j=1}^{s} z(j)}{r + sW}$$

$$P(s|s) = \frac{rW}{r + sW}$$

To solve this problem, we note that the model is

$$\mathbf{x}(k+1) = x(k), \qquad y(k) = x + v(k)$$

thus $A(k) = 1$, $G(k) = 0$, $Q = 0$. Also $H(k) = 1$.
For this case, (58) and (62) become

$$\hat{x}(k|k-1) = \hat{x}(k-1|k-1)$$
$$P(k|k-1) = P(k-1|k-1)$$

Also from (67)–(69) we obtain

$$K(k) = P(k|k-1)[r + P(k|k-1)]^{-1}$$
$$= \frac{P(k-1|k-1)}{r + P(k-1|k-1)}$$

$$\hat{x}(k|k) = \hat{x}(k|k-1) + K(k)[z(k) - \hat{x}(k|k-1)]$$
$$= [1 - K(k)]\hat{x}(k|k-1) + K(k)z(k)$$
$$= \frac{r\hat{x}(k-1|k-1) + P(k-1|k-1)z(k)}{r + P(k-1|k-1)}$$

and

$$P(k\,|\,k) = [1 - K(k)]P(k - 1\,|\,k - 1)$$

$$\frac{rP(k-1\,|\,k-1)}{r + P(k-1\,|\,k-1)}$$

It is clear that the new estimate $\hat{x}(k\,|\,k)$ is a linear combination of

1. The old estimate $\hat{x}(k-1\,|\,k-1)$ weighted by the variance of the new measurement
2. The new measurement $y(k)$ weighted by the variance of the old estimate

This is a consequence of the trade-off between the confidence we have in the old estimates and those in the new measurements.

On using the above relations in conjunction with the data $\hat{x}(-1\,|\,-1) = m$, $P(-1\,|\,-1) = W$, we obtain the desired results

$$\hat{x}(s\,|\,s) = \frac{rm + W\sum_{j=1}^{s} y(j)}{r + sW}$$

$$P(s\,|\,s) = \frac{rW}{r + sW}$$

These results show that

1. As we use new measurement records, the variance $P(s\,|\,s)$ of the estimation error decreases.
2. In the limit when $s \to \infty$, all traces of the initial conditions disappear, and we have

$$\lim_{s \to \infty} \hat{x}(s\,|\,s) = \sum_{j=1}^{s} \frac{y(j)}{s}$$

$$\lim_{\to \infty} P(s\,|\,s) = \lim_{s \to \infty} \frac{0}{W} = 0$$

which means that the estimate asymptotically approaches the arithmetic mean of the measurement records.

Example 8.5. In our third example, we consider the scalar process

$$x(k + 1) = ax(k) + w(k)$$

$$y(k) = x(k) + v(k)$$

with the standard assumptions, $\{w(k)\}$ is a zero mean Gaussian white sequence with constant variance q, $\{v(k)\}$ is a zero mean Gaussian white sequence with constant variance r, $x(0)$ is a zero mean Gaussian random

Optimal Design Methods

variable with variance $W(0)$, and a is a constant. We also assume that the two Gaussian white sequences and $x(0)$ are independent.

The linear minimum variance filter equation is

$$\hat{x}(k|j) = \hat{x}(k|k-1) + K(k)[y(k) - \hat{x}(k|k-1)]$$

since $H(k) = 1$ for all k. From (62) and (73), we obtain the results

$$P(k|k-1) = a^2 P(k-1|k-1) + q$$

and

$$K(k) = [a^2 P(k-1|k-1) + q][a^2 P(k-1|k-1) + q + r]^{-1}$$
$$= \frac{a^2 P(k-1|k-1) + q}{a^2 P(k-1|k-1) + q + r}$$

subject to the initial condition $P(-1|-1) = W(0)$.

Since $P(k-1|k-1) \geq 0$ by definition, then we can see that $P(k|k-1) \geq q$. This means that the variance of the system disturbance sets the performance limit on the prediction accuracy.

It is readily seen from the gain equation that $0 \leq K(k) \leq 1$ for most cases. Combining the gain equation with the error covariance equation, we arrive at

$$P(k|k) = rK(k)$$

which entails that $0 \leq P(k|k) \leq r$ for $k \geq 0$.

Suppose that $W(0)$ is very large ($\gg r$). Then the first measurement record $z(0)$ will reduce the filtering error variance from $W(0)$ to $P(0|0) \leq r \ll W(0)$.

Another point to note is that when $q \gg r$. In this case we see that $K(k) = 1$ and $P(k)r$ for all k. The interpretation of this is that the performance limit on filtering accuracy is now set by the measurement error variance.

On examining the asymptotic behavior of the error variance $P(k|k)$ with $q = 0$, we find that

$$P = \frac{ra^2 P}{a^2 p + r}$$

where $P = P(k|k) = P(k-1|k-1)$.

The above expression possesses two solutions

$$P = 0 \quad \text{and} \quad P = \frac{(a^2 - 1)r}{a^2}$$

We note that $P = 0$ is allowed only when $a^2 < 1$ since P is a variance.

To study the nature of the steady-state value P we define

$$\delta P(k \mid k) = P(k \mid k) - P$$
$$\delta P(k - 1 \mid k - 1) = P(k - 1 \mid k - 1) - P$$

By direct manipulation we obtain

$$\delta P(k \mid k) = \left[\frac{a^2 r}{(a^2 P + r)}\right] \frac{r}{r + a^2 P(k - 1 \mid k - 1)} \delta P(k - 1 \mid k - 1)$$

Now for $a^2 < 1$, it is readily seen that

$$\delta P(k \mid k) < \delta P(k - 1 \mid k - 1), \quad \text{for all } k$$

and for $p = 0$

$$P(k \mid k) < P(k - 1 \mid k - 1)$$

Thus we conclude that $p = 0$ is a stable equilibruim point for the filtering error variance whenever $a^2 < 1$.

Let us consider the case when $a^2 > 1$. Here we must consider both solution $P = 0$ and $P = (a^2 - 1)r/a^2$. For $P = 0$, it is easy to show that

$$\delta P(k \mid k) = \frac{a^2 r}{r + a^2 P(k - 1 \mid k - 1)} \delta P(k - 1 \mid k - 1)$$

In the case when $P(k - 1 \mid k - 1) = 0$ for sometime we see that $P(k \mid k) = a^2 \delta P(k - 1 \mid k - 1)$. This means that even if the filtering error variance becomes zero, it will not remain zero. Hence, $P = 0$ is an unstable equilibrium point. For the second solution, $P = (a^2 - 1)r/a^2$,

$$\delta P(k \mid k) = \frac{r}{r + a^2 P(k - 1 \mid k - 1)} \delta P(k - 1 \mid k - 1)$$

Since $a^2 > 1$, the above expression implies that $\delta P(k \mid k) < \delta P(k - 1 \mid k - 1)$. Consequently, $P = (a^2 - 1)r/a^2$ is a stable equilibrium point when $a^2 > 1$.

To summarize, the filtering error variance will converge to zero if $a^2 < 1$ and to $(a^2 - 1)r/a^2$ if $a^2 > 1$. Therefore, for sufficiently long filtering times, the state of the linear discrete time process can be determined exactly when $-1 < a < 1$, but can only be specified to within an error to within an error variance of $(a^2 - 1)r/a^2$ when $q = 0$.

Example 8.6. Consider a discrete-time process described by

$$x(k + 1) = x(k) + u(k), \quad k = 0, 1, \ldots$$

where $\{u(k), u(k - 1), \ldots, u(0)\}$ is a sequence of independent random variables. Assume that $x(0)$ is a random variable which is independent of the sequence $\{u(k), \ldots, U(0)\}$. Is the process Markovian?

To answer the question posed above, we must examine the conditional probability

$$P[x(k+1) | x(k), \ldots, x(0)]$$

By iterating the discrete model we obtain

$$x(k) = u(k) + u(k-1) + \cdots + u(0) + X(0)$$

In view of the independence assumption, it is clear that $x(k)$ is independent of $u(k)$. The same is true for $x(k-1), x(k-2), \ldots, x(0)$. We could therefore write

$$P\{x(k+1) | x(k), \ldots, x(0)\} = p\{x(k+1) | x(k)\}$$

and hence the process is Markovian.

Example 8.7. A scalar process is modeled by the difference equation

$$x(k+1) = ax(k) + v(k), \quad |a| < 1$$

where $\{v(k)\}$ is Gaussian white noise with zero mean and covariance q. The initial state $x(0)$ is $N(0, W_0)$ and it is uncorrelates with $\{v(k)\}$ for all k. It is required to examine the asymptotic behavior of the variance. What will happen when $E[x(0)] = m_0 \neq 0$? The variance of the process is given by (50) in Appendix D. Setting $A(k) = a$, $G(k) = 1$, $Q(k) = 1$, we obtain

$$W(k+1) = a^2 W(k) + q$$

Starting from W_0 at $k = 0$ and iterating, we obtain

$$W(k+1) = a^{2k+2} W_0 + q(1 + a^2 + \cdots + a^{2k})$$

$$= a^{2k+2} W_0 + q\left(\frac{1 - a^{2k+2}}{1 - a^2}\right)$$

Since $|a| < 1$ then

$$w(\infty) = \frac{q}{1 - a^2}$$

The assumption $x(0)$ is $N(0, W_0)$ implies that $m(k+1) = 0$ and we conclude that the process is a stationary Gaussian process. Suppose that $E[x(0)] = m_0 \neq 0$, then from (56) in Appendix D we have

$$m(k) = a^k m_0$$

which tends to zero as $k \to \infty$. Therefore, the condition $|a| < 1$ always yields an asymptotically stationary process (as $k \to \infty$) with a zero mean irrespective of the actual mean of the initial state.

Example 8.8. Here we derive the formulae that define a Gauss-Markov process in terms of the probability distribution of the initial state $p\mathbf{x}(0)$ and the transition probability density $p\mathbf{x}(k+1)|x(k)$.

Recall the assumptions made in Appendix D. Since the random vectors $\mathbf{x}(k+1)$ and $\mathbf{x}(k)$ are jointly Gaussian, it is sufficient to calculate the conditional mathematical expectation $E[\mathbf{x}(k+1)|\mathbf{x}(k)]$ and the conditional covariance matrix. For the conditional expectation we have

$$E[\mathbf{x}(k+1)|\mathbf{x}(k)] = A(k)\mathbf{x}(k) + G(k)E[\mathbf{w}(k)|\mathbf{x}(k)]$$

But $\{\mathbf{w}(k)\}$ and $\{\mathbf{x}(k)\}$ are independent by assumption so that

$$E[\mathbf{w}(k)|\mathbf{x}(k)] = E[\mathbf{w}(k)] = \mathbf{0}$$

hence

$$E[\mathbf{x}(k+1)|\mathbf{x}(k)] = A(k)\,\mathbf{x}(k)$$

For the conditional covariance matrix

$$E[\{\mathbf{x}(k+1) - E[x(k+1)|\mathbf{x}(k)]\}\{\mathbf{x}(k+1) - E[\mathbf{x}(k+1)|\mathbf{x}(k)]\}^t|\mathbf{x}(k)]$$

it is readily simplified, using the conditional expectation derived above, into

$$E[G(k)\mathbf{w}(k)\mathbf{w}^t(k)G^t(k)|\mathbf{x}(k)] = E[G(k)\mathbf{w}(k)\mathbf{w}^t(k)G^t(k)]$$
$$= G(k)Q(k)G^t(k)$$

When the matrix $G(k)Q(k)G^t(k) = D(k)$ is nonsingular, we will have

$$p\{\mathbf{x}(k+1)|\mathbf{x}(k)\} = \left\{\frac{1}{(2\pi)^{n/2}}\right\}\{\det[D(k)]\}^{1/2} \exp\left\{-\frac{1}{2}[\mathbf{x}(k+1)\right.$$
$$\left. - A(k)\mathbf{x}(k)]^t D^{-1}(k)[\mathbf{x}(k+1) - A(k)\mathbf{x}(k)]\right\}$$

Example 8.9. Consider the random process (43)–(45) of Appendix D subject to the noise processes $\{\mathbf{w}(k)\}$ and $\{\mathbf{v}(k)\}$ having known nonzero means. What are the expressions for the evolution of the mean and covariance of $\mathbf{x}(k)$?

Let the mean of the input noise process be $\mathbf{n}(k)$. By taking the

Optimal Design Methods

expectation of (43) of Appendix D, we obtain

$$E[\mathbf{x}(k+1)] = E[A(k)\mathbf{x}(k)] + E[G(k)\mathbf{w}(k)]$$
$$= A(k)E[\mathbf{x}(k)] + G(k)E[\mathbf{w}(k)]$$

which can be written as

$$\mathbf{m}(k+1) = A(k)\mathbf{m}(k) + G(k)\mathbf{n}(k)$$

The solution of this equation, given $\mathbf{m}(0)$, can be expressed as

$$\mathbf{m}(k) = \Psi(k,0)\mathbf{m}(0) + \sum_{j=0}^{k-1} \Psi(k, j+1)G(j)\mathbf{n}(j)$$

A comparison of this expression with (48) in Appendix D shows that the mean $\mathbf{m}(k)$ now depends on the random sequence $\mathbf{n}(0), \ldots, \mathbf{n}(k-1)$.

To calculate the covariance matrix we write

$$W(k+1) = E[\{\mathbf{x}(k+1) - \mathbf{m}(k+1)\}\{\mathbf{x}(k+1) - \mathbf{m}(k+1)\}^t]$$

It should be noted that

$$\mathbf{x}(k+1) - \mathbf{m}(k+1) = A(k)[\mathbf{x}(k) - \mathbf{m}(k)] + G(k)[\mathbf{w}(k) - \mathbf{n}(k)]$$

Combining the above two expressions, we arrive at

$$W(k+1) = A(k)W(k)A^t(k) + G(k)Q(k)G^t(k)$$

where $Q(k)$ is the covariance of $\{\mathbf{w}(k)\}$. This shows that the covariance of the process $\{\mathbf{x}(k)\}$ is unaltered.

8.4 LINEAR QUADRATIC GAUSSIAN CONTROL

In this section, we extend the LQ design problem to include stochastic disturbances.

8.4.1 Problem Formulation

The design problem is specified in terms of the linear model

$$\mathbf{x}(k+1) = \Phi\mathbf{x}(k) + \Psi\mathbf{u}(k) + \mathbf{v}(k)$$
$$\mathbf{y}(k) = C\mathbf{x}(k) + \mathbf{e}(k) \tag{79}$$

where $\mathbf{y}$ and $\mathbf{e}$ are discrete-time Gaussian white noise processes with zero mean and

$$E[\mathbf{v}(k)\mathbf{v}^t(k)] = R_1$$
$$E[\mathbf{v}(k)\mathbf{e}^t(k)] = R_{12}$$
$$E[\mathbf{e}(k)\mathbf{e}^t(k)] = R_2 \tag{80}$$

It is assumed that the initial state $\mathbf{x}(0)$ is Gaussian distributed with

$$E[\mathbf{x}(0)] = m_0 \quad \text{and} \quad \text{Cov}[\mathbf{x}(0)] = R_0$$

We consider that $R_0 \geq 0$, $R_1 \geq 0$, and $R_2 \geq 0$, and that the model (79), (80) is both reachable and observable. The cost functional to be minimized is given by

$$J = E\left\{ \sum_{k=0}^{N-1} [\mathbf{x}'(k) Q \mathbf{x}(k) + \mathbf{u}'(k) R \mathbf{u}(k)] \right\} \tag{81}$$

The admissible controls are assumed to be such that $\mathbf{u}(k)$ is a function of Y_{k-1} where

$$Y_k = \{\mathbf{y}(j), \mathbf{u}(j) \,|\, j \leq k\} \tag{82}$$

represents the data known up to the kth instant. This means that there is a computational delay of one sampling period. The *linear quadratic Gaussian* (LQG) control problem is stated as follows. Find an admissible control signal that minimizes the cost functional (81) when the system is modeled by (79) and (80). The design parameters are the matrices in the cost functional.

8.4.2 Solution Strategy

In terms of the model and cost matrices, we consider that the associated matrix Riccati equation

$$\begin{aligned} P(k) &= \Phi' P(k+1)\Phi + Q - G'(k) \\ &\quad \times [R + \Psi' P(k+1)\Psi]^{-1} G(k), \qquad P(N) = 0 \end{aligned} \tag{83}$$

where the matrix G is defined by

$$G(k) = [R + \Psi' P(k+1)\Psi]^{-1} \Psi' P(k+1) \Phi \tag{84}$$

has a solution $P(k) \geq 0$ and the matrix $[R + \Psi' P(k)\Psi]$. The solution to the LQG problem is given by

$$\mathbf{u}(k) = -G(k)\hat{\mathbf{x}}(k\,|\,k-1) \tag{85}$$

and the corresponding minimum value of the cost function is

$$\begin{aligned} J^* &= m_0' P(0) m_0 + \text{Tr}[P(0) R_0] + \sum_{k=0}^{N-1} \text{Tr}[P(k+1) R_1] \\ &= + \sum_{k=0}^{N-1} \text{Tr}[P(k) G'(k)[\Psi' P(k+1)\Psi + R]^{-1} G(k)] \end{aligned} \tag{86}$$

It is interesting to note that the difference in the minimum cost functions

Optimal Design Methods

of the LQ and LQG problems is due to the estimation of the state variables. The solution so obtained is often called the *separation theorem*. One consequence of the separation theorem is that the synthesis problem can be divided into two parts, which can be solved separately. First, the deterministic LQ control problem is solved, giving $G(k)$. Second, the state is estimated using the Kalman filter. A block diagram of the system with the optimal control law is shown in Figure 8.4.

An interesting feature of the LQG problem is that the solutions to the LQ control problem and the state estimation problem are very similar and in some sense they can be obtained from each other. This

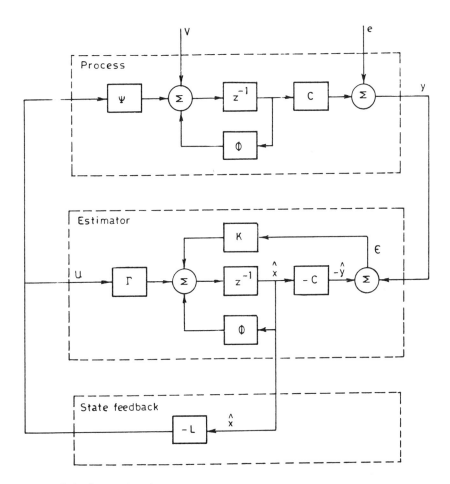

Figure 8.4 Separation theorem.

facet is called *duality*, that is, the state estimation problem is equivalent to an LQ problem by making certain substitutions among system matrices.

Given the system (79), (80) along with the control (85) and the Kalman filter

$$\hat{x}(k+1|k) = \Phi\hat{x}(k|k-1) + \Phi u(k) + K[y(k) - C\hat{x}(k|k-1)] \quad (87)$$

in terms of the estimation error $\tilde{x} = x - \hat{x}$, the closed-loop system can be written as

$$\begin{bmatrix} x(k+1) \\ \tilde{x}(k+1) \end{bmatrix} = \begin{bmatrix} \Phi - \Psi L & \Psi L \\ 0 & \Phi - KC \end{bmatrix} [x(k)\hat{x}(k)]$$

$$+ \begin{bmatrix} I \\ I \end{bmatrix} v(k) + \begin{bmatrix} 0 \\ -k \end{bmatrix} v(k) + \begin{bmatrix} 0 \\ -k \end{bmatrix} e(k) \quad (88)$$

It is obvious that the dynamics of the closed-loop system in (88) are determined by $[\Phi - \Psi L]$ and $[\Phi - KC]$, which correspond to the dynamics of the LQ control problem and the optimal filter, respectively.

8.4.3 Concluding Remarks

Optimal design based on state-space models has been discussed in this chapter. The LQ controller and the Kalman filter have been shown to have many interesting properties. The potential of the developed techniques hinges upon the availability of a full-order model that accurately describes the process dynamics and the ability to translate the design specifications into an appropriate cost functional. With the help of interactive computer facilities, one can develop an adequate design procedure for practical applications.

8.5 Problems

1. Given the linear model

   $$x(k+1) = \alpha x(k) + \beta u(k)$$

   with the cost functional

   $$J = \sum_{k=0}^{N} x^2(k)$$

   determine the control strategy $u(k)$ as a function of $x(k)$ which minimizes the cost.
2. Consider the model

Optimal Design Methods

$$x(k+1) = ax(k) + v(k)$$
$$y(k) = x(k) + e(k)$$

The equation is represented by

$$y(k) = [1 \quad 0]\mathbf{x}(k) + v(k)$$

where $v(k)$ is a zero mean Gaussian process with covariance $= 1$. The initial state is Gaussian with

$$m_0 = \begin{bmatrix} 95 \\ 1 \end{bmatrix}, \quad R_0 = \begin{bmatrix} 5 & 0 \\ 0 & 1 \end{bmatrix}$$

The measurement records of the position are $y(0) = 102$, $y(1) = 100$, $y(2) = 97.9$, $y(3) = 94.4$, $y(4) = 92.7$, $y(5) = 87.3$, $y(6) = 82.1$. It is required to estimate the position and the velocity of the falling body at $k = 6$. Obtain the corresponding with the standard assumptions and $a = 1$, $Q = 25$, $R = 15$, and $R_0 = 100$.

 a. Determine the asymptotic value of $p(k \mid k)$.
 b. Carry out a few computational cycles and hence detect the discrete instant at which the filter can be considered to be in a steady-state.
 c. Write the equation of the filtered estimate in a steady-state.

3. A discrete model of a noise-free system describing a falling body in a constant field is given by

$$\mathbf{x}(k+1) = \begin{bmatrix} 1 & 1 \\ 0 & 1 \end{bmatrix} \mathbf{x}(k) + \begin{bmatrix} -0.5 \\ 1 \end{bmatrix}$$

where x_1 is the position and x_2 is the velocity. Find the observation error covariance matrix.

4. For the scalar process

$$x(k+1) = \alpha x(k) + v(k)$$
$$y(k) = x(k) + e(k)$$

where all the standard assumptions hold with $v(k) \to N(0, q)$, $e(k) \to N(0, r)$ and $x(0) \to N(0, s_0)$, develop the filtered estimate equation by maximizing

a. $p[x(k+1) \mid Y(k+1)]$

b. $p[x(k+1) \mid Y(k)]$

and compare the results.

5. An inventory system is modeled by

$$x(k+1) = \begin{bmatrix} 1 & 1 \\ 0 & 0 \end{bmatrix} x(k) + \begin{bmatrix} 0 \\ 1 \end{bmatrix} u(k)$$

$$y(k) = \begin{bmatrix} 1 & 0 \end{bmatrix} x(k)$$

   a. With $Q = I$ and $R = r$, determine the steady-state LQ controller.
   b. Investigate how the closed-loop poles vary with the factor r.
   c. Obtain the state trajectories as function of r for $x(0) = [1 \quad 1]'$.

NOTES AND REFERENCES

A brief overview of the available results on state-space optimal design tools has been given. LQG control, LQ control, and optimal filters are the subjects of many books including Kwakernaak and Sivan (1972), Anderson and Moore (1971), Astrom (1971), and Anderson and Moore (1979). Main contributions to the development of the recursive optimal filters have been due to Kalman (1960), Bucy (1959), and Kalman and Bucy (1961). Other related work on optimal control can be found in Singh and Titli (1978), Bryson and Ho (1969). Different treatment and examination of least-squares estimation are carried out by many authors including Cox (1964), Eykoff (1974), Balakrishnan (1972), Sorenson (1970), and Bierman (1973).

9
Basic Hardware and Peripherals

9.1 INTRODUCTION

In recent years there has been a tremendous increase in the computing power available to engineers, managers, and scientists. This has been brought about by the advances in integrated circuit technology during the 1970s, which enable the implementation of a very large number of basic digital logic elements and units on a small area of a single wafer of silicon; hence the well-known name *silicon chip technology*. The development of large-scale integration (LSI) technology resulted in the fabrication of very high density and complex circuits. While the technology required to manufacture these chips is very advanced, the final product is highly reliable. This high reliability stems from the fact that a single circuit is performing the functions previously carried out by hundreds of components. As the number of components in a system increases, the chance of any one of them failing and hence the possibility of system failure also increases.

The development of microprocessors and microcomputers (micros for brevity) also brought about a revolution in engineering design. Micros are robust, small, and cheap enough to be used as components in a wide range of systems; industrial controllers and laboratory instrumentation are just two applications where mechanical, electromechanical and nonprogrammable electronic control systems are being replaced by micros.

As the use of mainframe computers became widespread, engineers and scientists learned how to program computers to solve purely numerical problems. Nowadays, however, the use of micros as integral parts of larger real-time systems represents a wide departure from this more "conventional" use of computers. To design and apply microcomputer-based systems in control engineering requires not only knowledge of the micros and their programming, but also of the remainder of the system. In this part of the book we provide a modest coverage of topics related to microcomputer-controlled systems. The material is divided into four chapters. Chapter 9 deals with the physical components of the system which are frequently termed the *hardware*. This includes the various building bricks and the associated peripherals that are fundamental to all engineering applications. Given the fact that a micro is a "do-anything" machine which can be programmed to perform any of a wide range of tasks, we present in Chapter 10 these programs or *software support*. Chapters 11 and 12 are devoted to typical case studies on industrial process control and on distributed computer control.

We start this section by identifying the three functions of microprocessor (microcomputer)-based systems: *processing*, *storage*, and *transmission*. These functions are then separated into three sections linked together. Figure 9.1 illustrates how these three sections within a microcomputer are connected in terms of the communication of information within the machine. The system is controlled by the *microprocessor unit* or *MPU* which supervises the transfer of information between itself and the memory and the input/output sections. It must be emphasized that the *processing* will be performed in the *MPU*, the *storage* will be by means of *memory* circuits and the *transmission* of information into and out of the system will be by means of special *input/output* (*I/O*) circuits.

The next section concerns the basic elements of the MPU. Broadly speaking, these contain:

1. Arithmetic and logic unit (ALU)
2. Instruction register
3. General purpose registers including an accumulator
4. Index register
5. Flag register
6. Interrupt vector register
7. Stack
8. Stack pointer register

Basic Hardware and Peripherals

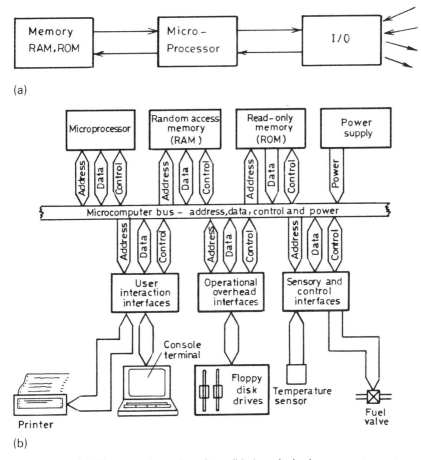

Figure 9.1 (a) Three sections of a micro. (b) A typical microcomputer system.

9. Program counters
10. Control unit

The MPU also contains flags which can be set or cleared as a result of the completion of specified operations.

9.1.1 Arithmetic and Logic Unit (ALU)

The principal function of a microprocessor is to manipulate data according to specified instructions. This function is actually performed by the arithmetic and logic unit. Usually the instruction is read from the memory, conveyed to the control section, and housed in the instruction register. Within the ALU, there is an accumulator in which one of the data items is stored and manipulated upon. A second data item can be held in a temporary register. The basic functions performed by the ALU include:

a. Binary addition
b. Binary subtraction
c. Complementing
d. Logical AND
e. Logical OR
f. Logical EXCLUSIVE OR

Other functions such as multiplication and division are carried out by repeated additions and subtractions. Once the data has been manipulated, the resulting data value replaces the original data value in the accumulator.

9.1.2 Registers

Registers are circuits or mechanisms which can hold information within a microcomputer by storing a fixed number of digits. There are various types of registers which, by and large, constitute the bulk of any microcomputer's circuitry. In the sequel we discuss some of the important types of registers and their role in micros.

General Purpose Registers

Here registers are treated as memory locations inside the microprocessor chip which can be utilized to keep data values. The vast majority of the microprocessor operations can be regarded as transfer of data between different registers. Individual registers have unique locations and are assigned unique numerical addresses. In this respect, microprocessor memory can be considered to consist of a number of registers. These registers might be general purpose or they might be used for specific functions. General purpose registers are used for high-speed access to and manipulation of data.

The accumulator is an example of a register with a specific function. This register is used to hold the data to be manipulated or the results of a data manipulation. Considerable data transfer takes place between the memory, external devices, and the accumulator. N-bit micros usually have N-bit accumulator registers.

Instruction Registers

This is an important element used during the execution of program instructions. Normally the program instructions, stored in the computer memory, are executed in a sequential fashion, one at a time. It is sometimes possible to jump to instructions in other sections of the program depending upon the satisfaction of specified conditions. The location of the required program instruction is determined followed by the retrieval of the actual instruction which is placed in the instruction register. This is followed by the decoding and interpretation of instructions to determine the operation to be performed by the arithmetic and logic unit. After the execution of the current, the next instruction is loaded into the instruction register and the whole process is repeated.

Index Register

This is a wide internal register, within the MPU, whose content is modified (subtracted from or added to) during the execution of an instruction to determine the memory location which will be acted upon by the instruction. Index registers are used during the indexed addressing operation.

Flag Register

Microprocessors also contain flags (in the form of binary switches) whose status is changed as a result of the execution of arithmetic or logical operations. For example, the status of overflow flags is altered as a result of the overflow of data stored in the accumulator. Similarly, a zero flag indicates when the data stored in the accumulator, following the execution of an instruction, has a zero value. The sign flag is used to indicate the positive or negative value of a number stored in the accumulator. Some microprocessors contain a large number of flags whose functions can be determined only by studying the relevant manuals supplied by manufacturers.

Interrupt Vector Register

One of the potential applications of microprocessors is for real-time data acquisition and process control. This forms a basic task for the microprocessor to be able to respond to external stimuli given their priority assignment. We should note that the demand for attention might come from one of a number of external devices. To this end, vector interrupt procedure is used to identify the external device which interrupted the normal execution of instructions and to transfer program control, after the current instruction has been executed, to the program section which will service the interruption device. It is also necessary to specify the return address so that the original program section can restart from the point at which interruption took place. The external device is identified by a data field or a vector.

9.1.3 Stacks

In the normal situations, the addressing schemes allow information to be accessed at known places in memory. Sometimes, however, it is convenient simply to put information to one side with the intention of recovering it later. In such cases the actual memory address used is irrelevant provided that information saved can be recovered when required.

If words are written sequentially into memory they can be read later in reverse sequence without the actual addresses used for storage being of importance to the program. The effect is quite similar to that obtained when documents are stacked up in a tray instead of filing them. Items can be found systematically in a stack of papers if they are removed one at a time in the reverse order to that in which they were placed on the stack. The idea of "stacking" information can be contrasted with that of "filing" it away at known memory addresses.

When data is stored on a *stack* it is customary to refer to it as being *pushed* on to the stack, and *pulled* or *popped* from the stack. A simple stack can be programmed using either indexed or indirect addressing and incrementing and decrementing the pointer (index or memory register) each time it is used, which is especially easy when autoincrement and autodecrement instructions are available. No matter how they are implemented, the key point about stacks is that they allow information to be stored in memory according to a *last in, first out* (*LIFO*) rule.

In applications, many microprocessor programs consist of a number of subroutines for implementing functions which are frequently used in the main body of the program. For this purpose, the microprocessor system uses a memory section, often referred to as *stack memory*, for storing subroutine addresses so that the main program can be executed from the point at which it was interrupted to process the subroutine program. The stack is also used during the processing of interrupts.

Using assembly language with the opcodes

        LD   LoaD register
        ST   STore contents of register

Basic Hardware and Peripherals

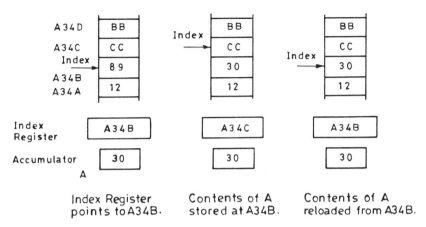

Figure 9.2 Implementing a stack using index register.

a stack can be programmed in the form

$$STA \quad 0, X+$$

to store the contents of the accumulator A at the address pointed to by the index register and move to the next address, and

$$LDA \quad -X$$

to move the index register back to the previous location and load the accumulator with the contents of that register. An example is shown diagrammatically in Fig. 9.2 where the index register points initially to the unused location A34B (the contents of the memory locations are arbitrary).

We noted previously that the data transfer to and from the stack takes place in a push-down/pop-up manner. For retrieval of stack data, the item which was pushed on the stack will be retrieved, that is, the last data item will pop up. Specifically, to retrieve the very first item of data, it will be necessary to access all the other data items first and in a sequential fashion. This is contrasted to the random manner in the case of random access memory.

The stack contains a number of memory locations. To keep track of the stack memory locations on which the next item of data can be written, a *stack pointer register* is used.

9.1.4 Control Unit

The microprocessor contains a control unit which directs and oversees the execution of various operations and synchronizes events. This control unit is connected to the clock which provides the signals for timing and control. An instruction is fetched from the memory, placed in the instruction register and decoded by the instruction decoder which contains the instruction set for the microprocessor. The decoded instructions are then executed. The address, in memory, of the next instruction is fetched from the program counter, the instruction is loaded into the instruction register and the whole process is repeated. This cyclical process of fetching and executing the instructions is repeated at very high speed, which varies from one microprocessor to another and also depends upon the complexity of the instruction to be executed. The time required for fetching and executing instructions depends upon the basic cycle time often referred to as the *micro-cycle time*.

In a microprocessor system, a master clock is used to synchronize the movement of data. A time reference is required for various input/output operations and for obtaining access to memory. This is achieved by using a high-speed pulse generator. Some micros include clock circuits on the chip while others use quartz crystals, connected to the clock chip, to generate time pulses.

9.1.5 Program Counter

The micro uses a special purpose register called the *program counter* (*PC*) to keep track of where it is in the program. The program counter register always contains the address of the next instruction to be carried out, and as the program executes the address which the *PC* contains, it is automatically incremented to point to successive instructions.

Microcomputer instructions are carried out in two phases called *fetch* and *execute*. During the first phase the program counter provides addresses for the binary opcode and operand information to be fetched from memory and in the second phase the micro acts upon this information and executes these instructions. The program counter is so called because its contents are incremented by one each time a word is fetched from memory, and most of the time it continues in this mode with the micro fetching the next instruction when it has finished executing the current one. Note that because the program counter has been used to fetch all the bytes of the current instruction into the microprocessor by the time that it starts to execute the instruction, the program counter will point to the start of the next instruction, while it is executing the current one. This is illustrated in Fig. 9.3. The program counter is used whenever instructions are fetched from memory and ensures that they are executed in numerical sequence.

Basic Hardware and Peripherals

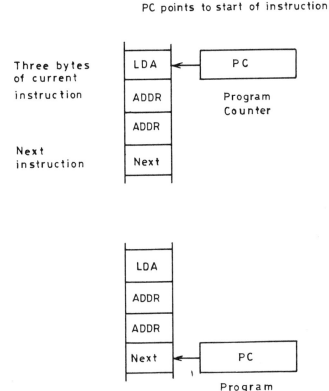

Figure 9.3 Operation of program counter.

9.2 MICROPROCESSOR CHARACTERISTICS

Microprocessor choice greatly affects overall system characteristics, so it is important to understand the performance of the microprocessor and to identify the relevant characteristics during the system evaluation stage.

9.2.1 Purpose

We have learned that micros are general purpose devices that can perform almost any task if given enough external support circuitry and processing time. However, they have built-in features that make them better suited to certain applications. The two major purposes are electronic data processing (EDP) and control. In this context, EDP refers to tasks requiring extensive arithmetic operations. But a control application may indeed require some EDP, making an EDP-type microprocessor well suited to the task.

One way to judge the microprocessor purpose is to look at such characteristics as (1) bit width, (2) instruction set, and (3) support hardware and software.

The very narrow word width of 4 bits is indicative of a controller. Arithmetic and ASCII character manipulation are difficult to implement using such a narrow word, and quadruple-precision arithmetic is necessary to represent even a comparatively small number such as 23754. On the other hand, a 4-bit word width would prove adequate for many control applications. For example, up to 16 traffic lights can be represented by a 4-bit code.

Micros with broader widths usually indicate an EDP orientation. This applies to all 16-bit processors since they are actually derived from minicomputers.

The instruction set of micros gives a clue to their purpose. To be suited for EDP tasks, an instruction set must allow arithmetic shifts and accommodate two's complement arithmetic.

Support hardware and software lend more evidence to the purpose of a microprocessor. A simple controller chip will not have a broad range of support chips such as floppy disks, controllers, memory popping units, and one-chip modems. A more sophisticated processor has a broad base of EDP-type hardware and software support.

9.2.2 Bit Width

The bit width of a microprocessor is simply defined as the number of parallel lines contained in the data bus. In general, the bit width has a great effect on system complexity and capability. Data and instructions are usually stored in a memory as wide as the bit width of the processor. The advantage of a microcomputer with a wide word width is that the microcomputer can handle a much wider range of arithmetic values before resorting to inefficient multiple-precision arithmetic. It can also have a much larger set of single-word instructions. The results of having these features are higher memory bit widths, wider data buses, and connectors, and usually wider bit widths on the interfaces tied to the bus.

A 4K memory for a 16-bit microcomputer takes twice as many RAMs as that of a 4K memory for an 8-bit machine. It is wise to keep the bit

Basic Hardware and Peripherals

width as low as is reasonably possible in an application because part counts, especially in the memory area, are much lower for narrow bit-width machines.

It is known that microprocessors with 4-bit word widths are almost exclusively defined for control applications. The 8-bit micro can be designed for either EDP or control, but in most cases it is designed to be general purpose enough for both. Double precision arithmetic is fairly efficient on these devices, and 16 bits of precision is adequate for most EDP work. Microprocessors with 16-bit word widths are almost exclusively used for data processing in which more than just control functions are required.

9.2.3 Bit-Slicing

Some micros, notably the high-speed bipolar types, are *bit-sliced*. Large bit-width microcomputers can be built from few 4- or 8-bit processor "slices".

Bit-slicing, especially with 4-bit elements, is primarily used for thermal reasons. Bipolar large-scale integrated (LSI) circuits draw a lot of current and tend to run hot. A 16-bit or even an 8-bit processor would generate too much heat for a single package to dissipate.

Bit-sliced processors are usually more like LSI building blocks than self-contained processors with strictly defined I/O protocols and instruction sets. By varying the number of bit slices and changing the control ROM (that contains a microprogram for the control sequence), the desired microcomputer bit width and even instruction execution method may be changed.

Bit-sliced microprocessors are used in custom, high-performance applications. Due to the bit-sliced microprocessor's versatility, systems built with them can be made to emulate more common computers efficiently. Bit-sliced micros are therefore being used extensively in the construction of minicomputers.

9.2.4 Processing Speed

Processing speed is the rate at which a microprocessor executes the application program, and this depends on three basic specifications: the clock rate of the microprocessor; the number of cycles required to execute a given instruction; and the instruction repertoire itself. To see the significance of these factors and the manner in which they interrelate, we must have a common understanding of the terms and their functional contributions.

Processor Clock Rate

The clock rate is defined as the frequency of the clock input to the micro; the number of clock pulses produced per second. Since the clock is the governor of all timed operations within a system, it follows that a high-rate clock permits more operations to be performed within a given period; but a high-rate clock coupled with low-rate peripherals translates to interface complexities.

Acquisition/Execution Rate

The acquisition and execution rate of a micro may be expressed in microcycles; the number of cycles or operational steps required to perform a given instruction. A microcycle consists of one or more clock cycles. Most metal-oxide-silicon (MOS) microprocessors require many microcycles to execute one instruction. Typically, one microcycle might be used to fetch the instruction, one or two more might be used for data access, and several more for the actual execution operation of the acquired instruction. The number of microcycles required by an instruction is affected by the addressing mode and the instruction complexity. For example, a simple add may take 14 microcycles, while a multiply may take 52 on the TMS 9900 16-bit micro.

Instruction Repertoire

The types of instructions a micro can execute determine its suitability to a task. Instructions should be evaluated on the basis of what they can do, not how many there are.

The number of instructions a micro can perform depends on the manufacturer and therefore cannot be unified. The instruction set should be oriented toward the kind of processing: In control applications, particular attention should be paid to I/O instructions and in EDP, the data manipulation instructions (arithmetic shifts, complement instructions, arithmetic branches) should weigh heavily in the choice.

The question asked is how system speed is determined? A true measure of how fast a program will perform a given task is how much time it takes to execute a total program. This figure is the number of clock cycles needed to execute a program multiplied by the microprocessor clock rate.

9.2.5 Common Microprocessor Families

In this section we provide brief descriptions and evaluations of a few common microprocessor families. No attempt has been made to cover

Basic Hardware and Peripherals

all micros since the market is dynamic and new devices appear constantly.

Intel 8080/8085/8086

The Intel 8080 was the first micro to gain wide acceptance in the microcomputer field and did, in fact, help create the field. From an architectural and feature standpoint, the 8080 is quite primitive and improved processors are usually designed into new products. But due to its wide acceptance, multiple sources, and large line of support chips, the 8080 is here to stay. The 8080 instruction set reflects a control nature in this microprocessor and possesses a few data processing features. A block diagram of the 8080 micro is shown in Fig. 9.4.

Among the relevant features are:

1. The architecture contains six 8-bit registers that may be used individually or in pairs for 8- and 16-bit operations.
2. It is capable of stack operations by incorporating a 16-bit stack pointer to keep track of the push-down stack.
3. It has asynchronous vector interrupt and direct memory access (DMA) capabilities.
4. The 8080 has about 100 instructions, with length varying between 1 and 3 bytes each.
5. There are four basic addressing modes:
   a. Direct-mode addressing allows direct loading or storing of the accumulator.
   b. Immediate addressing allows the loading of any register or register pair.
   c. Implied addressing is used in operations needing no memory reference.
   d. Indexed addressing allows the contents of register pairs to be used as 16-bit pointers.

By far, the 8080 is currently one of the best supported micros in regard to interface chips. Software support is equally diversified.

An upgraded version of the 8080 is the Intel 8085 which incorporates a built-in clock and system controller. It is software compatible with the 8080 and contains two additional instructions. The 8086 is the first micro to use the new silicon-gate short channel HMOS process. This process makes the 8086 faster as well as more reliable than a similar product fabricated from regular NMOS. Having an expanded register set enables numerous processing operations to be implemented. It is actually two processors in one package.

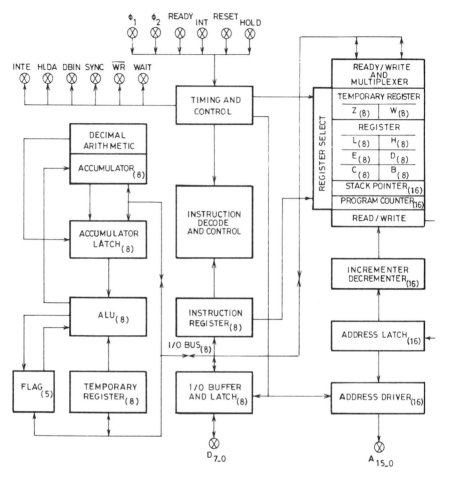

Figure 9.4 Block diagram of 8080 micro.

Zilog Z80

The Zilog Z80 is a greatly enhanced upgrade of the 8080. Enough similarity is maintained to allow 8080 programs to be used and additional instructions (which correct the lack of arithmetic capability) are included in the instruction set. More than twice the number of internal registers are used and two independent index registers enhance the addressing capabilities. Improved functions have been integrated into the chip yielding a dual-purpose nature. The Z80 qualifies for both EDP and control applications.

Basic Hardware and Peripherals

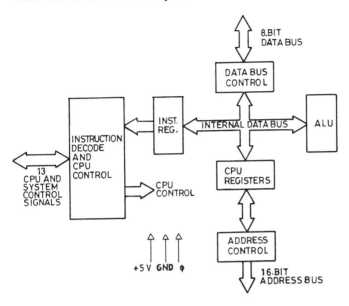

Figure 9.5 Block diagram to the Zilog Z80 (courtesy of Zilog, Cupertino, CA).

Among the relevant features are:

1. It contains eighteen 8-bit registers and four 16-bit registers. Two accumulators and flag registers are also provided. Figure 9.5 illustrates the Z80 structure.
2. Many additional instructions have been added, eliminating most of the 8080's arithmetic and data processing shortcomings.
3. From a software viewpoint, I/O is a bit simpler with the Z80 than 8080 since any register can be written to the output or loaded through the input instruction. A comparison of 8080 and Z80 I/O methods is shown in Fig. 9.6. One very unique Z80 feature is its simultaneous I/O capability.
4. The Z80 CPU has a dynamic RAM refresh capability.

Motorola M6800

The M6800 is another one of the most widely used microprocessors. This family of chips has design features that make it very desirable in control applications. It features total two's complement arithmetic as well as control capability. The 6800 therefore does very well in EDP-type applications. The particular features are:

330   Chapter 9

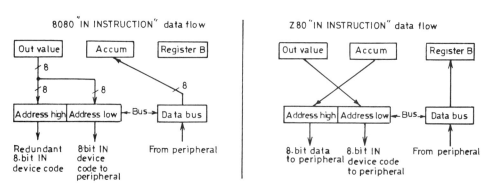

Figure 9.6 A comparison of 8080 and Z80 "IN INSTRUCTION" data flow.

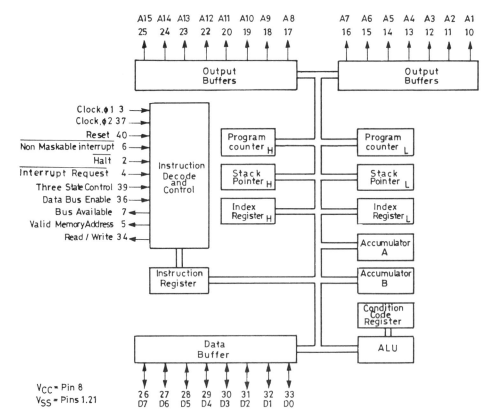

Figure 9.7 Architecture of the 6800 micro (courtesy of Motorola Semiconductor Products, Inc.).

Basic Hardware and Peripherals

Table 9.1 Some Data of Microprocessors

Item	8080	Z80	6800	9900
Manufacturer	Intel, AMB	Zilog, Mostek	Motorola, AMI	Texas Inst
Device technology	NMOS	NMOS	NMOS	NMOS
Data word width	8-bits	8-bits	8-bits	16-bits
Addressing range	65,536 words, external	65,536 words, external	65,536 words, external	32,768 bytes
Instruction width	8-bits	8-bits	8-bits	8–16-bits
Instruction count	78	158	72	
Clock frequency	0.5–4 MHz	5 kHz–4 MHz	DC–2 MHz	0.5–4 MHz
Package	40-pin DIP	40-pin DIP	40-pin DIP	60-pin DIP
Power requirement	12 V at 40 mA 5 V at 60 mA −5 V at 10 mA	5 V at 90 mA	5 V at 100 mA	5 V at 75 mA 12 V at 40 mA −5 V at 100 mA

1. It can be viewed as a functionally balanced microcomputer building-block set.
2. The 6800 is memory-oriented and its architecture follows the philosophy of using a low-speed clock with many actions per clock cycle. Most 6800 instructions execute in 2, 3, or 4 clock cycles (versus 8 or 9 for the 8080).
3. Data can be manipulated with two 8-bit accumulators, and a 16-bit index register is available for address manipulators.

The 6800 instruction set's direct addressing mode and extensive conditional branch capabilities allow the majority of the instructions being performed to be one or two bytes long. This reduces memory requirements and decreases memory access, thereby increasing processor performance. These factors make the 6800 desirable in real-time applications in which execution speed is important. Figure 9.7 shows the 6800 structure. One basic feature is that the 6800 relies on memory-mapped I/O and therefore no independent I/O channel or I/O instructions are provided.

TMS 9900

One of the first 16-bit one-chip microprocessors was Texas Instruments' TMS 9900, which is produced as one large chip in a 64-pin package. It was initially intended to be a central processing unit that could successfully compete with minicomputers in scientific applications. Unlike other micros described earlier, the TMS 9900 is a 16-bit unit that runs with a maximum clock rate of 3 MHz. The architecture follows the philosophy of high-speed clock with many small operations per cycle. Its main features are:

1. A memory-oriented machine whose structure consists of a program counter, status register, and workspace pointer
2. Three I/O methods to facilitate versatile communications
3. A high-order language that eases real-time programming tasks

In Table 9.1, some of the relevant technical data are summarized. We must emphasize that the selection of micros for a typical application is a design problem in itself.

9.3 MEMORY

Computer processors are usually composed of three functional pieces: the central processor, the input/output unit, and the storage unit or

Basic Hardware and Peripherals

memory. This division results from the concept of sequential computation and the need to store commands and data within the computing machine. Because memory is such an integral part of computer systems, its technology has evolved and become at least as complex as processor technology.

The present generation of microcomputers can be classified as cell-addressable single-memory machines. A cell-addressable memory is one that accepts a processor-issued address and returns a program or data word from the location in memory corresponding to that address. The term single-memory machines refers to the way in which a processor accesses memory. It has the advantage of having less hardware.

9.3.1 Memory Hierarchies

In computer systems, the problem of memory storage capacity versus memory speed and accessibility has confronted the designers for some time. It has been solved using *memory hierarchies*. Our purpose here is to examine some of the memory types. The MPU section of a computer system typically uses a high-speed memory to temporarily store programs and data while it is being processed. This is called a short-term memory or *working store*. Working stores usually range in size from a few bytes to a few tens of kilobytes in most microcomputer systems. This amount of working storage is enough to store the programs and data for most micro applications.

A few hundred kilobytes of programs and data are accessed often enough in most applications to require a fairly quick storage unit to hold this data. Medium-term storage units serve this purpose. Disk drives, charge-coupled devices, and budded memories work well as medium-term storage media.

Huge blocks of data that will be referenced periodically require a bulk storage media. Magnetic tape and large disk packs are ideal for this long-term storage requirement. Long-term storage units are capable of storing hundreds of megabytes of data quite slowly but at a very low cost.

The three types of storage (short-term, medium-term, and long-term) constitute a memory hierarchy, as demonstrated in Table 9.2. At the top of the hierarchy, a fourth level is added to represent high-performance processors.

In terms of memory hierarchies, data must be searched for and read-in from the long-term storage devices, put into the medium-term devices and ranged into the short-term memory as needed. After the data

Table 9.2 Characteristics of Memory Hierarchies

Speed	Type Processor	Typical Device
Less than 250 ns	Ultrahigh-speed memory	RAM, HMOS, RAM
250 ns to 2 μs	Short-term memory (high-speed working store)	MOS, RAM, magnetic core
10 μs to 10 ms	Medium-term memory (medium-speed, medium capacity)	Floppy disks, bubble memory, drum memory
2–100 ms	Long-term memory (low-speed, high capacity)	Magnetic tape, cards, laser memories

processing is complete, all the data has to be sent back down the hierarchy to the long-term storage devices.

9.3.2 Working Store

Short-term working store media has the important characteristics of fast access time and random-access capability. Semiconductor RAMs, ROMs, and magnetic core memories are the most common.

The availability of RAM has made the micro possible. Without low-cost, short-term memory, microprocessors have very little cost advantage over any other CPU. A RAM consists of two functional blocks, the memory cell array and the peripheral interface circuitry as illustrated in Fig. 9.8. Through address lines, the particular cell can be examined by I/O circuitry for read or write operation.

The memory cell array is the heart of the memory unit. There are many approaches to the design of the individual memory cells; the most common of which are:

1. Static flip-flop cells
2. Transistor-base-charge dynamic cells
3. Pseudostatic charge-pumped cells

A RAM peripheral circuitry consists of:

1. An address decoder that takes half of the memory address field and divides it into many single row-select lines
2. A decoder that divides the other half of the field into many column-select lines

Basic Hardware and Peripherals

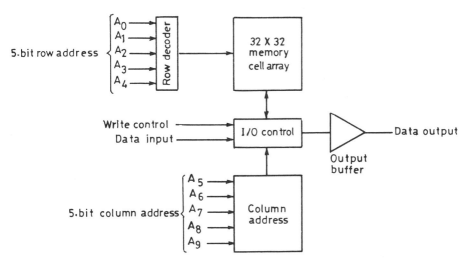

Figure 9.8 Memory peripheral blocks.

The most important RAM features are:

1. Memory type, static versus dynamic
2. Memory size
3. Memory configuration, that is, the way in which memories are organized
4. Memory speed represented by memory access time and memory cycle time (read time plus read recovery time)
5. Device technology covering transistor-transistor logic (TTL), integrated injection logic (I^2L), V-groove semiconductor technology yielding VMOS and MOS families.

Sometimes it is desirable to have a nonvolatile memory store within a computer to hold often used programs. If a program can be present at the moment a micro is turned on, the need for program loading and the associated load delay time are eliminated. A permanent, nonvolatile program is also useful for storing a small program consisting of common system utility programs and an initialization program or a bootstrap program. Read-only memories (ROMs) and programmable read-only memories (PROMs) are the most popular nonvolatile storage devices. Other versions include erasable PROMs (EPROMs) and electrically alterable ROMs (EAROMs).

A ROM is similar in design to a RAM, with the primary difference

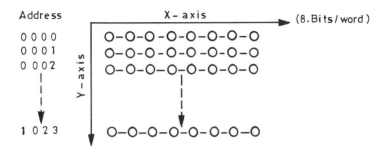

Figure 9.9 A 1024 × 8 word-organized 2-dimensional ROM array.

in the type of memory cells. A two-dimensional, word-organized ROM array is displayed in Fig. 9.9.

9.3.3 Medium Capacity Storage

Storage devices of medium capacity are used temporarily to store blocks of data that are used often enough to require fast random access but are too big to fit into a short-term working store, at least all at once. Devices included in the medium-term mass-store category are nonvolatile media such as floppy disks, magnetic-bubble memories, and solid-state disks built from charge-coupled device technology.

The floppy disk is a flexible, paper thin disk spinning at 360 rpm in its jacket. It contains a number of concentric tracks (77 tracks for a full size floppy and 35 tracks for a minifloppy) that are accessed by radially moving read-write heads onto the proper track using a stepping motor or magnetic "voice coil" linear motor. The disk drive has two indications that it is on the proper track. Because stepping motors are used to move the heads, the step movements are counted and the step count indicates the proper track. Secondly, a track identifier code is written at the beginning of the first sector on any given track.

The tracks on the floppy disk are divided into 10 to 26 pie-wedge sectors. Small blocks of data (typically 128 bytes by 8-bits) are stored serially on each sector along with a few bytes of preamble data and empty data gaps to keep sectors well isolated from one another.

On the other hand, bubble memories are magnetic devices that store data as magnetically oriented domains in a sheet of magnetic material; but unlike most magnetic media, the magnetic material stays stationary and magnetic domains or bubbles move around within the magnetic material under the influence of a rotating magnetic field. By forming a long loop of magnetic bubbles using a bubble generator/cater and a magnetic field, a large "shift register" memory is set up.

Bubble memories combine some of the best features of both magnetic and semiconductor memory storage. They are large-capacity nonvolatile storage devices and the area in which the actual data is being stored requires no power to hold the data.

Finally, charge-coupled devices fall into the same class of medium-term storage because it too is a serial shift register. They store data as charges on a row of either T-shaped or chevron-shaped charge storage cells. By applying the proper voltage levels to these cells, charges can be moved around; a circulating shift register is thus formed.

Unlike bubble memories, charge-coupled devices are volatile in character and much faster because they move charges rather than magnetic domains. However, they do not pack the same density of storage into a given area. To make memory access time quicker, many short shift registers and decoding logic are often incorporated into a charge-coupled device chip.

9.3.4 High Capacity Storage

In terms of memory hierarchies, high-capacity storage devices reside at the lowest levels since they are capable of storing vast amounts of data in a somewhat permanent, nonvolatile form at a very low cost. Access speed and random-access capabilities are of secondary importance for long-term storage devices; memory storage size is the primary goal. In regard to long-term storage for microcomputers, high-capacity and medium-capacity devices are usually one and the same. This situation is brought about by the nature of micros.

The three most common forms of long-term storage in microcomputer systems are floppy disks, magnetic tape (usually cassette), and paper tape.

We conclude this section on memory types and hierarchies by noting that the remaining task is that of interfacing with the micro system. Selection of components and the matching between them are among the major design issues to be examined in implementation. We shall shed some light on these points later in this chapter.

9.4 INPUT AND OUTPUT METHODS

Two important areas of microcomputer systems that are often overlooked are the control structure of micros and the I/O methods used to get data in and out of the closed microcomputer-memory system.

This is due to the fact that the efficiency and overall performance of micros are greatly affected by the control logic and accordingly most manufacturers prefer to keep this unique design feature proprietary. Generally speaking, control logic is somewhat invisible to a computer programmer. However, control logic and the associated I/O protocols

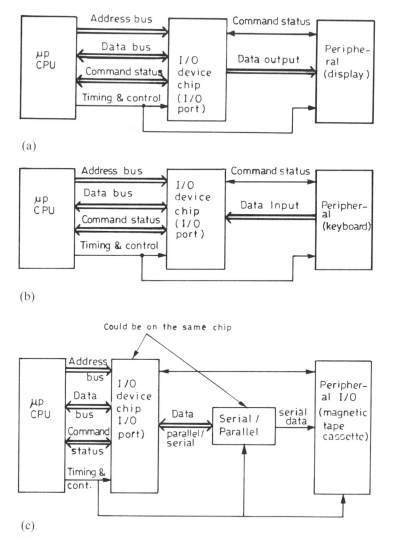

Figure 9.10 I/O interface between micro and peripherals. (a) Block diagram for an output transfer. (b) Block diagram for an input transfer. (c) Block diagram for I/O peripheral with serial/parallel sense conversions.

Basic Hardware and Peripherals

of micros play the largest role in communicating with the outside world. A single-port I/O chip for an output device such as a display is shown in Fig. 9.10(a) in block diagram form. A similar setup for an input peripheral, such as a keyboard is shown in Fig. 9.10(b). The setup for a peripheral that has both input and output capabilities is shown in Fig. 9.10(c). This section is devoted to clarifying concepts and methods of an I/O unit.

9.4.1 Data Transfer

An input or output operation is defined as the act of selectively transferring data to or from a selected peripheral device. We now illustrate two simple cases: (1) single-bit output transfer, and (2) single-bit input transfer. With reference to Fig. 9.11(a), the output device consists of a bit-storage element (say a rising-edge-triggered register), a light-emitting diode (an indicator) and the bit-generator (micro). The micro presents valid data and an output cycle clock line to the register. At the midpoint of the data-valid intervals, the control unit raises the output cycle clock line to logic 1 as shown in the timing diagram. As a result, the data is transferred from the micro to the register and the diode lights up.

Figure 9.11(c) depicts a single-bit input cycle. In this case, the input clock signal issued by the micro is used to sample the data at the D input of the register. The data becomes stable at the output end after a short time from rising the edge of the input clock. The control unit provides all the proper timing signals for the data transfer and usually transfers the valid data on the line to one of the registers in the micro where it can be accessed by the programmer.

Extending these cases to parallel data transfer can be achieved in a logical way as can be seen in Fig. 9.11(b) and (d).

9.4.2 Serial I/O

Providing several bits of data to an output register and supplying an input clock to chop the data is really a fast and simple way of transmitting data to a peripheral device. In practice it is not recommended to have eight or more data lines plus a clock line extending out to all the peripherals in a system. It is advantageous in these cases to replace parallel data transfer with serial data transmission (SDT). This SDT is concerned with the process of breaking bytes of data down into single bits and conveying them to the peripheral devices one at a time. Indeed this is not a simple matter and it involves some problems. The first concerns the parallel/series data conversion. Some form of shift regis-

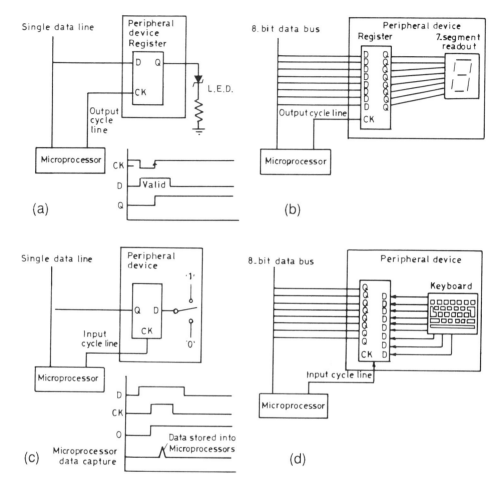

Figure 9.11 Simple I/O data transfers. (a) Single bit output transfers. (b) Parallel output transfers. (c) Single bit input transfer. (d) Parallel input transfer.

ters implemented by micros under software control can be constructed to receive the parallel bits and deliver them sequentially and vice versa. A second problem arises when the task of determining which bit of the transferred byte is needed. One common approach is to synchronize the transmit and receive shift registers initially. Then the circuitries of the transmit and receive ends should keep track of the byte boundaries in the bit stream. This type of serial transfer is called *synchronous communication*. It is fair to admit that the practicality of synchronous

Basic Hardware and Peripherals 341

communication depends, to a large extent, on the ability of the transmitter and the receiver to stay synchronized after initialization. The registers of both the transmit and receive ends can be initialized through a character matching process. Usually a clock signal is needed to maintain synchronized serial data transmission.

9.4.3 Common I/O Methods

Every micro has its own way of applying data transfer principles. This section deals with some common methods of input and output.

The Data Channel

In the standard architecture of computers, the memory interfaces with the processor through one interface and to the peripherals through a different interface. Separate processor instructions are set aside for memory reference and I/O operations. This kind of an I/O scheme is called *data channel input/output*. Micros have optimized I/O schemes in order to fit a whole computer into a 40-pin integrated circuit (IC). Some micros have updated the channel-oriented I/O schemes and others have introduced an I/O request line to save additional devices.

Memory-Mapped I/O

Hardware optimization and simple-to-use instruction sets have become fundamental features of computer systems. One typical example lies in approaching the I/O and memory interface process logic simultaneously rather than independently. A memory-mapped I/O interface is nearly identical to a memory interface, but instead of using RAMs or ROMs, input and output registers are incorporated. Like a memory interface, the peripheral interface must contain a complete N-bit address recognizer (for N-bit address lines) as well as buffers and associated logic.

9.4.4 Communication Buses

The signals coming out of and going into a microprocessor chip are adequate to communicate with any peripheral or memory device controlled by the microprocessor, but the signals rarely are sent directly to interfaces and memory. Instead, additional logic is used to form a standardized communication bus (the microprocessor bus) to which memory and peripherals may be interfaced. Microprocessor buses offer many advantages over irregular connection of microcomputer interfaces. It is sufficient to mention features like modularity, high fan-out, standardization, and circuit protection. First, we define a microcomputer bus as a set of address, data, control, and power lines arranged in a standardized manner and operating under a strict set of data communication rules. Power lines are required to provide enough excitation to various interface logic and peripherals. Reliable computer systems entail the use of good solid ground for all circuits in the system, particularly when a wide range of voltages for operation is employed.

Two approaches are taken in distributing power on the bus. In one, an external, regulated power supply places a voltage of precisely the proper level on the power lines. In the other approach, on-board voltage regulators are used.

Most micros have bidirectional data lines that permit transfer of data to and from memory and peripheral devices and save I/O pins on the IC package.

Address lines from the microprocessor are usually set to high-current drivers to drive address lines on the micro bus.

The control lines are dependent on the processor driving the bus. Some of the common control lines are: clock lines, initialize lines, memory control lines, interrupt lines, halt and wait lines, and direct memory access (DMA) status lines.

9.4.5 I/O Transfer

It is important to determine when to start a data transfer within the microprocessor. Two methods are in common use: the examination of device status under program control (polling) and peripheral-initiated program interruption (interrupt driven I/O).

Polling

To help the microprocessor in keeping track of the data transfer, it is customary to have a separate input register built into the interface which the processor can use to obtain status information about new data submitted to the interface.

By getting into a programmed loop that repeatedly examines or polls the status register waiting for some input command (key to be pressed for instance), the micro can be made to read the new data, and so on.

Interrupt-Driven I/O

Following the idea that the action of a peripheral marks the beginning of a data transfer, it seems more reasonable to have the peripheral device "inform" the processor when it is ready with new data. This is

Basic Hardware and Peripherals 343

precisely what an interrupt-driven I/O means: When a peripheral device has data to transfer, it lets the processor know.

An interrupt system must be incorporated in the control structure of a microprocessor if this I/O initialization method is used. It causes a program to suddenly halt, and diverts execution to a separate program which inputs or outputs the new data.

Finally we discuss the idea and operation of interface devices.

9.5 INTERFACE COMPONENTS

To complete the description of hardware devices, we examine some of the interface components that are in common use in micros.

9.5.1 Driver Circuits

Receiver and driver circuitry is used on all input and output lines on flip-flops, latches, and registers. The main problem is that of properly matching the individual units together, without too much loss or noise pickup. A driver is an output device capable of generating a standardized voltage or current that other parts can use.

Transistor-Transistor Logic (TTL) Drivers

A TTL driver consists of one or more transistors that switch voltage levels or apply current to represent logic values. Generally speaking, the driver is characterized by:

1. Its high and low (voltage) output levels.
2. The number of receivers it is capable of driving (fan-out).
3. Its switching speed.
4. Its rise time.
5. Its noise threshold.

Among the TTL driver circuits are:

1. Open-collector driver.
2. Totem-pole driver.
3. Tristate driver.

MOS Drivers

MOS and CMOS systems draw much less power and work with much lower currents than TTL systems, particularly when run at low frequencies. These are a variety of circuits built around MOS units.

However, MOS provides low-power characteristics, a factor which one should watch in system design.

Emitter-Coupled-Logic (ECL) Drivers

ECL is a current-oriented logic as opposed to TTL, which is voltage-oriented. Since it is nonsaturating, ECL is a high-speed logic. As a result, a bus can be built with ECL drivers.

9.5.2 Receivers

A receiver is an input device capable of converting a signal from a driver into a signal that is usable within the chip with which it is associated. Internal voltage levels and current requirements vary widely from chip to chip, and it is the receiver's job to match the internal requirements to the signals arriving at the chip's input pins.

The important receiver characteristics are:

1. Input current level
2. Input voltage level
3. Circuit immunity to environmental noise

The above characteristics are to be considered when applying TTL, MOS, ECL, or even I^2L.

We should mention that factors related to driver circuits carry over to receivers as well. All the foregoing information put emphasis on the functional layout rather than on the electrical details. This will be our approach in the remaining section.

9.5.3 Data Acquisition

Data acquisition is the process of taking analog signals from the real world, preprocessing them, converting them to digital data and finally bringing the resulting digital data into the computer memory. Figure 9.12 illustrates an overall picture of a data acquisition system. Physical parameters are first converted to electrical signals by a transducer. An amplifier is used to guarantee the voltage level for processing. Then follows an active filter to remove unwanted high- and low-frequency signal components. Special conditioning circuits are used as necessary for signal compression, multiplying, and squaring. The sample-and-hold unit samples the voltage level of the input at a specific instant of time and holds it constant at its output. So the analog-to-digital (A/D) conversion circuitry can sample a steady voltage level. The A/D unit converts the stable voltage level to a digital value corresponding to the

Basic Hardware and Peripherals 345

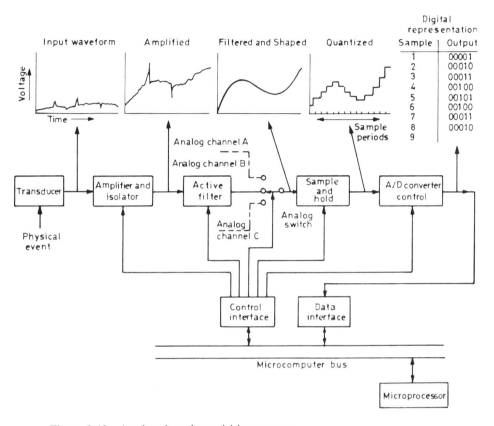

Figure 9.12 A micro-based acquisition system.

input voltage. This data can be put through some interface parts and sent to a micro.

The micros or discrete control circuitry must control the signal acquisition and processing circuits to insure that:

1. Proper analog inputs are selected.
2. Data is sampled at the proper time.
3. Data is held long enough for the D/A converter to make a valid conversion.

The most important parameters of a sample-and-hold circuit are:

1. Acquisition time (the time between start of sampling and stable output)

2. Aperture time (the time it takes for the sampling switch to fully open)
3. Aperture uncertainty time (the variation characteristics of the aperture time)
4. Decay rate (the change in output voltage resulting from capacitor discharge per period of time)
5. Feed through (the amount of input signal that manages to leak through to the output when in the hold mode).

Among the A/D conversion schemes are:

1. Dual-slope integrator
2. Successive approximation
3. Parallel conversion

Converting a digital word to a proportional analog value is a necessary task and is accomplished by the digital-to-analog (D/A) converters. Weighted-current source circuits are typical examples of D/A converters having fast and accurate performance.

To complete the interface components, micro systems employ devices which convert electrical energy to mechanical energy or vice versa. Such devices are called transducers. There are numerous types, each of which depends on the application at hand. Examples include pressure-sensing, flow-rate, solenoids, and stepping motors.

9.5.4 Some Remarks

It should be noted that assembling a microcomputer system for a particular application should include the tasks of:

1. Interfacing the various components
2. Selecting and matching the proper devices
3. Building the system using acceptable design and construction techniques

The information needed to carry out these tasks is usually available from diverse sources including manufacturer's specification sheets, journal articles, industry papers, and textbooks. Indeed, the information gathering is generally a time-consuming process. It can be minimized, however, with the aid of some software capabilities; a subject that will be discussed the next chapter. An in-depth study of each unit or component will eventually help in simplifying the design effort.

9.6 PROBLEMS

1. State the basic functions of
   a. Memory
   b. CPU
   c. Instruction decoders
2. Briefly describe the difference between a volatile and a nonvolatile memory.
3. For EPROMS
   a. How is the previous information erased?
   b. Is it possible to erase only one word and save the rest?
   c. Can the chip be erased while still inserted in the circuit board?
4. Name a few advantages and some disadvantages that electrically alterable ROMS (EAROMS) have over erasable PROMS (EPROMS).
5. What role do the I/O interfaces play in a microsystem? What functions do they perform?
6. What are the two modes in which the serial I/O interface unit can possibly operate? Briefly explain each mode.
7. Briefly define the following operating parameters of an A/D converter.
   a. Input range
   b. Resolution
   c. Conversion time

NOTES AND REFERENCES

The main topic of this chapter has been the hardware components of microprocessors and microcomputers. For further reading, one can use one of several references depending on the selected topic. A general overview of the digital computer world can be found in Hilburn and Julich (1976), Evans (1981), and Leventhal (1978). More detailed information on the internal structure of micros can be found in Sloan (1980). Input/output techniques are discussed thoroughly in Stone (1982) and Buzen (1975). A wide coverage of topics related to interfacing of all components in microsystems is achieved in Artwick (1980). An overall treatment of micros architecture and systems can be found in Khambata (1982). The books by Kochhar and Burns (1983) and Craine and Martin (1985) represent a good transition between exposure of micros information and micros applications.

10
Software Support

10.1 INTRODUCTION

Early computer systems were developed with an emphasis on hardware and components; the software was developed later to support the hardware functions to yield desirable performance. In recent years, however, system hardware is designed to meet software constraints and demands. The present chapter is devoted to the study of computer system software and to examine the main tasks and requirements of different programs.

System software can be classified into a number of levels:

1. *Control software*, which includes system control software and application control software. System control software is designed to let the system perform its specific functions such as support program development, control program execution, and structure memory. This entire set of programs is called the *operating system*. Application control software is designed to control and monitor the process in the broadest sense.
2. *Support software*, which includes text editors, a link editor, and language processors such as Assembler, Fortran.
3. *User-written software*.

Software Support

A computer operating system will be examined more closely in the subsequent sections. The heart of the operating system is the real-time executive, which performs numerous functions including:

1. Structures and allocates memory
2. Controls program execution
3. Manages peripherals
4. Detects a system malfunction
5. Monitors power failures
6. Provides memory and bulk storage

User-developed application software can be written in different languages depending on the type of computer. In some cases, computer manufacturers employ some kind of "fill in the blanks" method to define databases. Many control and monitoring algorithms are standard on the system and all the user does is select parameters such that the algorithms fit his needs.

Support software is a collection of programs that supports the user in his communication with the operating system. It provides valuable tools for efficient use of and communication with the computer. Most modern systems have text editors, which make file access and modification very easy. Other utilities include Fortran, Pascal, and Assembler language processors; a link editor that links data, files, and programs to produce an executable module; a console system editor for the creation of graphic, tabular, and bar graph displays; and a software maintenance system (SMS) that documents all the software and gives the user the ability to maintain it. The importance of the SMS stems from the fact that computer systems use hundreds of programs and files. Indeed, bad documentation will eventually lead to inefficient use of the computer system.

10.2 APPLICATION PROGRAMS

The application program source code, whether written in assembly language or a high-level language, must be translated into object or machine code to be executed by the microcomputer. Assembly language and high-level language programs are translated into machine code by using assemblers and compilers/interpreters respectively. In what follows we will briefly consider program development in the four types of code: machine, hex, assembler and high-level.

10.2.1 Machine Code

Simply stated, the machine code is the actual code executed on the micro. Each machine code instruction consists of binary strings of 0s and 1s, because that is the only format in which data and instruction can be stored in the registers, and subsequently manipulated by the arithmetic and logic unit. Frequently, the machine code is referred to as the object code.

When developed, the machine code is closely related to the architecture of the particular microprocessor/computer. Therefore, machine code application programs written for one type of microprocessor cannot be used directly on another type of microprocessor system. A typical binary instruction looks like 00101011 or 10111001. At first glance, there is an obvious difficulty in understanding and remembering this code which is obviously a drawback. However, experienced programmers can write concise and quite efficient programs for simple applications. It is also important to keep track of the memory locations and the programmer must have an adequate knowledge of the computer hardware.

We have to admit that despite the efficiency and fast execution of machine code programs, they are time-consuming since the actual programming, testing, and debugging process is extremely tedious. Numerous errors could take place quite easily in a machine code program since it is usually lengthy. The situation is further compounded since it is not allowed to include any comment statement in the program layout. In short, it is quite difficult to document, amend, and maintain machine code programs.

In the early days, computer users, specialist engineers and scientists wrote application programs in machine code before it became possible to write programs in hex code, assembler, and high-level languages. At the present time, it is very hard to find application programs that are actually written in machine code since this method of programming does not offer any significant advantages over other programming methods.

10.2.2 Hex Code Programming

A logical candidate to replace machine code is to use hexadecimal numbers (HEX code) for application programming. HEX code has a base of 16 and makes use of the numeric digits 0 to 9 and letters A to F to form the primary factors. Since $16 = 2^4$, thus four binary digits (bits) can be combined and represented by an equivalent HEX code. For illustration, the relationship between binary, decimal, and hexadecimal representation of numbers between 0 and 15 is given in Table 10.1.

Like the machine code, HEX code is also related to the architecture of the microprocessor/computer and therefore HEX code programs are

Software Support

Table 10.1 Relationship between binary, decimal and hexadecimal numbers*

Decimal	Binary	Hexadecimal
0	0000	0
1	0001	1
2	0010	2
3	0011	3
4	0100	4
5	0101	5
6	0110	6
7	0111	7
8	1000	8
9	1001	9
10	1010	A
11	1011	B
12	1100	C
13	1101	D
14	1110	E
15	1111	F

*The binary number 00111011 corresponds to 3B in HEX code and 59 in decimal code.

not transportable. It should be clear that there is a one-to-one relationship between a machine code binary instruction and its HEX equivalent. Although easier to write and interpret when compared to machine code, the hexadecimal code is still difficult to remember and it does not really provide any clue to the operation that will be carried out, nor does it directly distinguish between an instruction and a data value. In light of this, very short programs can be written in HEX code and then converted into machine code by using a translation program which is available for use with the vast majority of microprocessors. Indeed, this conversion is necessary since the microprocessor executes only machine code in a direct way. Such a translator can be resident in the read only memory or the programmable read only memory of a microcomputer system. Today, almost every simple kit incorporates a HEX code translator for evaluation purposes. While useful for understanding the micro programming process at its least sophisticated level, the hexadecimal code is not really practical for writing lengthy and intricate programs. Comment statements cannot be included in the program, resulting in numerous problems, and debugging of HEX code programs is quite difficult.

10.2.3 Assembly Language Programmng

The task of writing application programs can be simplified considerably by using mnemonics to represent each basic instruction: this constitutes the code of assembly languages. Typical examples of such mnemonics are

ADD, LD, INC, DEC, SUB, STR, ACC, etc.

The use of such easily understandable codes is indeed a significant improvement over machine code and HEX code. We should note that most assembly language statements have at least one equivalent machine code instruction while some will have more than one machine code equivalent. Also, labels can be used to represent variables. It should be remarked that while the code of assembly language is again peculiar to a particular micro, it does give a good indication of the operation that will be executed by the microcomputer. For example, the code **ADD** appears to express an addition operation to be carried out while **LD** refers to a loading operation. Once again, unless the instruction set as well as the mnemonics used are identical, assembly codes written for one micro system cannot be used directly on another micro system. The system programmer must understand the mode of operation of the micro system.

Like HEX code, conversion of the assembly language program (source code) into its machine code equivalent (object code) is attained via a translator (see Fig. 10.1).

A translator (sometimes called assembler) translates the assembly input code into machine code and allocates memory locations to program instructions. The assembler usually comes with the micro system. The memory locations are allocated sequentially starting from 0. The user has the freedom to use labels and symbols instead of specifying the resources to be used by the programmer and comment statements can also be included. Assembler programs require more memory than the hexadecimal translator. In view of the cost reduction in chip technol-

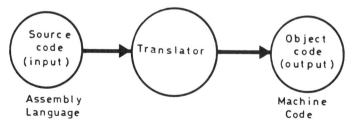

Figure 10.1 Translation of assembly language into machine code.

Software Support

ogy and the convenience of using symbols and labels, the extra expense involved is worthwhile. It is to be expected that the execution time in the case of assemblers is more than required for HEX progams.

There are other features in an assembler of significance to the user of a microprocessor/computer system. One way to categorize assemblers is into absolute assemblers and relocatable assemblers. An absolute assembler always starts at a single fixed memory location or one of a number of fixed memory locations. Specific memory locations are referred to, and the program must be loaded into those locations. An alternative is to use a relocatable assembler. No absolute memory locations are specified and the assembler starts from a base address to allocate the memory locations. The user can also decide the order in which various sections of a program will be assembled and loading of the program in a different order will not cause any problem. The relocatable assembler will simply allocate different memory locations to the program.

Assemblers can be of the self-assembling variety or they can be cross-assemblers. The self-assembler is run on the microcomputer system on which the application program will be executed. A cross-assembler is a program which is used on a host machine other than the target system which represents the micro on which the resulting machine code will be executed. The host machine is usually a bigger computer system, e.g., a mini- or a mainframe computer system which will accept the application program, written in the assembler code of the target machine. Cross-assemblers translate the assembly language source code into the machine code of the target machine. The host computer system will output the machine code in a form acceptable to the target machine. It might be in the form of a punched paper tape, stored on cassette or directly burnt into a PROM or EPROM. Cross-assemblers implemented on large machines, are often better than self-assemblers implemented on small machines.

The assembler operates by translating each mnemonic code into the appropriate binary code instruction. It also assigns memory locations to each instruction and operand, and substitutes numerical addresses in place of the symbolic address. Tables, stored in the memory, are used during the assignment of numeric values to mnemonic codes and symbolic addresses. All the possible mnemonic codes are built into the structure of the assembler. It can also check the format of the address labels as well as the delimiters which must be used to ensure that the format of the statement corresponds to the allowed format. Thus the assembler is able to check any syntax errors and references to other labels and then output any error messages. The use of such error detecting facilities, which are not available on hexadecimal translators, greatly simplifies the task of assembling, testing, and debugging the applications program. However, the assembler cannot detect any logical errors which can only be identified by comparing the program output with the expected output. This necessitates a thorough testing of every application program.

An assembler might be a one-pass or a multi-pass assembler. Most microprocessor/computer systems make use of two-pass assemblers, i.e., each application program must be passed twice through the assembler in order to produce the machine code. The first pass through the assembler results in the creation of a table in which all the symbols used in the program are collected. Errors are also detected at this stage. These symbols are then assigned numerical values during the second pass and this results in the generation of the machine code which may be executed on the microprocessor. Some assemblers require more than two passes to produce the machine code.

Macro-instructions can be used on some assemblers. A macro-instruction consists of a number of basic instructions and can be particularly useful when some sections of code are repeatedly used in the program. This facility allows the programmer to write the macro-instruction of that section of the program only once. Wherever the macro-instruction is specified in the program, the assembler will replace it by the specified group of instructions. The macro-instruction can for all practical purposes become part of the assembly language repertoire.

An assembly language statement can contain up to four fields. These are the label field, the mnemonic code field, the address field and the comment field. The mnemonic code field is mandatory for all statements while other fields can be blank. Delimiters such as commas, blanks, and semi-colons are used to separate different fields.

The label field is optional and used to provide a reference, if necessary, to the instruction address. Symbols are preferred over direct numerical values. The assembler program will assign the label a memory location in which that instruction is stored so that the programmer does not have to worry about memory locations and this label can be referred to in other sections of the program. It is only necessary to assign labels to statements to which a program can jump from another instruction. Subsequent modification of programs is also easy if it becomes necessary to add or delete statements. The assembler will then relocate the instruction to another memory location and make the required adjustments to other statements.

The label field is followed by a delimiter and the mnemonic code field which symbolically specifies the operation to be carried out. Every mnemonic code is translated by the assembler into appropriate binary machine code. The mnemonic code might be an instruction or operation which will actually be executed. Alternatively, it might be a pseudo-operation which provides information to the assembler. A good example of a pseudo-operation is the END statement used to indicate that the program is complete.

A delimiter separates the mnemonic code field and the operand or address field which provides the additional information, if any, required by the mnemonic code field. It may contain the symbolic address of the memory location, a register, an input/output port or a data value. The operand field data can be provided in many different ways; for example, as a decimal value or as a hexadecimal code.

The operand or address code is followed by a delimiter and an optional comment field which can be used to describe the action of that particular statement. The main objectives of including comment statements are to improve the readability of the program, and improve the quality of program documentation. Comment statements are not translated into the object code. It is desirable that the comments should be kept brief but understandable.

Many application programs, particularly those involving direct interface between the microprocessor/computer and a machine or process, are written in assembly language. Compared to machine code programming, the preparation of assembly language programs is very much easier and does not require as much time. They are more efficient than high-level languages, and the resulting object code is compact, making them suitable for high volume applications. However, to perform the same task they do require a much higher number of individual instructions, which cannot be easily followed, than a program written in high-level language.

10.2.4 High-Level Languages

High-level languages are general purpose application-orientated programming languages. They are not orientated towards any particular microprocessor or microcomputer, as is the case with assembly languages. These high-level languages are easy to program and the way in which the actual application program is executed is transparent to the user. Thus, the programmer does not have to worry about housekeeping problems such as the contents of registers or individual memory locations. All these tasks are carried out automatically so that the programmer can concentrate his attention on the logic of the program, i.e., how a particular task is to be carried out. These languages are not generally related to a particular system and it should theoretically be possible to use a given application program on another computer system which has facilities for translating the program into machine code. Compilers are computer system programs used for translating a high-level language program into machine code which can subsequently be executed on a microprocessor/computer. If a high-level language compiler contains any facilities which can only be implemented on a particular computer, then the programs cannot be transported from one machine to another.

The preparation of an application program, in a suitable high-level language, is easy and less time-consuming when compared to an assembly language program. Several studies have shown that the coding of a high-level language program requires only 10% of the time required for the coding of an equivalent assembly language program. However, the general purpose nature of the high-level languages results in the generation of a large amount of machine code to be executed by the computer. Whereas an assembly language instruction has an equivalent machine code instruction, a high-level language instruction results in a large number of binary machine code instructions. The number of translated machine code statements is often far greater than the number of statements required to program the same task in the assembly language. Thus, a large amount of memory will be required to store the object code which has been generated by a high-level language compiler. Obviously the user is having to pay for the privilege of using the convenience offered by high-level language statements. It is to be emphasized that computer memory is becoming relatively cheap whereas the assembly language programmers are often highly paid. Thus, a compromise has to be made between the cost of writing the program and the cost of memory required to store the program. For on-off applications, it is best to write the program in a high-level language, while programs to be burnt into memory and used in high-volume applications should be written in assembly languages.

An even more important factor is the speed of execution of the program. Since the binary machine code, generated by the high-level language computer, contains more instructions, it will also require more time for execution. In many microprocessor/computer-based systems, the execution speed is an important factor, since direct interfacing between a process or machine and a microprocessor is involved and it is necessary to respond to events as they take place.

Another difficulty with some high-level languages is the lack of good input/output facilities. Some applications require direct control over the input/output ports as well as the facilities available in the microprocessor/computer architecture.

It is sometimes possible to overcome these difficulties by writing

Software Support

the program sections which require direct input/output control or fast execution speed in the assembly language. These sections, written in assembly language, are subsequently combined with the remainder of the program written in a high-level language to produce the overall application program which can then be translated into the binary machine code. Such facilities are often necessary for the development of microprocessor/computer programs and many development systems, used for the testing and debugging of microprocessor/computer systems, contain these facilities. Thus, the programmer can get the benefit of writing the vast majority of a program in the convenient high-level language, yet has control over input/output operations and can influence the speed at which the program is executed.

The ability to write a program in a high-level language clearly depends upon whether or not a compiler exists for translating the application program into the machine code to be executed on the target microprocessor system. Some microcomputers do not have self-compilers. This difficulty can be overcome by using cross-compilers on a host computer system. The cross-compilers accept a high-level language application program and produce machine code for the target microprocessor/computer system on which it will be executed. As with cross-assemblers, the machine code can be generated in a format suitable for the target machine. Convenient formats are magnetic tape, cassettes or burning the machine code into the read-only memory.

Two passes are required to translate a high-level language application program into machine code. The compiler has to perform more tasks than an assembler program. It must allocate memory locations and registers, decide on the use of registers for particular tasks and the need to save their contents, provide links between different sections of a program and generate the machine code to be executed. During the first pass a table containing all the symbols and addresses used in the program is created. This table is subsequently used during the second pass and the binary machine code is generated. External functions, kept in the run-time library, are also linked to the application program. The compiler will detect any errors in the syntax of program instructions and will also ensure that no reserved words have been used and that the application program obeys all the rules. If the rules are not obeyed, then an error message will be output so that the programmer can correct his source code which must then be re-compiled. However, the compiler cannot detect any logical errors. The programmer is responsible for detecting and correcting such errors and the only way to detect such errors is to test the computer program thoroughly using different sets of test data, execute the binary code, and compare the program output values with the expected output values. It is worth noting that while a program is being compiled, an actual execution of instructions takes place. The program compilation may be followed by its execution. The binary machine code can also be stored on a suitable secondary mass storage device, such as a floppy disk, and executed at a later stage.

Popular examples of compiled high-level languages are FORTRAN, COBOL, PASCAL, PL/1, ALGOL and C. When used on the micro systems, small versions of these high-level languages are usually needed and are required to have relevant features such as input/output interface, real-time control facilities, and portability. PASCAL is a good example of a standard high-level language which is portable and compilers are available to implement it on a very wide range of microprocessor/computer systems. Updated versions of high-level languages which are structured and expanded are now available on the market, including ADA, C, MODULA-2, and COBOL. It is generally advisable to use a high-level language which has structured programming facilities, since these help to reduce errors in application programs and also provide for a sound and thorough testing of individual modules.

An alternative to compiled high-level languages is to use an interpretive high-level language. Interpreters are used to translate a high-level language statement to the machine code. This process is repeated every time an instruction is encountered. Source statements are not translated in one go into the object code and saved for subsequent execution. Interpreters are rather slow and inefficient so the time required to execute a program written in an interpretive language is much higher than that required to execute a compiled language program. The interpreter as well as the applications program must be resident in the computer memory at the same time. However, they do offer some advantages. They are very interactive. For example, the execution of a program can be interrupted at any stage. The program can then be modified and its execution resumed from the point at which it was interrupted or from any other point in the program. Syntax errors in any program can be detected immediately after the statement is input to the computer system. It is only necessary to learn a few instructions and programs can be easily written. Interpretive languages can also be used in immediate mode as in desk top calculators.

The most popular interpretive computing language is BASIC (Beginner's All Purpose Symbolic Instruction Code). Designed at Dartmouth College, it was originally used for teaching purposes. However, it is now in widespread use for all types of applications and the vast majority of available computer systems have BASIC interpreters. Because of its small size and simplicity, it has been implemented on the whole range of available microcomputer systems. It is very easy to learn and use. The major drawback of BASIC, apart from its slow speed, is the fact that different versions have been implemented on different computer systems, even within the range of computers of a particular manufacturer. It is highly unlikely that a complex application program written in BASIC for one computer system could be used on another computer without some modifications. However, BASIC does provide a good introduction to computing and the availability of inexpensive microcomputers has further increased its popularity.

The use of an interpretive language for dedicated data acquisition and process control applications poses a further problem. As remarked earlier, the execution of an interpretive application program requires that the interpreter should be resident in memory at the same time as the program, since it is not possible to generate binary code which can be burnt into a ROM, PROM, or EPROM. Thus, memory will be required to store the interpreter and this will result in additional cost. This is clearly not desirable for systems used in large volumes. Some computer system suppliers have started supplying compiled and interpretive versions of BASIC. Thus a program can be input, tested and debugged in interpretive mode. The compiler can be used to produce the equivalent binary machine code which can then be burnt into ROM, etc., and executed subsequently.

The choice of a language for developing application programs depends upon many factors. Most programs for commercial and scientific applications are written in high-level languages such as COBOL, BASIC, FORTRAN, C, and PASCAL, while assembly languages provide the most popular method of writing programs for industrial applications. This is partly due to the need to write efficient and time critical programs which do not require too much memory space. Efficiently compiled high-level languages suitable for process control and other real-time work have started to emerge. It is highly likely that these high-level languages will increasingly be used for industrial applications, particularly if the computer system manufacturers and users could reach agreement on a standard high-level language. Substantial improvements in programmer productivity can be achieved by using a standard industrial high-level language which is also portable. The portability is particularly important in the microprocessor/computer field in which last year's latest product quickly becomes obsolete as newer systems with enhanced facilities become available. It is highly desirable that programs written for use on one microcomputer system can be executed on any new range of equipment that is introduced.

10.3 OPERATING SYSTEMS

As their name implies, operating systems are developed to assist the operator in running a batch-processing computer; they have been developed to support both real-time systems and multi-access on-line systems.

The traditional approach to providing a system to meet a wide range of requirements has been to incorporate all the requirements inside a general purpose operating system. The typical arrangement is illustrated in Fig. 10.2. Access to the hardware of the system and to the I/O devices is through the operating system. In many real-time and multiprogramming systems, restriction of access is enforced by hardware and software traps. The operating system is constructed, in these cases, as a monolithic monitor. In single-job operating systems, access through the operating system is not usually enforced, but it is good programming practice to do so; additionally, it facilitates portability in that the operating system entry points will not be changed in different implementations. In any general purpose system there will be some facilities which will be required in a particular application. The tra-

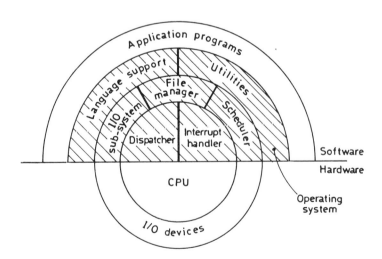

Figure 10.2 General purpose operating system.

Software Support

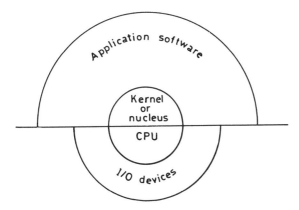

Figure 10.3 Minimal operating system.

ditional approach is to provide a mechanism which allows a limited range of modifications; this process is known as system generation (SYSGEN).

In addition to supporting and controlling the basic activities, operating systems typically provide various utility programs, e.g., loaders, linkers, assemblers, and debuggers, as well as run-time support for high-level languages.

In recent years there has been considerable interest in developing an approach in which only a minimum kernel or nucleus of the operating system is provided on to which additional features can be added by the applications programmer writing in a high-level language. This structure is shown in Fig. 10.3; it is only the nucleus or kernel to which access is limited. In this type of system the distinction between the operating system and the application software becomes blurred; it enables all the operating system functions to be modified or designed to suit the application.

10.3.1 Classification

There are many different types of operating system; the position has further expanded with the development of ROM-based "monitors" which are frequently provided by manufacturers. These monitors have grown out of bootstrap loaders and now provide additional support features for terminals and simple I/O operations. The various types of monitors/operating systems have been categorized by Mellichamp (1983) and are summarized in Table 10.2. The major emphasis in

Table 10.2 Classifications of operating systems

System	Key characteristics	Memory size (K words)
Terminal interface monitor	Is used only by the simplest single application dedicated computers; it will handle process interrupts and allow the user to start a program, display or alter memory locations, set breakpoints and load or dump programs	2
Input/output monitor	Handles its own initialization and control as well as communication between itself, system programs, user programs, and a simple I/O subsystem	2
Keyboard monitor	Is a single-user operating system which requires some form of bulk storage, typically disk drives; it includes all the facilities of an I/O monitor plus routines to accept and act on console commands as well as the ability to modify I/O assignments dynamicallly	3–6
Background/ foreground monitor	Is a dual program executive that includes all the facilities of the keyboard monitor and also controls processing and I/O in a time-shared or interrupt-driven environment.	4–12
Multi-programming executive	Includes all the facilities of the background/foreground monitor with the additional capability for concurrent execution of (up to) several hundred foreground tasks.	>8

monitors, as in the single-user, single-job operating system described below, is in communication with the user and in providing simple support for the input/output devices.

In the following sections operating systems of the following types

Software Support

will be briefly described:

1. Single-user, single-job
2. Foreground/background
3. Multitasking

10.3.2 Single-User (Single-Job) Operating System

As an example of a single-user, single-task, disk-based operating system, the CP/M 80 system of Digital Research will be described. This system is available for 8080 and Z80 based computer systems.

CP/M consists of three major sections:

1. Console command processor (CCP)
2. Basic input/output system (BIOS)
3. Basic disk operating system (BDOS)

The relationship between the various sections of the operating system, the computer hardware and the user is illustrated in Fig. 10.4.

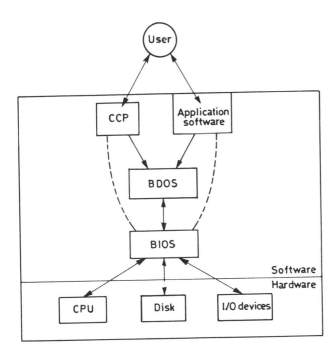

Figure 10.4 Block diagram of CP/M.

The console command processor provides a means by which the user can communicate with the operating system from the computer console device. It is used to issue commands to the operating system and to provide the user with information about the actions being performed by the operating system. The actual processing of the commands issued by the user is done by the BDOS which also handles the input and output and the file operations on the disks. The BDOS makes the actual management of the file and input/output operations transparent to the user. Application programs will normally communicate with the hardware of the system through "system calls" which are processed by the BDOS.

The BIOS contains the various device drivers which manipulate the physical devices; this section of the operating system may vary from implementation to implementation as it has to operate directly with the underlying hardware of the computer. For example, depending on the manufacturer the physical addresses of the peripherals may vary, as may the type of peripheral, and the type of controller used for the disk drive. All these differences will be accommodated in the coding of the BIOS.

On starting the computer, the three subsystems which reside on the disk are loaded into the memory. The computer must, therefore, have some means of booting the operating system from a systems disk. The normal arrangement is to provide a small bootstrap loader in a ROM chip. Following system boot, the memory allocation is as shown in Fig. 10.5. The combination of BDOS and BIOS is frequently referred to as the FDOS (functional disk operating system) and these two units have to remain in memory. Since the CCP functions cannot be called from a user program, the CCP area can be overwritten by a user program provided that CP/M is reloaded when the program terminates.

Full details of the commands and logical device drivers are readily available in the manual issued by Digital Research. The commands are briefly listed below:

1. **DIR (afn)** Display a list of all filenames which satisfy the ambiguous filename **afn**. If the parameter (**afn**) is not supplied all the filenames on the current drive are listed. Various switches can be added to the parameter to display, e.g., the size of each file.
2. **ERA (afn)** All files which satisfy the ambiguous filename **afn** are erased. All the files on the disk can be erased by setting (**afn**) equal to *.*.
3. **REN (new) = (old)** If the file (**old**) exists and is no file with the name (**new**) then the file (**old**) is renamed as (**new**). If a file with

Software Support

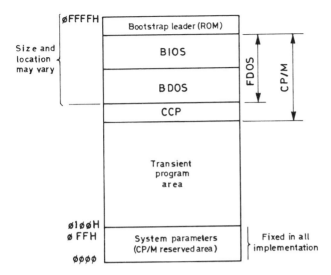

Figure 10.5 Typical memory allocation for CP/M.

the name (**new**) exists then the command is aborted. If the file (**old**) does not exist then the filename is displayed on the console followed by a question mark.
4. **SAVE** (n) (ufn) Starting at the address 0100H, n blocks of 256 bytes of memory are written to the unambiguous filename ufn. Note if the file (ufn) already exists, the contents of the file will be overwritten.
5. **TYPE** (ufn) The contents of the file (ufn) are listed on the console.
6. **USER** (n) This is a facility to enable separate file areas to be maintained on the disk. Any integer value in the range 0 to 15 can be entered.
7. (ufn) Entering an unambiguous filename of a file which has the extension .COM is interpreted as requesting the file to be loaded into memory and executed. The file will be loaded into the (transient program area) TPA and run. Note the file extension .COM need not be entered. If the (ufn) does not have an extension .COM-the command is ignored.

In addition to the direct command, various utility programs are provided including an editor, assembler, and debugger; these programs are loaded into the TPA as required and executed in the same way as user programs.

User programs are run in the transient program area which starts at location 0100H. The first page of memory (000H to 0FFH) is used by CP/M to store system parameters and the information stored in this area is shown in Table 10.3.

Recall that the operating system has the fundamental task of supporting user programs while they are running. In this regard, the CCP facilities are not sufficient as they support developing user programs. For this purpose, the facilities provided by BDOS are essentially concerned with operating the input/output devices attached to the system.

Two types of input/output devices are supported:

1. Sequential (data is transferred one byte at a time)
2. Blocked (data is transferred in fixed length blocks or records of 128 bytes: these are assumed to be disk drives in CP/M).

In dealing with input/output devices, CP/M considers devices to be "logical" or "physical". Logical devices are software constructs used to simplify the user interface; user programs perform input and output to logical devices, the BDOS connects the logical device to the physical

Table 10.3 Use of page 0 by CP/M

Location	Contents
00H–02H	Contains jump instruction to warm start entry point in BIOS; this allows the use of a standard warm start entry of JMP 0000H regardless of actual location of BIOS
03H	Contains IOBYTE used in connecting logical devices to physical devices
04H	Current default disk drive number (0 = A, 1 = B, etc.)
05H–07H	Contains a jump instruction to the BDOS entry point, hence CALL 05H to use BDOS functions. Since the BDOS entry point is also the lowest memory location used by CP/M, locations 06H and 07H give the lowest memory address used by CP/M
08H–27H	Standard interrupt locations not used by CP/M
30H–37H	Interrupt location 6, not currently used but reserved for future use by CP/M
38H–3AH	Entry point to the debugger from programmed breakpoints
3BH–3FH	Reserved for future use
40H–4FH	Scratch area for customized BIOS
50H–5BH	Reserved for future use
5CH–7CH	Default file control block, produced for a transient program by the console command processor
7DH–7FH	Reserved for future use

Software Support

device. The actual operation of the physical device is performed by software in the BIOS.

Sequential devices

CP/M supports four sequential logical devices, for historical reasons they are named CON:, RDR:, PUN:, and LIST:. (Note that the colons are part of the names.) They have the following characteristics:

1. CON: This is the "console" device used to communicate with the user.
2. RDR: This is a general purpose serial input device which was originally intended for a paper-tape reader.
3. PUN: This is a general purpose serial output device which was originally intended for a paper-tape punch.
4. LST: This designates the listing device. It is normally used to direct output to a printer, but can be used to send information to a mass storage device other than a disk.

Associated with each logical device are one or more logical device drivers. These are:

1. CON: a. CONIN Input one character at a time from the console input device.
   b. CONOUT Output one character at a time to the console output device.
   c. CONST Examine the console input device to test if an input (a character) is pending.
2. RDR: READER Read one character form the serial input device.
3. PUN: PUNCH Output one character to the serial output device.
4. LST: a. LIST outputs one character at a time to the list output device.
   b. LISTST tests the list device to determine if it is ready to receive a character or if it is still busy.

CP/M makes provision to map the logical devices onto 12 physical devices; the actual mapping is controlled by the bit pattern in the IOBYTE which is stored in location 0003H in the memory. The arrangement of logical and physical devices is shown in Fig. 10.6. Illustrated is the CON: device which is assumed to be connected to a VDU and a keyboard. It should be noted that it is only at the BIOS level that information regarding the physical characteristics of the device, e.g., address on I/O bus, data-transfer rates, is required. At the logical device level, all that is required is to know, for example, that CON: is a sequential device capable of input and output.

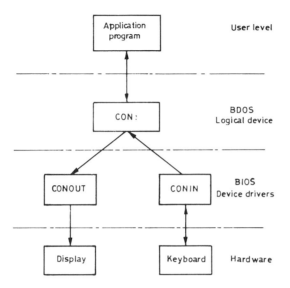

Figure 10.6 Logical and physical I/O devices.

Blocked Devices

Up to 16 physical devices which use blocked format can be supported and CP/M assumes that they will be disk drives. Only one logical driver is supported. This driver assumes that data is stored in blocks of 128 bytes; if the physical device uses a different block size, the physical device driver in the BIOS has to split the physical blocks into 128-byte logical blocks (or if the block size is smaller, group the blocks).

Associated with the driver is file control block (FCB) which is located in the first page of memory and holds information about the file which is to be accessed by the driver. The format of the FCB is shown in Table 10.4.

It is the user's responsibility to put the correct information into the FCB before using any of the disk operations available in the BDOS. Note that if several files are to be accessed by the user program, the program must provide storage for each FCB; if only one file is to be used then the FCB area in locations 005CH to 007CH can be used (this is the default area used by the console command processor).

Table 10.4 File control block format for CP/M

	0	1	2	3	4	5	6	7	8	9	A	B	C	D	E	F
00	DN					FN					FT		EX	0	0	RC
10								DM								
20	CR	RN														

DN Disk drive reference — values are in the range 0 to 16 where 0 refers to the current drive (i.e., the drive whose number is stored in location 0004H) and 1 to drive A, 2 to drive B, etc.

FN These locations contain the first eight characters of the file name, the entry is left justified and filled with spaces

FT The three characters of the file name extension are stored in these locations. The high bits of bytes 09 and 0AH are used to specify the attributes of the file. Normally these bits are set to 0: if the high bit of byte 09 is set to 1 then the file is read only; if the high bit of byte 0AH is set to 1 then the file is a system file which cannot be listed by the DIR command

EX, RC and DM These are used by BDOS to hold information about the location of the file on the disk and should not be changed by the user

CR This contains the current record to be accessed by the driver; it is normally set to 0 by the user on entry in order to start reading from the beginning of the file.

RN This is optional; if used it contains the number of a record which is to be read, i.e., it provides for random access to records

System Calls

Both the sequential and blocked operations provided by the BDOS are called in the same way. Locations 0005H to 0007H contain a jump instruction to the start of the BDOS and the appropriate operation is selected by loading the C register with a number corresponding to the operation required and using an instruction CALL 5; any parameters required are passed in the various CPU registers. By using this approach, the BDOS can be located in different areas of memory for different implementations of CP/M and yet application programs which use the facilities solely through calls to the BDOS will run on the different systems.

Example 10.1 (Use of System Calls). The assembly coding given below shows the coding required to transfer a character to the CON: output.

```
; Example on the use of BDOS calls
;
; Define symbols
;
BDOS    EQU  5   ; address of entry point for BDOS system calls
WRCH    EQU  2   ; function number for CON: output
;
; output character 'X'
;
        LD E,  'X'        ; parameter to be passed
        LD C,  WRCH       ; BDOS function number
        CALL BDOS         ; call BDOS
```

The various functions which are available in a typical BDOS for CP/M are listed in Table 10.5; the actual number of functions provided will vary with the version number of CP/M being used, but it should be noted that there is compatibility of function number between the different versions of CP/M.

In addition to the BDOS routines listed in Table 10.5, the user can also gain access to the BIOS routines, since the start of the BIOS is usually stored in location 0000H and the first locations in the BIOS contain a jump table to the various routines. The standard BIOS entry points are listed in Table 10.6.

Although it is not a real-time multitasking operating system the CP/M system has been dealt with at some length since it illustrates some of the features of a traditional operating system. The emphasis

Software Support

Table 10.5 System (function) calls to BDOS

Function number	Name	Action
0	RESET	Re-initializes the disk system
1	RDCH	Waits for ASCII character to be typed at console and returns it in A
2	WRCH	Sends character in E to console and waits until character has been accepted
3	TAPIN	Waits for character from READER, returns it in A
4	TAPOUT	Sends character in E to logical PUNCH device
5	PRNTER	Sends character in E to logical LIST device
6	RAWIO	If on entry E contains 0FFH then this is treated as an input request and the console input register is examined; if a character is present it is returned in A; if no character is present then the value 0H is returned; if the value in E is not 0FFH then the character in E is output; this routine is used to get and send the full ASCII character set; functions 1 and 2 trap certain control characters and interpret them
7	GETIO	IOBYTE value is returned in A
8	SETIO	Value in E is transferred to IOBYTE
9	PRSTR	Sends a string character by character to console until $ is encountered; the address of the first character in the string is held in DE
10	RDLINE	Reads characters from console and places them in buffer with address in DE until CR is entered
11	CSTAT	Returns a 0H in register A if no character is available at the console; it returns non-zero if character is available
12	VERNUM	Returns the version number of CP/M in registers HL
13	RSTDSK	All disk drives are reset
14	SETDSK	Selects the disk drive whose number is in E
15	OPEN	Must be used before attempting to access an existing file; DE contains a pointer to the FCB
16	CLOSE	Close an existing file; DE points to FCB
17	SRCHF	Searches for a file with the ambiguous file name contained in the FCB. DE contains pointer to FCB. The search starts at the beginning of the directory
18	SRCHN	Searches for a file as in SRCHF, but search begins at previous match
19	DEL	Delete file; DE contains pointer to FCB

Table 10.5 *Continued*

Function number	Name	Action
20	RDSEC	Read sector into current disk buffer (default buffer starts at location 0080H and is 128 bytes long); DE contains pointer to FCB
21	WRSEC	Write sector (128 bytes) from current buffer to disk; DE points to FCB
22	CREAT	Create new file, DE points to FCB
23	REN	Rename file, old name is in bytes 1–11 of FCB, new name is in bytes 16–26 of FCB
24	LOGV	Returns in HL information regarding which drives are on-line and which are off-line
25	CURD	The current disk drive number is returned in A
26	CURB	Sets the current disk buffer to the address contained in DE
27	ALLOC	Returns information in HL about the allocation vector for the current disk
28	WRPR	Sets current disk in write protect mode
29	GETR	Returns information about which drives are currently set to read only
30	SATT	Sets file attributes
31	DPAR	Returns address of disk parameter block
32	SUSR	Allows display and setting of current user number
33	RDRN	Read a random record from file to disk buffer
34	WRRN	Write a random record
35	CFS	Return number of records in a file
36	SRR	Returns current random record number
37	RSDR	Reset several drives as specified in value in DE
40	WRZ	Similar to **WRRN** except that the first time a previously unallocated group is written, all records in the group except the one being written are set to 00

is on support for the disk file system and the input/output devices. Access to the system functions is by means of subroutine calls and information is passed in the CPU registers of the machine. Because of the latter, the functions cannot be called directly from most high-level languages and hence there is a degree of isolation between the operating system and a programmer using a high-level language. The isolation is deliberate: it is an example of "information hiding". The connection between the high-level language and CP/M is made by the compiler

Software Support

Table 10.6 BIOS entry points

```
;
; BIOS jump table
;
BIOS:        JP  CBOOT      ; cold boot entry point
WBOOTE:      JP  WBOOT      ; warm boot entry point
             JP  CONST      ; console status
             JP  CONIN      ; console input
             JP  CONOUT     ; console output
             JP  LIST       ; list device output
             JP  PUNCH      ; punch device output
             JP  READER     ; reader device input
             JP  HOME       ; restore disk drive
             JP  SELDSK     ; select disk drive
             JP  SETTRK     ; select track number
             JP  SETSEC     ; select sector number
             JP  SETDMA     ; select buffer address
             JP  READ       ; read sector
             JP  WRITE      ; write sector
             JP  LISTST     ; list device status
             JP  SECTRAN    ; translate skewed sector
                            ; number
```

writer through the provision of run-time support routines which convert the CP/M virtual machine into the virtual machine described by the high-level language.

The isolation is not complete in that it is possible to call assembly-coded routines from high-level languages and to pass parameters between the high-level language code and the assembly code; this does require some detailed knowledge of the system, however. Again, information hiding is used in that the details of the physical implementation on the CPU and of the IO devices are hidden within CP/M and hence operations are performed on the CP/M virtual machine.

10.3.3 Single Foreground/Background Operating System

It is possible to convert CP/M-based machines into a simple foreground/background system. In order to do this some form of real-time clock must be available on the computer — a suitable clock can be set up by using the Zilog CTC chip — and it must be possible to connect the clock chip to the IRQ interrupt line on the CPU chip. In most CP/M systems the interrupts are not used, hence the clock interrupt will be the only one on the system. If there are other interrupts, the CTC chip must be connected to the interrupt daisy chain. Some systems provide a "type-ahead" facility which uses interrupts from the keyboard to input characters to a buffer.

Strictly speaking the successful conversion of CP/M to a foreground/background system using interrupts depends on how the BIOS has been implemented for a given system. In particular it may not be possible to use the disk file system if the BIOS does not protect critical disk operations, i.e., sector reading and writing, by disabling interrupts during such operations.

The TPA part of the memory can be divided into two parts: one part will contain the program which will be entered when an interrupt occurs (the foreground) and the other part will contain the program which will run in the normal way (the background). In order to divide the memory in this way a linker (or loader) program which provides control over the locations in which the code is located and which allows several segments of code to be combined is required. A suitable linker is the Microsoft L80, used in conjunction with the M80 assembler and FORTRAN 80.

To make use of this approach the tasks to be performed must be divided into two categories: time critical and nontime critical. The time critical tasks are run inside the interrupt routine as the foreground and the other tasks are run in the background. Examples of the former include the acquisition of the plant data, output of plant control, and calculation of the control outputs. These would have to be gathered together as the foreground task; the display to the operator and the input from the operator become the single background task. It is possible to write the bulk of both tasks in a high-level language as is shown in the skeleton programs given below.

Example 10.2 (Simple Foreground/Background System).

Foreground task
```
;
;   Interrupt routine forming main body of foreground task
;
    PUBLIC IRHO            ; make address of interrupt routine
                             available to external programs
;
    EXT   SR,RR,CON,TIME   ; external routines, addresses will
;                            be provided by the linker
```

Software Support

```
IRHO:    CALL SR          ; save all registers
         CALL CON         ; control routines (could be
                            FORTRAN routines)
         CALL TIME        ; clock routine
         CALL RR          ; restore all registers
         EI
         RETI
         END
```

```
         Background task
C
C        Body of background task
C
C        Initialize routine for setting up the system
C
                 CALL COMVAR (p1, p2, ... , pn)
C
C        Above is a routine which transfers the addresses of
         variables
C        p1, p2, ... , pn to the foreground program, it is only
         required if
C        the control routines are written in assembler
C
                 SECS = 0
                 DISPTM = 0
C
C        Set up interrupts and clock interval
C
                 CALL INTP (IBASE, ICHAN, ICT)
C
C        Main body of background
C
100              CALL TIME (SECS)
                 IF SECS > DISPTM THEN 200
                 CALL KEYB (STATUS)
                 IF STATUS THEN 300
                 GOTO 100
C
200              CALL DISPLAY
                 CALL TIME(SECS)
```

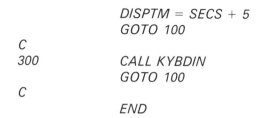

```
                DISPTM = SECS + 5
                GOTO 100
     C
     300        CALL KYBDIN
                GOTO 100
     C
                END
```

It should be noted in the above program example it is assumed that the subroutine CON, SR, RR, INTP, COMVAR, and TIME are located in the foreground partition as is shown in Fig. 10.7 and that their location has been determined by the programmer. In fact, provided that the routine used to set up the interrupts (INTP) finds within itself space for the interrupt response table, loads the address of this table into the Z80 I register, into the CTC chip, and also places the address of IRH0: in the table, then the allocation of memory can be left to the linker. If allocation is left to the linker then there will be no identifiable contiguous foreground or background memory areas.

It is also assumed in the above that the controller parameters are held in COMMON and hence are available to both the keyboard input subroutine (KYBDIN) and the control routine (CON). If the control routine is written in assembler in order to get faster execution, the background program is allowed to determine the location of the storage area for the parameters and the address of the actual locations is transferred to the foreground control subroutine using COMVAR.

A wide range of foreground/background monitors have been developed; perhaps the best known is the DEC RT/11 operating system. Some provide little more than the CP/M modification described above, i.e., they allow interrupt routines to be used and guarantee that critical operations within the operating system will be protected from disturb-

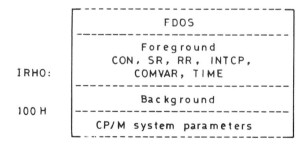

Figure 10.7 Example of foreground/background partition.

Software Support

ance caused by the use of interrupts. Others provide extensive facilities, including the ability to have multiple tasks in both the background and foreground with the ability to roll tasks in and out of memory. This type are perhaps better described as multitasking operating systems with only two priority levels. Many systems provide some form of memory protection barrier between the foreground and background systems, either through software or hardware mechanisms. The provision of memory protection is particularly desirable in systems which allow program development and debugging to be a background activity.

Foreground/background systems often restrict the use of particular devices either to the foreground or to the background, but not to both, i.e., a printer may be allocated either to the foreground or to the background, but not to both; if a printer is required for both, two printers must be provided. Where such restrictions are not imposed, the problem of IO device-sharing becomes similar to that which occurs with multitasking operating systems.

10.3.4 Real-Time Multitasking Operating Systems

It has been indicated that the natural way to structure a typical computer control system is in the form of a number of different tasks which all apparently run in parallel. The implementation of a design based on this approach is made easier if an operating system which supports multitasking can be used. The traditional real-time operating systems are based on the assumption that all the tasks in the system will be executed on a single CPU or processor. In what follows the same assumption will be made.

Confusion can arise between multi-user or multiprogramming operating systems and multitasking operating systems. The function of a multi-user operating system is illustrated in Fig 10.8. The operating system ensures that each user can run a single program as if they had the whole of the computer system for their program. Although at any given instance it is not possible to predict which user will have the use of the CPU or even if the user's code is in the memory, the operating system ensures that one user program cannot interfere with the operation of another user program. Each user program runs in its own protected environment; a primary concern of the operating system is to prevent one program corrupting another, either deliberately or through error. In a multitasking operating system it is assumed that there is a single user and that the various tasks are to cooperate to serve the requirements of the user. Co-operation will require that the tasks communicate with each other and share common data. This is illustrated in Fig. 10.9. In a good multitasking operating system the way in which

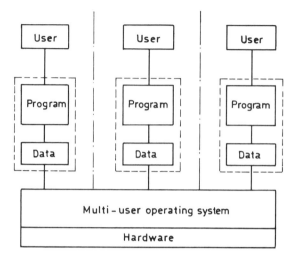

Figure 10.8 Multiuser operating system.

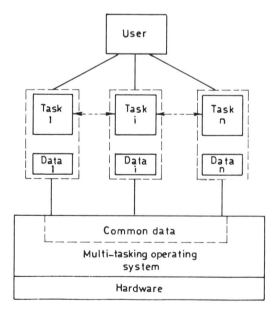

Figure 10.9 Multitasking operating system.

Software Support

tasks communicate and share data will be regulated such that the operating system is able to prevent inadvertent communication or data access (arising through an error in the coding of one task) and hence protect data which is private to a task. (Note that deliberate interference cannot be prevented — the tasks are assumed to be co-operating).

A fundamental requirement of an operating system is to allocate the resources of the computer to the various activities which have to be performed; in a real-time operating system this allocation procedure is complicated by the fact that some of the activities are time critical and hence have a higher priority than others. There must therefore be some means of allocating priorities to tasks and of scheduling allocation of CPU time to the tasks according to some priority scheme.

A task may use another task, i.e., it may require certain activities which are contained in another task to be performed and it may itself be used by another task. Thus tasks may need to communicate with each other. The operating system therefore has to have some means of enabling tasks, either to share memory for the exchange of data or to provide a mechanism by which tasks can send messages to each other. In addition, tasks may need to be invoked by external events; hence the operating system must support the use of interrupts.

Similarly, tasks may need to share data and they may require access to various hardware and software components; hence there has to be a mechanism for preventing two tasks from attempting to use the same resource at the same time.

In summary, a real-time multitasking operating system (RTMTOS) has to support the resource sharing and the timing requirements of the tasks and the functions can be divided as follows:

Task scheduling
Interrupt handling
Memory management
Code sharing
Device sharing
Inter-task communication and data sharing

In addition to the above the system has to provide the standard features such as support for disk files, input/output device support and utility programs. The typical structure is illustrated in Fig. 10.10. The file manager is the equivalent of the BDOS and the input/output subsystem (IOSS) the equivalent of the BIOS which were described in the section on CP/M. The resource management and allocation module is mainly concerned with the management of the memory and with memory protection if used. The command processor is the equivalent of the

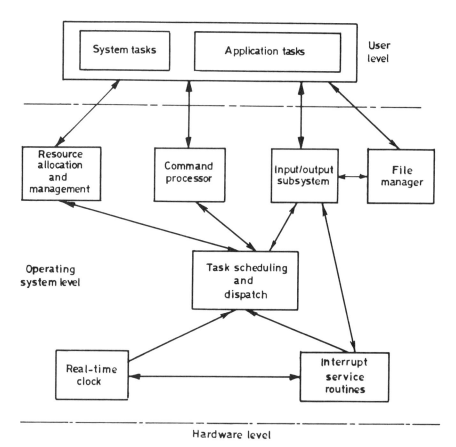

Figure 10.10 General outline of real-time operating system.

console command processor of CP/M, but also handles the system calls from the tasks. The overall control of the system is provided by the task scheduling and dispatch module which is responsible for allocating the use of the CPU. This module is also referred to as the "monitor" and as the "executive control program" (or more simply the "executive"). At the user level in addition to application tasks a box labeled "systems tasks" is also shown since in many operating systems some operations performed by the operating system and the utility programs run in the memory space allocated to the user or applications — this space is sometimes called "working memory".

Further details on the various parts of the operating system described

Software Support

above are beyond the scope of this chapter and can be found in Mellichamp (1983) and Lister (1979).

10.4 SOFTWARE DEVELOPMENT TOOLS

A number of other tools are required to help develop application programs. A loader is required to load the application program. An editor is necessary to modify source programs. A linker is used to join together different sections of a program and a debugger helps in the process of producing an error free program. Monitor programs are required to supervise the operations of the microprocessor. A file management system is used to store programs, data files and a directory for tape and disk systems. Large microcomputer-based systems with a number of peripherals operate under the control of microcomputer operating systems which combine all these tools. The facilities available in these tools are important during the application program development stage.

10.4.1 Loaders

These are system programs used to transfer programs, stored on an external medium, to the computer memory. The program might be stored in the form of a program file on floppy discs or cassettes, in which case the disk or cassette operating system will be used to load the program into the computer memory. Often, however, the microcomputer may not have a cassette or disk operating system. In such cases a bootstrap loader, a basic software utility tool provided by the manufacturer, can be used to load the application program code, stored on paper tape, into the computer memory. The microcomputer front panel or a monitor stored in the PROM can be used to control the program loading. Absolute loaders are required to load the object code, generated by absolute assemblers, which refer to specific memory locations. To ensure that the object code is loaded correctly, it is necessary to provide the base address. The base address can be included in the program source code which provides the required information to the absolute assembler and will be treated as a pseudo operation (i.e., it will just provide information). An alternative is to use the microcomputer front panel for providing the base address.

A relocatable loader is used for loading the code, generated by relocatable assemblers, anywhere in the memory since the memory locations are relative rather than absolute. Relocatable loaders are better than absolute loaders and also make it possible to load assembled programs separately. Different sections of the code can be loaded separately in a sequence and any memory location problems can be resolved at the loading time. If it is necessary to load the program sections in a different sequence, the base address can be changed and taken into account. It is not necessary to generate new assembly code as would be the case with absolute assemblers and loaders.

Some loaders have sophisticated facilities such as memory mapping, i.e., a map of the memory contents, and provide error messages when a program cannot be loaded due to problems such as data transmission errors. It is also possible to link loaders with text editors so that the code is modified and a new correct, file is generated.

10.4.2 Editors

Most application program source codes, when first created, contain errors. The errors might be in the syntax of statements. Such errors are relatively easy to detect. The more difficult errors are in the logic of the program. In both cases, facilities are required to amend the source code. The necessary amendments can take the form of addition/deletion of lines or individual characters. Text editors can be used to modify existing program files and create new ones and they are also useful for creating totally new files input via the keyboard. In a disk- or cassette-based microcomputer system, these new files can be stored on the floppy disk- or cassette. Alternatively, the modified programs can be punched out on paper tape or a listing can be produced on a printer. Text editors are essential tools for developing programs since most programs have to be modified a number of times before correct versions are finally produced.

The minimum desirable features in a text editor used on microprocessor/computer systems are the loading of programs into memory; listing of program sections; editing of one line at a time; replacement/deletion/addition of individual characters; search for program lines which contain a particular string of values so that editing of lines with these characters is simplified; and output of the complete program or a section of it on to an output device such as a paper tape punch, printer, cassette or a floppy disk.

10.4.3 Linkers

It is often necessary to combine two or more program segments which have been compiled or assembled separately into one complete program so that an absolute binary image of the complete program can be created and subsequently executed. When an error is detected in a

Software Support

program module, it is often better to just re-compile or re-assemble that particular section and link it with the remainder of the program. A table of the variables which are referred to in different sections of a program is created and a unique memory location is allocated to them. Two passes are required to create the binary image of the complete program. Linkers, link loaders or link editors are programs used to link object code modules. As discussed earlier, some modules of a program, used on microcomputer systems, are written in assembly language to facilitate high-speed execution and better control over input/output peripherals, while the convenient high-level languages are used to write the nontime critical program modules. Link loaders are necessary to join the assembler generated code with the compiled binary code.

10.4.4 Debugger

It is frequently necessary to modify the object code so that an error free version can be produced and executed. Debuggers are programs used to monitor and control the execution of a program, and patch the object code. They can be used for the specification of break points at which program execution should be interrupted; printing out the values of variables or contents of memory locations; modification of the memory contents and variable values; execution of programs starting at particular memory locations and trapping of particular memory locations or variables. It is thus possible to test whether the object code can cause long-term problems. If too many patches have been made to a program and it becomes necessary to re-assemble or re-compile the source code, which has not been modified, then errors, which have previously been corrected in the object code, could reappear. It is therefore necessary that whenever the object code is modified for any reason, an equivalent correction should also be made to the source code. A record should also be kept of the amendments to the source and object codes.

10.4.5 Monitor

Software modules used, at a rather crude level, for monitoring and controlling the operation of microprocessor systems and associated peripherals are described as monitors. They can be used, for example, to load programs, execute the program from its starting point or a specified memory location, stop its execution at a particular point or after a break point has been reached or a particular type of instruction has been executed, and provide information about the contents of memory locations or registers which form part of the central processing unit.

Monitors are also used to schedule the tasks to be carried out by various elements of the microcomputer system. Some microcomputer system suppliers will provide a software monitor for their configuration. In other cases it is necessary either to use the front panel switches for monitoring or the user must write a crude monitor himself. In general it is better to select a system with a built-in software monitor resident in a ROM, PROM or EPROM. In such cases the system is ready for use as soon as the power is switched on.

10.4.6 File Handling/Management

All floppy disk-based and some cassette-based microcomputer systems include file handling/management facilities. A file can be a high-level language source program, an assembly language source program, the binary object code of a program or a collection of data used in a program.

Data files are divided into records and the access to these records can be sequential, indexed sequential or random. In a sequential file system all the records are accessed in a serial fashion. Thus to locate a record halfway through the file, all the previous records must be read first. No algorithm or index is used to determine the location of individual records.

In an indexed sequential file system, the file handler maintains an index of the location of each record as specified by its key. Thus when access to a particular record is required, its location on the disk can be looked up in the index, and the record is then read into the memory. It is no longer necessary to read all the records which precede the one required. An indexed sequential filing system is better than a simple sequential filing system. If required, an indexed sequential file can also be accessed sequentially.

In a random file system, an algorithm is used to define the relationship between the record key and the location of the record on the disk. These files can be accessed in a random manner by specifying the key. Sequential access to a random file is not possible.

Files can be loaded into the computer memory from secondary storage devices such as a floppy disk or a cassette, as and when required, by issuing a simple command. On disk-based systems a directory is kept of all the files stored on the floppy disk. The directory itself is an example of a file. By issuing a simple command, the user can get a list of all the files in his account or on the complete disk. Each file is allocated a unique identity and it is usually possible to overwrite an existing file, i.e., replace the old version by a new one. While this

Software Support

facility is useful, it can and often does result in the inadvertent loss of a file which might be required at a later stage. To avoid such happenings, some file management systems allow a file to be locked and it cannot then be overwritten unless the user unlocks it first. Such a facility is highly desirable to avoid unintentional loss of a file as considerable effort is usually required to recreate a lost file.

The actual storage of a file on a floppy disk involves the allocation of the required amount of space which is usually divided into a number of fixed length blocks of 512 bytes. Depending upon the size of the program or data file, a complete number of blocks is allocated to a file. The blocks may be contiguous or noncontiguous. A contiguous system is to be preferred since the file will occupy adjacent blocks and the time required to load the file into memory will be less than that required for a noncontiguous file. In the case of files stored on noncontiguous blocks, a pointer indicates the location of the next block on which the file is continued. Although inefficient for disk access purposes, the noncontiguous method of storing files does make better use of the available storage space. This is particularly true when an existing file is deleted, but the space vacated by it is not big enough to store a new file. The file handler will continue to look for a number of contiguous blocks which may not be there. Thus over a period of time, a number of small contiguous blocks will become available. The total space available in a number of such contiguous blocks, spread all over the disk, may be far greater than that required to store a new file, yet it cannot be accommodated because these blocks are not contiguous. This difficulty can be overcome by reorganizing the disk at suitable intervals and creating contiguous blocks which can be used subsequently. In the case of a noncontiguous file handling system, a file can be used subsequently and a file can be stored as long as there is enough empty storage space anywhere on the floppy or hard disk.

The following facilities are desirable in a file handling system used on microcomputer systems.

1. Creation, deletion, and copying of all files
2. Allocation of meaningful names, consisting of a number of characters, to files
3. Closing and opening of files
4. File protection to avoid accidental erasure
5. Listing of complete directory or files whose names contain a specified character string
6. Initialization or formatting of disks
7. Error handling
8. Diagnostic routines for indicating bad disk blocks

These file handling facilities are extremely useful in a program development environment and are found on most disk-based operating systems and microprocessor development systems.

10.4.7 Microcomputer Operating System

Many of the currently available microcomputers can be run under the control of microcomputer operating systems which provide an interface between the hardware and the user. As described earlier, an operating system consists of a set of program modules, many of which have been discussed earlier, for allocating and controlling automatically the resources available on the microcomputer system. The operating system might be cassette-, floppy disk- or hard disk-based. A disk operating system is better than a cassette operating system. Only part of the operating system is resident in the RAM at any one time.

In addition to the software development tools already discussed, an operating system includes routines for automatic handling of the input and output peripherals. The operating system also supervises the various tasks carried out on the computer. Interrupts are used for controlling the flow of information and synchronizing events.

Major user benefits of microcomputer operating systems include the facility to carry out relatively complex tasks by issuing simple one or two word commands, automatic loading of programs and/or data from peripherals, automatic output, as and when required, of values to related peripherals such as printers, paper tape, and floppy disk, and scheduling of the sequence in which the programs will be executed.

Desirable features of a microcomputer operating system are:

1. Starting up the system automatically or by issuing a simple user command after the power has been switched on.
2. Simple methods of using peripherals such as floppy disks, cassettes, printers, paper tape readers/punches.
3. File handling/management.
4. High-level language compilers/interpreters.
5. Assemblers.
6. Text editors.
7. Loaders and link-loaders.
8. Program chaining.
9. Control of program execution.
10. Debugging routines.
11. Methods for handling hardware interrupts.
12. Possible communication with other similar systems to develop a microcomputer network. Alternatively it should be possible to

interface the microcomputer to a host system used at a higher level. This facility is extremely useful in many manufacturing environments and can be used to transfer data, to be used for further analysis, to a higher level system.

While all these facilities, in particular the last one, may not be available in a single operating system, they are highly desirable for using microcomputers in a flexible manner.

10.5 INTERFACING

10.5.1 Introduction

The successful implementation of a microcomputer-based system requires that means be found to link a process to the microcomputer. This is true in all cases except where the data are being input via a visual display unit for routine nonreal-time processing and output via the visual display unit and line printers. In real-time process data acquisition systems, the sensors used to detect the current state of a process have to be interfaced directly to the microcomputer. Similarly, if the microcomputer is being used to effect direct digital control (DDC), then the actuators must be interfaced to the microcomputer system. The interfacing procedure can be represented in conceptual terms by Fig. 10.11. In some cases it is necessary to amplify the current or voltage output by the microcomputer so that the actuators can be operated.

Careful interfacing of the microcomputer to the sensors and actuators is required so that real-time data acquisition and control can be carried out. Specialized and rather expensive interface circuitry and actuators are necessary. The interface requirements for all the available sensors and actuators are not identical. Thus, for example, the millivolt output from a thermocouple must be amplified to bring it to the microcomputer range of $+5$ V, and the amplified analog voltage must be converted into digital format, using analog to digital converters, before being input to the microcomputer. A light emitting diode (LED) requiring a very small amount of voltage and current to drive it could be directly interfaced to the microcomputer. The application of the microcomputer to control the power input to a heating element would necessitate the use of a silicon-controlled rectifier; similarly, the operation of pneumatic signals before being applied to the actuator. Thus the interfacing requirements for sensors and actuators vary from one element

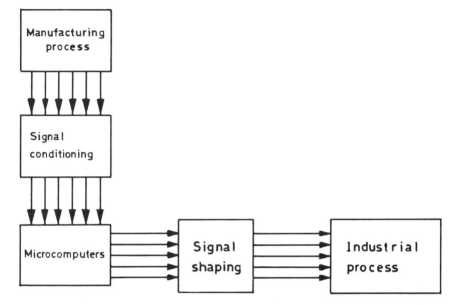

Figure 10.11 Overview of the microcomputer interfacing with an industrial process.

to another. The characteristics of the microcomputer used also have to be taken into account.

Most microcomputers have limited numbers of pins for input/output purposes. It is, therefore, necessary to use a multiplexer which transfers data from a number of lines (sensors) into one line for connecting to the microprocessor input port. Similarly a de-multiplexer is necessary for transferring the data from one output port to a number of lines connected to actuators or other peripheral devices. Multiplexers and de-multiplexers are effectively time-sharing devices which simplify and reduce the cost of interfacing a process to the microcomputer.

In an overall sense, the inputs, the outputs, the microcomputer, and the interface circuitry requirements can be illustrated by Fig. 10.12. While this diagram illustrates the schematic of microcomputer/process interfacing, the actual configuration will vary depending upon the number of inputs and outputs, and the accuracy/resolution requirements of converters. Similarly, the characteristics of multiplexers have to be taken into account. Suitable sensors are not available for monitoring many of the variables of interest which must then be inferred. Under these circumstances it becomes necessary to use sample and hold

Software Support

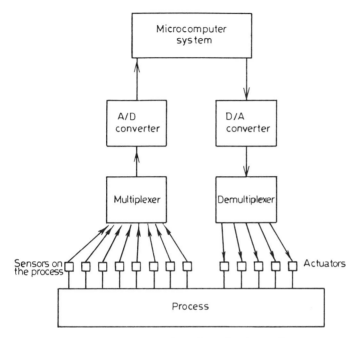

Figure 10.12 Microcomputer interface circuit requirements.

devices to ensure that the two operative variables from which the value of another variable is to be inferred, were all sampled at the same instant of time. The cost of many sensors can be considerable. Similarly, the fitting of new actuators to existing machines, in order to implement a suitable control system, can be very expensive. Without the availability and interfacing of suitable sensors and actuators, the microprocessor/computer cannot be used for on-line data acquisition and process control. The cost of the microcomputer and associated computer peripherals will be much lower than the cost of the sensors, actuators, and interface circuits. It is therefore essential that, wherever possible, the sensors, actuators, and interface circuits should be selected to minimize the overall system cost.

While a given microprocessor chip, manufactured in large quantities, can form the basis of a large number of microcomputer configurations, the same is not true of the interface logic circuits. A large number of interfacing chips exist, but they have to be configured to suit the application under consideration. It is difficult to standardize interface circuits since the requirements vary from one sensor to the next, and from one actuator to the next. Standard circuits can, however, be used for related functions of analog to digital conversion (ADC), digital to analog conversion (DAC), multiplexing and de-multiplexing.

10.5.2 Microprocessor I/O Ports

All microprocessors have input/output ports which can be used to transfer information between the central processing unit and external devices such as floppy disks, cassettes, visual display units, and line printers. The microcomputer supplier usually provides the interface circuitry required for connecting floppy disks and cassettes to the microcomputer. Standard interfaces, such as RS 232 or a suitable parallel interface, are used to connect VDUs and line printers to the microcomputer. Further input/output ports are available to interface the process with the microcomputer. These ports contain buffers for holding data until they are required by the microprocessor or the peripheral to which the data is to be output. These input/output ports are connected to the data bus. The microcomputer can address the input ports and transfer data from the port to the data bus which is bidirectional, i.e., it can transfer data to and from the microcomputer. Control lines (circuitry) are required to select the port from which data are to be input or to which data are to be output so as to avoid any conflicts. This is achieved by selecting a particular input/output port and allowing the data transfer to take place. The high speed at which data transfer takes place inside a microcomputer means that a number of ports can be addressed over a short period of time. The transfer of data between the central processing unit (CPU) and the peripherals can take place either in parallel mode (8 or 16 bits at a time) or in serial mode (1 bit at a time). Different techniques such as polling, interrupted input/output, and direct memory access (DMA), are used for communications between the microcomputer and the peripherals. These modes all have their applications and it is possible to use all of them, in a given microprocessor, for different types of data transfer. The actual transfer of data might be one data value at a time or it might be a continuous transfer of a block of data. Programs are needed to acquire the data from the peripherals or sensors and store them on secondary (mass) storage devices or output the data to other storage media such as magnetic tape.

10.5.3 Parallel I/O Ports

The task of data transfer between the microprocessor and the outside world (sensors, actuators or peripherals) can be simplified by using large-scale integrated (LSI) chips which can be programmed to suit a

Software Support

particular set of requirements. Typically a chip used for parallel data transfer contains two or more 8-bit ports which can be configured to input data to the microprocessor or output the data to a peripheral device. A register is used to define the direction in which the data is to be transferred by the corresponding bits of the port. The microprocessor can be programmed to define the data transfer direction.

The port and the data direction register associated with it contain 8-bits each. The direction of data transfer is specified by setting all the 8-bits in the data direction register to 0 for input and to 1 for output. It is also possible in some systems to define the direction of data transfer for each individual line or bit by setting the bit appropriately in the data direction register. The high degree of flexibility of the input/output chips makes them suitable for a wide range of applications. Thus, for example, if it was necessary to use the first four lines for input and the remaining four for output, the first four bits of the data direction register would be set to 0 (zero) and the remaining four to 1 (00001111). A subroutine program can be used to initialize the contents of these registers. These chips carry out additional functions such as "hand shaking", data transfer to and from the bus, interrupt and status control. Chip select lines are used to select the input/output chip by means of a suitable code and distinguish it from other registers in the random access memory and read only memory. The programmer can also address any of the individual registers in the interface chip directly.

A number of peripheral interface adapters (PIA) or programmable peripheral-interfaces (PPIs) are available and can be used for programmable input/output in parallel mode. Typical examples are the Motorola 6821 PIA chip for the Motorola 6800 family of microprocessors, and Intel 8255A for Intel 8085 microprocessors. Detailed information and the facilities available in these chips can be obtained by consulting the manufacturers' manuals.

It is necessary to carry out the following tasks for successful input/output of data:

1. Decoding of the address
2. Determination of the status of the device to which data are to be transferred
3. Buffering of data and its multiplexing since there are two or more ports
4. Transfer of data
5. Control functions

The instruction repertoire of most microprocessors includes instructions for transferring data, under software control, from the input ports to the accumulator, and from the accumulator to the output port. The actual transfer of data is carried out under program control.

10.5.3 Serial I/O Ports

In many microcomputer applications, it is expensive and impractical to use parallel lines over long distances. One of the problems, frequently encountered over long distances, is that all eight bits of data required to define a character might not arrive at the same time. Furthermore, many of the peripherals used with microcomputers, for example teletypewriters and visual display units, do not have 8-bit parallel interfacing facilities. Such devices can only transfer data in serial mode, one bit at a time. While it would be possible to use only one bit of the port for serial mode data transmission, a more common procedure is to use a LSI chip for converting the serial data to parallel mode. Similarly, converters can be used to convert parallel data into serial format. It is also possible to carry out serial data transmission under program control. The program required for serial data input and output is relatively simple. However, this procedure tends to be slow and it is better to use hardware converters.

The universal asynchronous receiver/transmitter (UART) is an example of a LSI chip which can be used to carry out parallel to serial and serial to parallel conversions. The serial data transmission of characters involves the use of start and stop bits which are not required for parallel data transmission. Thus the UART must take parallel data, store it in a register, and then add start, stop, and parity bits before transmitting the data in serial mode. Similarly, the conversion of serial data to parallel format requires that the start, stop, and parity bits be removed. These functions are performed by the transmitter, receiver, and control sections of the UART. The transmitter takes the parallel data, holds it in a register and adds the required number of start, stop, and parity bits as specified by the programmer. The data is then transmitted to the receiver which carries out the reverse function of removing the start, stop, and parity bits. Functions such as parity checking and overrun errors can also be carried out by UARTs. A start bit indicates that a new character is to be transmitted so that the receiver can accept the bits at the appropriate sampling rate, which can be selected by the user. The stop bits indicate that a complete character has been transmitted. The double buffered facility of UARTs makes it possible to keep two characters in the transmitter, one that is being transmitted and one that is being loaded into it. High data transmission rates of the order of 200 kbaud can be achieved by using the available UARTs. The actual rate at which data is to be transmitted can be selected by the user.

Software Support

The control section of the UART is used for the synchronization of events. Control lines or signals are used to indicate that a complete character has been received or that the transmitter buffer is empty and ready to accept a new character.

UARTs are relatively inexpensive LSI circuits, compatible with most microprocessor systems, which simplify the process of serial data transfer and interfacing.

Synchronous data transmission is possible by using a universal synchronous receiver/transmitter (USRT). These devices synchronize the transfer of data and provide better performance than UARTs since it is not necessary to use start and stop bits. The transmitter and receiver are synchronized by the transmitter sending a synchronizing character which is recognized by the receiver. From then on synchronization is maintained by transferring data at the exact specified rate. In the event of the nonavailability of data in the transmitter, null or dummy characters are transmitted. It is thus possible to transfer data continuously. USRTs are only necessary for high-speed data transmission.

Chips, referred to as universal synchronous-asynchronous receiver/transmitters (USART), provide for both synchronous and asynchronous serial transfer of data. Intel 8251 USART is an example of a universal receiver/transmitter which can transfer data in synchronous and asynchronous modes. Other sophisticated interfacing circuits are being developed. For example, most UARTs require additional circuitry for decoding the microprocessor address and control signals. Chips which carry out this decoding, in addition to the normal functions of the UARTs, are also available and simplify the process of interfacing the UART to the microprocessor. A typical example is the Motorola asynchronous-communications interface adapter (ACIA) 6850 used with Motorola microprocessors.

10.5.4 Data Acquisition Modules

The actual transfer of data between the microcomputer and external devices is commonly carried out using three different methods:

1. Polling
2. Interrupts
3. Direct memory access

These methods can be used individually or in combination, thus forming the core of data acquisition modules.

Polling

Polling routines are used for programmed input/output purposes. A number of peripheral devices attached to the microcomputer are continuously polled at specified time intervals to determine whether the device wishes to input data to the microprocessor. Similarly, for the transfer of data to an output device, the state of its buffer is checked to determine if it is empty and ready to receive the data. A software program is required to check the state of the peripheral devices. Hardware flags are used to indicate the send or receive condition of the peripheral. If a device is ready to transfer the data, then an appropriate software routine is used to service that particular device so that data transfer can take place. The service routines first save the accumulator contents before carrying out data transfer so that the execution of the polling routine can be resumed from the point at which it was interrupted. It is necessary to ensure that none of the data from any of the peripherals are lost while the polling routines or any of the peripheral device servicing routines are being executed. The probability of losing data is very small since most peripherals are slow devices. Other techniques, such as direct memory access (DMA), can be used to achieve high-speed data transfer from a given peripheral.

Polled input/output is a very simple and common method of data transfer which does not require any additional logic circuitry. The only requirement is for an efficient program which is executed at very high speed.

Interrupted Input/Output

The polling procedure is a form of time-sharing whereby the facilities of the computer are shared between a number of users. The computer continually polls each of the terminals, and serves a terminal for a specified period of time before moving on to the next one. This procedure is inefficient because system overheads are incurred in polling all the peripheral devices irrespective of the actual need. The time available for the processing of instructions is reduced significantly. Furthermore it may not be possible to respond to the real-time requirements of process control systems in which a particular device might require urgent attention while the processor is servicing another input/output device. In some circumstances the actual data might be lost. These difficulties can be overcome by using the hardware interrupt input/output systems. The output of the microcomputer system can be increased substantially by using such interrupts, since an input/output device is serviced only after it indicates that it is ready to send or receive data. This request

Software Support

for input/output can occur at any time, i.e., in an asynchronous mode, and the device indicates this by setting up an interrupt flag. The microprocessor acknowledges the interrupt, completes the execution of the current instruction, saves the content of the registers and then services the requesting device by transferring control to an appropriate input/output routine. Once the peripheral device has been serviced, the microprocessor resets the flag to indicate that it has been serviced, the control is transferred back to the original program which resumes execution from the point at which it was interrupted.

Some real-time control applications involve the use of critical programs which must not be interrupted during their execution. Some of the interrupts are trivial while others have to be serviced. For example, data input can be delayed while alarms must be raised immediately. Different types of interrupts, maskable and nonmaskable, can be used to overcome these difficulties. Software programs are used to enable or disable maskable interrupt. Thus, if the maskable interrupt has been disabled, the microprocessor will ignore the interrupt request. A nonmaskable interrupt has to be acknowledged and serviced by the microprocessor.

The interrupt used in a microprocessor system might be a fixed interrupt or a vectored interrupt. A fixed interrupt requires relatively simple hardware. The flag is set on the interrupt line to indicate that service is required. If there is only one device attached to the interrupt line, then the control will be transferred to a fixed memory location which will provide the program for servicing that device. When a number of devices are attached to a given interrupt line, each with its own servicing routine, then the peripheral responsible for the interrupt has to be identified. Thus a polling routine must be used to detect the peripheral requiring service and then determine the memory location at which the program for servicing that device starts. Multiple interrupt lines can be used to overcome these difficulties.

Interrupts with multiple lines provide the addresses of different memory locations to which the program should transfer in response to an interrupt on a particular line. It would be necessary to use as many lines as the number of interrupts or resort to polling routines if more than one device is attached to the same interrupt line.

A preferred alternative is to use a vectored interrupt whereby the interrupting device is directly identified. This device identification can be used to look up the starting memory location of the service routine for that device. Some vectored interrupts not only identify the interrupting device, but also provide the starting memory location of the service routine to which the program should jump. Thus the address of the memory location, stored in the peripheral controller, is placed on the data bus and used for transferring the control.

A system with only one device which can interrupt the normal work of the microcomputer is a very simple one and the particular interrupt can be readily serviced. In practice a large number of devices can provide interrupt requests to the microcomputer. Under normal circumstances, it is likely that two or more devices may provide an interrupt request at the same time. It then becomes necessary to decide the priority to be allocated to individual interrupt requests. A number of procedures can be used for priority allocation purposes.

Priorities can be allocated to peripherals attached to a single interrupt line by using a simple daisy chain procedure. Once interrupted, the microcomputer sends a signal to the first device in the daisy chain. If it is the interrupting device, then it will provide the memory address of its service routine and the signal will not be passed on to the other devices in the daisy chain. If, however, the first device did not cause the interrupt, then the message will be passed on to the next device. This procedure is repeated until the interrupting device is located. Clearly the first device in the daisy chain will have highest priority, followed by the next one and so on. Once an interrupt request has been acknowledged and is being serviced all the other interrupts either have to be disabled or higher priority interrupts must be allowed to break into the current service routine.

An alternative to daisy chains is to use priority interrupt circuits which identify and service the device with highest priority. Level 0 means highest priority, followed by level 1, level 2, and so on. These priority interrupt circuits identify up to 8 interrupt levels by means of a 3-bit code which can be inspected by the microprocessor. Devices requiring higher priority are connected to interrupt levels, for example, 0, 1, and 2, with higher priorities. Since events such as power failure must be identified quickly and alarms must be raised as soon as possible, these subsystems are attached to higher priority interrupts. Similarly, other peripherals requiring fast response can be allocated higher priority than slow-speed peripherals. Facilities exist for the programmer to mask selectively one or more interrupt levels. When an interrupt is being serviced and another interrupt takes place, then the priority of the new interrupt is compared with the priority of the existing interrupt. A higher priority interrupt is allowed to suspend the servicing of the current interrupt, otherwise the lower priority interrupt will have to wait until the higher priority one has been serviced.

Some of the currently available priority interrupt circuits carry out far more than the identification of the highest priority device which

Software Support

requires attention. They will also provide the address of the memory location to which the control should be transferred. If two interrupts occur simultaneously, the device serviced by the microprocessor program has the higher priority.

Direct Memory Access

While interrupted input/output throughput rate is higher than the throughput rate achieved by the use of polling techniques, it may not be high enough for many applications requiring fast data transfer. Direct memory access (DMA) techniques bypass the central processing unit and substantially increase the data transfer rate which can be as high as the memory cycle time allows. The other limiting factor is the speed of the peripheral device. This technique can be used to write data required at high speeds or for transfers between the memory and the mass storage devices attached to the microcomputer. DMA involves the isolation of all devices, other than the single device to be used for data transfer, from the memory while the data transfer between the memory and the appropriate peripheral is taking place. Special purpose DMA controllers are used to achieve this high-speed block data transfer. The direct memory access controller requires the use of the address and data buses to carry out the transfer.

Direct memory access operations can take place in different modes. It is possible to suspend the normal operation of the microprocessor completely for the period of time during which direct memory access operations are being carried out. In this particular type of direct memory access, the peripheral wishing to carry out high-speed data transfer informs the DMA controller by means of an interrupt signal. The controller has to obtain control of the data and address buses before the high-speed data transfer can take place. This is achieved by sending a HOLD signal to the microprocessor which suspends its operation after executing the current instruction. Control of the data and address buses is transferred to the DMA controller which is also given the address of the memory location at which the block starts and the number of words to be transferred. Data is then transferred between the memory and the external device. During this data transfer period, the microprocessor does not carry out any other operations. This type of DMA operation, referred to as the visible or burst mode, is very fast and frequently used in microcomputer systems. However, the speed of operation of the microprocessor is reduced.

Another mode of direct memory access used in larger systems involves the stealing of cycles during which the microprocessor is carrying out other tasks which do not require access to the memory and the data bus is not being used. Data is transferred one byte at a time. The cycle stealing can be transparent to the microcomputer in the sense that the normal operations of the processor are not suspended. This mode of DMA operation requires that the CPU and the external device must not attempt to gain access to the memory at the same time.

The vast majority of the available microprocessors have DMA facilities and suitable LSI-based DMA controller chips are obtainable. Sophisticated systems make use of dedicated microprocessor chips as DMA controllers.

To summarize, the three input/output methods discussed here have their own advantages and disadvantages, and are suitable for connecting to different types of devices. The programmed input/output is suitable for use with fast devices which are regularly providing data to the microcomputer. Interrupted input/output can be used with slow peripherals such as teletypewriters. High-speed data transfer between the memory and external storage devices, such as floppy disks, can be achieved by using direct memory access.

Modules which contain the complete circuitry required to implement a multiplexer, instrumentation amplifier, sample and hold circuit, and subsequent analog to digital conversions are also available. They include facilities for using the software to specify the gain of the instrumentation amplifiers so that different gain levels can be specified for different channels. Typical modules can be used to acquire data from 8 or 16 channels. These modules are compatible with many microcomputers and simplify the task of interfacing sensors to a microcomputer. Such modules are cheaper than the combined cost of the discrete components which would otherwise be required and provide a ready made solution to many data acquisition problems. Higher throughputs, if required, can be achieved by using other data acquisition systems which have separate circuits for handling analog data from each input channel.

10.6 DISCUSSION

All microcomputer-based systems, irrespective of the applications, require software support in the form of suitable application programs. These programs contain the instructions necessary to carry out the task under consideration. The development of suitable programs, particularly for dedicated applications involving direct interface between microprocessors and production machines/processes, requires very considerable effort on the part of skilled programmers. This software development process can be simplified by using appropriate development tools such as high-level language compilers, assemblers, cross-com-

Software Support

pilers, cross-assemblers, monitors, operating systems, universal or dedicated microprocessor development systems, in-circuit emulators and logic analyzers.

We have briefly described the main features of operating systems, which are usually specific to a particular computer or range of computers. Further details of the material presented are available in the reference list.

NOTES AND REFERENCES

The subject matter of this chapter has been diverse and machine-dependent. We have made every effort to stress the major items and to examine the underlying principles and features. What has been covered in the foregoing sections is by no means complete, and can be considered a guided tour along the basic elements of software methodologies for real-time and computer-controlled systems. For further reading and more elaborate studies it is recommended to consult Bowles (1977), Chamberlain (1980), Gladstone (1979), Imbriale (1980), Kallis (1979), Lee (1977), McIntyre (1978), Motorola (1979), Ogdin (1978), and Wintz (1980) on topics related to software development tools. Technical issues and applicatons of operating system concepts are found in Allworth (1981), Brookes et al. (1985), Cassell (1983), Comer (1985), Johnson (1984), Kaisler (1982) and Lister (1979). Relevant points pertaining to the interface of micros and engineering applications have been discussed in Barney (1985), Tzafestas (1983), Wolfe (1980) and Kochhar and Burns (1983).

11
Industrial Process Control Systems

11.1 INTRODUCTION

All industrial process are affected by outside disturbances and have to be controlled to ensure that the product quality lies within the required acceptable limits. Automatic control of continuous processes dates back to the early 1950s. Over the past twenty years computer systems have been continually adapted to process control, however, with an emphasis on continuous processes. Later on, in the late 1960s, the programmable logic controller (PLC) was developed in response to the automotive industry's requirements. In the early 1970s, PLCs were first applied in control of batch processes in the chemical industry. At that time (late 1960s/early 1970s), a limited number of computer systems that offered a batch control software package was available.

A major breakthrough came in the early 1980s when computer manufacturers developed batch control systems for mini- and microcomputer systems. Until recently, there were a number of differences between a PLC and a computer system:

1. A PLC has a simple operating system, and only limited security checking is done. To bring the security checking of both hardware and software functions within a PLC at the level of a computer

Industrial Process Control Systems

would at least require hardware modifications and major (expensive) software modifications.
2. A PLC has a simple and specific application language in the form of a ladder diagram or Boolean algebra. This implies a limited instruction set.
3. The operating system of a PLC is not accessible to a user. Higher level programming languages cannot be used.

The scope of batch computer applications is usually different from the scope of a PLC application. A computer can be used for trending and documentation of process variables, report generation, more sophisticated control through supervisory (Fortran or Pascal) programs, and so forth, but until recently these functions were not available in PLCs. The latest development is that computer manufacturers develop interfaces between PLCs and computer systems. PLCs are then used for sequences which are repeated continually; the computer monitors and controls the PLCs by initiating sequences and performs the data collection and information processing. The advantage of this approach is that the computer is freed from several tasks and thus can be used for other tasks. PLC is also an order of magnitude less expensive than a computer system. One should keep in mind, however, that the security and integrity checking of a PLC is limited, although continuous improvements are being made in this area.

Another recent development is that PLC manufacturers have designed interfaces to personal computers providing color graphics and documentation facilities. Modules are now available to interface PLCs to mass storage devices.

Since 1984 there have been PLC-based systems on the market which perform all the functions of a conventional programmable controller, with additional capabilities previously found only in mainframe process control computers. They are characterized by a design architecture that combines the features of a programmable controller (ruggedness, high-speed scan ability, low-cost I/O, ease of programming) with the most desired features of a process control system (communications capability, higher level programming languages, large memories for data storage).

In typical industries complex process control systems can have as many as a thousand control loops, some of which are interacting, i.e. any change in the value of a single input or control variable affects two or more output variables. Analog controllers simply attempt to maintain the value of a single variable near to its set point.

Computers are used in many different modes for process control purposes. At the very lowest level, they are suitable for direct digital control (DDC), i.e. the replacement of conventional analog controllers. Sophisticated adaptive computer control systems attempt to optimize the overall performance of the whole process, taking into account the interactions between variables and the constraints imposed on the complete operation of the process, in order to maximize the profit. Computer based optimal start-up procedures are implemented to minimize the time over which the process achieves a steady-state condition. Similarly, the computer is used to shut down a process plant in a safe manner.

The objective of this chapter is to discuss the computer-controlled systems used in industrial processes and examine the application of various control algorithms. Emphasis will be placed on introducing the terminology and practices of typical process control schemes. This will eventually take the form of cases studies.

11.2 FUNCTIONS AND BENEFITS OF A BATCH PROCESS CONTROL

For the purpose of demonstration, we summarize below the basic functions of the software needed in a batch process control system:

1. Discrete (on/off) control. This deals with the checking of contact inputs and the driving of contact outputs. When there is a discrepancy between the actual state of a contact and its desired state, one or more actions can be taken:
   a. Generating an alarm message
   b. Invoking an alarm/service subroutine which has been coded by the user and deals with the discrepancy
   c. Shutting down part or all of the process.
2. Interface with the continuous control package. For example, a blower is started (contact-output activated) upon low flow (analog value).
3. Sequence control (start/stop pumps, open/close valves, etc.). Sequence control is usually handled by the batch executive, the main program that controls the execution of other programs and interfaces with status tables which contain process unit information related to the batch type of operation.
4. Performance of calculations.
5. Notifying operator of alarm conditions.
6. Providing an operator interface to the process.
7. Reporting and logging.
8. Recipe handling.

A unit can be running in different operating modes. Different batch control systems use a different terminology, although the following modes are common:

1. *Automatic.* The sequence is carried out under control of a user-written program without operator intervention. This is the normal operating mode.
2. *Manual.* Process action is carried out by the operator from and via the operator interface.

Depending on the batch process control software, more operating modes may exist. When a unit runs in automatic mode, it can be in different states:

1. *Normal.* In this state, sequence control runs without a problem.
2. *Hold.* In this mode, sequence control is halted as a result of predefined error conditions, and the unit is brought to a predefined safe state.

The major benefit of automation of batch processes is increased safety. Because of high-frequency (typically two seconds or less) scanning of contact inputs, such as the status of all valves and pumps in the process, a discrepancy will be determined immediately. Upon detection of a discrepancy, predefined action will be taken, and, in case of an emergency, the process will be brought to a safe state. Possible errors in manual actions are eliminated, leading to increased personnel and plant safety.

It is difficult to convert increased safety into an economic gain in terms of k$/year. It is sometimes necessary to operate under computer control for two or more years before a real comparison with past operation without a computer can be made and exact credits can be identified.

The second most important benefit of automation of a batch process is increased production through a reduction in cycle time. Considerable savings are possible, especially when many steps are involved.

Another important benefit is increased consistency of product quality. In a competitive environment, it is extremely important to produce a consistent product. In a decade of overproduction, quality is more important than quantity. Off-spec incidents can usually be reduced by a factor of four, which can be translated into an economic gain.

A last benefit that should be mentioned is flexibility in charging through the use of multiple recipes. Especially when many different products are made, a situation exists which is prone to errors. By having a recipe for each product grade, the required ingredients are added in appropriate amounts, react at the right temperature, and so forth.

11.3 PROCESS CONTROL CONCEPTS AND ACTIONS

11.3.1 Process Control Concepts

Process control basically involves holding the machine or process control variables at values that maintain the product quality. In a computer process control system, the process data acquired by the transducer are input to the computer. The actual control of a process might be open loop or closed loop.

In open-loop mode, the value of an input variable is set to a given value in the expectation that it will result in the desired value of the output variable. There is no automatic comparison of the actual output value with the desired output value to determine the changes, if any, that should be made to the input setting.

The overall performance of a process is substantially improved by operating it in closed-loop mode whereby the actual value of the output variable is compared with the desired output value to determine the changes required in the input control variables.

All the control actions are dependent on the actual state of the process. The implementation of a closed-loop system requires the following facilities.

1. Instruments or transducers to monitor the actual state of the process
2. Comparators to compare the actual values of process variables with the set point values
3. Means of determining the control action to be implemented
4. Actuator or other means of implementing the control action

The process loop might be closed by a human operator who interprets the process output to determine the deviation between the desired and actual outputs. He uses this knowledge of the deviation to make the adjustments necessary for correcting the process output. A computer used to acquire the process data from suitable transducers may also be used to make the comparisons and display the necessary control actions that should be carried out. The actual changes to the input settings are made by the human operator. Figure 11.1 illustrates such a manual or computer-aided closed-loop system.

It is highly unlikely that a human operator will be able to make all the necessary changes to the large number of input variables which might be present in a complex process. The closed-loop control of many

Industrial Process Control Systems

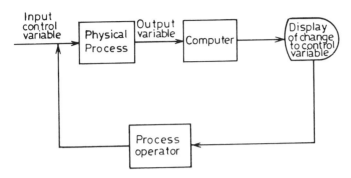

Figure 11.1 A computer-aided process control system.

process loops can be carried out by using suitable automatic control mechanisms which determine the deviation between the actual and desired values of a process variable and automatically alter the input to the process to maintain it around the desired value. A microcomputer along with a suitable actuator is ideal for carrying out all these tasks of data acquisition, comparison and subsequent closing of the process loop.

The implementation of a closed-loop control system on a microcomputer requires that the commands or signals used for this purpose must be compatible. Thus, the data input to the microcomputer must be in digital format. As will be shown later the vast majority of the sensors used in the process industry have an analog output which must be amplified and converted into digital format using analog to digital converters. Similarly, the digital signals output by the microcomputer must be converted into analog format and amplified to operate the actuators.

Apart from the microcomputer system configuration requirements for implementing closed-loop control, particular emphasis has to be placed on the accuracy with which the sensors measure the process variables. The accuracy of control actions carried out by the microcomputer is better than that of analog controllers but depends critically on the accuracy with which the process data are measured.

Open- and closed-loop control systems are not totally exclusive. They are often used together to control individual processes. Closed-loop systems may be used for the normal control of a process whereas open-loop control may be used to override a specified set point and alter the state of a process.

11.3.2 Computer Control Actions

The control actions carried out by a process computer vary according to the algorithms (strategies) used. Commonly used control actions are:

1. ON-OFF control
2. Proportional (P) control
3. Integral (I) control
4. Derivation (D) control
5. Three-term (PID) control

We now discuss each action in the following sections.

ON-OFF Control

This is the simplest type of action which can be implemented by using a computer or a conventional analog controller. As implied by the term ON-OFF, the control element, such as a heater, can be in two states of being fully turned on or fully turned off. There are no intermediate states. It is a discontinuous control system. Perhaps the simplest example is the ON-OFF control of home heating systems. Thermostats are used to operate the heater around the set point temperature value. When the actual room temperature is below the desired set point value, the heater is fully switched on, and the room temperature begins to rise. The heater continues to operate until the room temperature, as measured by a thermostat, exceeds the set point value

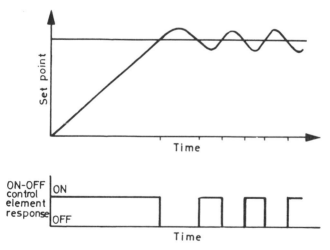

Figure 11.2 Set point ON-OFF control.

Industrial Process Control Systems

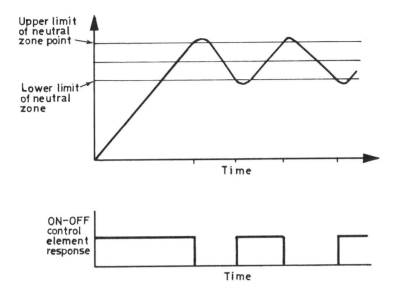

Figure 11.3 ON-OFF control system with a neutral zone.

when it is switched off. Loss of heat to the surroundings results in a temperature drop and when the actual temperature falls below the set point value, or the lower limit of a neutral zone in which no action is taken to avoid frequent cycling, the heater is switched on again. This process is continuously repeated. Figures 11.2 and 11.3 illustrate single set point and neutral zone ON-OFF control systems. While easy to implement and relatively inexpensive, such a control system does not satisfy the requirements of many industrial processes. The response of the control element does not take any account of the magnitude of the error between the desired and actual values of the controlled variables.

All ON-OFF control systems, irrespective of whether they are based on set point control or neutral zone control, are subjected to cycling and overshooting as illustrated in Figs. 11.2 and 11.3. Microcomputers, if used as part of a control system configuration, can provide much better control than the ON-OFF control illustrated here.

Proportional Control

The control of a process is improved to some extent by using a proportional control system whereby the correction signal used varies in direct proportion to the deviation between the actual and desired values of a process variable. Thus, the output of the controller is proportional to the error signal. This can be represented mathematically by the equation

$$U = K_p e \tag{1}$$

where U = control action, K_p = constant of proportionality or controller gain and e = difference between the desired and actual values of the process variable.

When the actual value exceeds the desired value, the deviation e is negative and the control action to be implemented by using a correcting signal is also negative. The actual value of the correcting signal is obtained by the multiplication of the error signal and the gain constant. Thus a large deviation between the actual and desired signal values results in an increased control action. As the actual value approaches the desired value, the correcting signal and hence the control action is reduced. Figure 11.4 illustrates the variation in error signals and the control actions.

Small values of K_p result in a small correcting signal and hence sluggish dynamic response to the error changes. This sluggish response is improved by increasing the value of K_p. However, too large a value of K_p can lead to instability and oscillatory response.

A certain amount of control element input, for example the energy input to heater elements, is required to maintain the process variable

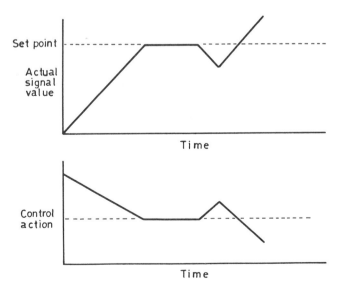

Figure 11.4 Proportional control system action.

Industrial Process Control Systems

at its operating point. Thus, (1) can be modified to provide this control action. The modified algorithm is

$$U = K_p e + K \tag{2}$$

where K is a constant and determines the control action when the error signal or deviation e is equal to zero. Thus the control system used for the implementation of proportional action on its own should maintain the control action to this value K whenever there is zero error between the desired and actual values of the process variables. Proportional action is carried out around this particular value of the input set point which corresponds to a particular value of the load. Any change in the load will require an appropriate adjustment in the value of K.

It is easy to program microcomputers to implement proportional control action. Since microcomputers only handle discrete data, (2) must be modified for use in discrete mode. This can be represented as follows:

$$U_n = K_p e_n + K \tag{3}$$

where U_n = control action after the nth sample has been taken, e_n = error after the nth sample.

Using a high level language, or assembly language for fast processing, it is simply a matter of subtracting the actual value of the variable from its desired value, followed by multiplication and addition operations. The correcting signal value, representing the control action, is then output to the actuator. Only a small number of instructions is required to implement the control action. Hence a microprocessor based proportional control system is relatively fast.

A major disadvantage of proportional control systems is the need to reset the value of the constant K whenever the load is changed. The control system will continue to produce correcting signals as long as the deviation e has a finite value. Better control is achieved by incorporating an integral term into the control algorithm used.

Integral Control

Integral control of a process loop involves integration of the error signal over a period of time. The rate of change of the correcting signal is proportional to the error. This can be mathematically represented as

$$\frac{dU}{dt} = K_i e \quad \text{or} \quad U = K_i \int e \, dt \tag{4}$$

where K_i is the gain of the integral controller or the integral action constant.

In this mode the controller will continue to take control action as long as a control error exists and any offset errors, caused by load changes, are eliminated. The correcting signal as determined by (4) is progressively increased until the error has been corrected. Expression (5) shows the discrete representation of the integral control action

$$U_n = K_i \sum_{i=0}^{n} e_i \Delta t \tag{5}$$

where Δt = sampling interval.

The effect of integral action is illustrated in Fig. 11.5. A positive error results in an increasing value of the integral control action while a negative error leads to a decrease in the correcting signal value. The value of the correcting signal remains constant for the zero values of the error signal, as shown in Fig. 11.5.

Integral control action is also referred to as reset control since the effect of load changes can be automatically counteracted without the need to reset values manually. This control mode is often combined with proportional control to produce a two-term, proportional-integral (PI) control algorithm. A major disadvantage of integral control is its tendency to overshoot or cause oscillations which can lead to instability.

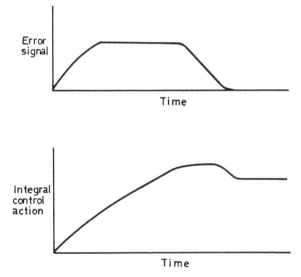

Figure 11.5 Effect of integral control action.

Industrial Process Control Systems

Microcomputer implementation of integral control action is slower than the implementation of proportional control since a number of multiplication and addition operations have to be carried out. However, only one multiplication and addition operation is required after a new sample has been obtained.

Derivative Control

In this mode of control, also referred to as anticipatory control, the corrective action is proportional to the rate of change of error. In mathematical terms, the derivative control action can be represented by

$$U = K_d \frac{de}{dt} \qquad (6)$$

where K_d = derivative control gain or derivative control action constant. Expression (7) is the discrete form, necessary for microcomputer implementation, of the derivative control action.

$$U_n = K_d(e_n - e_{n-1})/\Delta t \qquad (7)$$

The importance of derivative control lies in the fact that the control action is varied to respond to the rate at which the error is changing. The error is corrected rapidly before it becomes too large. The deriva-

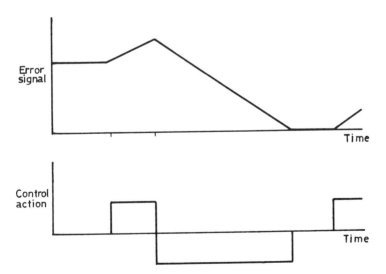

Figure 11.6 Effect of derivative control action.

tive control term has the effect of damping the error and stabilizes the system performance. Use of derivative control in combinatlon with proportional control allows the proportional control factor to be higher than would be possible otherwise. The speed with which derivative control action is implemented enables the control system to follow quickly any step changes in the set point. Figure 11.6 illustrates the effect of derivative control action.

No control action is taken as long as the error signal remains constant. Any change in the error signal results in a correcting signal whose value depends upon the derivative control gain and the rate of change of error. As illustrated in Fig. 11.6 a constant rate of increase in the error signal results in a constant control action.

Derivative control action is seldom used on its own, since a constant value of the error does not result in a corrective control signal. It is therefore necessary to combine derivative control action with proportional or integral control actions. Three-term controllers which combine proportional, integral, and derivative control actions are frequently used in processing industries.

Three-term (PID) Control

The overall control of a process loop is substantially improved by using a correcting signal whose value, and hence the control action, has been determined by combining the algorithms used for proportional, integral, and derivative control. This mode is referred to as three-term or PID (Proportional-Integral-Derivative) control. Three-term control can be represented mathematically by

$$U = K_p e + K_i \int e\, dt + K_d \frac{de}{dt} + K \tag{8}$$

$$= K_p \left(e + \frac{K_i}{K_p} \int e\, dt + \frac{K_d}{K_p} \frac{de}{dt} \right) + K$$

$$= K_p \left(e + \frac{1}{T_I} \int e\, dt + T_d \frac{de}{dt} \right) + K \tag{9}$$

Conventional analog controllers usually specify the integral and derivative action times which are more meaningful to a control engineer than the proportional constants K_i and K_d. Equation (8) has been modified to equation (9) in order to make use of the integral time constant T_I, which is a ratio of the proportional gain constant and the integral gain constant, and the derivative gain constant T_d which is a ratio of the derivative control gain and proportional gain. Equation (9) is suitable for implementing conventional analog control action. Implementation of three-term control on a microcomputer requires that the equation be modified and represented in discrete format. Equation (10) is the discrete form representation of the control algorithm after the nth sample has been obtained.

$$U_n = K_p \left[e_n + \frac{1}{T_I} \sum_{i=1}^{n} e_i (\Delta t) + \frac{T_d(e_n - e_{n-1})}{\Delta t} \right] + K \qquad (10)$$

It follows that for the $(n - 1)$th sample

$$u_{n-1} = K_p \left[e_{n-1} + \frac{\Delta t}{T_I} \sum_{i=1}^{n-1} e_i + \frac{T_d}{\Delta t} (e_{n-1} - e_{n-2}) \right] + K \qquad (11)$$

On subtracting (11) from (10) we obtain

$$U_n - U_{n-1} = K_p \left[(e_n - e_{n-1}) + \frac{\Delta t}{T_I} e_n + \frac{T_d}{\Delta t} (e_n - 2e_{n-1} + e_{n-2}) \right]$$

$$= K_p \left\{ \left[1 + \frac{\Delta t}{T_I} + \frac{T_d}{\Delta t} \right] e_n - \left(1 + 2 \frac{T_d}{\Delta t} \right) e_{n-1} + \frac{T_d}{\Delta t} e_{n-2} \right\}$$

$$= K_p \left[1 + \frac{\Delta t}{T_I} + \frac{T_d}{\Delta t} \right] e_n - K_p \left[1 + 2 \frac{T_d}{\Delta t} \right] e_{n-1} + K_p \frac{T_d}{\Delta t} e_{n-2}$$

$$= A_0 e_n - A_1 e_{n-1} + A_2 e_{n-2} \qquad (12)$$

where A_0, A_1, and A_2 are constants defined by the following relationships

$$A_0 = K_p \left(1 + \frac{\Delta t}{T_I} + \frac{T_d}{\Delta t} \right)$$

$$A_1 = K_p \left(1 + 2 \frac{T_d}{\Delta t} \right)$$

$$A_2 = K_p \frac{T_d}{\Delta t} \qquad (13)$$

Rewriting (12) in the form

$$U_n = U_{n-1} + A_0 e_n - A_1 e_{n-1} + A_2 e_{n-2} \qquad (14)$$

to determine the control action after the nth sample has been obtained.

The above representation of the three-term control algorithm is easy to implement on a microcomputer. It is only necessary to store the last three values of the error signal and the previous control action. Simple multiplication and addition operations are necessary to calculate the control action which must be taken after a new sample has been obtained and the error signal calculated. Symbol U_n represents the direct digital control signal value after the nth sample, for example, the speed of a motor. Equation (15) represents an incremental direct digital control algorithm suitable for implementation on a microcomputer based control system.

$$U_n = A_0 e_n - A_1 e_{n-1} + A_2 e_{n-2} \qquad (15)$$

where U_n is the change to the previous value of the input control variable, for example the increase or decrease in the motor speed.

Equation (14) is a particular form of a more general control algorithm of the form

$$\begin{aligned} u_n &= B_1 u_{n-1} + B_2 u_{n-2} + \cdots + A_0 e_n + A_1 e_{n-1} + A_2 e_{n-2} \\ &= \sum_{j=1} B_j u_{n-j} + A_0 e_n + A_1 e_{n-1} + A_2 e_{n-2} \end{aligned} \qquad (16)$$

In an actual implementation of such a digital control algorithm, it is necessary to select the values of coefficients $A_0, A_1, A_2, \ldots, B_1, B_2, B_3, \ldots$ so as to tune or optimize the control and ensure that oscillation, instability, and offset errors are minimized without too frequent cycling. It is easy to implement these algorithms on a microcomputer since all the constants and previous values of error signals and control actions are stored in the computer memory.

Direct digital control of process loops offers very considerable advantages over conventional analog controllers. It is possible to use a control algorithm which takes full account of the dynamic characteristics of individual loops whereas conventional analog controllers can implement only the simple algorithms, usually in the form of a three-term controller, built in by the designer. Changes in the characteristics of the process are catered for easily and at low cost after the original design of the process controller. Redesign of conventional controllers is a difficult, time consuming and expensive business. The use of the microcomputer and the software required for the implementation of the direct digital control algorithm offers very considerable flexibility to the control system designer.

11.4 SUPERVISORY CONTROL

The use of control algorithms discussed so far is based on the assumption that the process loops are to be individually controlled to a set

Industrial Process Control Systems

point. No account is taken of the interaction between different process loops. The values of set points may be based on the a priori knowledge of the process or may be determined experimentally to optimize the process performance. This procedure does not take account of the changes in the behavior of the process and the subsequent need to alter the set points of process loops. Another problem is the interaction between different process loops. The overall performance of a complete process is substantially enhanced by using a supervisory control system in which the set points are calculated by a supervisory control program stored in the computer memory. This program may be regarded as a simple model of the process. It is then possible to coordinate the overall control of the process. The actual implementation of a supervisory control system may be carried out using one or more microcomputers. For simple processes with only a limited number of loops, a single microcomputer may be used. The control of a large number of process loops will require a number of microcomputers for direct digital control and a coordinating microcomputer in which the supervisory control program is executed. Figures 11.7 and 11.8 illustrate the two microcomputer configurations which may be used for combined direct digital control and supervisory control.

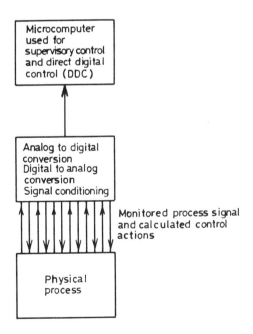

Figure 11.7 A simple supervisory computer control.

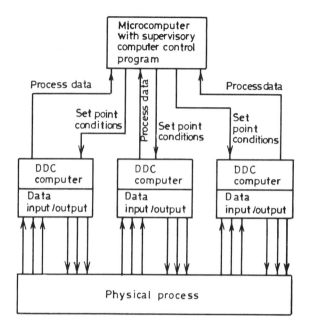

Figure 11.8 Implementation of supervisory control system.

Supervisory computer control takes many forms ranging from simple program control to cascade control. The various forms of supervisory computer control can be divided into the following categories.

1. Ratio control
2. Program control
3. Sequence control
4. Cascade control

These control strategies are now considered in detail.

11.4.1 Ratio Control

This control mode is a very simple example of the supervisory computer control of processes. The controller is set to a value which bears a constant relationship with the measured value of another variable. It is particularly useful for controlling blending operations in which different items must be mixed in a constant proportion. Formulation of drugs, for example, requires the use of such a control system. Another example is the need to maintain a constant air-fuel ratio in combustion operations.

Industrial Process Control Systems

Considerable savings are possible by using a ratio control system for burning fuel in jet engines.

A microcomputer is ideal for the implementation of a ratio control system. It is necessary to monitor the value C_1 of a control variable and multiply it by M_1, the constant ratio, which determines the proportion in which individual items must be mixed or blended. The microcomputer calculates C_2, obtained by the multiplication of C_1 and M_1, which becomes the set point for the second variable. Figure 11.9 illustrates a simple ratio control system.

Actual implementation of the microcomputer based ratio control system involves sampling the control variable C_1 at regular intervals. Denoting the nth sample of control variable C_1 by $C(K, 1)$, the nth value $C(K, 2)$, of the controlled variable C_2 is given by the relationship

$$C(K, 2) = M_1 C(k, 1) \rightarrow C(K, 2) = M_1 C(K - 1) \qquad (17)$$

It is better to use an incremental controller in which it is only necessary to determine the adjustment which must be made to the second control variable. $C(K, 2)$ can also be written in the form

$$C(K, 2) = C(K - 1, 2) + C(K, 2) \qquad (18)$$

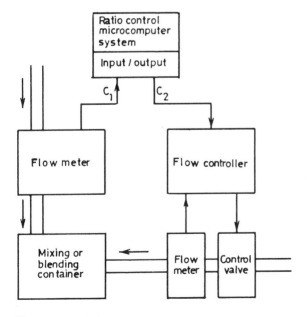

Figure 11.9 Microcomputer-based ratio control system.

where $C(K, 2)$ represents the incremental adjustment to the second control variable after the nth sample has been obtained.

Thus

$$C(K - 1, 2) + C(K, 2) = M_1 C(K, 1)$$

or

$$C(K, 2) = M_1 C(K, 1) - C(K - 1, 2) \qquad (19)$$

The above algorithm enables a quick calculation of the incremental adjustment to the second controlled variable. It is possible to extend the above ratio control system to cover more than two variables.

11.4.2 Program Control

The start and shut down of many complex manufacturing processes used in the chemical industries has to be gradual to ensure safe operation. In some cases, for example, the operation of high-temperature furnaces used in the steel and chemical industries, the start up and shut down procedure can last a number of days to take the process through the complete cycle. Program control is used for this purpose. Microcomputers are suitable for implementing program control since the time periods of particular phases can be stored in the memory. Following the completion of a particular phase, the microcomputer issues commands to alter the set points of individual loop controllers which might be conven-

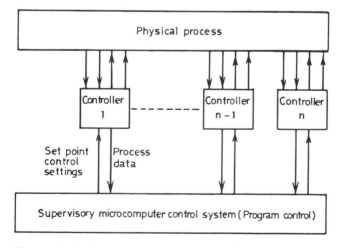

Figure 11.10 Supervisory program control system configuration.

Industrial Process Control Systems

tional analog or microprocessor based. A supervisory control program is used to determine the set point conditions, over a period of time, of individual controllers. Program control systems are also used to guide the process from one set of operating conditions to another, so that a new product, requiring different operating conditions, can be produced. In all cases, it is desirable to minimize the time taken, and hence the cost, to change the state of the process. Constraints on the values of input and output variables have also to be taken into account. Dynamic programming and other optimization techniques can be used to determine the best process path which must be taken to reach the new state. Figure 11.10 illustrates a supervisory control system configuration for implementing program control of processes.

11.4.3 Sequence Control

In this control mode, an extension of program control, the individual controllers and the actuators associated with them are switched to ON and OFF states. The time period over which an individual controller remains in the ON or OFF state is long compared to the process time lag. Sequence control is not carried out on a strict time period basis, characteristic of program control. Sensors are used to make measurements of the actual process data and the control action is taken after a particular condition has been satisfied and the required action has been carried out.

A typical example of a computer based sequence control system is in automatic assembly line operations. The line must not be moved

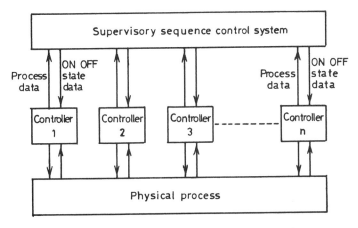

Figure 11.11 Microcomputer-based sequential control system.

until the previous operation has been completed. Another example of a sequential control system is in the formation of glass bottles. Quality problems may cause one or more of the glass bottle forming sections or moulds to be out of action. A sequential control system is required to transfer the glass melt to other sections until the faults have been corrected. In other processing industries, heaters are switched ON and OFF to vary the heat input so as to bring the temperature up to the required level. Microcomputers, interfaced to the process via suitable sensors, are particularly suited to the implementation of sequence control systems. The input data are analyzed to determine whether a particular condition has been satisfied when the required sequence control action can be implemented. Figure 11.11 illustrates a microcomputer based sequential control system.

11.4.4 Cascade Control

In most processes the process operator is concerned with the control of output or controlled variables such as pressures, temperatures and flow rates, so that the quality of the final product lies within the desired limits. This process control may be achieved by controlling the heat input, motor speeds or material feed rates which are basically machine variables rather than process variables. The process operator hopes that particular values of the input or control variables will result in the desired values of process variables. Feedback control is achieved by changing the control setting of one or more variables. This control can have the effect of introducing disturbances into the process if more than one process output variable is affected by the changes in any of the input control variables. Such a situation can best be illustrated by considering the plastics extrusion process in which the product quality depends primarily on the melt temperature and pressure at the die end. Melt temperature is affected by the temperature of the barrel, divided into a number of zones and controlled by individual loop controllers set to different values, and the speed at which the screw rotates inside the barrel. Pressure is primarily, but not solely, dependent upon the screw speed. Changes in barrel zone temperatures also have an effect on the melt pressure. Both pressure and temperature depend upon the properties of the material being processed. Material properties vary from one batch to another in a random manner. The process and associated control is schematically illustrated in Figure 11.12.

Any changes in melt pressure can be counteracted by altering the screw speed which also disturbs the melt temperature. This error in melt temperature can be corrected by changing the set points of the

Industrial Process Control Systems

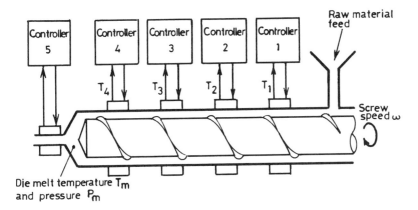

Figure 11.12 Extrusion process and associated control system.

barrel zone temperature controllers. The use of a number of independent controllers for individual loops, affecting a given process condition, results in considerable interactions between their control actions. This often leads to a cycling of process conditions making it necessary to reduce the controller gains to such low values that they only enable very long term averaging control. The objective of the control system

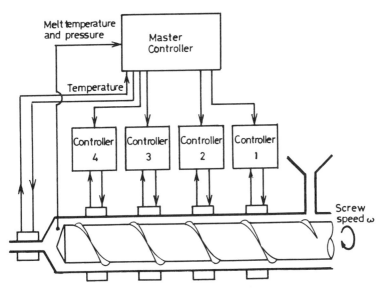

Figure 11.13 Cascade control system for the plastic extrusion process.

is to minimize the time taken to correct this error in temperature without degrading the polymer in any way. A cascade control system, illustrated in Figure 11.13, can be used to alter the settings of the individual zone temperature controllers to achieve the desired result in an optimal manner. Thus a cascade control system can be characterized by a master controller which is used to alter the set points of slave controllers.

In such a cascade control system, the die melt temperature is set up on a master controller. The temperature settings of the slave controllers, which control the temperature profile of the extruder barrel, are continuously adjusted to obtain the desired melt temperature condition. A model, showing the relationships between the temperatures of the individual controllers and die melt temperature, is required to impement a cascade control system. Such a system shows considerable improvement over the simple individual loop control systems.

The implementation of a cascade control system using conventional analog hardware is expensive since controllers are required at two levels. Conventional controllers do not allow the implementation of a sophisticated control system strategy. Apart from the advantage of inherent flexibility, the microcomputer implementation of cascade control systems is cheaper and better. The rudimentary model used for cascade control purposes can be altered in the light of the experience of operating the process. To be effective, the cascade control system must be fast without introducing instability and oscillations.

11.4.5 Some Practical Considerations

A large number of functions have to be carried out to implement a microcomputer based process control system. Some of these functions may be carried out on hardware while others involve the use of software. Typical computer hardware based process control systems would consist of:

A suitable 8-bit or 16-bit microcomputer
Battery backed random access memory
Power supply
EPROMs
Input signal conditioning circuits such as amplifiers
Multiplexer
Analog to digital converters
Real-time clock
PIAs

Industrial Process Control Systems

Serial interface circuits
Digital input circuits
Digital to analog converters
Demultiplexer
Output amplifiers
LED displays or visual display units

The above items form the core of the microcomputer hardware required for process control. Other items can be added to this list to suit the needs of the process to be controlled. The characteristics of all the hardware items used must satisfy the process control requirements.

Software requirements can vary depending upon the type of system used. In a single board computer system, the required application program in machine code would be resident in a suitable EPROM. Such a single board computer (SBC) configuration would be suitable for first level direct digital control, based on proportional, integral, derivative or some combined control strategy, or individual process loops.

Implementation of more sophisticated process control-systems would require the use of a larger microcomputer system configuration. Such a system configuration, in which high-level language programs are used, would require a suitable system executive, system monitor or a complete operating system. There might also be a need for high-level language compilers/interpreters and assembler programs. A visual display unit would be used to display the current process conditions while a hard copy device, such as a slow thermal printer, might be used to print out summary or exception conditions data. Floppy disks or cassettes or cartridge could be used to store programs and/or the process data for off-line analysis. Control of complex processes would involve the use of a hierarchical computer system configuration. Such a configuration is also necessary for the implementation of optimal and adaptive process control systems. The microcomputers at the first level are used for data acquisition and direct digital control. Supervisory control and data analysis functions are performed by the micro- or minicomputer used at the second level. It is possible to take this configuration two stages further whereby the computer systems used for supervisory control or data analysis purposes are linked to computers at two higher levels used for production scheduling and control, and the overall business management. While only a few configurations of this type are in current use, the trend is to develop distributed process control and data processing systems which satisfy the requirements of all the people in a manufacturing company. Communications between different computer systems are necessary to eliminate data duplication and redundancy. Various distributed processing system configurations will be discussed in the next chapter. A detailed discussion of production control and management information systems is outside the scope of this book.

11.4.6 Software for Process Control

Most conventional control systems simply consist of specialized hardware which is used to implement the required control actions such as ON/OFF control, proportional control, and three-term control. Such systems are usually based on the application of analog control techniques. These systems are necessarily restricted and inflexible.

Many digital control systems are based on the combined use of hardware and software. Application programs are required to implement the required control action which may be far more sophisticated than simple ON/OFF control, proportional control or three-term control. It is possible to use different, flexible combinations of hardware and software for designing and implementing almost any type of computer control system.

Apart from interfacing the process with a microcomputer, the most time consuming and expensive part in the implementation of a computer process control system is the development of the necessary software. Development of real-time data acquisition and control software is more demanding than the development of software used for routine data processing applications. Implementation of microcomputer based systems requires synchronization of events, and quick response. The actual response requirements depend upon the process time constants.

Much of the currently used data acquisition and control software is written in assembly languages or machine code. The cost of developing such software is substantially reduced by using suitable high-level industrial languages which are now available for use on microcomputers. It is thus possible to develop a library of programs which can be used on any of the microcomputers for which a suitable compiler, for the selected high-level language, is in existence. Apart from the paramount functional requirements of the language used, it must be efficient, simple to use, easy to learn and standardized. Portability is particularly important since different implementations of a language do not permit the use of the application program on another microcomputer system. High-level industrial languages such as PASCAL, CORAL, RTL-2 and control BASIC (a subset of BASIC) are suitable for process control applications. Another language which offers considerable potential for future applications is ADA. Several other special purpose process control languages have also been developed.

Dedicated microcomputer systems, suitable for process control purposes, are also available. Some of these systems minimize or eliminate the need for conventional programming. It is possible to purchase a range of standard EPROM based modules for performing individual functions such as data acquisition, three-term control, proportional control. A complete process control system can be implemented by linking together the individual modules required. While such a system is not as flexible as a system which has been completely programmed by the user, the time and effort required to implement an EPROM module based system is reduced.

11.4.7 Microprocessor Based Programmable Controllers

The design, development and implementation of a customized microcomputer based process control system is an expensive and time consuming process. While such a system might provide a large number of benefits, the implementation cost and effort required is often a stumbling block for many companies. It is possible to achieve a high proportion of the potential benefits by using a microprocessor based programmable controller. These controllers differ from the programmable logic controllers which are only suitable for ON/OFF, sequence or interlock control in applications such as materials handling and machine tool control. Many of the microprocessor based programmable controllers can be interfaced with continuously varying analog signals. Proportional control, integral control and three-term control facilities are also available.

Other facilities include communication with computers or other programmable controllers, hierarchical computer control and automatic fault detection. In general, such microprocessor based control systems are suitable for automating processes which require sequential and/or logical actuation of various devices.

Microprocessor based programmable controllers incorporate many useful facilities such as a good operator interface, easy expansion of input/output modules, ease of repair and maintenance, communications capabilities and simplified programming. It is possible to integrate many programmable controllers into total factory systems so that the production data can be used in the production planning and control, accounts and sales departments. Similarly the manufacturing set points, used in a programmable controller, can be altered under control from a centralized control system at a higher level.

Microprocessor based controllers incorporate the following typical modules.

1. Memories in the form of RAM, EPROM or EAROMs. These memories can be expanded as required up to a maximum capacity.
2. Timer units.
3. Ac and dc voltage outputs with various switching capabilities.
4. Analog to digital converters and digital to analog converters with or without multiplexing.
5. Thumb wheel input devices.
6. Modules for handling binary coded decimal input and output data.
7. Addressable power failure sensors to reset or inhibit process outputs.
8. Input/output simulators.
9. Real-time clocks.

The built-in system software is normally resident in an EPROM which gives a user the opportunity to extend the range of control functions by replacing an existing EPROM by a new one. User programming of the tasks necessary to control a given process or machine takes the form of flow charts or ladder diagrams which are input via a simple keyboard. Conventional programming techniques are not used. The application program is stored in Electrically Alterable Read Only Memory (EAROM) so that any necessary amendments can be carried out by the user himself without the need to use PROM or EPROM programmers. Facilities are normally provided to input, verify, alter, and fix the program into the EAROM. Random access memory is used for the storage of current process conditions. These controllers can be programmed to provide alarms whenever exceptional conditions are encountered.

The user has the opportunity to select one or more input modules and output modules for use on his system. Input variables might be digital with different levels of ac and dc voltages. Alternatively, it is possible to use analog input modules with normal signal levels of 4–20 mA or 0–10 V. Similarly a number of digital and analog output modules are available for interfacing with different types of actuators such as solenoids, proportional valves, and stepping motors.

A number of such programmable controllers are available. Their capabilities are being enhanced to make them suitable for a wide range of applications. Major advantages of using such programmable controllers, for suitable applications, include the ease and flexibility of usage, simplified programming which is well within the capability of most engineers who do not possess computer programming skills, suit-

Industrial Process Control Systems

ability for use in noisy industrial environments, ease of interfacing with the process and built-in fault diagnosis.

11.5 ADVANCED CONTROL

Ninety percent of the control loops in the chemical and petrochemical industry work well by using simple proportional, integral, and derivative (PID) control. The PID controller has been used for decades, first in pneumatic instrumentation, later in analog instrumentation, and today in digital instrumentation and computer control systems.

In some cases the PID controller does not perform so well, and it may be necessary to study and analyze these situations. Personal computers offer the control engineer an excellent tool to study the behavior or controlled and uncontrolled processes. Several computer-aided design packages with tremendous flexibility to analyze virtually anything the control engineer can think of are on the market.

This section provides concise account of some of the modern control methods for practical use in process industry. These methods include:

1. Dead-time compensators
2. Inferential control
3. Feedforward control
4. Constraint control
5. Adaptive control
6. Multivariable control
7. Self-tuning control

Our focus will be on the orientation of the methods to specific cases.

11.5.1 Dead Time Compensators

One of the most common control problems in the process industry today is control of a process with significant dead time. Control of such systems will usually involve a model-based approach. The two techniques that will be discussed in this chapter are the Smith Predictor and the Dynamic Reconciliator dead time compensators. These will be applied to single input single output (SISO) systems and in subsequent sections the problem of feedforward, decoupling, and a fully interactive 2×2 system will be addressed. The self-tuning regulator (STR), which has rapidly gained in popularity in recent years and which not only deals with the problem of dead time but also changing process parameters, e.g., process gain of a fixed-bed reactor changes with the deactivation of the catalyst, will also be discussed briefly later on.

Dead time compensator algorithms are typically written at the supervisory (computer) level and output to the basic regulatory level. In other words, they are usually applied to cascade systems. Generally the process model is assumed to be first order with dead time although this is not a restriction to the applicability of the dead time compensation techniques. However, this type of process model is simple and is suitable for many processes in the chemical and petrochemical industry. As a rule of thumb, dead time compensation techniques are applied whenever the dead time exceeds twice the time constant ($\theta \geqslant 2t$).

Smith Predictor

The control block diagram for a SISO Smith Predictor control strategy is shown in Fig. 11.14. More specifically, this is an example of using the reflux flow rate to control the impurities in the overhead of a distillation tower via an analyzer feedback loop. Dead time can arise, e.g., from analyzer cycling time, holdup in the reflux drum (analyzer located on the reflux or distillate line), and so forth.

In Fig. 11.14, the Smith Predictor predicts the concentration, $(C_{k+\theta})$, θ time slots ahead and subtracts the present predicted concentration C_k. This value is then subtracted from the actual analyzer reading C resulting in the compensated value C^* that will be used in the analyzer PI controller. The predicted concentration $C_{k+\theta}$ is calculated using the Laplace transform

$$C_{k+\theta} = \frac{\hat{K}_p}{1 + \hat{\tau}_p S} R_k \qquad (20)$$

When we discretize this equation and use absolute values for $C_k + \theta$

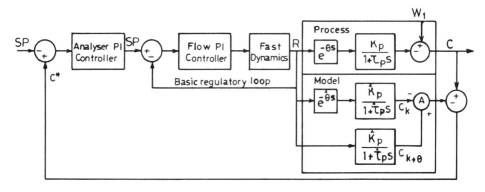

Figure 11.14 Smith predictor control block diagram. W_1 = Load.

Industrial Process Control Systems

and R_k the result is

$$C_{k+\theta} = H(C_{k+\theta-1}) + (1 - H)\hat{K}_p R_k + \text{bias} \tag{21}$$

or using Z-transform

$$C_{k+\theta} = \frac{(1 - H)\hat{K}_p R_k z^{-1}}{1 - Hz^{-1}} + \frac{\text{bias}}{1 - H(z^{-1})} \tag{22}$$

where

$C_{k+\theta}$ = impurity concentration predicted θ time slots ahead
$H = \hat{\tau}_p/(\hat{\tau}_p + \Delta t)$
Δt = sampling time
R_k = present reflux flowrate
z^{-1} = backward shift operator, $C_{k-1} = z^{-1}C_k$
bias = $(C_{ss} - KR_{ss}) \times (1 - H)$
C_{ss} = impurities concentration at steady-state
R_{ss} = reflux flow rate at steady-state

The present predicted concentration is calculated according to

$$C_k = \frac{\hat{K}_p e^{-\hat{\theta}_s}}{1 + \hat{\tau}_p s} R_k \tag{23}$$

or in discretized form

$$C_k = HC_{k-1} + (1 - KH)\hat{K}_p R_{k-\theta} + \text{bias} \tag{24}$$

Using Z-transform we obtain

$$C_k = \frac{(1 - H)\hat{K}_p z^{-\theta/\Delta t - 1}}{1 - Hz^{-1}} \cdot R_k + \frac{\text{bias}}{1 - Hz^{-1}} \tag{25}$$

The dynamically compensated concentration C^* can be calculated from the foregoing equations by subtracting equation (23) from (20) and then subtracting the actual impurity concentration $C_{k,\text{actual}}$

$$C^* = C_{k,\text{actual}} - \left[\frac{\hat{K}_p}{1 + \hat{\tau}_p s} R_k (1 - e^{-\theta s})\right] \tag{26}$$

By discretizing (26) we arrive at

$$C^* = C_{k,\text{actual}} - [H(C_{k+\theta-1} - C_{k-1}) + (1 - H)\hat{K}_p(R_k - R_{k-\theta})] \quad (27)$$

which can be put into Z-transform format as

$$C^* = C_{k,\text{actual}} - \left[\frac{(1 - H)\hat{K}_p z^{-1}}{1 - Hz^{-1}}[1 - z^{-\theta/\Delta t}]R_k\right] \quad (28)$$

Because the concentration of the impurities is dynamically compensated, the analyzer PI controller can be tightly tuned without producing the oscillatory response typical of noncompensated feedback loops with large dead time.

Whenever the Laplace transform of a transfer function is discussed, input and output measurements are assumed to be expressed in terms of deviation variables (absolute value minus steady-state value). If absolute values are used, which is highly desirable from a practical viewpoint, then an additional term (bias) will be introduced in the discretized equations (21), (22), (24), and (25). However, it is not necessary to determine its value since it is cancelled at the summation point A, shown in Fig. 11.14.

There are many ways in which the Smith Predictor block diagram can be equivalently represented. Another simpler representation is shown in Fig. 11.15. However, implementing the control strategy as shown in Fig. 11.14 allows the user to "fine-tune" the process model parameters on-line by comparing C_k against the actual analyzer reading. Once this comparison is favorable, the control engineer is assured of a good process model and thus good feedback control. The analyzer PI controller can then be tuned as if no dead time exists in the analyzer feedback loop.

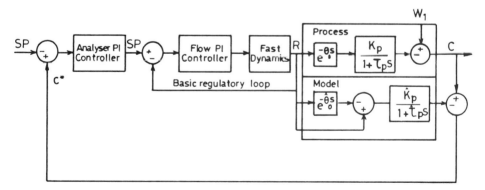

Figure 11.15 Alternative Smith predictor control block diagram.

Although a first order transfer function plus dead time model is assumed, the Smith Predictor technique is by no means limited to this special case. The method holds equally well for higher order models with dead time. A first order plus dead time model is utilized here because most processes in the process industry can be adequately described by such a model. As stated before, the analyzer PI controller can be tuned as if the process had no dead time. It should be noted that the signs (+ and −) shown in the summation junctions of Fig. 11.14 and 11.15 are specific only to the preceding example. Different situations will require different summation junctions, requiring each situation to be analyzed in detail. The following tuning guideline works well: Controller gain K_c is equal to the inverse of the process gain K_p and the integral time is equal to the process time constant $\hat{\tau}_p$. Consider the model

$$\frac{Y}{U} = \frac{K\bar{e}^{\theta s}}{1 + \hat{\tau}_s} = KG \qquad (29)$$

where

$$G = \frac{e^{-\theta s}}{1 + \tau s}$$

θ = process dead time

τ = process time constant

K = process gain

Y = output deviation variable = $y - y_{ss}$

y = absolute value of output

y_{ss} = steady-state value of the output

U = input deviation variable = $u - u_{ss}$

u = absolute value of input

u_{ss} = steady-state value of the input

As noted previously, for absolute values, equation (29) then becomes

$$y = KGu + \text{bias} \qquad (30)$$

where bias = $(y_{ss} - Ku_{ss})(1 - H)$ if equation (30) is discretized via backward difference approximation, H is as defined before. The bias is not calculated directly but is estimated from (see equation (30))

$$\text{bias} = y - KGu \tag{31}$$

The control equation is derived from the static version of equation (30)

$$u_{sp} = \frac{y_{sp} - \text{bias}}{K} \tag{32}$$

which leads to the SISO control structure shown in Fig. 11.16 in which

u = absolute value of the manipulated variable (e.g., flow or temperature)

y = absolute value of the controlled variable (e.g., concentration or conversion)

L_d/L_g = Lead-lag block with gain = 1

With reference to Fig. 11.16 and equation (31)

$$\text{bias} = y_k - y_k^*$$

where y = absolute value of the output and y^* = model output using

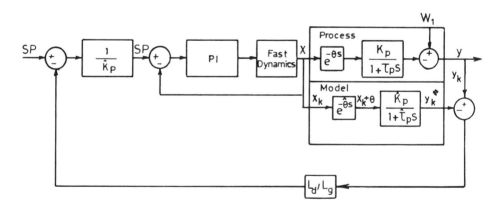

Figure 11.16 Dynamic reconciliator control block diagram.

the absolute value as the input. The value of y^* is calculated as follows

$$y^* = \frac{K_p e^{-\theta s}}{1 + \tau_p s} u$$

which can be written as

$$y^* + \tau_p \frac{dy^*}{dt} = K_p u(t - \theta)$$

or, when using the backward difference approximation

$$y_k^* = y_{k-1}^* H + (1 - H) K_p u_{k-\theta/\Delta t - 1} \tag{33}$$

where $H = \tau_p/(\tau_p + \Delta t)$, Δt = sampling time, and $u_{k-\theta/\Delta t - 1}$ = absolute value of input that has been delayed by θ units of time.

The lead-lag block in the feedback path allows the user to fine-tune the feedback controller. For example, if the process output (y_k) is noisy, a first order lag would be appropriate. On the other hand, if disturbances have slow dynamic, a lead-lag with more lead than lag would be appropriate. Extensive simulations gave the following guidelines for tuning the lead-lag:

$$A = \max(\theta, \tau) \tag{34}$$

$$B = \min(\theta, \tau) \tag{35}$$

$$\tau_{\text{lag}} = 0.5 A \tag{36}$$

$$\tau_{\text{lead}} = 0.5(A + (2 - \theta/\tau) B) \tag{37}$$

It was found that, in many cases, control performance is satisfactory without using the feedback lead-lag.

As in the case with the Smith Predictor, process model parameters can be fine-tuned on-line by comparing y_k^* with actual output y_k^*. Also, the signs (+ and −) shown in the summation junctions of Fig. 11.16 are specific to a process where an increase in the input x will result in an increase in the output y (positive gain).

Processes with higher order model plus dead time are handled in a similar manner. The only difference is in the form of the discretized equation (33) that is used.

Some Guidelines

Although these dead time compensation techniques are relatively simple to understand and implement, they have nevertheless proven themselves very effective in the process industry. The question of when a particular technique should be used over the other is basically a matter of personal preference. Table 11.1 is a comparison list for the Dynamic Reconciliator and the Smith Predictor.

Some practical guidelines for implementation are listed below:

1. Before attempting to close the supervisory feedback loop, ensure that the predicted output value (C_k or $y^*t = k$) is a good fit of the actual and the estimate values over the same time period. The same approach for updating the model parameters should be taken if process conditions have changed significantly.
2. If an analyzer is used in the feedback path, there is the possibility of unknowingly adding a dead time that will be less than the analyzer update time (the time it takes for a new reading of the same component to be determined) to the process. This depends on when the application was started. Ideally, the control action should be taken immediately after the analyzer is updated. This condition can be approached by forcing a controller tag to execute at a much faster rate (about four times as fast) than the update time, but outputting only on the execution following an update.
3. For reasons similar to those presented above, test runs to identify process models should be started immediately after the pertinent analyzer reading has been updated.
4. Tune a Smith Predictor PID controller as if no dead time exists in the process.

Example 11.1 A dead-time compensator is used in control of a process with dead time; the process input is controlled by a secondary loop. The process is a distillation tower, the controlled variable is the concentration, and the control variable is the temperature. Both primary and secondary controllers in the Smith Predictor Strategy are PI controllers. The control structure is shown in Fig. 11.17.

The process in the secondary loop uses the following data:

Industrial Process Control Systems

Table 11.1 A Comparison of the Dynamic Reconciliator and the Smith Predictor

Smith Predictor	Dynamic Reconciliator
PID (proportional-integral-derivative tunings are required)	Aside from the lead-lag block, which is often omitted from the implementation, no tuning constants are required)
Control performance is adequate with a poorly tuned basic (regulatory) controller	Works well with a poorly tuned basic (regulatory) controller
Supervisory setpoint changes are less likely to "bump" the process	Supervisory setpoint changes can "bump" the process, especially if the process gain is small ($\ll 1$). This can be avoided by various means, e.g., rate of change clamp to the setpoint of the basic controller, "filtering" the supervisory setpoint changes via a first order lag, etc.
Requires parameter update when process conditions change significantly	Requires parameter update when process conditions change significantly

$$K = 1.0, \quad \tau = 4 \text{ min}, \quad \theta = 1 \text{ min}$$

Using a control tag processing frequency of 1 min, H becomes equal to $4/(4 + 1) = 0.8$.

The process equation becomes

$$T_k = 0.8 T_{k-1} + (0.2 u_{k-2} \times 400) + w_{k-1} \quad (38)$$

The constant 400 is introduced to size controller output between 0 and 1, which could correspond to 0 and 100% valve opening.

The process in the primary loop is characterized by the following data:

$$K = 0.02, \quad \tau = 10 \text{ min}, \quad \theta = 13 \text{ min}$$

The process equation is

$$C_k = 0.91(C_{k-1} - 4) + [0.09 \times 0.02(T_{k-14} - 200)] + 4 \quad (39)$$

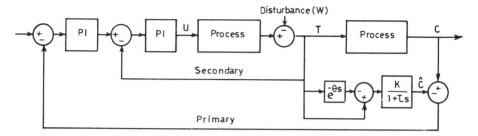

Figure 11.17 Smith predictor control for simulation example.

The secondary loop uses a PI controller. Good controller tuning parameters are a gain $Kc = 3.0$ and the integral time $\tau_i = 10$ min. The error is defined as

$$e_k = SP_k - T_k \tag{40}$$

and the controller equation as

$$u_k = u_{k-1} + \frac{Kc}{400}\left(e_k - e_{k-1} + \frac{1}{\tau_i}e_k\right) \tag{41}$$

The division by 400 brings u_k in the range $0 \ldots 1$ (corresponding from 0 to 100% valve opening).

The primary loop uses a PI controller; the algorithm is similar to equation (41).

The Smith Predictor uses the process model and the filtered process input using the following equation:

$$C_k = 0.91 C_{k-1} + [0.09 \times 0.02(T_{k-1} - T_{k-14})] \tag{42}$$

The control strategy using dynamic reconciliation is shown in Fig. 11.18. The primary and secondary process are the same as for the Smith Predictor. Also, the secondary controller is the same as before. The primary "controller" is $1/K = 1/0.02$.

The equation of the lead-lag element in the feedback path is

$$y_k^* = 0.8 y_{k-1}^* + 0.2 y_k + 0.8(y_k - y_{k-1}) \tag{43}$$

where the coefficient 0.8 determines the amount of lead action. The

Industrial Process Control Systems

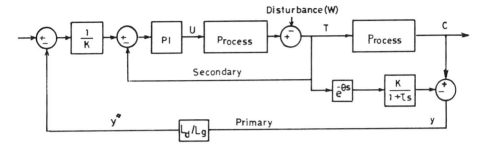

Figure 11.18 Dynamic reconciliation applied in simulation example.

lead action is adapted to the disturbance pattern: for slow changes the lead should be increased; for step changes the lead action is not required at all.

11.5.2 Inferential Control

In inferential control, a control scheme uses an inferential variable as the controller PV. The inferential variable can be either measured or calculated and usually infers another variable which cannot be measured or is difficult to measure. A well-known example is found in distillation columns where temperature changes reflect composition changes. If a reliable analyzer is not available, it may be possible to use the temperature to infer the composition. But even if the analyzer is available, temperature changes can often be detected well before composition changes, and, therefore, it offers an advantage to include the temperature measurement in a control scheme. Another advantage of using the temperature (inferred concentration) measurement is that, if the analyzer is temporarily out of service, control can still continue using the inferred concentration.

Inferential variables do not have to be used only for feedback control — they can be used in feedforward control and multivariable control as well. If, for example, the stripping section temperature of a distillation column increases (as an indication of changing bottom composition), this temperature could be used as a feedforward to an overhead controller that adjusts the reflux in order to avoid a change in overhead composition. An inferential variable does not necessarily have to be calculated from one measurement — more measurements could be involved. Figure 11.19 shows the concept of inferential control.

The controller structure could be a simple PI control algorithm, or, in the case of a major dead time, a dead time compensator as discussed

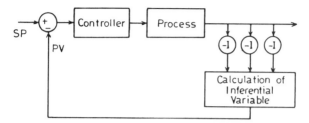

Figure 11.19 Inferential control scheme.

in the previous chapter could be used. Table 11.2 gives some examples of inferential control. It should be mentioned that selection of inferential variables is not always obvious (it may require a good process understanding), and often extensive data collection and testing is required to find a unique relationship between the inferred and actual variable.

The main reason for controlling an inferential variable is that the variable that should be controlled cannot be measured directly or there is not sufficient economic justification for its measurement. Analyzers, for example, are still relatively expensive and require a considerable amount of maintenance, which may force us to infer the required measurement. Also process conditions, e.g., measurement is difficult and it is easier to infer the measurement.

Table 11.2 Examples of Inferential Control

Variable to be Inferred	Inferential Measurement
Distillation tower composition	Tray temperature(s), pressure and flow(s)
Flooding in distillation towers	Differential pressure, tower overhead vapor flow, calculation
Internal liquid flow in distillation tower	Internal reflux rate
Distillation tower pressure	Reboiler and condensor duties
Reaction rate	Reactor temperature difference
Fuel gas heating value	Specific gravity
Mooney	Viscosity
Heat exchanger fouling	Overall heat transfer coefficient
Tank level	Flow rates
Composition in in-line blending	Composition and flow rates of individual components

Industrial Process Control Systems

In many light ends distillation towers, temperature can be correlated to concentration by means of a simple relationship:

$$c = -a_1 T + b_1 \qquad (44)$$

in which c = component concentration, T = temperature and a_1, b_1 are constants.

In high purity towers often a logarithmic relationship is found

$$\log c = a_2 T + b_2 \qquad (45)$$

It may also be possible to correlate product composition to external tower variable such as

$$\log c = \frac{a_3 Q_h - a_4 IR}{F} + b_3 \qquad (46)$$

in which Q_h/F = tower heat input per unit feed and IR/F = internal reflux per unit feed.

If a relationship such as equation (44) is found, the tray temperature could be used as an inner loop of a cascade control structure that eliminates disturbances quickly because the temperature responds to disturbances much faster than the analyzer does. The analyzer could then be used as the outer loop of the cascade to reset the setpoint of the inner loop. This is illustrated in Figure 11.20.

If a relationship such as equation (45) is valid in a certain situation, then control of temperature (as inferred composition) is even more useful and could lead to better control performance. In a basic control system, a logarithmic function may not be available. If the straight analyzer reading would then be used for control, we would have to control a system with a variable process gain. From equation (45) this

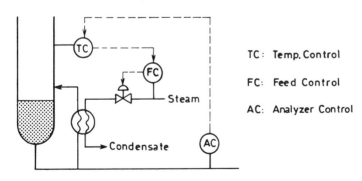

Figure 11.20 Inferential control in distillation tower.

gain can be derived:

$$K_p = \frac{dc}{dT} = -a_2 e^{-a_2 T + b_2} = -a_2 c \qquad (47)$$

When a disturbance comes in, resulting in an increase in temperature, the process gain decreases. It is known that a process with a variable gain is difficult to control by using simple PID control. If the temperature was controlled, however, disturbances would mostly be eliminated by the temperature controller. Straight analyzer to temperature control could function satisfactorily since temperature and concentration changes would be restricted to a limited operating range, and the process gain would vary only slightly.

The selection of an inferential variable for control purposes is not always obvious. In some cases, however, a process model consisting of energy, mass, and partial mass balances may give an indication which variable(s) to select. Consider a simple example. Assume an ideally mixed adiabatic reactor for which the following static partial mass balance may be written:

$$F(c_{in} - c_{out}) = Vr \qquad (48)$$

with V = reactor volume, c = reactant concentration, F = volumetric flow, r = reaction rate, and the static energy balance:

$$F \rho c_p (T_{in} - T_{out}) = -Vr \Delta H \qquad (49)$$

with ρ = density, C_p = specific heat, T = temperature, and ΔH = heat of reaction, negative for exothermal reactions.

Combining equations (48) and (49) and using the definition for conversion

$$C = 1 - \frac{c_{out}}{c_{in}} \qquad (50)$$

it can be shown that

$$C = \frac{-\rho c_p}{C_{in} \Delta H} (T_{in} - T_{out}) \qquad (51)$$

If the reactor inlet concentration was known and it was possible to measure and calculate the other variables, the reactor conversion could be inferred from the above equation. Although in reality models may be more complex, it may be possible to reduce model complexity in a certain operating region.

Another important source for inferential variable selection is past experience. Past experience often has no physical background but has been obtained by monitoring and analyzing the process. The role of operators in this area shoud not be underestimated. An empirical model can be just as good as a model based on mass and energy balances.

In what follows it will be illustrated, by means of a simple example, that the selection of an inferential measurement should be based not only on static considerations but also on dynamics. Equation (51) showed a unique relationship between conversion and temperature difference. Assume that the temperature difference has to be controlled by manipulating the inlet temperature. It can be seen from equation (51) that maintaining a constant $T = (T_{out} - T_{in})$ results in a constant conversion. Although this is statistically true, dynamically there are some problems. When the inlet temperature is changed stepwise, the outlet temperature will lag behind as shown in Figure 11.21.

Although in Figure 11.21 it was assumed that a one degree change in inlet temperature results in a one degree change in outlet temperature, this is usually not the case in chemical reactors due to the reaction that takes place. If the relationship between a change in inlet and a change in outlet temperature is, for example

$$\delta T_{out} = 1.2 \frac{e^{-3s}}{1 + 2s} \delta T_{in} \qquad (52)$$

then the change in temperature difference is

$$\delta(\Delta T) = \left[1.2 \frac{e^{-3s}}{1 + 2s} - 1 \right] \delta T_{in} \qquad (53)$$

The response is shown in Figure 11.22.

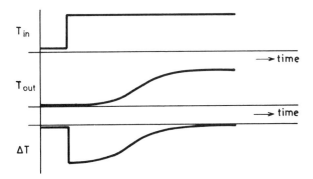

Figure 11.21 Temperature response of chemical reactor.

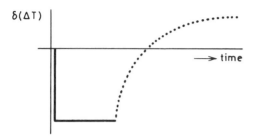

Figure 11.22 Temperature difference response.

Since this is a nonminimum phase response, control may be difficult. Model-based control would have to be used in order to make tight control possible. A control scheme that uses this concept is shown in Figure 11.23.

The *PV* can be written as:

$$PV = \left[\frac{1.2e^{-3s}}{1+2s} \delta T_{in} - \delta T_{in} + 1.2\,\delta T_{in} - \frac{1.2e^{-3s}}{1+2s} T_{in} \right] \frac{1}{1+2s}$$

$$= \frac{0.2\,\delta T_{in}}{1+2s} \tag{54}$$

In other words, the *PV* indicates the future value of the temperature difference via a first order lag, hence it can now be used to predict final reactor conversion. When considering inferential measurements, dynamic considerations should not be ignored.

Example 11.2. In this section some examples of inferential control will be discussed in more detail. The first example in Table 11.2 was distillation tower composition control by controlling tray temperature. In binary columns at constant pressure there is a unique relationship

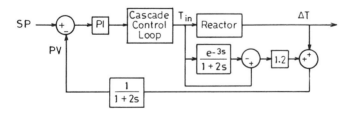

Figure 11.23 Model-based control to eliminate non-minimum phase behavior.

Industrial Process Control Systems

between temperature and composition. Although most industrial columns are not binary, it is often still possible to select a tray temperature that correlates fairly well with a key composition.

In an effort to minimize energy consumption in distillation towers, many control schemes use pressure minimization. In case of upsets, however, pressure can vary considerably. In that case it is advisable to compensate the temperature measurement for changes in pressure:

$$T = T_{\text{actual}} + \frac{dT}{dP}(P_{\text{ref}} - P_{\text{actual}}) \tag{55}$$

in which dT/dP = the inverse of the slope of the vapor pressure curve in a reference point, P_{ref} = a reference pressure, e.g., an average operating pressure, and T = the temperature.

dT/dP can be determined in actual plant operation by selecting two pressures close to the normal operating pressure and determining the corresponding temperatures. It is obvious that the selection of the tray location is of ultimate importance in correlating temperature to concentration.

In order to estimate flooding in distillation towers, the differential pressure is often used as an indicator. Once the pressure difference has reached a certain level, however, flooding has usually started and vapor and/or liquid flow have to be decreased considerably in order to return to normal operation. To make the pressure drop indicator more sensitive, it would be better to measure pressure drop above and below the feed tray. Flooding usually starts either above or below the feed tray. When the upper part starts flooding, the pressure drop in the lower part will decrease even initially, and, thus, we could detect flooding earlier than when there was only one pressure measurement available.

Another indicator for flooding is a tower mass balance. In a flooding situation more mass is entering than leaving the tower. Also, here the problem is that the tower is usually flooded before it is detected. An indicator that was found to be very useful is the tower overhead vapor flow, or the sum of reflux and distillate flow. A maximum flow limit has to be determined empirically. An advanced control strategy could take appropriate action (e.g., reduce the reflux flow) once a predetermined maximum value is exceeded.

Internal liquid flow in distillation towers can be controlled by manipulating internal reflux to the tower. Internal reflux control will minimize changes in fractionation due to condenser disturbances. It is often applied in towers with flooded condensers where the reflux can be considerably subcooled. The degree of subcooling will affect the internal liquid flow in the tower. The expression for the internal reflux flow can be derived from an energy balance around the top of the tower: internal reflux is equal to the external reflux plus the amount of overhead vapor condensed by the subcooled reflux and therefore:

$$ER = \frac{IR}{1 + (c_p/\Delta H)(T_{OH} - T_R)} \qquad (56)$$

in which IR = internal reflux, ER = external reflux from reflux drum to tower, c_p = liquid specific heat, ΔH = heat of vaporization, T_{OH} = overhead vapor temperature, and T_R = reflux liquid temperature.

In a cascade control scheme, the internal reflux setpoint IRSP could be manipulated by an analyzer controller. Two things should be pointed out when applying internal reflux control. First, the overhead vapor temperature should always be greater than the reflux liquid temperature. An application should verify this condition before calculating equation (56). Second, there may be instances where the internal reflux controller may show an inverse response. If the heavy components in the feed increase, T_{OH} will increase before the reflux temperature changes, and the reflux will initially decrease. This, however, is not desirable since we would like the reflux to increase. The analyzer controller will increase the reflux once more heavies are detected in the overhead. To get around the problem either an average but constant T_{OH} could be used or a feedforward strategy using a feed analyzer should be implemented.

Control of distillation tower pressure is usually done by feedback control to the heat input (e.g., steam) or condenser duty (e.g., cooling water flow). Figure 11.24, however, shows a situation where pressure control would benefit from feedforward of reboiler and condenser duties. The pressure controller of tower one would be helped considerably if the heat input to tower two and the condenser duty of tower one are used in a feedforward loop. The pressure response of tower one is

$$\delta P = G_1 \delta F_{S1} + G_2 \delta F_{S2} - G_3 \delta F_{C1} \qquad (57)$$

in which δP = change in pressure, δF = change in flow, and G = transfer function.

In order for δP to be equal to zero, equation (57) has to be rewritten as:

$$\delta F_{S1} = \frac{G_3}{G_1} \delta F_{C1} - \frac{G_2}{G_1} \delta F_{S2} \qquad (58)$$

Equation (58) would be the model for the feedforward compensator.

The last example of inferential measurement that was mentioned was the use of specific gravity in inferring the heat of combustion. For

Industrial Process Control Systems

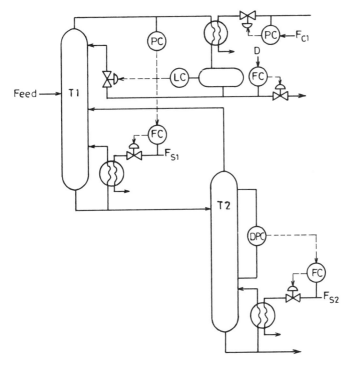

Figure 11.24 Distillation process in which heat and condenser duties can be used in pressure control.

many hydrocarbons the heat of combustion is linearly dependent on the density. Thus, density can be used as an indicator of the heat of combustion. If a change in density is measured, the flow could be adjusted to maintain a constant heating value of the gas.

11.5.3 Feedforward Control

Feedforward control can be usefully applied whenever a process disturbance can be measured. The general concept of feedforward control is shown in Fig. 11.25.

Assume the process output is affected by a disturbance via transfer function G_1 and by a control variable via a transfer function G_2. The disturbance is added to the control signal via a transfer function G_{FF}. Ideally the process output y does not change when the disturbance comes in, therefore the control u does not have to change. A change in y can be written as

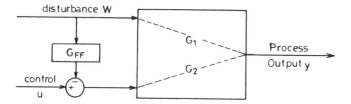

Figure 11.25 Concept of feed-forward control.

$$\delta y = G_1 \delta w - G_{FF} G_2 \delta w \tag{60}$$

which gives with the above conditions

$$G_{FF} = \frac{G_1}{G_2} \tag{61}$$

Not in all cases can a useful expression for G_{FF} be derived. If, for example, the path G_1 is faster than G_2, feedforward action always comes too late. This can be easily seen from the following example:

$$G_1 = \frac{e^{-3s}}{1 + 30s}, \quad G_2 = \frac{e^{-5s}}{1 + 30s} \tag{62}$$

Then

$$G_1/G_2 = e^{2s} \tag{63}$$

which cannot be realized using physical components. Experience has shown that the best thing that can be done is to use a lead-lag for G_{FF} with the lead larger than the lag. This, however would give an initial kick to the control variable.

Example 11.3. Consider the control configuration of a distillation tower shown in Fig. 11.26.

An analyzer controller is manipulating a relatively small distillate flow. If only the level in the reflux drum was manipulating the reflux flow, and if the reflux drum was also large, then the reflux would be adjusted extremely slowly, resulting in poor analyzer response. The response, however, can be improved by adding feedforward action of the distillate flow to the reflux flow. As soon as the distillate flow changes, the reflux is adjusted, hence the time lag of the reflux drum will be virtually eliminated. Feedforward control is used in a true form, since any increase or decrease in the distillate flow is passed on to the reflux flow.

Industrial Process Control Systems

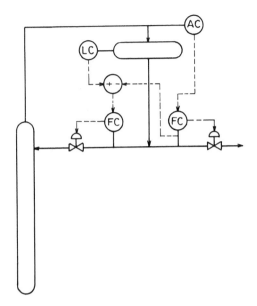

Figure 11.26 Feed-forward control in distillation tower.

In general, feedforward control in a system without a major time delay is usually simple to implement; however, if time delays are present and are compensated for in the feedback loop, then an added degree of complexity is introduced, making feedforward control more difficult to implement. This is illustrated in the following example.

Example 11.4. Consider a situation whereby it is not possible to find a temperature that could be used as an inferential measurement for the overhead impurity in a distillation tower. The results of test runs that have been performed indicate that the transfer function from reflux to analyzer reading can be put in the form

$$\frac{\delta A}{\delta R} = -\frac{0.03 e^{-18s}}{1 + 19s} \tag{64}$$

in which δA = change in analyzer reading and δR = change in reflux.

Because of the relatively large time delay, it is convenient to use a Smith Predictor as a dead time compensator. The control tag will run every two minutes. By discretizing (s^4) it can be written in Z-transform as:

$$\frac{\delta A}{\delta R} = \frac{-0.003 z^{-10}}{1 - 0.90 z^{-1}} \tag{65}$$

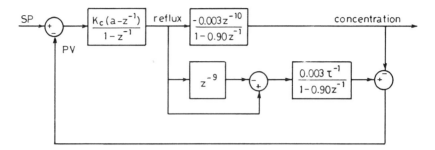

Figure 11.27 Analyzer control using Smith predictor.

The Smith Predictor control structure is shown in Fig. 11.27. The controller gain is shown as K_c; a is a function of the integral time τ_i:

$$A = 1 + \frac{2}{\tau_i} \tag{66}$$

It can be shown that control works well, except for large feed changes. Therefore, it was decided to use feedforward control in order to improve control performance further. From another test run the transfer function from feed to analyzer was found:

$$\frac{\delta A}{\delta F} = \frac{0.015 e^{-20s}}{1 + 13.3s} \tag{67}$$

or in z-notation

$$\frac{\delta A}{\delta F} = \frac{0.0021 z^{-11}}{1 - 0.86 z^{-1}} \tag{68}$$

It is clear that the feedforward action should be added between the output of the controller and the input to the process. Incremental feedforward will be used rather than absolute feedforward since in a steady-state situation feedforward should not contribute to control. The purpose of feedforward is to have such an impact on the process that the analyzer reading does not change, and, therefore, the feedback control loop does not have to act. For that to be the case the model part in Figure 11.27 should not be affected by feedforward control and the addition of feedforward is therefore just before the process but after the junction where the reflux is fed into the Smith Predictor. The feed also has to be dynamically compensated. G_{FF} can easily be calculated from Fig. 11.28.

Industrial Process Control Systems

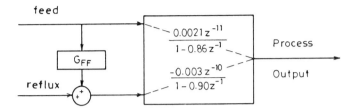

Figure 11.28 Determination of feed-forward action.

$$\frac{0.0021z^{-11}}{1 - 0.86z^{-1}} = G_{FF} \frac{0.003z^{-10}}{1 - 0.90z^{-1}}$$

or

$$G_{FF} = 0.7z^{-1} \frac{1 - 0.90z^{-1}}{1 - 0.86z^{-1}} \qquad (69)$$

It can be seen that equation (69) represents a one-cycle delay and lead-lag function. The lead-lag part can be written as

$$\frac{y}{x} = 0.7 \frac{1 - 0.90z^{-1}}{1 - 0.86z^{-1}} \qquad (70)$$

or

$$\begin{aligned} y_k &= 0.86 y_{k-1} + 0.7 x_k - (0.7 \times 0.90 \times x_{k-1}) \\ &= 0.86 y_{k-1} + (0.5 \times 0.14 x_k) + 0.63(x_k - x_{k-1}) \end{aligned} \qquad (71)$$

Let us look again at equations (64) and (67). Let $K_1 = 0.03$, $\theta_1 = 18$, $\tau_1 = 19$, $K_2 = 0.015$, $\theta_2 = 20$, and $\tau_2 = 13.3$. Then it can be seen that the discrete approximation of the lead-lag element has the form

$$y_k = H_g y_{k-1} + K(1 - H_g) x_k + H_e(x_k - x_{k-1}) \qquad (72)$$

which when compared with (71) yields $H_g = 0.86$, $K(1 - H_g) = 0.5 \times 0.14$, $K = 0.5$ and $H_e = 0.63$. The definition of the respective variables in (72) are

$H_g = e^{-\Delta t / \tau_2}$ is the lag-element gain

$H_e = K\tau_1/(\tau_2 + \Delta t)$ is the lead-element gain

$K = K_2/K_1$ is the transfer function gain

The control scheme with feedforward action can now be constructed and is shown in Fig. 11.29.

The scheme may need some explanation. The major reason to use an incremental feedforward is that in a steady state situation the feedforward control signal does not contribute to the process input:

$$F_k^* = (1 - z^{-1})F_k^*$$
$$= F_k^* - F_{k-1}^*$$

in which F_k^* is the dynamically compensated feed at time k. In a steady-state situation $F_k^* = 0$. However, if the feed F^* is multiplied by $(1 - z^{-1})$, it also has to be divided by it in order to leave F^* unchanged and avoiding additional dynamics. This division together with the addition of the feedforward to the feedback signal is shown in the dotted box in Fig. 11.29. At this point, the following relation is valid

$$\Delta F_k^*/(1 - z^{-1}) + u_{FB.K} = u_{P.K} \tag{73}$$

which can be expanded into the form

$$u_{P,K} = u_{P,K-1} + [u_{FB,K} - u_{FB,K-1}] + [F_K^* - F_{K-1}^*] \tag{74}$$

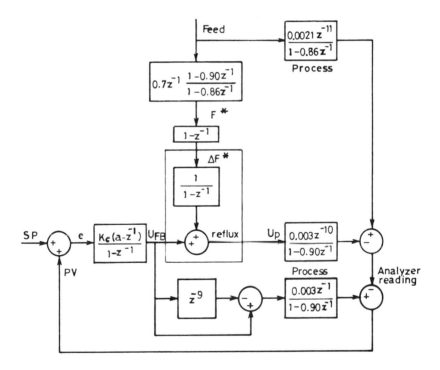

Figure 11.29 Smith predictor with feed forward.

Industrial Process Control Systems

and the corresponding terms are interpreted as: $u_{P,K}$, current value of the reflux; $u_{P,K-1}$, previous value of the reflux; $[u_{FB,K} - u_{FB,K-1}]$, change in flux due to feedback; $[F_K^* - F_{K-1}^*]$, change in reflux due to feedforward.

It is important to emphasize the practical value of (74) in actual implementation. Although $u_{FB,K} = u_{SP,K}$ the input to the Smith Predictor, the implementation will assume the form

$$u_{SP,K} = u_{SP,K-1} + [u_{FB,K} - u_{FB,K-1}] \tag{75}$$

where $u_{SP,K}$, $u_{SP,K-1}$ are the current and previous inputs to Smith Predictor, respectively.

The advantage of the use of equations (74) and (75) in any practical application is that limit checks can be made before feedforward and feedback incremental control signals are used. This is the necessary security checking part of any advanced control application. Also this form of implementation provides a simple way of incorporating feedforward control in feedback loops that are dynamically compensated.

11.5.4 Constraint Control

So far, control in which there is no active constraint has been discussed. It is not uncommon, however, to encounter situations where constraints become violated during process operations. Design of a control scheme should take these situations into account, and the result is often a more complex control scheme than if there were no constraints. If constraints are not taken into account, they may be violated, which may result in serious loss or damage to equipment and the environment. Two types of constraints can be distinguished:

1. Hard constraints
2. Soft constraints

Each of these types of constraints will be discussed in more detail. Once the supervisory control scheme is designed and all constraints are properly dealt with, there may still be a number of degrees of freedom that are not used for control. These remaining process variables can then be used for optimization.

Hard Constraints

Hard constraints are constraints that should not be violated because of operational reasons or cannot be violated because of physical reasons. A typical example of a hard constraint is a valve position which cannot exceed 0 and 100%. This is a physical constraint that, upon violation, can result in loss of control. If the controller were to continue to output once the valve has reached a limit, the controller would saturate or wind up. Most modern electronic controllers have antiwind-up circuits that inhibit integral action once the valve reaches the open or closed position. In a cascade system the primary controller can also be protected against wind-up by monitoring the deviation between secondary setpoint and process variable. An effective way of preventing wind-up is to maintain a constant secondary setpoint as long as the deviation between secondary setpoint (SP) and measured value (PV) is larger than a predefined limit. Only changes are allowed that move the setpoint away from the constraint.

Most supervisory control systems offer standard control algorithms that inhibit integral action in the primary controller as long as the secondary controller is in a wind-up state (see Fig. 11.30). It is of ultimate importance to use this feature if available. If it is not available, it is better to write your own control algorithm and build in the anti-wind-up protection.

In the case where feedforward control action is added to the output of the primary controller, the situation is somewhat more complex. Also in this case, however, the deviation between secondary setpoint and measurement could be used to arrest output from the primary to the secondary controller.

Another example of hard constraints are equipment constraints. Some protection can be obtained by limiting the controller output between high and low limits although changes in process operation

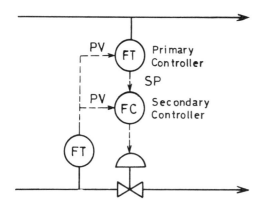

Figure 11.30 Feedback of secondary measurement to prevent windup.

Industrial Process Control Systems

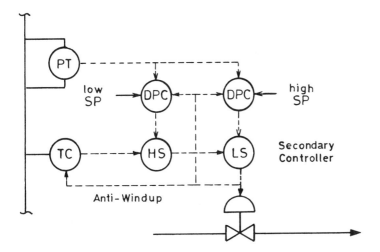

Figure 11.31 High and low selectors in protecting equipment.

may change the relationship between valve position and equipment constraint. It will therefore be better to measure the constrained variables and use them in a control scheme. An example is a distillation tower where the reboiler is on temperature control. The tower vapor flow should remain within a low and high limit in order to avoid weeping or flooding. Both phenomena are rather irreversible, e.g., when a tower starts flooding, operating conditions have to be changed considerably in order to return to a normal operating mode. A control scheme that protects against high and low vapor flow is shown in Fig. 11.31. If the tower starts flooding, the differential pressure controller with the high setpoint will reduce the steam flow and take over control from the temperature controller.

Another example of a hard constraint is the surge point of a compressor. It is not acceptable to exceed the surge limit, and, as such, the compressor has to be protected against the flow becoming too low.

Soft Constraints

Soft constraints are limits that can usually be violated for short periods of time without causing serious problems. An example is the product of a distillation tower which is blended with other components and is then sent to a hold-up tank. Minor variations in tower product quality can be blended out and will be further smoothed out in the

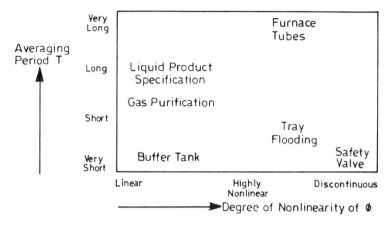

Figure 11.32 Classification of constraints.

hold-up drum. Another example is the tube temperature in furnaces. A higher than usual tube temperature is often allowed provided the period over which it occurs is short. Soft constraints are approachable from both sides of the operating region in contrast with the compressor surge constraint, which is always approached from one side of the constraint. In general, a constraint can be characterized by an expression of the form

$$\int_0^T \phi(x)\,dt \le c_{\max} \qquad (76)$$

in which ϕ is a linear or nonlinear function of one or more process variables x; T is an averaging period or the period of optimization and $c_{\max}$ the limiting value. Some constraints classified according to equation (76) are shown in Fig. 11.32. In the graph, the degree of nonlinearity ϕ is plotted versus the averaging period. In this graph, various phenomena occurring in chemical plants can be plotted. It should be noted that when the value of T is equal to zero, the horizontal scale no longer has significance.

Single Boundary Control

Process operation is often optimal near a constraint. In that case there is the desire to maintain the process near the constraint despite process upsets. Figure 11.33 shows the diagram for control of a single variable near a constraint or single boundary constraint.

Industrial Process Control Systems

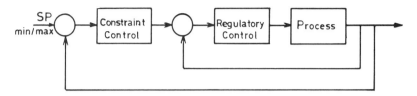

Figure 11.33 Single boundary control.

This diagram looks very much like a cascade control system. The constraint controller can be an integral controller only, which slowly drives the regulatory setpoint. The changes in the regulatory setpoint are so slow that the regulatory control loop has sufficient time to adjust to the changes in its setpoint. The integral time should be set as:

$$\tau_i = K_p \Delta t \tag{77}$$

in which K_p = process gain and Δt = tag execution interval.

The tag execution interval should be least equal to the dead time plus three times the time constraint:

$$\Delta t \geq \theta + 3\tau \tag{78}$$

in which θ = process dead time and τ = process time constant.

It is obvious that this type of constraint control will give a very slow response although it is applied, for example, in tower pressure minimization as shown in Fig. 11.34. A valve position controller (VPC) adjusts the pressure setpoint. The intention is to hold the condenser control valve in its fully open position (on the maximum constraint) in order to minimize the pressure and consequently minimize energy consumption. The pressure controller (PC) has proportional and inte-

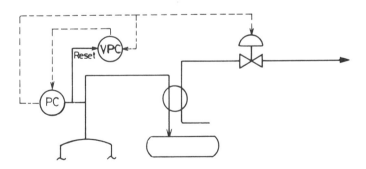

Figure 11.34 Tower pressure minimization.

gral action in order to eliminate process upsets quickly. The integral valve position controller moves the pressure setpoint to its optimal long-term value. Wind-up of the VPC is prevented by using the pressure measurement as reset feedback.

Constraint control often creates a very nonlinear objective function. To make this clear, consider that control of the acetylene concentration is 25 ppm and the actual concentration drops to 15 ppm, there is obviously the need to bring it back to setpoint. However, if the concentration increases to 35 ppm, the situation is more serious and the concentration has to be brought back to target much faster in order to avoid off-spec material.

This creates a situation where even control action may become non-symmetric: upon a negative offset, the controller acts faster than on a positive offset. In the literature there are at least ten different nonlinear controllers proposed to improve control near a constraint. From practical experience it was found that all of these controllers offer only limited success. One controller that was found to be useful for the situation just outlined was one in which the gain was tripled for positive offsets. Another interesting approach is given by Westerlund et al. (1985). In this case the authors do not try to develop a nonlinear controller but update the controller setpoint based on past process values. A process is operated against a soft constraint, i.e., only during a certain period is the controlled variable allowed to be outside the desired quality limit y_M. The setpoint of y at time k is denoted $y_{SP,k}$. A natural choice for the setpoint is given by

$$y_{SP,k} = y_M \pm y_{M,k} \tag{79}$$

where $y_{M,k}$ is a "confidence" interval. The $\pm$ sign corresponds to an upper or lower quality limit. The confidence interval can be given by

$$y_{M,k} = F_{s_{H,k}} \tag{80}$$

where F is a constant factor, or more precisely, it is obtained from the density function of the stochastic process $H(k)$, and s_H is the estimated standard deviation of $H(k)$:

$$H(k) = y_k - y_{SP,k-1} \tag{81}$$

If slow variations in the estimation of H are allowed, the variance of H can be approximately computed by giving old values of $H - (k)$, an exponentially decreasing weight:

$$S_{2H,k} = \frac{\sum_{i=0}^{k} \lambda^{k-i} [y_k - y_{SP,k} - 1]^2}{\sum_{i=0}^{k} \lambda^{k-i}} \tag{82}$$

where λ is a weighting factor, $0 < \lambda < 1$. Equation (82) can be approximated by

$$S_{2H,k+1} = S_{2H,k} + (1 - \lambda)[y_{k+1} - y_{SP,k}]2 \qquad (83)$$

It is important to mention that (79), (80) and (82) constitute a simple method for on-line updating of the setpoint. The value of F in (80) can be obtained from the statistical distribution of y.

Multiple Boundary Control

The previous section presented two techniques of driving a process variable against a constraint. However, sometimes more than one process constraint can limit process operation and more advanced control is required. This section will deal with that situation. The process variables of interest should be monitored or calculated. A detailed plant model can be used where necessary to estimate constraints which cannot be measured directly. The control system must determine which of the process variables is the active constraint and limits plant operation. Plant production can then be maximized by adjusting the predetermined manipulated variable in order to drive the plant against the active constraint. A detailed process analysis is usually required to determine which process variables have to be considered. Three approaches to the constraint control problem will be discussed:

1. Steady-state approach
2. Dynamic constraint control using a single PI controller
3. Dynamic constraint control using multiple PI controllers

Steady-State Approach. If it is not required to control the process tightly near the constraints, a steady-state approach will suffice. In this case the manipulated variable is increased by a fixed amount Δu, after which the control system allows the process sufficient time to settle out. If no constraints are violated, the manipulated variable is again increased by the same amount. This procedure is repeated until the process variable reaches a constraint. If a constraint is violated the control system decreases the manipulated variable by a fixed amount, which should be about twice as much as the change made when moving toward the constraint.

It is obvious that this type of constraint control is not very tight because it ignores dynamics completely; however, a large portion of the credits of constraint control can be captured this way with relatively minor effort.

Dynamic Constraint Control Using Single PI Controller. This situation is very similar to the one discussed in the previous section. It can be used here when the dynamics of the different constraints are similar. Figure 11.35 shows the approach. Each constraint has a dedicated integral-only controller. As in Figure 11.33, the PV of the controller is the constraint variable. Since the constraints can be imposed upon different process variables they are normalized according to

$$\Delta m_i = \frac{y_i - y_{m,i}}{K_{P,i}} \tag{84}$$

in which Δm_i = change in manipulated variable, required to bring the ith constraint variable to its limit in one time step, y_i = current value of the ith constraint variable, $y_{m,i}$ = constraint value of the ith constraint variable, $K_{P,i}$ = process gain = $\delta y_i / \delta u$, and u = manipulated variable.

All values of Δm are compared and the minimum is selected because any larger value of Δm would violate one or more of the constraints. The integral action of each integral-only controller is calculated using equation (77). The selected output Δm_i is the PV of the PI controller. The setpoint of this controller is zero. The purpose of this is to output the selected Δm_i slowly to the process, so that when a constraint is violated, the process will not be upset. The single PI constraint controller is simple to implement and proven to be quite adequate in situations where the dynamics of the different constraints are similar.

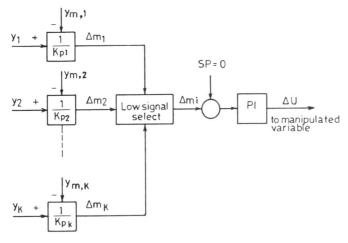

Figure 11.35 Dynamic constraint control using a single PI controller.

Industrial Process Control Systems

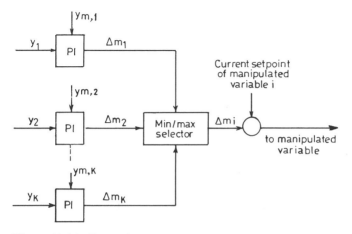

Figure 11.36 Dynamic constraint control using multiple PI controllers.

Dynamic Constraint Control Using Multiple PI Controllers. When the constraints have different dynamics, this approach is required for good constraint control. The scheme is shown in Figure 11.36. In this case each constraint has a separate PI controller calculating Δm_i. A selector picks the largest Δm_i if the process can still be driven toward the constraint. The PI controllers have proportional action based on PV rather than error. The output Δm_i is added to the current setpoint of the manipulated variable. A drawback of this approach is that it is rather sensitive to noise. If one or more of the signals y_i to y_k is noisy, the best thing to do is to eliminate the source of the noise or filter it out.

11.5.5 Adaptive Control

All real processes are affected by noise. Hence the model, or the equations describing the process, are not always very accurate. Variations in the physical properties of the materials being processed mean that the actual process varies over a period of time. Many processes operate in an uncontrolled environment. For example, the room temperature or humidity might change. Under these circumstances the optimal values of input control variables, determined from a suitable process model or earlier experimental data, will no longer be optimal. It is therefore necessary to update the physical process model to take account of the changes in process properties and operating conditions. This updating can take the form of changes in the values of process coefficients. Adaptation is particularly necessary if the actual process is operating outside the linear region which might have been assumed in the development of the process model. Computer implementation of adaptive process control systems is schematically illustrated in Figure 11.37.

In such adaptive control systems, the microcomputer has the role of data acquisition, direct digital control and the comparison of process model and actual outputs. On-line updating and optimization of the process models is complex requiring very considerable data processing power which is beyond the capabilities of most microcomputers. It would be necessary to use a hierarchical computer system configuration, which makes use of micro- and minicomputers. The minicomputer would be used for updating the physical process model and its subsequent optimization. Updating of the process model involves measurement of the input control variables, process output variables and other physical properties. These data are subsequently used to calculate the process performance which is compared with the desired performance to determine the action which must be taken. A number of techniques are available to develop and adapt the models used for adaptive process control. If the process model has been updated, it must then be optimized to determine the new values of input control settings. In some cases, it might be necessary to change the control strategy completely

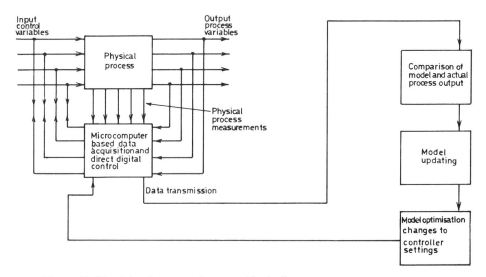

Figure 11.37 Adaptive control system block diagram.

Industrial Process Control Systems

instead of making minor adjustments to the controller settings. Under these circumstances, the computer must be programmed to make decisions about the occasions on which the controller settings may be adjusted or a complete change in control strategy is required. The deviations between the actual process performance and the model performance must be statistically significant before a decision is made to adopt either of these procedures. In either case, the optimization of the process model must be followed by the hardware implementation of the necessary changes so that the process continues to operate in an optimal manner.

11.5.6 Multivariable Control

The PID controller uses the process input as the control variable and the process output as the controlled variable. In a process with multiple inputs and multiple outputs, however, one will usually not find a situation where one control variable can be paired with one controlled variable without interactions from other control variables.

A simple example is the blending of one component with an inert flow. If the concentration in the final blend changes, the flow of the component can be changed in order to control the composition. However, this will change the total flow of blended material, which requires the inert flow to be changed as well. The control problem cannot simply be treated as two independent control loops. A number of techniques are available to solve the problem (Fig. 11.38(a)). By applying the technique shown, independent control loops are obtained. In the literature, this type of control is often called "noninteractive control". The major advantage of this approach is that the control scheme is visible to the operator since it is broken down into independent control loops. Failure of one control loop does not affect the rest of the control strategy.

Another approach is to use "interactive control" (Fig 11.38(b)). All process outputs now have an impact on all process inputs. This type of control is more complicated from an operator point of view. The method requires a process model, giving the relationship between all process inputs and outputs (state variables):

$$u_{1,k} = a_{11}x_{1,k-1} + a_{12}x_{2,k-1} + \cdots + b_{11}u_{1,k-1} + b_{12}u_{2,k-1} + \cdots$$

$$u_{2,k} = a_{21}x_{1,k-1} + a_{22}x_{2,k-1} + \cdots + b_{21}u_{1,k-1} + b_{22}u_{2,k-1} + \cdots \quad (85)$$

in which $x_1, \ldots, x_n$ are state variables and $u_1, \ldots, u_n$ are control variables.

Equation (85) can be written in matrix notation as

$$x_k = Ax_{k-1} + Bu_{k-1} \quad (86)$$

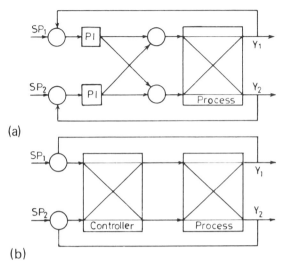

Figure 11.38 (a) Noninteractive control. (b) Interactive control.

The process model is usually combined with a measurement model:

$$y_k = C x_k \tag{87}$$

in which y is the measurement vector. The feedback control law can be written as

$$u_{k+1} = G y_k \tag{88}$$

in which A, B, C, and G are matrices.

Optimal control theory can be applied using a quadratic cost function. The result is a calculation of the gain coefficients in the gain matrix G. The process description can be easily modified to include process and measurement noise. The solution for the gain matrix G depends strongly on these estimates as well as on the weighing coefficients in the quadratic cost function.

11.5.7 Self-Tuning Control

Our interest here is to present a brief summary of self-tuning regulators since they can also be used in model-based control in which the dead-time has to be eliminated.

Self-tuning regulators (STR) can be particularly useful in processes with changing parameters, e.g., fouling of equipment or decay of catalyst. The process may have a dead time, however the value of it should

Industrial Process Control Systems

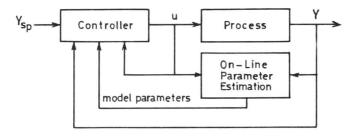

Figure 11.39 Control using self-tuning regulator.

not change. A self-tuning regulator has the structure as shown in Figure 11.39. In this technique a process model is assumed, and a recursive least squares estimation algorithm is used to update the model parameters on-line based on current and past process input and output values. After the model parameters are updated, the new control output is calculated.

Example 11.4. Assume that a particular process model can be described by

$$y_k = \alpha_0 y_{k-2} + \beta_0 \nabla u_{k-2} + \beta_1 \nabla u_{k-3} \qquad (89)$$

in which y_k = predicted deviation between process output and setpoint, y_{k-2} = measured deviation between process output and setpoint, two time steps ago, $\nabla u_{k-2} = u_{k-2} - u_{k-3}$, u_{k-2} = process input two time steps ago, u_{k-3} = process input three time steps ago, and α, β = model parameters.

This model is now used to find the prediction y_{k+2}:

$$y_{k+2} = \alpha_0 y_k + \beta_0 \nabla u_k + \beta_1 \nabla u_{k-1} \qquad (90)$$

The target is to control or setpoint, hence the best estimate is $y_{k+2} = 0$; therefore

$$\alpha_0 y_k + \beta_0 \nabla u_k + \beta_1 \nabla u_{k-1} = 0 \qquad (91)$$

resulting in

$$u_k = 1 - \frac{\beta_1}{\beta_0} u_{k-1} + \frac{\beta_1}{\beta_0} u_{k-2} - \frac{\alpha_0}{\beta_0} y_k \qquad (92)$$

which shows that the current control input depends on the current process output and the previous two control inputs. The parameters α_0, β_0, and β_1 can be updated via a least squares estimation technique.

11.6 SENSORS AND ACTUATORS

11.6.1 Introduction

The implementation of any microprocessor/computer based control system in the industrial environment requires that the data relating to the state of a process be input to the system. Suitable sensors or transducers are required to measure the process variables. Once the process measurements have been input to the microcomputer system, the measurement is inferred by using a suitable calibration curve and the calibrated value compared with the desired value of the variable. In a closed-loop system, any differences between the setpoint and actual values are converted into the control action which is implemented by using suitable output transducers or actuators such as relays, valves, solenoids, power transistors, thyristors, and stepping motors. In open- as well as closed-loop systems, the process data are frequently displayed so that the process operator is aware of the actual values. The display might be on a cathode ray tube (visual display unit) or it might be in the form of a light emitting display (LED) or liquid crystal display (LCD). A detailed discussion of instrumentation technology is beyond the scope of this book. However, the principles of the commonly used transducers are briefly discussed.

11.6.2 Sensors

The sensors or transducers form the first and perhaps the most important link in the development of a data-acquisition and process control system. It is not always possible to measure every single variable of interest. Many variables have to be inferred by measuring variables, for which appropriate sensors exist, and combining the data, by using suitable analytical or empirical relationships. Sometimes it might be necessary to examine the basic physical phenomena in order to design a suitable specialized sensor. Leaving aside such specialized sensors, the main requirements are for measuring pressure, temperature, flow, level, displacement, velocity, acceleration, force, torque, vibration, and weight. In each case the mechanical, thermal or optical measurements must be converted into an electrical signal which can be input to the microcomputer after suitable conversion into digital format. The following techniques are commonly used for measuring the above variables.

Pressure Transducers

A vast majority of process control systems involve the measurement of liquid or gas pressure. Different types of transducers are required

Industrial Process Control Systems

to measure low or vacuum pressures and high pressures. Some pressure sensors are only suitable for the laboratory environment while others can withstand the harsher industrial environments. The pressure measured might be absolute, gauge or differential. Pressure monitoring involves the measurement of the movement of the sensing element due to the application of the pressure. The absolute movement of the sensing element will depend upon the pressure on both sides of the sensing element. Thus if the pressure on one side is atmospheric and the pressure being applied to the other side is not atmospheric, the measurement will indicate differential pressure. Absolute pressure transducers measure the pressure relative to perfect vacuum.

It is possible to use mechanical, capacitive or piezo-electric sensors for measuring the pressure.

One of the most commonly used components in a pressure transducer is the diaphragm whose displacement, caused by the application of pressure, is measured and subsequently converted to the proportional pressure or electrical signal. Depending upon the pressure applied, the diaphragm will extend or contract, and this extension or contraction is detected by a force balancing mechanism. Bellows, consisting of a single metallic piece, are also used to monitor differential pressure which result in a physical movement. This physical movement can be calibrated and subsequently used to record the pressure.

A number of techniques can be used to convert the mechanical displacements discussed above into proportional electronic signals. Commonly used techniques are based on the application of strain gauges, variable capacitance, linear variable differential transformers and piezo-electric effects. The electronic sensors which make use of above techniques provide improved accuracy and can be used in conjunction with microprocessor/computer based systems. They also cover a wide range of pressures and are more flexible. The decline in the cost of electronic circuitry has resulted in the availability of low cost electronic pressure sensing devices. Recent technological advances have resulted in the fabrication on a single chip of all the circuitry required for Wheatstone bridge arrangements, signal conditioning, and amplification. The low cost of such electronic based pressure transducers will make them increasingly popular.

In electronic sensors based on strain gauges, the application of pressure results in a strain on a gauge made from metals or semiconductor materials such as silicon and germanium. This strain causes an equivalent change in the resistance which can be measured using a suitable Wheatstone bridge arrangement. The output from a Wheatstone bridge circuit will consist of a few millivolts and the actual value is proportional to the pressure being measured. There is a linear relationship between the changes in strain and resistance.

Capacitive pressure transducers are based on the principle that any movement of elastic dielectric elements results in a change in capacitance which can be detected by a bridge circuit and converted into an appropriate pressure level. Pressures up to 10 k lb/inch^2 (6.7×10^7 N m^{-2}) can be measured and the accuracy is approximately 0.5%.

A linear variable differential transformer (LVDT) consists of a variable transformer whose movable core is connected to the element to which pressure is being applied. A single primary excitation coil and two symmetric secondary coils are used to detect the changes in the voltage caused by the movement of the core. Electronic circuits are used to provide a dc voltage level proportional to the movement of the core which is in turn related to the applied pressure. The normal range of LVDT pressure transducers is 30 lb/inch^2 (2×10^5 N m^{-2}) to 10k lb/inch^2 (6.7×10^7 N m^{-2}).

Piezo-electric materials generate an output voltage when a force is applied to them. This effect has been used to design piezo-electric pressure transducers which are particularly suitable for measuring differential pressures. It is possible to measure pressures up to 10k lb/inch^2 (6.7×10^7 N m^{-2}) with an accuracy of approximately 1%.

A number of factors, apart from the accuracy and range, have to be taken into account during the selection of a suitable pressure transducer, e.g., it must be able to cope with the temperatures to which the transducers will be subjected. The frequency response of the transducer will determine its ability to detect dynamic changes. Its sensitivity to vibrations and shock will be important in many industrial environments. High output levels obviate the need to use amplifiers before the data is input to the computer. The long term reliability and freedom from drift are other important factors and the cost of the transducer is obviously a consideration. However, the transducer selected must be able to measure the pressure accurately.

Temperature Sensors

It is often necessary in industrial environments to monitor accurately temperatures in the range $-200\,°C$ to $3000\,°C$. In some cases it is possible to use sensors which come into contact with the body of material whose temperature is being measured, while noncontact sensors must be used in other situations. It is not possible to use a single sensor to cover the whole temperature range of $-200\,°C$ to $3000\,°C$ which might be of interest. Simple transducers, such as bimetallic strips

Industrial Process Control Systems

and glass thermometers, cannot be used in conjunction with microcomputer systems which require a proportional electrical signal output. It is therefore necessary to use other sensors such as thermocouples, thermistors and pyrometers for monitoring the temperature.

Thermocouples in their simplest form consist of two wires of dissimilar metals joined together by soldering or welding to form a junction. Heating of the junction results in the generation of a very small amount of electromotive force in the millivolt range whose actual value depends upon the temperature of the body or material. Different metals, some of them very expensive, are used to form the thermocouples to cover the required range. Typical materials used for temperature values in the range of 0–1000 °C include copper, constantan, chromel, and iron. Higher temperatures can be measured using chromel/alumel and platinum-rhodium/platinum thermocouples. These latter materials are very expensive and the length of the wires made from these materials has to be kept as short as possible. Characteristics of most thermocouples include relatively low cost, large measurement range, small sensor size, fast response to temperature changes, good repeatability and sensitivity of measurement, nonlinear voltage temperature relationship requiring accurate calibration, good accuracy, the possibility of picking up noise and the need to avoid temperature gradients.

Thermistors are low cost semiconductor temperature sensors made from oxides of metals, such as copper, chromium, cobalt, magnesium, manganese, nickel, and titanium, which can be used to monitor temperatures in the approximate range $-100\,°C$ to $300\,°C$. Within this range the thermistors are highly sensitive to any temperature changes. The temperature variations are accompanied by a change in the resistance of the thermistor and this change can be converted into an output voltage by using a suitable condition of thermistor characteristics; this is necessary since the relationship between resistance/voltage and temperature is highly nonlinear. Advantages of thermistors include small sensors which can be fabricated in many shapes such as beads and discs, very good accuracy of the order of 0.1% of full scale, good repeatability, no need for cold junction compensation which is necessary in the case of thermocouples, high degree of sensitivity to temperature changes and relatively low cost. Disadvantages include the narrow temperature range and nonlinear relationship between temperature and resistance.

The operation of thermocouples and thermistors relies upon direct contact between the transducer and the body whose temperature is to be monitored. In some situations, for example when monitoring the temperature of glass and steel furnaces, such direct contact is not possible for a variety of reasons and it is necessary to use other noncontact measurement techniques. They can also be used to measure the temperature of nonstationary bodies. Radiation pyrometers, optical pyrometers and thermopiles are examples of such noncontact temperature measuring devices.

Some pyrometers, referred to as total radiation pyrometers, are sensitive to the radiation of all wavelengths while partial radiation pyrometers only react to particular wavelengths. The heat radiated by the body, whose temperature is to be monitored, is focused on a transducer by using a suitable optical system. The transducer used to detect the temperature might be thermal, for example a thermopile, which consists of a number of thermocouples connected in series. A thermopile absorbs the heat of radiation from the hot body, and the difference between the temperature of the hot and cold junctions represents the output. Such devices are examples of total radiation pyrometers. Their response time is rather long. They can be used for continuous monitoring of temperatures up to 3500 °C. Alternatively it is possible to use photon-detectors which are only sensitive to certain wavelengths and produce an output due to the effect of the heat energy on electrical charges stored in materials such as silicon, or cadmium sulphide. At certain wavelengths the molecular structure of the particles changes very rapidly. It is thus possible to detect dynamic changes in the temperature of objects moving at high speed. Such pyrometers, referred to as photo-electric pyrometers or continuous optical pyrometers, can be used to monitor temperatures in the range of 450–2000 °C. These measurements are continuous and can be input to a microcomputer system after the continuous analog value has been converted into digital format using suitable analog to digital converters.

Many pyrometers are portable units which require an operator to monitor the temperature of the hot body and are unsuitable for use with any automatic control system. Optical pyrometers are of this particular variety. Although more accurate than total radiation pyrometers, the optical pyrometer can only be used in the visible range at temperatures between approximately 700 °C and 3000 °C. Disappearing filament optical pyrometers are frequently used. These are based on the comparison between the images of the selected radiation wavelength from the hot body whose temperature is being monitored and the radiation of the same wavelength from a suitable source of reference. The heat or current input to the reference source and hence its temperature can be calibrated. The current input to the reference source is varied, until the two images cannot be distinguished. It is possible to monitor the temperature by calibrating the current/temperature relationship.

The response characteristics of all pyrometers are highly nonlinear

Industrial Process Control Systems

and they are only suitable for monitoring temperatures in specified ranges. Full scale accuracy lies in the range of 0.5–2%. They provide good repeatability but are rather expensive systems.

A number of factors have to be taken into account during the selection of a pyrometer for monitoring the temperature of a given hot body. These include the desired temperature range, accuracy required, material from which the body is made, the distance between the body and the pyrometer, whether the body is moving or stationary, and the need for continuous or discrete monitoring of the hot body temperature.

Other transducers used for monitoring temperature include quartz crystals and resistance bulb thermometers.

Level Transducers

The measurement of liquid levels is necessary in most processing industries. Similarly the levels of oils, other lubricants and coolants have to be monitored and maintained within required limits to ensure the efficient running of machinery. It will be increasingly necessary to measure the liquid levels automatically as the complexity of machinery used in manufacturing processes increases and microcomputer based condition monitoring and control systems become commonplace. A number of techniques can be used to achieve this automatic monitoring of liquid level. The simplest type of liquid level monitoring and control involves the use of ON/OFF switches, while other transducers enable the continuous monitoring of the actual liquid level.

The ON/OFF type of level monitoring can be carried out using techniques such as beam breaking, float type level switches, capacitance level detectors and conductivity switches. Beam breakers provide the simplest method of checking whether the liquid level has reached a certain point. All that is required is a beam of light which is transmitted and picked up by a receiver on the opposite side. As soon as the level of material reaches the required point, the beam is broken and the further supply of material can be stopped. As soon as the level falls below the required point, the beam is no longer broken and the supply of material can be resumed. The simplicity of this technique makes it ideal for the ON/OFF type of level control.

Float type level switches follow the liquid level; when a reference point is reached, an automatic alarm is generated. Magnetic, pneumatic or microswitches can be used to determine whether the reference point has been reached. This information can be easily input to the microcomputer by using an interrupt.

Conductive level probes are suitable for detecting high or low liquid levels. Two probes are required to monitor both the low and high levels. They are based on the simple principle that for high level monitoring, a circuit can normally be left open. The circuit is closed when the liquid comes into contact with the probe. A relay can then be operated and information input to the microcomputer system.

Continuous level monitoring is carried out using techniques such as the measurement of capacitance, differential pressure, ultrasonic or sonic level detection and radioactive sensing. Capacitance measurement techniques involve the formation of a variable capacitor by using a probe as one electrode and the metal wall of the container as the second electrode. The dielectric constant depends upon the level of liquid in the container. A bridge arrangement and a high frequency crystal can be used to monitor the capacitance which is directly proportional to the liquid level.

The level can also be monitored using the measurement of differential pressure between the bottom of the tank containing the liquid and the top of the tank, which gives an accurate indication of the head of liquid. Pressure transmitters, discussed under the heading of pressure measurement, can be used to provide the required data to a microcomputer system.

Ultrasonic level monitoring devices are suitable for continuous as well as reference point measurement of the material level. A beam, with a frequency between the top end of the sonic range and the ultrasonic range, is generated and transmitted in bursts from a transmitter located above the material. The beam is reflected by the material surface and collected by a receiving transducer. The time taken to receive the reflected echo is a measurement of the material level in the container. Mounting of the transmitter above the material has the advantage that the transducer does not come into contact with the material. The accuracy of measurement can be improved by mounting the transducer underneath the liquid. However this method will involve contact between the transducer and the material. It is also possible to use the same transducer for monitoring the level in a number of containers. Available circuitry can be used to transmit the level measurement data to the microcomputer system. These transducers can also be used to detect the interface between solids and liquids.

Other level measurement techniques include the use of silicon transmitters, displacement level sensors, bubblers, diaphragms, glass and magnetic gauges, microwaves, thermal sensors, and resistance tape sensors.

Factors to be taken into account during the selection of a suitable level monitoring sensor include the material whose level is to be moni-

Industrial Process Control Systems

tored, the nature (hazardous or otherwise) of the environment, the pressure and temperature ranges in which the sensor will have to operate, the accuracy required, the need for ON/OFF or continuous measurement, the level measurement range, the cost of the sensor, the reliability of the sensor and its maintenance requirements.

Flow Transducers

It is frequently necessary to measure the absolute and differential flow rates of fluids in processing industries. Such flow measurements are also required to monitor and control the flow rate of fuels such as gases and liquids, and coolants in situations where a given amount of heat transfer has to be achieved. A large variety of flow meters is available and can be used to monitor the flow rate. The techniques used to monitor the flow rate vary. Some of the available techniques are only suitable for the measurement of the flow rates of liquids or gases while others can be used for both. Special purpose flow meters are required to monitor the flow rates of viscous fluids and slurries. The flow can be measured in terms of the mass flow rate, volume flow rate or the velocity, although the last measurement is not very common.

Orifice plates mounted in the flow path provide the simplest method of monitoring the flow rate. The orifice plate restricts the area through which the fluid can flow and this area restriction increases the downstream flow velocity and results in a differential pressure across the orifice plate. The differential pressure can be measured and converted into the appropriate flow rate. Orifice plates are simple and relatively inexpensive, but result in a high pressure loss and are not suitable for highly accurate monitoring of liquid flow rates.

While orifice plates provide a sudden constriction to flow, venturi tubes are carefully shaped to provide a smooth and gradual decrease in diameter. Once the minimum diameter has been reached, it is smoothly increased to its original size. This smooth change results in a much smaller head loss than in the case of an orifice plate. The differential pressure across the venturi tube is monitored and converted into the flow rate.

Other differential pressure flow meters include flow nozzles which lie somewhere between the orifice plate and the venturi tube. Although cheaper than venturi tubes, their head loss is only slightly lower than for orifice plates. Such nozzles can be installed in existing mains and the lack of sharp edges makes them suitable for use with abrasive liquids and high velocity mains.

Pitot tubes are used to monitor the difference in pressure between that due to the impact made by the flowing fluid on a cylindrical tube and the static pressure. The differential pressure, proportional to the square of the flow velocity, is measured and converted to the flow rate. Pitot-venturi tubes, comprising a Pitot tube and a venturi meter, generate higher differential pressure than a Pitot tube. These Pitot-venturi tubes are used to monitor the flow velocity and hence the flow rate.

A rotameter is an example of a variable area flow meter. It consists of a uniformly tapered tube, made out of a glass or metal tube, in which a float is allowed to move freely. The float movement is proportional to the rate of fluid flow and the float stabilizes when the forces exerted by it and the fluid flow are equal. Glass tubes are only suitable for a visual indication of the flow rate while metal tubes can be used to provide signals which can be transmitted electrically. The flow rate monitoring accuracy of rotameters varies between 0.5% and 2% of full scale.

Displacement Measurement

Accurate measurement of angular and linear displacements is necessary for the automatic control of machinery used in many manufacturing processes. The simplest method of displacement measurement involves the monitoring of the effect of displacement on a potentiometer. Resistance changes produced by displacement can be converted into a proportional voltage, by applying constant current, which can then be input to the microcomputer system. Linear variable differential transformers are used to generate a dc voltage output which bears a linear relationship, within specified limits, to the displacement, which might be angular or linear.

Variations in capacitance can also be used to monitor the displacement. The relative linear movement of two parallel metal plates, separated by air which acts as a dielectric, results in a change in capacitance which can be measured. Alternatively it is possible to vary the exposed area which also changes the capacitance. This change in capacitance is converted into a voltage which is input to the microcomputer system.

Displacement can also be measured by monitoring the effect of moving a permeable core on the inductance of a coil. A suitable circuit can be used to convert the inductance into an appropriate electrical signal suitable for input to a microcomputer system.

Angular displacement can be monitored by counting the number of voltage pulses generated by the teeth of a wheel as they move past an inductive pick-off. This measurement technique is an example of the

Industrial Process Control Systems

digital transducers which are increasingly being developed for use with computer based monitoring and control systems.

Optical shaft encoders are other examples of digital transducers used for displacement monitoring. Tachometer encoders are suitable for sensing the movement, while incremental encoders can be used to monitor the direction of motion as well as actual displacement. Absolute position can be measured using absolute encoders. Such optical encoders typically make use of a suitable light source, a grated glass disc, photo sensors, and an optics system. The binary output from the transducer is amplified and can be input to the microcomputer directly or it can be converted into an analog voltage, proportional to the displacement, by using a suitable electrical circuit.

ON/OFF switching position sensors, based on solid state techniques, are available and can be used to change the output signal state at a particular point. Such sensors can operate on the principle of a broken beam, produced perhaps by a LED system, or proximity sensing.

Factors to be taken into account when selecting a suitable displacement sensor include resolution, accuracy, cost and the ease with which the sensor can be mounted and used in conjunction with a microcomputer system.

Velocity, Vibration, and Acceleration Measurement

Many automatic control systems used in modern machinery require the measurement of linear and angular velocities and accelerations. Velocity is the first derivative of displacement and acceleration the second derivative of displacement. Electronic circuits can be used to differentiate the displacement after it has been measured by a suitable displacement transducer. Similarly if the acceleration can be measured by a suitable transducer, then its output can be integrated to provide the velocity.

Linear velocity over short distances can be measured by detecting the time taken by a moving object to move past suitable reference points. Vibration amplitudes are relatively short and the vibration velocity can also be monitored using a permanent magnet, which moves in the field of a coil. The relative motion induces a voltage proportional to the velocity of vibration.

Angular velocity of a rotating shaft is used, wherever possible, to deduce the linear velocity which is the product of the angular velocity and radius of the shaft. Tachometers, electro-mechanical, electrical, optical or electronic, can be used to monitor the angular velocity. Electro-mechanical tachogenerators produce an output voltage proportional to the shaft speed. The accuracy of measurement of the angular velocity is improved by using digital techniques. The digital output can be input to the microcomputer without requiring the application of any analog to digital converters.

Digital tachometers do not require any direct contact with the rotating shaft. It is only necessary to attach a suitably coded disc to the shaft. The number of voltage pulses detected by the transducer, during a given period of time, is proportional to the shaft velocity. Accuracy is improved by increasing the number of voltage pulses generated during one complete rotation of the shaft.

The angular velocity can also be monitored using multi-pole magnets fitted to the rotating shaft. As magnets move past a detector, a pulse is generated by the closing and opening of a reed switch and the pulse frequency is proportional to the angular velocity.

Voltage pulses induced in a given time period by a toothed wheel, mounted on the rotor shaft, as it moves past a coil wound around a permanent magnet sensor provide another method for measuring angular velocity.

Accelerometers are used for measuring acceleration. They are particularly required for monitoring the acceleration of vibrating bodies. Popular accelerometers are based on the application of strain gauge and piezo-electric techniques. The strain gauge accelerometers are based on the measurement of changes in strain gauge resistance caused by the application of force due to acceleration. The change in resistance is proportional to the acceleration and can be monitored using a suitable Wheatstone bridge circuit.

Piezo-electric crystals are also used for measuring acceleration which results in the generation of a strain in the crystal and hence an output voltage. The voltage produced is proportional to the acceleration. These transducers can be use to monitor acceleration in the frequency range 5–20 kHz. Piezo-electric based accelerometers are less sensitive than strain gauge based accelerometers. The principles discussed above are used for vibration measurement in terms of displacement, velocity and acceleration.

11.6.3 Actuators

Any microcomputer based control system requires that the data input via sensors be compared with the desired output and action be taken if the actual output value is different from the desired output value. Thus, it might be necessary to close or open a valve, input additional

Industrial Process Control Systems

power to a heater or increase the speed of a dc motor. Actuators are required to implement the necessary control action. A very wide range of actuators is used in the manufacturing industries. Commonly used actuators are now considered briefly.

1. Solenoids are ON/OFF devices which convert output signals into linear motion and are often used for controlling pneumatic or hydraulic cylinders. The low electrical signal output from the microcomputer is first amplified and then used to operate the solenoid. Being digital devices, these solenoids can be easily interfaced to the microcomputer without any need for digital to analog (D/A) conversion. The solenoid motion is relatively limited and it is often necessary to use this motion for operating valves or switches and thereby achieve the required longer via pneumatic/hydraulic cylinders or motors.

2. Relays, mechanical or solid state, are also ON/OFF devices which are frequently used in many control systems. Reed relays, being mechanical devices, take a relatively long time to respond to the electrical signal. Small voltages, typically 5 V, and currents, typically 10–15 mA, are required to operate these relays. Many miniaturized mechanical relays are available in the market. Solid state relays, capable of performing the functions of mechanical relays, are also available. They do not suffer from the contact arcing problem encountered with mechanical relays. The small voltage and current output by the microcomputer can be used for direct operation of the relay. The digital (ON/OFF) nature of relays makes them suitable for direct interfacing with the microcomputer.

3. Hydraulic cylinders are used to provide high magnitude linear and rotary motion. Large forces can be generated using high pressure oil, a pump unit being used to generate the required high pressures. An increase in the amount of motion needed results in the use of large hydraulic cylinders which tend to operate slowly. The high stiffness of the system means that good position control system can be provided. Solenoids are used for interfacing hydraulic cylinders with microcomputers.

4. Pneumatic cylinders make use of the normal air supply and are used for providing motion. It is necessary to remove any oil particles and filter the air before it is used for operating the pneumatic cylinder. The stroke length is restricted due to the compressibility of the air; long stroke cylinders have relatively low stiffness. The force that can be exerted is also small. The main advantage of using pneumatic cylinders lies in the fact that the actuation speed for small cylinders can be very fast. High precision control of dimensions is difficult. The mechan-

ical movement can be restricted by placing mechanical stops on the traverse. Solenoids are generally used to operate pneumatic cylinders in a fashion similar to that used for operating hydraulic cylinders.

5. Stepping motors are very useful actuation devices for achieving point-to-point control. Such control is required, for example, in the case of numerical control systems. Stepping motors are digital devices and there is no need for digital to analog conversion when they are used in conjunction with computer systems. Control signals are first generated by the microcomputer to determine the direction of rotation of the motor. Each pulse results in an angular motion which varies from one stepping motor to another. The number of pulses generated, during a given time period, defines the actual speed of rotation. Thus the angular velocity can be altered by increasing or decreasing the timing of pulses. Accurate rotational speed is achieved by counting the number of pulses until a pre-set limit is reached. Pulse to motor rotation is an exact ratio making it possible to achieve precise rotational speed. The stepping motors can be operated in an open-loop mode. Alternatively they can be used in conjunction with digital rotation encoders for very precise positioning systems.

Stepping motors are expensive, when compared to other types of motors, and require the use of relatively complex circuits. The power output is also relatively low. For this reason they are restricted to small table sizes on numerical control machines. Careful selection of drives is necessary for achieving high torques at high speeds. Maximum torque is developed at slow speeds to enable the start-up of heavy inertia items.

6. Field control dc motors are used in many manufacturing industry applications such as large robotic systems in preference to the pneumatic cylinders and stepping motors used in small robots. Such dc motors are also used as actuator elements in many process control systems. These motors make it possible to achieve a high power output at high speeds as well as a high power to weight ratio. The interfacing of dc motors with microcomputers requires the application of digital to analog conversion techniques.

Accurate positioning, as required for numerical control systems and robotic systems, necessitates the use of rotary and linear position encoders. Digital tachogenerators can be use for velocity feedback. Position and velocity control can be achieved by using a cascade control system. Field control dc motors with an output of up to 2 horse power (1.5 kW) are used for position and velocity control systems.

7. Armature control dc motors are used in general purpose systems producing a wide range of power outputs.

Industrial Process Control Systems 477

8. Ac motors are relatively inexpensive and are used in comparaively low power systems. High power induction motors, when used in conjunction with thyristor frequency conversion circuits, are suitable for variable speed applications.

9. Hydraulic motors also have a wide range of power output and provide good performance particularly at low speed and high torque. The interfacing of a hydraulic motor to the microcomputer requires the use of a digital to analog converter whose output is connected to a proportional valve. The valve provides a hydraulic signal which is proportional to the voltage signal output by the microcomputer system.

11.7 DISCUSSIONS

Improved process control algorithms for industrial processes are necessary for the production of quality goods, scrap minimization, and reduction in the amount of energy used. Convential analog controllers only enable the control of individual process loops. It is usually difficult to take account of the substantial interactions between different process variables.

Microcomputers are making a very substantial impact on the control of processes. They are suitable for the flexible direct digital control of process loops. The control algorithms used for individual loops can be timed to suit their requirements which is not possible with conventional hardwired analog controllers. At a higher level of sophistication, it is possible to use microcomputers for supervisory process control based on the application of ratio control, program control, sequence control, cascade control, and feedforward control algorithms. Control and adaptive control of complex processes offers very considerable advantages over simple direct digital control or supervisory control. However, the implementation of these controls systems for complex processes requires far more processing power than is normally available on microcomputer systems. It is therefore necessary to use more powerful host minicomputers for the required data analysis while microcomputers are used at a lower level for direct digital control purposes.

Considerable time and effort is required for the implementation of user designed microcomputer based process control systems. It is possible to reduce the time and effort for many applications by using EPROM based program modules for individual functions. An alternative is to use microprocessor based programmable controllers.

The implementation of microprocessor/computer based systems requires the use of a wide range of sensors, actuators and auxiliary devices. Sensors are external to the microprocessor/computer and are used to obtain data about the state of a process. It is necessary to convert the physical measurement into an electrical signal which can be input to the microcomputer. Amplifiers are required to amplify the low level sensor signal to the level required for input to the microcomputer. A vast majority of the sensors provide continuous analog outputs and require analog to digital converters for transforming the continuous analog value into the binary digital format.

Actuators form the final element of many control systems. Some actuators, such as solenoids, are digital devices while a majority of the actuators requires analog outputs for operation. It is often necessary to use auxiliary devices such as amplifiers, thyristors and digital to analog converters to connect the microcomputer to the actuators.

NOTES AND REFERENCES

In this chapter, elements and components of computer controlled systems in industrial processes have been reviewed and their functional operations have been discussed. A more detailed account of the related subjects can be traced in Amrehn (1977), Roffel and Rijnsdorp (1981), Mehta (1983), Ghosh (1980), Hanus (1980), Roffel et al. (1986), Bartman (1980), Badavas (1984), Bristol (1986), Westerlund et al. (1985), Kochhar and Parnaby (1977), Bristol (1979), and Kendler and Lutte (1979). Further discussion on sensors and actuators can be found in Jones (1979), Syndenham (1980) and Woolvet (1977), among many others.

12
Distributed Digital Control Systems

12.1 INTRODUCTION

Most of the computer control systems implemented during the 1960s and 1970s were centralized whereby a single large process computer was used to acquire data from one or a number of processes and control a large number of process loops. This was the subject of the previous chapter. Computers and their associated peripherals were relatively expensive and it was necessary to utilize the available computing power efficiently. The cost of interfacing the computer to one or more processes was relatively high due to the need to use long and costly screened cables which are necessary in a typical production environment. In such centralized computer process control systems it is necessary to bring all the cables back to the central control room. This approach clearly has its own advantages and disadvantages. On the positive side it is possible to achieve overall coordination and optimization of the process. Large amounts of process data can be stored on associated peripherals such as hard disks and subsequently analyzed to improve the process performance. It is necessary to use only one set of expensive peripherals such as disks, printers, and plotters. Process optimization can only be carried out by analyzing past data, developing/validating process models and continuously updating them. Data from different sections of a process are required in order to infer the values of variables which cannot be measured directly. Large amounts of computing power might be required to solve relatively complex equations for inferring the values of variables or for determining the changes to input process variables which can be manipulated to counteract the effect of process disturbances. Large computers, with multiprogramming facilities, or even minicomputers operating in foreground-background mode can carry out these necessary calculations while concurrently performing the input/output functions necessary in computer process control systems. Another significant advantage has been the availability of some packages, with extensive facilities, for computer control of different processes. Many mainframe computer system manufacturers provided such computer software packages in order to promote the sale of computer hardware suitable for process control. Considerable progress was made in the field of computer process control systems using such centralized computer systems.

While offering the benefits outlined above, the approach of using centralized computers for process control purposes suffers from a number of disadvantages. First and foremost is reliability. Any computer used for controlling processes must be highly reliable since any system breakdown will result in a disruption of production and complete shutdown of the process, which could prove to be extremely expensive. A highly reliable system can only be built by building a considerable redundancy element into it. One possible approach is to have a standby computer system, ordinarily used for development work, and use it for process control purposes in the event of the breakdown of the main unit.

The cost of a centralized computer process control system is obviously an important consideration. Mainframe and large minicomputers used for centralized process control are still relatively expensive. Such computers are produced in small quantities and hence have to carry large overheads. The provision of a standby computer to be used in the event of the breakdown of the main unit can be a costly exercise. Cabling and interfacing costs are also high. A number of terminals might be required to keep shop floor process operators and supervisors informed about the state of the process. Computing power is not available at the point of actual use. Furthermore, it is difficult to ensure the integrity of the process database.

There has been, in recent years, very rapid development of computer control in application to industrial systems, such as steel, petrochemical, and electric power. Factors contributing to this development include: (1) increasing pressures on industry to improve productivity, conserve

Distributed Digital Control Systems

energy, satisfy environmental constraints; (2) advances in computer technology and software which have made modern, highly sophisticated multicomputer, microprocessor-based distributed control systems economically feasible, reliable, and available; and (3) extensive progress in the development of concepts and theory for large scale systems analysis and design.

In general, an industrial system involves the confluence of materials, energy, labor, and equipment in a complex sequence of operations and processes carried out under prescribed conditions. Performance depends on a variety of factors including; technological design of the system and its components; (2) the nature of resources available and environmental constraints; and (3) the choice of operating conditions, allocation of resources, scheduling of operating sequences, etc. The purpose of the control system is to make the best decisions with respect to (3) within the constraints imposed by (1) and (2). Thus, the term control is employed to include not only the traditional process control functions, but also real-time applications of information processing and decision making, such as production planning, scheduling, optimization, operations control, etc. The common characteristic underlying control, in the sense, is the basing of actions, responses, decisions, etc. on information describing the current state of the system (and its environment) as interpreted through appropriate models.

Many factors contribute to the need for more effective control of modern industrial systems: (1) the need for more efficient utilization of resources (e.g., energy, water, labor, materials) because of increasing cost, limited availability, or both; (2) demands for higher productivity to meet more intense competition from abroad; and (3) more stringent requirements concerning product quality, environmental impact, and human safety because of government regulations and greater consumer awareness.

Industrial systems are inherently complex and large scale. They are characteristically multivariable, nonlinear, time varying, and subject to disturbances and constraints of various kinds. Effective control involves consideration of dynamic couplings among the system components, multiobjective decision making under uncertainty, man-machine interactions, etc.

Different control strategies were examined in Chapter 11 to tackle the control problem of a typical industrial process. In this chapter, we follow a different route by providing alternative control-oriented architectures that exploit the recent developments in distributed information processing capabilities. First, the features and advantages of distributed processing techniques are demonstrated and then the implementations of these techniques are addressed in terms of system applications.

12.2 DISTRIBUTED PROCESSING

The availability of powerful micro- and minicomputers has made it possible to implement computer control systems which do not require the facilities of a large powerful mainframe computer system with expensive peripherals. Such distributed processing systems are suitable for use in the process control as well as the general business applications environment. It is also possible to combine process control systems with manufacturing management and business computer systems. In such systems the computing power is available at the point of use. Instead of using a large monolithic mainframe computer for performing a large number of unrelated tasks, a number of mini- or microcomputers are used to carry out the required tasks. In the case of a process control system, the limit is reached if a separate microcomputer is used for controlling each individual process loop, as was the case with conventional analog controllers, or for controlling a small number of loops close to the process. However, this does not imply that each microcomputer is used in a stand-alone fashion. Efficient use of micro/minicomputers and the data associated with them demands that such microcomputers should be linked together or to a host system at a higher level. In a large and complex system, there might be a number of levels at which different types of computer systems are used depending upon the type of functions which must be performed. The type of configuration used will depend upon the characteristics of the production environment in which one is operating as well as the requirements of the overall control system. It is highly likely that future computer systems, in process control as well as general business applications, will make an ever increasing use of distributed processing concepts, simply because small computer systems are produced in very large quantities and are relatively inexpensive. The computing power provided by linking together such inexpensive computers is far greater than that available from a mainframe computer which costs as much. It is important to distinguish between different types of distributed processing systems, some of which may not be suitable for the typical process control environment.

12.2.1 Elements

Any distributed processing system will consist of the following essential elements.

Distributed Digital Control Systems

1. The hardware in the form of micro/mini/mainframe computers on which the necessary data manipulations will be carried out.
2. The data, whether acquired from a process or stored on mass storage devices such as disks.
3. The interfaces necessary for connecting the various computers.
4. The operating system or networking system required to control and share the processing of data on different system elements.

The way in which the data is actually processed is also sometimes considered as an element of a distributed processing system. In some distributed processing systems, an individual application program can only be executed on a particular computer while other systems allow an application program to be executed on any of the computers which form part of the network.

Typical distributed processing systems can have a horizontal structure or a vertical (hierarchical) structure.

12.2.2 Hierarchical Processing System

An alternative often preferred to a horizontal distributed processing system is the vertical distributed processing system. The overall functions to be performed by the system are divided and a number of systems are connected in a hierarchical fashion with a strict definition of the tasks to be performed by each one of them. The complexity and/or sophistication of the computer configurations used at different levels of the system vary according to the tasks to be performed by them. Simple microcomputers without expensive peripherals, or programmable controllers would be used at the lowest level for data acquisition and closing one or more process loops. Alternatively the microcomputers at the lowest level might be used simply for condition monitoring. The application program logic would vary depending upon the tasks to be performed. It is not necessary to provide direct operator interface at this level. Such an operator interface would be provided via a higher level computer system. Similarly the task of process coordination and optimization would be carried out by the computer system at a higher level. In addition to data acquisition and control of individual process loops, the microcomputers or controllers at the lowest level would also transmit the required data to the coordinating computer at a higher level. For example, in the case of a condition monitoring microcomputer system, very little data normally have to be transferred to a coordinating computer at a higher level. Thus if the condition being monitored is within the desired range, then it is not necessary to transmit data to the higher level computer. As the condition becomes unsatisfactory, the microcomputer at the lowest level would use the interrupt facility to transfer data to the coordinating or supervisory computer system at the higher level. The supervisory computer would then output an alarm and print or display it on the operator console.

Mass storage facilities are generally provided on the supervisory computer system and can be used to keep a log of alarms, significant process conditions and trends as well as a summary of process data. Thus it is not necessary to provide expensive peripherals with each

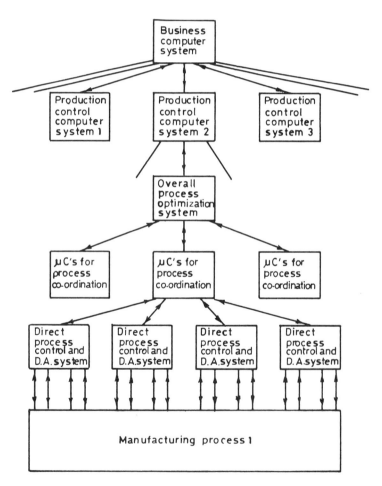

Figure 12.1 Hierarchical processing system (μC = Microcomputer; D.A. = Data acquisition.

Distributed Digital Control Systems

microcomputer system. In the case of microcomputer systems used for direct control of process loops or for data acquisition, the frequency of data transfer to the higher level supervisory computer would be very much higher than that for condition monitoring purposes. Complex processes, for example glass bottle production lines, might require the use of a number of microcomputers for coordinating different sections of the process and for providing direct process interface. For such large processes it would be necessary to have another minicomputer at a still higher level for overall process optimization purposes and to provide process settings for different sections. In a large factory, with a number of manufacturing processes or systems, a number of such coordinating minicomputers would be used and these will provide data to a production planning and control computer system. In turn a number of such production control computer systems might provide data to an overall business control system which provides links between the production data and other functions such as overall business planning, accountancy, payroll, marketing, research and development. Such a computer configuration is illustrated in Fig 12.1.

The development of such a complex distributed processing system, covering a large number of processes is evidently a long-term activity requiring considerable thought and planning. However, this does not imply that the complete system would be implemented simultaneously. In fact any attempt to do this will only result in disastrous consequences.

12.2.3 Horizontal Processing Systems

A horizontal distributed processing system is one in which a number of computer systems are linked together in the form of a network, and share the overall load of the tasks which must be carried out on the whole computer system. The possible network configurations are described in the following pages. Further distinctions can be made between horizontal distributed processing systems depending upon the way in which the load is shared between different computer systems. Clearly the best form of horizontal distributed processing would be one in which any of the data processing tasks can be carried out on any of the network computers. This would provide the best form of load sharing so that the application program can be loaded into the memory of any of the computers, executed until interrupted for any reason followed by subsequent processing to completion. The results of processing will then be returned to the user. It is necessary that the load sharing be transparent to the user since in a complex computer system, he cannot be expected to know the best system on which his application program should be executed. The access to such load sharing systems can be further improved by making the data as well as application program storage transparent to the user. This implies that the data and application programs might be stored on different systems. The user simply specifies, from his terminal, the application program and the data files to which the program will require access, without worrying about the computer on which they are stored. The network operating system will have to take over the responsibility, in the extreme case, for transferring the application program and data, stored on devices attached to different computers, to the CPU of the computer on which the data will be manipulated and subsequently transfer the results of data manipulation to yet another system to which the user terminal is attached. Further complications arise when different computers, although part of the network, are controlled by different operating systems. A very complex and sophisticated network operating system would be required to schedule the work and transfer information between such different computer systems. Such distributed processing systems are still in their infancy and are not really suitable for the data acquisition and process control environment.

Process control systems and other systems used in a typical manufacturing environment involve data acquisition from sensors and subsequent closed-loop control via actuators attached to individual micro/minicomputers. The tasks to be carried out by these mini/microcomputers can be clearly defined and an optimum configuration can be specified. The type of true horizontal distributed processing discussed above does not offer any benefits in this instance. In fact it will only introduce extra complications and cost. However, this form of distributed data processing will provide considerable benefits in a general manufacturing/commercial environment in which it is difficult to predict accurately the data processing requirements which might be there at different times.

A different and simplified form of horizontal distributed processing allows the load to be shared between different processors without involving transfer of application programs and large amounts of data. This type of distributed processing is suitable for the data acquisition and process control environment. The overall data processing and control functions are analyzed, partitioned and allocated to different processors. Thus each processor has to carry out only a specified set of functions. The application programs required to perform these functions will be resident in the memory of that processor or on a mass storage device attached to it. Similarly the data to be used by these application programs will either be stored on the attached secondary

Distributed Digital Control Systems

storage device or directly acquired from the process by means of suitable sensors. In a process control environment, that processor will be responsible for data acquisition, calibration, and for carrying out any necessary preprocessing. This pre-processed data will then be used to determine the required control action which must be effected via the actuators interfaced to the micro/minicomputer. Each of these processors will also be responsible for performing any necessary optimization calculations for that section of the process and implementing the required changes to process settings. Any form of planning calculations will also have to be carried out on the local processor. A separate operator interface and logging device would be required for each processor. Process coordination can be achieved by transferring, via communications lines, the necessary small amounts of data to other processors which need them. It is difficult to carry out overall process optimization since none of the computers has complete information about the state of the overall process.

12.2.4 Benefits

Distributed processing systems offer some very considerable benefits over the conventional concepts of centralized systems. In what follows, the major benefits are categorized:

1. Provided that the overall system structure has been carefully designed, the system implementation can be carried out in a modular fashion thereby providing incremental system growth. Thus one can start at the lowest level by using simple microcomputers or microcontrollers for data acquisition and control of individual process loops. At the same time, computers can be used, in stand-alone mode, at higher levels for applications such as production and management control. The interlinking of different computers, micro, mini and mainframe, to provide an integrated computer-aided management and process control system, can then be carried out in a number of stages. Successful integration will only be achieved by careful planning and ensuring good interfaces and protocols between different computer systems. It is easier to develop and implement systems in a modular fashion than to attempt to develop a centralized and complex computer control system.

2. The cost of developing such distributed systems is lower than the cost of a centralized system which performs the same functions. Small computers are relatively inexpensive and, when linked together, they provide more computing power than a large mainframe system of the same cost. It is only necessary to purchase computers required for

immediate implementation whereas a large centralized system may not be adequately utilized until all the functions have been developed and implemented. Technological breakthroughs mean that at a later time when another module is to be implemented, it is possible to make use of a better processor, for example a 16-bit microprocessor or a 32-bit processor with considerably enhanced inbuilt facilities instead of the conventional 8-bit microprocessors/computers currently in use. The cost of microcomputers is still falling and taking into account high rates of inflation, the real cost of implementing a module at a later stage might be much lower.

3. The cost of upgrading the system will also be lower. Replacement of a large centralized mainframe computer system is a very expensive exercise. Small, relatively inexpensive systems can be easily upgraded, if necessary, without incurring heavy expense. If the centralized computer system has reached the limit of its processing capability, it may not be possible to accommodate extra functions without incurring heavy expenditure or deterioration of system response time which can prove critical in process control systems.

4. Each processor carries out a clearly defined set of functions and system growth would be achieved by incorporating additional processors. The tasks to be carried out by individual processors are relatively simple. Complex operating systems, with sophisticated multi-programming, time-sharing and real-time interrupt facilities, may not be necessary at the lower levels. Application programs are easy to develop, since they must only perform a limited number of functions. The application software required for handling a large number of real-time functions, as would be necessary for a centralized system, is very complex. Large application programs are difficult to develop and test, and some of the errors are never detected until a particular combination of values is encountered. In the case of a distributed processing system, each application program carries out only a limited range of functions and can be developed easily. If the tasks to be carried out by a number of processors are identical, or if there are only small differences, as in the case of data acquisition and three-term process control, a library of basic programs can be maintained and individual programs customized, as necessary, to suit the particular requirements. The testing of individual application programs, developed at different times, is easy. New programs and systems can be independently tested whereas in a centralized system the new application programs must be tested in conjunction with existing programs.

5. A distributed processing system is inherently more reliable than a centralized processing system. Each small system has a small number

Distributed Digital Control Systems

of components and hence is more reliable. It is easy to provide back-up for small inexpensive micro- and minicomputers used for real-time functions. The probability of the simultaneous failure of a large number of individual processors is extremely low. It is thus possible to keep in stock a small number of microcomputers which can be used to replace any faulty microcomputer systems. Another method of providing a back-up for real-time data acquisition and process control microcomputers is to use the coordinating or supervisory micro/minicomputer in the event of any breakdown.

6. Hierarchical computer control systems allow for the sharing of expensive peripherals such as floppy or hard disk drives and printers. Thus it is not necessary to use floppy disks with each microcomputer used for data acquisition and process control. In some cases, it may not even be necessary to use floppy disks with each of the coordinating microcomputers for each process system, if the small amount of data required for analysis are transferred to and stored on peripherals attached to the overall process coordinating minicomputer system. Similarly, visual display terminals or hard copy printers are not required at the lowest level. However, a terminal would be required for use with each of the coordinating microcomputers so that the process operators/supervisors are aware of the state of the process.

7. There is a substantial reduction in the cost of interfacing the process to the computer system. It is no longer necessary to use large lengths of screened cable to feed the sensor data to the centralized computer system. The first level microcomputers or micro-controllers interfaced to the process will be in the immediate vicinity of the process, requiring short cable lengths. A number of sensors might be interfaced to one first level microcomputer. However, only one cable is required to interface the first level microcomputer to a coordinating microcomputer. Similarly, only one cable is required to link the production control computer system, interfaced in turn to a number of lower level microcomputers, to the manufacturing management/business computer system. It is highly likely that fiber optic systems will be used in future for data transmission to eliminate noise and reduce errors.

8. Distributed processing systems are more flexible than a centralized processing system. The current configuration can be easily altered, within limits, to suit changed requirements. Microcomputers with firmware programmed to carry out specified discrete functions are available in the market. Such microcomputers can be readily configured to satisfy the overall system requirements. Large centralized computer systems involve a considerable amount of application programming which often results in a delay in system implementation.

9. Distributed processing systems allow the duplicate storage, if necessary, of critical data. For example, critical data relating to process conditions might be stored on the floppy disks attached to a supervisory micro/minicomputer as well as on hard disks attached to a production control computer system.

10. The management of computer operations can be simplified. Line managers can have control over the computers used in their areas while at the same time providing a link to the data used and created in other areas. This is perhaps the main advantage of distributed data processing compared to decentralized processing in which stand-alone and independent processors do not communicate with each other.

11. In a process control environment, the use of microcomputers/ controllers, placed in the immediate vicinity of the process, results in a minimization of data losses and errors. In a centralized system, process data can be lost over long distances or errors can be introduced. Considerable system overhead is required to recover the lost data.

12.3 MULTI-PROCESSOR SYSTEMS

An alternative form of distributed processing involves the use of multi-processor systems. For some real-time data acquisition and process control system applications, the speed of many individual microcomputer systems is inadequate. In such circumstances it becomes necessary to use more powerful and high speed 16- and 32-bit computer systems. Many applications also require that some data manipulations should be carried out simultaneously or in parallel. Digital computers, whether micro or mainframe, are serial devices which can only execute one instruction at a time. These difficulties can be overcome by using tightly coupled multi-processor (a number of processors) systems which share a common bus. Each microprocessor, forming part of the multi-processor system, can simultaneously carry out the required tasks, such as input/ output and data manipulations. The results of individual data manipulations can be combined, if necessary to perform the overall function. Although such multi-processor systems might appear, in the first instance, to be similar to distributed processing systems, there are a number of significant differences in the way in which these two configurations are designed and implemented. In a distributed processing system, as discussed earlier, the total system functions are partitioned and the tasks to be carried out by each of the processors are clearly defined and fixed. The distributed processing systems do not share a common bus. Each processor carries out a specified task and might

Distributed Digital Control Systems

transfer summary data or exception conditions data to a higher level system for process coordination and optimization. Most distributed processing systems are loosely coupled and the communications between the first level microcomputers, directly interfaced to various processes or sections of a process are via the supervisory computers at a higher level. Each of these microcomputers has its own memory. Multi-processing systems consist of a number of tightly coupled processors in which data are exchanged between processors via a common bus system, over relatively short distances. The memory is shared between different processors. It is then possible to carry out parallel processing of data and increase the amount of computer power available to carry out the required tasks. This approach enables the user to have access to powerful computer systems, based on the use of a number of inexpensive microcomputers, at a fraction of the cost required to provide the same amount of computing power on a centralized mainframe computer system. This approach might soon be the most practical method of providing increased computing power. Technological developments have resulted in a shortening of the distances between the elements of a large-scale integrated circuit and further miniaturization, while continuing to develop, will soon reach the theoretically possible limit.

A typical multi-processor system would comprise two or more microprocessors, of comparable power, with common memory, access to peripherals, and controlled by a single operating system. Clearly the operating system required to control the system would have to be very complex for resolving access and synchronization problems.

12.3.1 Interconnection of Processors

As discussed in Chapter 10, the mode of operation of a computer, whether micro or mainframe, consists of repetitive serial execution of instructions, whereby an instruction is fetched from memory and decoded, followed by the fetching of data from memory and the subsequent execution of the instruction. Thus a number of basic computer cycles are required to execute a single instruction. One can imagine that the processing power of any computer can be substantially improved if these basic functions are carried out in a parallel rather than sequential manner. Such techniques are commonly used in assembly line or flowline manufacturing environments, whereby a number of manufacturing operations are carried out in parallel on the products which are flowing past different work stations. The use of such techniques results in substantial productivity improvements. By the same token if it becomes possible to execute an instruction, while another is being decoded and the data are being fetched, the total time required for execution would be substantially less. Just as conveyer belts are used on assembly lines to move the next item after a given operation has been completed on the previous work station, a "pipeline" type of system can be used to execute the next instruction, since it has already been decoded and the data has been fetched. In such systems all the sequential functions required to execute an instruction are divided and carried out simultaneously on a number of separate units. The process is represented schematically in Fig. 12.2.

It is thus possible to carry out four tasks in parallel. By considering four sequential instructions, for example numbers 2, 3, 4, 5, the parallel processing at a given instant would involve the following:

1. Execution of instruction 2
2. Retrieval of data for instruction 3
3. Decoding of instruction 4
4. Fetch instruction 5 from memory

The execution of instruction 2 will be followed by immediate execution of instruction 3 since it has been previously fetched from memory and decoded. Similarly, the data for instruction 3 has also been retrieved. The use of pipelining techniques improves the throughput of the computer system without having to use faster processors.

Different techniques can be used to implement various types of multiple processing of operations. It might be achieved internally within a central processing unit as has been done in the case of the Intel 8086 microprocessor. This approach can be schematically represented by Fig. 12.3. In this internal multiple processing system approach the four processes of fetching an instruction, decoding it, fetching data and executing an instruction are carried out in parallel so that as soon as the previous instruction has been executed, the next one is ready for execution. The overall throughput rate clearly depends upon the speed of the slowest of the four modules. This speed can be improved by subdividing the slowest element or carrying out the functions of the slowest element in parallel. While increasing the throughput power of the

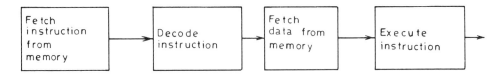

Figure 12.2 Pipelined processing of instructions.

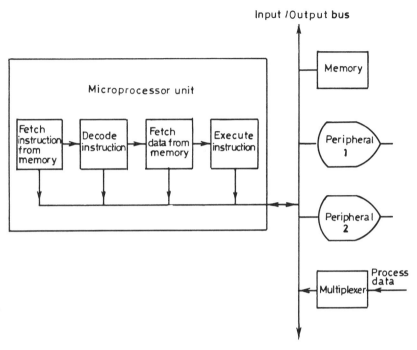

Figure 12.3 Schematic representation of parallel processing within a central processing unit.

microcomputer system, this approach does not permit the simultaneous execution of a number of instructions on a number of data values. The only practical way to achieve this is to use a number of processors and link them together. The required linking of processors can be achieved in different ways. The processors might be strongly coupled or weakly coupled. Alternatively, processors may not be coupled, but a processor is used to carry out input/output operations which involve frequent interruptions. An example of this is the use of a front end processor linked to peripherals as well as the host mainframe computer system.

Strongly Coupled Systems

A system, consisting of a number of processors, is considered to be tightly coupled if the processors share a common memory and peripherals. Thus any of the processors can access any of the memory locations and execute the necessary instructions. Clearly this access by different

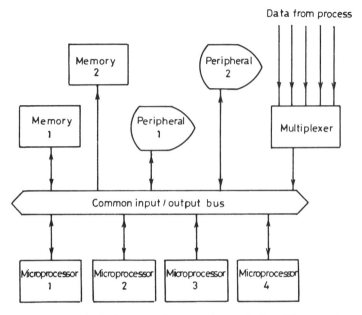

Figure 12.4 Block diagram of a strongly coupled multi-processing systemn.

processors to common memory cannot be simultaneous. It is therefore necessary to devise procedures for synchronizing events as well as deciding on the priority of access. A sophisticated operating system is required to control the operations of such systems. Multiple processing systems can be considered to be tightly coupled. Figure 12.4 illustrates a tightly coupled multi-processor system. Different techniques used for tight coupling of systems are discussed later in this section.

Weakly Coupled Systems

Loosely coupled systems do not have access to a common memory. Thus each processor must have its own memory and input/output peripherals. The functions of loosely coupled systems are clearly defined. Each processor must have its own operating system which might be identical for a number of processors. Hierarchical distributed processing systems are loosely coupled. Figure 12.1 can be considered to be a schematic representation of a loosely coupled system and shows that communication links are required to connect the microcomputers into a network. These networks can be used to interconnect equipment from a number of manufacturers. An advantage of networks is that the

Distributed Digital Control Systems

failure of any system element connected to the network does not result in the failure of the complete system which can be extremely important in a general manufacturing as well as the process control environment. Processors can be interconnected in the form of a star network, ring network, matrix structure or hierarchical network and these will all be discussed in more detail later.

12.3.2 Linking Techniques

A number of different techniques can be used to link the processors to form the required overall system configuration. The available techniques can be divided into the following categories.

1. Processors with common bus
2. Processors with multi-port common memory
3. Processors arranged in the form of a star network
4. Processors arranged in the form of a loop or ring network
5. Procedures arranged in the form of a hierarchical network
6. Processors arranged in the form of a matrix structure

These coupling technologies can be used to design many different types of system configuration to suit the actual requirements.

Processors with Common Bus

Figure 12.4 is a schematic illustration of the connection of a number of processors to a common input/output bus and common memory. Free access to common memory as well as input/output ports is available to any of the processors connected to the system. All the processors are master processors which operate independently. Hence a sophisticated system is necessary to achieve the required synchronization of events and decide on the priorities of different tasks. The system must have a knowledge of the units attached to the bus as well as their addresses.

This type of linking of a number of processors is relatively simple. The actual transfer of information involves the bus, the unit that wishes to transfer information and the unit which will receive the information. All the units must be available if the transfer is to take place. Thus the transferring unit has to determine first whether the bus is available. Nonavailability of the bus means that transfer cannot be initiated. If the bus is free, then the status of the receiving unit must be determined. Transfer can be initiated only if the receiving unit is free and can receive the data.

The overall system can be easily re-configured by adding or removing units attached to the bus. Interconnection costs are low for small systems and it is possible to transfer information at high speeds.

While this approach is relatively simple, it does suffer from a number of disadvantages. The major disadvantage is the overall reliability of the system which is only as good as the reliability of the weakest unit. The breakdown or malfunction of the bus interface of a single unit will result in an overall system failure. Furthermore, the use of this approach can result in serious bottlenecks if the speed of data transfer over the bus is low and less than the data transfer rate required for access to input/output ports and memory. Common bug configurations are only suitable for small systems.

These disadvantages can be overcome by using a number of different

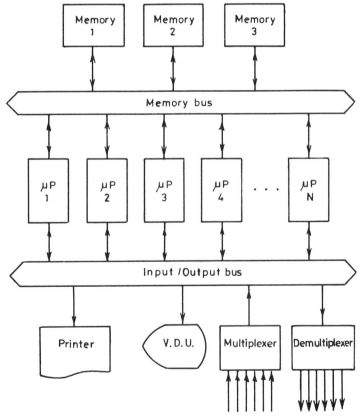

Figure 12.5 Block diagram of a system with a memory bus and a separate input/output (μP = Microprocessor).

approaches. For example, each of the processors can have a small amount of separate private memory in addition to the memory shared between different processors. An alternative is to use two separate buses, one for common access to memory and the other for common access to input/output peripherals. Such a system is schematically represented in Fig. 12.5; this system would be much faster than a system with a single bus. This approach can be extended to include further buses.

Processors with Multi-port Common Memory

Some of the problems associated with memory access and synchronization can be resolved by using common memory systems with a number of ports. At a simple conceptual level, this process can be illustrated by Fig. 12.6, for a single common memory connected to four processors.

The memory ports can be allocated different priorities by building in the appropriate logic at the interface between the memory and the processor. Thus it is possible to easily re-configure the system in order to allocate a higher priority to a particular processor, by simply attaching it to a higher priority port. However, it is not possible to attach additional processors if the memory does not have an additional port. It is also possible to attach input/output units directly to the common memory

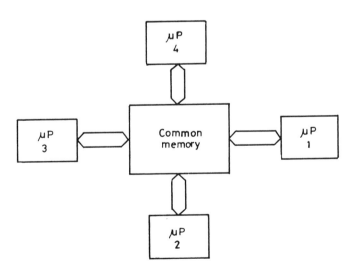

Figure 12.6 Schematic illustration of processors connected to a multiport common memory system (μP = Microprocessor).

units. Multiple buses are used to interface the common memory units to the processors and input/output ports. Thus a system with four processors, three memory units and three input/output units can be configured as shown in Fig. 12.7. This approach offers many advantages including protection of some data items by treating memory as private to particular processors. The use of multiple buses means that high data transfer rates can be achieved to enable the configuration of a powerful computer system. Synchronization problems are minimized by allocating priority to each memory port. For these reasons, this particular approach is often used in practice for configuring computer systems.

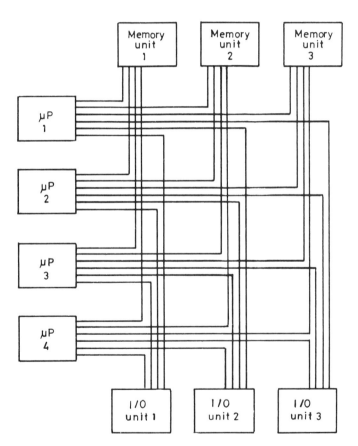

Figure 12.7 A multiport memory system involving three memory units with four ports connected to four processors also linked to three input/output units (μP = Microprocessor; I/O = input/output).

Distributed Digital Control Systems

The major disadvantages of this approach are the cost of multi-port memory and the difficulty in re-configuring the system if the number of memory ports is inadequate for the new configuration. The cost of cabling and connecting is also high particularly when the system is expanded. Maximum system configuration has to be specified at the start.

A variation on the multi-port memory approach is the complex cross-bar switch system which makes use of switches to enable the transfer of data between memory processors and input/output units. The memory unit has only one port and is connected by separate buses to the processors and input/output units. This approach can be used for the simultaneous transfer of information. The synchronization and priority problems are resolved by using a complex switching system. All the system elements can be interconnected. It is possible to make a number of simultaneous mutually exclusive connections. System expansion to include further processors and input/output ports is possible, since it is only necessary to provide additional cross-over points and high rates of data transfer can be achieved. However, it is necessary to provide circuitry for priority allocation, switching and controlling at each point whereas in a multi-port memory system this circuitry is only required at the ports.

Processors within a Star Network

A typical two-level star network is shown in Fig. 12.8. It consists of a master processor and a number of slave processors. The master processor is the center of the network. Individual slave processors only communicate with the master processor. Thus, if two slave processors need to communicate with each other, this can only be achieved via the centralized master processor. Independent serial links are used to interface the master processor to slave processors. The master processor may simply be used for handling communications or it might perform communications tasks in the foreground mode and other applications related, but not time critical, tasks in the background mode. While the star network system is relatively simple, severe problems occur if the master processor breaks down. The breakdown of any of the communications lines also results in problems. The failure of a communications line or a slave processor will only affect that part of the system. Network reliability can be improved by providing standby communications links. However, it is not possible to share communications lines between different processors. Bottlenecks can occur if the level of communications between processors is high and the centralized master processor is unable to cope with this level of activity.

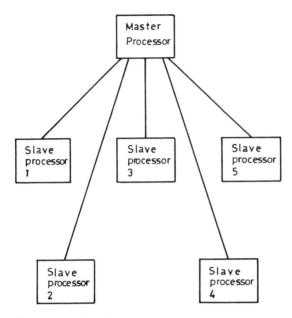

Figure 12.8 Typical star network configuration.

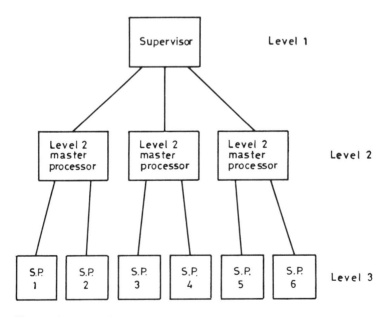

Figure 12.9 Multi-level hierarchical system (S.P. = Slave processor).

Distributed Digital Control Systems

Star configurations are suitable in circumstances in which slave processors mostly communicate with the master processor and the level of communications between slave processors is low. In some star networks, particularly those involving the use of microcomputers, the application processor is linked to the master processor via a communications processor to minimize the interruptions.

Star networks can also be used to implement hierarchical system in which computers are used at more than two levels. Thus the master processor at one level may be a slave processor for the next higher level. A typical multi-level system is schematically represented in Fig. 12.9. Such systems can be used for linking computers, via communications lines and concentrators, at a number of sites.

Processor within a Ring Network

Ring or loop networks are increasingly being used for the local interconnection of a number of microcomputers. Figure 12.10 shows a typical ring network system.

The processors forming part of the network can be interconnected by means of a serial unidirectional link in the form of a loop. It is necessary to transfer the information until the destination is specified at the start of the message. Thus the intermediate processors only need to decode the address, instead of decoding the whole message. The processor for which the message is destined can remove it from the ring or pass it along if the message is also meant for another processor. It is thus possible to send a message to all the processors in the network. Data can be transferred in digital format and it is not necessary to use

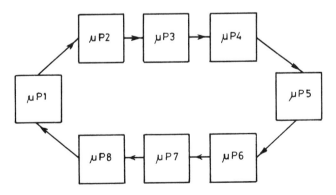

Figure 12.10 Ring network interconnection of processor (μP = Microprocessor).

modems, etc. Implementation of ring network is relatively simple and inexpensive.

Such ring networks are only suitable for data transmission over short distances and applications which are not time critical. Data have to be transferred unnecessarily from one processor to another until the destination is reached and the rate of information transfer is restricted to the speed of the slowest node. Problems are caused by the malfunction of any of the networked processors. Diagnostic procedures can be used to check periodically that the processors are performing satisfactorily and in the event of a breakdown, the faulty processors can be bypassed. Reliability of ring networks can be improved by providing double loops.

Any of the ring networked processors may also be connected to a host processor as shown in Fig. 12.11. In such configurations, the networked processor can also transmit messages originated from the

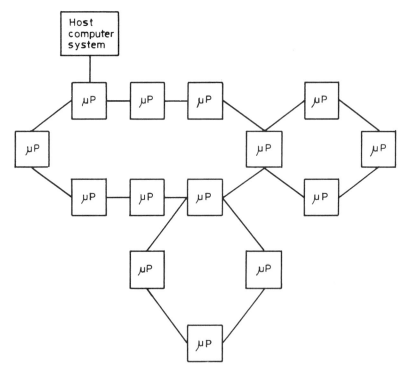

Figure 12.11 Implementation of multiple loops and use of a host computer system (μP = Microprocessor).

Distributed Digital Control Systems

host system. Furthermore a processor may form part of more than one network, as illustrated in Figure 12.11.

Processors within a Hierarchical Network

This is the most common and traditional form of networking in which a number of processors with varying capabilities and processing powers are linked together. Figure 5.1 illustrates the full-fledged form of hierarchical network that might be implemented in a manufacturing organization. At the lowest level, a number of simple microprocessors or microcontrollers are used for data acquisition and closing of process loops. The functions to be carried out by individual processors become increasingly complex as the higher levels of network are reached. Each higher level processor receives and coordinates information from a number of lower level processors. The amount of data transferred from lower level processors to the next higher level is condensed since all of the information used at one level of a manufacturing organization is not required at the next higher level. For example, in the case of microcomputers used for automatic testing of products, it is only necessary to pass on information about the number of products accepted and rejected along with the causes of rejection, to the computer used at the higher level for production planning and control purposes. Similarly, the production control computer need not transfer the reasons for rejecting individual items to the computer used at the higher management level. Thus the information is usually condensed before being passed on to a computer at the next higher level. The transmission of every bit of data created at the lowest level of the network to the highest level computer would result in severe communications problems. Computers at the higher levels of a network are used to carry out complex tasks involving extensive processing of data. It is therefore necessary to use powerful computers at the highest level. Thus a typical hierarchical configuration might consist of microprocessors at the lowest level, minicomputers at the middle level and a powerful mainframe computer at the highest level. The processors can be selected to suit the requirements.

These hierarchical configurations can be considered to be implementations of star networks. It is possible to use ring networks for implementing a hierarchical system configuration, although this is not usual.

The reliability of such hierarchical networks is a very important consideration. Simple and hence reliable processors are used at the lowest level, whereas the higher level systems, controlling a section of the network or the complete network, are complex and less reliable.

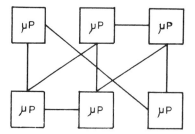

Figure 12.12 Matrix (mesh) network (μP = Microprocessor).

The breakdown of a single serial link between two processors at higher levels means that a part of the system is disconnected from the rest. This breakdown can result in difficulties in the case of applications involving coordination/optimization of processes. Redundant communications links can be built into the network to overcome these difficulties.

Many of the networked processors in a system with several levels of hierarchy may not be doing very much more than simple transfer of information. The response time of a system with a large number of hierarchical levels can be relatively long. However, such hierarchical configurations can be used to achieve high throughput rates and will continue to be implemented for networking processors on a number of sites and performing unrelated or related functions.

Processors Arranged in the Form of a Matrix Structure

Processors can be networked in the form of a matrix or mesh, as illustrated in Fig. 12.12, whereby each mode (processor) is connected to a number of processors. While improving system performance and reliability, the high cost of providing such additional connections is a deterrent. Furthermore, the development of communications protocols in a system without a properly defined structure can cause problems.

12.3.3 Advantages

Multiple processing systems offer many advantages over the conventional systems. Depending upon the type of system configuration implemented, the following benefits can accrue to the user.

1. Multi-processing systems operate under the control of a single operating system, allocating and sharing resources as necessary to perform the required tasks.
2. Compared to a single centralized system, a multi-processing system

offers a faster response since the frequency of interruption of a single processor is not as high.
3. The cost of building a multiple processing system is much less than the cost of a centralized system with equivalent computing power.
4. A multi-processing system is not as complex as a single centralized system since it is built from a number of individual and hence simple processors. Centralized processors built from a large number of components are necessarily complex and less reliable.
5. The level of availability and reliability of a multiple processing system is higher than that of a centralized system.
6. The multiple processing system can be adapted to suit changed requirements. Hence it is easier to provide for an increase or change in system demands.
7. Multiple processing systems with shared memory simplify the task of communications between different processors. All the data are transferred over the common bus. This reduces the cost of developing the communications network.

While multiple processing systems do offer these significant advantages, a price has to be paid because of the need for complex operating systems.

We now direct attention to the application of the hierarchical systems control approach (Lefkowitz, 1977; Mesarovic, 1970; Mahmoud, 1977 and Findeisen, 1970) which provides a conceptual framework for organizing the integration of the many diverse decision-making and control functions that affect system performance. Simply stated, the hierarchical approach is predicated on replacing the initially complex control problem by a set of sub-problems which are more easily solved and implemented. Compensations for model approximations and interaction effects are effected through the coordinating efforts of a supremal control unit. It is significant to observe that the hierarchies provide motivation for orderings with respect to time scale, degree of aggregation, frequency of control action, and other attributes that are to be considered at the system design stage. The hierarchical structure also plays an important role with respect to organizing the flow of information through the system and in providing the mechanisms for effective utilization of feedbacks for control and decision making. Now, all aspects of information processing, data gathering, process control, on-line optimization, operations control–advancing even to real-time scheduling and production planning functions–may be included in the range of tasks to be carried out by the computer control system. This has made possible the realization of integrated systems control in which all factors influencing plant performance are taken into account in an integrated fashion–recognizing the couplings, interactions, and complex feedback paths existing in the system–to achieve an overall optimum performance.

12.4 INTEGRATED SYSTEMS CONTROL

The problems of realization and implementation of an integrated systems control are generally formidable because of the complexity of the production processes, the variety of constraints to be satisfied, the nonlinear, time-varying dynamics, etc. Multi-level and multi-layer hierarchical control approaches provide rational and systematic procedures for resolving these problems. In effect, the overall systems control problem is decomposed into more easily handled sub-problems; the sub-problem solutions are coordinated by a higher level controller so as to assure compliance with overall objectives and constraints (Lefkowitz, 1977; Mesarovic, 1970; Mahmoud, 1977).

With respect to decision making and control, we distinguish three basic elements of the system: plant, controller, and information processor.

12.4.1 Plant

The plant denotes the controlled system and may refer variously to a production unit, a processing complex, or even an entire company, depending on the level of control being considered. We assume that the plant is governed by causal relationships, that is, its behavior relevant to our control objectives may be described (in principle) by a set of input-output relationships.

We assume further that some of the plant inputs are free to be selected by a decision maker or controller so as to influence the plant's behavior in a desired direction. Thus, we may classify the variables associated with the plant as follows:

1. *Disturbance inputs.* These are inputs that are independent of the control and cause the system to deviate from desired or predicted behavior, hence motivating control action. In general, disturbances represent the interactions of the plant with other plant units and with the environment, that is, load changes, weather changes, etc. A special class of disturbance, called contingency occurrences, refers to events that occur essentially at discrete points in time, that is, a component has failed or a unit is taken off line. Often a contingency event signals that the system is no longer operating according to assumptions implied by the current control model and that, as a result, it is necessary to

Distributed Digital Control Systems

modify the structure of the system, go into a new control mode, or develop some other non-normal response.

2. *Controlled inputs* (also referred to as decision variables). These inputs are the results of the decision-making process carried out by the computer/controller. They are determined so as to compensate for the effects of disturbances by either directly or indirectly modifying the relationships among the plant variables, e.g., by changing the energy balance in the system.

3. *Outputs*. These are variables of the plant which (1) are functionally dependent on the designated input variables, and (2) are relevant with respect to the performance measure on which control of the plant is based.

4. *State*. In the sense used here, the state relates to energy or material storages in the system, the status of production units, and any other factors necessary to the identificaton of the appropriate input-output relationships (models) to be applied.

12.4.2 Controller

The purpose of the controller is to generate the controlled inputs to the plant which tend to maximize performance consistent with the constraints imposed on the system. The constraints define the region of feasible or acceptable plant operation. They may characterize actual technological limits of the equipment, e.g., maximum loads, capacity limits, etc. Constraints are also imposed to ensure the safety of operating personnel or the security of the production means (e.g., temperature limits on a furnace to minimize deterioration of the refractory lining). Finally, we impose constraints to ensure that various "quality" requirements are met, e.g., product specifications, effluent pollution restrictions, etc.

The control functions may be performed by man, by machine (computer), or more generally, by an integration of human operators, schedulers, and planners with the computer control and data management system. The functions performed by man include those requiring judgments that cannot be standardized, or decision processes that have not been adequately established, or coordinations that involve the integration of a great many factors whose subtleties or nonquantifiable attributes defy computer implementation. The functions performed by computers are essentially those where the tasks are routine and well defined and where the operating standards are quantified and established. Thus, the main planning and coordination functions are carried out by humans, with computers providing the basic information on which the operator's judgment and decisions are based. The computer is involved in the gathering, processing, and dissemination of data, the distribution of operating instructions, and the implementation of controls and operations at the technological level. In addition, the responsibility of responding to contingency occurrences and special requirements rests generally with the human component of the system.

12.4.3 Information Processor

The underlying assumption in the achievement of integrated control is that the controller acts on the basis of (real-time) information concerning the state of the plant, external inputs, etc. Major functions of the information system include:

1. Data gathering and processing, e.g., data-smoothing, noise-filtering, prediction and extrapolation, etc.
2. The monitoring of system status for contingency events to determine whether diagnostic and/or corrective responses are to be initiated
3. The storage and retrieval of operating instructions, standards, parameter values, and other information required for the functioning of the system.

A block diagram of the relationship of the controller to the plant is given in Fig. 12.13. Current values of the output variables y (for feedback actions) and some of the disturbance variables z (for feedforward actions) are transmitted to the controller by means of sensors or measuring devices. The raw information set x may be further processed (filtered, smoothed, transformed, etc.) by the information processor. The controller generates its outputs according to current information concerning y and z, in relation to the input r, which defines the desired behavior of the plant, e.g., provides the setpoint values at which certain output variables are to be maintained. Finally, the controller must communicate its decisions/actions to the plant, and this is the role of the actuators. The elements represented in Fig. 12.13 are basic to all control and decision-making functioning and are embedded within each of the hierarchical control structures to be described below.

The overall complex problem may be decomposed according to various criteria; these include:

1. Decomposition according to control function: functional multilayer control hierarchy
2. Decomposition according to sub-system classification or system structure; multi-level control hierarchy

Distributed Digital Control Systems

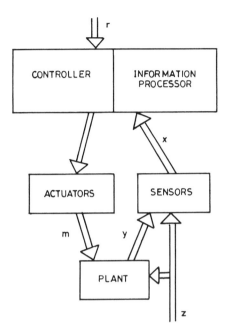

Figure 12.13 Basic control configuration.

3. Decomposition according to time scale: temporal multilayer control hierarchy

The multi-layer and multi-level hierarchies many also be characterized as "vertical" and "horizontal" control structures, respectively.

12.4.4 Functional Multi-layer Control Hierarchy

The functional control hierarchy is characterized by the diagram in Fig. 12.14 in which four categories of control function are identified, namely: direct interaction, supervisory, adaptive, and self-organizing functions.

First Layer

The first or direct layer function constitutes the interface between the controlled plant and the decision-making and control aspects of the system. An important characteristic of the first layer, therefore, is its ability to interact directly with the plant and in the same time scale.

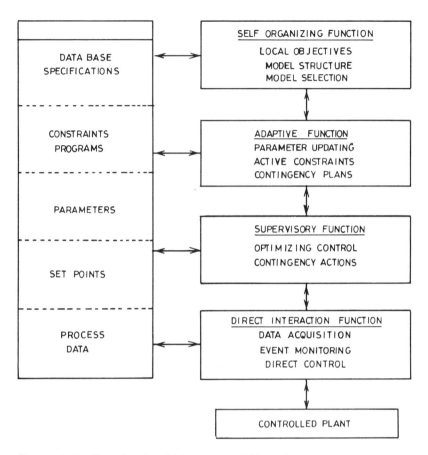

Figure 12.14 Functional multi-layer control hierarchy.

We distinguish three sub-functions:

1. *Data aquisition.* Responsible for providing the controller with the necessary data to carry out its control functions, including data from plant sensors and operator inputs. Data may be processed (smoothing, averaging, normalizing, linearizing, transforming, etc.) before being input to the control system for storage and subsequent utilization.
2. *Event monitoring.* Responsible for detection of discrete event, occurrences which affect the control problem. The event may (1) cause the controller to initiate an action or response, (2) signal the completion of a previous task or operation assignment, (3) signal

Distributed Digital Control Systems

the introduction of new parameter values, or (4) signal a change in operating mode. We note that the event occurrence may be signaled by a functional evaluation or pattern classification of plant data, e.g., the determination of whether the current operating state lies within a specified region.
3. *Direct control.* Responsible for implementing the decisions generated by the higher layer functions of the controller. Specifically, the direct control function implements the target or strategy defined by the second layer function through direct actions on the plant. The direct control functions may imply algebraic computations, logical operations, table look-up or various combinations. The controller output may evoke a single act or operation in response to the supremal signal, or a train of actions, each triggering the next.

Second Layer

The second layer or supervisory function is concerned with the problem of defining the immediate target or task to be implemented by the first layer. In the normal mode, the objective may be control of the plant for optimum performance according to the assumed mathematical model. Under emergency conditions, different objectives may take precedence through implementation of the appropriate contingency plan. In general, there may be a number of operating modes or topologies identified for the system; each then may have a different mathematical model through which information describing the current state of the system is transformed into directives applied to the first layer function.

In the conventional process control application, the second-layer intervention takes the form of defining the setpoint values for the first layer controllers. In the discrete formulation, the output of the supervisory function may be a specified or "next state" to be implemented by the direct controller through a predetermined sequence of actions.

Third Layer

The third layer or adaptive function is concerned with the problem of updating algorithms employed at the first and second layers. The adaptive layer may intervene in the operation of the lower layers in the following ways:

1. Updating of parameter values associated with the first and second layer control algorithms, say by least-squares fitting of the underlying mathematical models to observed plant behavior.

2. Updating of parameters associated with the event-monitoring function. Of particular interest here is the sensing (and generation of an event notice) of the transition of the plant from one operating mode to another.
3. Specification of the constraints to be applied in the second layer optimization problem formulation. In security control, for example, these constraints derive from the set of contingencies against which the system is to be made secure. More generally, the imposed constraints may reflect changes in plant topology occasioned say, by the removal from service of a piece of equipment.
4. Development of contingency plans, i.e., alternative procedures for second layer implementation, to be invoked when the plant degenerates to an emergency mode.

A common and distinguishing feature of the third layer function is that its actions are a reflection of operating experience over a period of time. The actions are discrete, taking place at predetermined intervals of time or, more generally, in response to event occurrences (e.g., operator inputs).

Fourth Layer

The fourth layer or self-organizing function is concerned with decisions relevant to the choice of structure of the algorithms associated with the lower layers of the hierarchy. These decisions are based on overall considerations of performance objectives, priorities, assumptions of the nature of the system relationships and input patterns, structuring of the control system, coordination with other systems, etc. The fourth layer intervenes directly in the operations of the lower-layer control functions via the mode selection function. Intervention is implemented through the executive program to schedule and control programs at the lower layers.

Application 1. An example of the application of the foregoing multi-layer control structure is provided by a catalytic reactor process (Lefkowitz and Schoeffler, 1972). The process inputs are controlled as continuous functions of time; however, at discrete points in time, the "normal" operating mode is disrupted to go into a "regeneration" mode for the purpose of re-sorting catalyst activity. Thus, a scheduling problem is superimposed on the continuous problem.

The direct control function is concerned with the task of controlling the process variables, e.g., pressures, temperatures and flow rates according to the trajectories of setpoint values defined by the supervisory control function. This function is implemented by means of conventional feedback control loops, with perhaps some feedforward considerations. The local objectives are essentially the maintenance of the controlled variables at their respective setpoint values within acceptable tolerances.

The determination of setpoints at the second layer is based on a model of economic performance appropriate to the mode of operation. Thus, in the normal operating mode we may determine values for the control variables which will tend to maximize product yield consistent with system constraints and specifications on product quality. In the catalyst regeneration mode, we may want to operate the plant so as to minimize the duration of the regeneration period, i.e., the period during which product is not being produced. There are at least two distinct tasks assigned to the third layer. The first relates to the updating of selected parameters of the lower layer control algorithms to take care of the effects of normal variations in operating conditions, catalyst activity, etc. The second task relates to the criterion function for switching from the normal operating mode to the regeneration mode. The conditions for switching may be determined through solution of a scheduling problem whose objective is to maximize an overall profit function in which is embedded the optimization model used at the second layer.

The fourth control layer has the responsibility of selecting the operating mode and, consequently, the programs to be used by the lower layer control functions. In particular, we note in this example that a transfer of mode requires extensive changes in the control structure; these are to be coordinated by fourth layer intervention.

Attributes of the multi-layer structure are:

1. The structure provides a natural hierarchy in which each layer has a priority of action over the layer below. In general, information passes up the hierarchy via the common data base; the results of decision making and evaluation proceed down the hierarchy either via the data base or via the computer executive program.
2. The layers of the hierarchy represent different kinds of control functions, hence require different kinds of computation and information processing algorithms. This means that we can do a better job of tailoring the hardware and software to the specific needs of the subproblem associated with each control layer.
3. The layers of the hierarchy can be designed to respond to disturbance inputs and/or discrete events having different frequency characteristics. This opens up the trade-off possibilities in more effective utilization of equipment, time-sharing of applications, bandwidth characteristics, and priority considerations.

A variation of the multi-layer control hierarchy described above is that shown in Fig. 12.15. In this formulation of the hierarchy, four levels are distinguished (Lefkowitz, 1966):

Level 1. Specialized dedicated digital controllers
Level 2. Direct digital control
Level 3. Supervisory control
Level 4. Management information

This approach is characterized by a stronger orientation toward the technological level of control, to the practical aspects of digital computer implementation and to the man-machine interface.

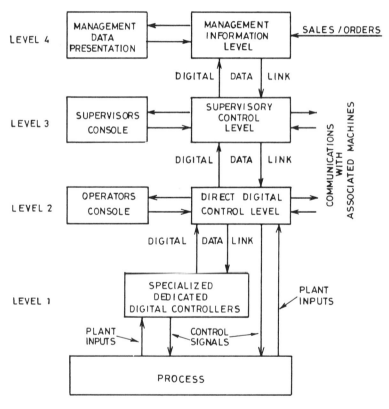

Figure 12.15 A digital control hierarchy.

12.4.5 Multi-level Control Hierarchy

In the multi-level control approach, the overall plant system is decomposed into subsystems, each with its own local controller. In this scheme:

1. The first-level controllers compensate for local effects of the disturbances, that is, maintain local performance close to the optimum while ensuring that local constraints are not violated.
2. The second-level controller modifies the criteria and/or the constraints for the first-level controllers in response to changing system requirements so that actions of the local controllers are consistent with the overall objectives of the system.
3. The second-level controller compensates for the mean effects of variations in the interaction variables.

In effect, the sub-system problems are solved at the first level of control. However, since the sub-systems are coupled and interacting, these solutions have no meaning unless the interaction constraints are simultaneously satisfied. This is the coordination problem that is solved at the second level of the hierarchy. A schematic of the multi-level structure is shown in Fig. 12.16 with two levels represented explicitly.

A variety of coordination schemes have been described in the literature (Mahmoud, 1977). The methods are similar in the sense that they serve to motivate an iterative procedure for solution of an optimization problem wherein a set of local sub-problems are solved at the first level in terms of a set of parameters specified by the second level. They may differ, however, in their applicability to a specific problem, in the computation requirements, convergence speed, sensitivity to model error, incorporation on-line, and other considerations.

As far as the plant is concerned, it is only the final result of the iterative process that is important. Thus, the entire multi-level structure is internal to the computational block generating the optimum control. However, in the on-line application, the computation depends on the current value of the disturbance input and this changes with time.

The decomposition of the overall system into sub-systems may be based on geographical considerations (i.e., relative proximity of different units), lines of managerial responsibility (e.g., steel-making shop and rolling mill in a steel works), or on the type of equipment (e.g., distillation tower and reactor in a chemical plant). In general, however, the plants are designed so that these divisions correspond to lines of weak interaction, i.e., through the incorporation of various "buffer" or control mechanisms, the resulting sub-systems are partially de-

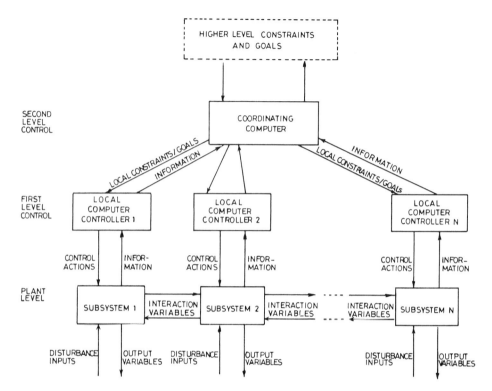

Figure 12.16 Multi-level computer control hierarchy.

coupled so that interaction effects tend to be small and/or only slowly varying with time.

The multi-level approach leads to the following advantages:

1. A reduction in the total computational effort because of less frequent second-level action.
2. A reduction in data transmission requirements because: (1) most of the control tasks are handled locally, (2) much of the information required at the second level consists of averaged aggregated data, and (3) the upper-level action takes place at lower frequency.
3. A reduction of development costs for the system by virtue of the fact that the models, control algorithms, and computer software can be developed in a step-by-step, semi-independent fashion. By the same token, the problems of system maintenance, modifying and debugging programs, etc. are considerably simplified.
4. There is increased system reliability because (1) a computer mal-

function at the first level need only affect the local subsystem, and (2) the system can operate in a sub-optimal but feasible mode for some time in the event of a failure of the second-level computer.

Application 2. A particular relevant application of the multi-level approach is in the electric power industry where the power generation and distribution system is designed as an interconnection of semi-independent sub-systems (Fink, 1971; DyLiacco, 1967). Thus, there is a natural decomposition induced by technological considerations at the generating unit level, geographical considerations at the generating station level, ownership boundaries at the company level, etc. This decomposition is illustrated by the multi-level hierarchical structure of Fig. 12.17. Here, S_j^i denotes the jth sub-system at the ith level; where $i = 1, 2, 3, 4$ corresponds, respectively, to the generating unit, the power station, the utility company, and the regional/power pool levels. Associated with each sub-system S_j^i is a controller/decision maker C_j^i. The

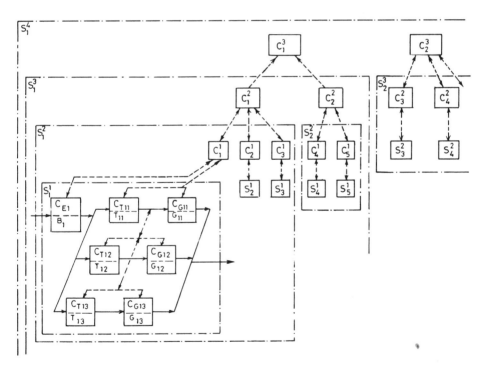

Figure 12.17 Multi-level control structure in electric power system.

system S_j^i is, in turn, comprised of sub-systems $\{S_k^{i-1}:kEI_j^i\}$ where I_j^i denotes the set of controllers informal to C_j^i, i.e. C_j^i serves as coordinator for controllers C_k^{i-1}, kEI_j^i. Note that $\{S_j^1\}$ represents the set of sub-systems at the physical plant level; each sub-system at this level may itself represent an interconnection of production units, each with its own local controller. For example, Fig. 12.17 shows S_1^1 expanded to include steam generator (B_1), turbines $\{T_{1j}\}$, and electric power generators $\{G_{1j}\}$, with controllers denoted by C_{B1}, C_{Tij}, respectively.

The power generation industry has more than two decades experience with direct digital control. In what follows, we shall describe some advanced control techniques used in Boiler-Turbine-Generator (BTG) units.

Boiler Control

The problem of boiler control involves the setting and maintenance of the boiler outputs of steam flow, pressure and temperature to their desired values. We thus need to employ both setpoint and regulatory control systems. For both types of control, the input quantities of air, fuel, and water are adjusted to obtain the desired outlet steam conditions. The requirements to be met by the regulatory controllers are that the outlet steam conditions are maintained close to their desired values under the steady-state conditions and that the steam conditions recover quickly without excessive oscillations when a system disturbance (load change) occurs.

A major control system in modern boilers is the combustion control system which is used to control the fuel or air input or firing rate to the furnace in response to a load index representing a demand for the level of fuel input. The demand for firing rate is to match an increase in load demand which can be met through increased steam flow. The boiler outlet pressure is generally used as an index of imbalance between fuel energy input and energy withdrawal in the output steam. A great variety of combustion control systems incorporating fuel flow and air flow controls have been developed over the years to fit the needs of particular applications.

Associated with the air-combustion control is the separate control action for the fuel supply. The fuel feed rate is controlled through a feeder controller working in conjunction with the air supply regulator as to maintain a fuel-air ratio. The desired fuel-air ratio has to be determined in order to set up the mill control loop. The following measurements are used: primary air flow; mill outlet-inlet differential pressure; and mill outlet temperature.

Distributed Digital Control Systems

To maintain the correct level of water in the drum of a boiler requires close control of three main functions involved in steam generation, i.e., water level, steam flow, and feedwater flow. The correct measurement of the true level of water and the correct control action to be applied during load changes is essential in very large boilers mainly because of the relatively small drum capacity which requires fast-acting and accurate control to keep water levels within gauge limits on varying loads.

Another common control system in all modern boilers is the feedwater control which regulates the flow or water in order to maintain the level in the drum between the desired limits. The three-element feedwater control system is shown in Fig. 12.18. It consists of a cascade feedback-feedforward control loop which maintains water flow input equal to the feedwater demand. The drum-level feedback-controller applies proportional control to the error between the desired set point and the measured drum-level signal. The sum of the drum-level error signal and the steam-flow signal is the feedwater signal. The feedwater signal is compared with the water flow signal and the difference is the combined output of the controller. Proportional-integral action is incorporated for valve regulation/pump speed control.

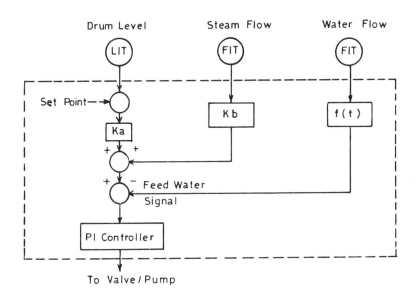

Figure 12.18 Three-element feedwater control.

In any modern thermal power station, it is of great importance to keep a very close control of the steam temperature at the boiler outlet, since excessive temperature has a bad effect on the metal parts of the boiler and turbine. Superheat and reheat control have to be considered for a load change from no load to full load. There are several methods for controlling the system temperature, their use has to be decided with due regard to boiler design and the requirements of turbine starting.

The current practice in 500 MW units is to use a number of parallel modulating control systems, the major control loops being designed to give a stable control action for maximum load changes. The improvements in performance through control algorithms such as three-term PID controller are significant and have been presented in the literature. The major control loops for direct-digital-control (DDC) of a boiler are

Feedwater (three-element)
Air flow and furnace pressure
Fuel flow
Primary air pressure
RH and SH steam temperature
Minor loops (turbine oil temperature, etc.)

More recently, the application of distributed digital control to boiler control problem has been examined in order to improve reliability and minimize the overall cost. The six major loops listed above have been placed under distributed control employing microprocessors at the JH Campbell Plant, Unit 3, Michigan, USA. Such schemes generally include operator display, logging facilities, alarm functions, and supervisory control.

Integrated BTG Control Systems

In a boiler-following control system (Fig. 12.19), the demand for a load change goes directly to the turbine control valves to reposition them to achieve the desired load. Following the load change, the steam pressure goes down at the throttle and the boiler feedwater and firing rate control systems restore throttle pressure to its normal operating value. The load response is very rapid, since the stored energy in the boiler is used to provide the initial load change. In a turbine-following system (Fig. 12.20), the turbine-generator is assigned the responsibility of throttle pressure control while MW control is the responsibility of the boiler. When an increase in load is demanded, the boiler control increases the feedwater and combustion rates which in turn starts raising

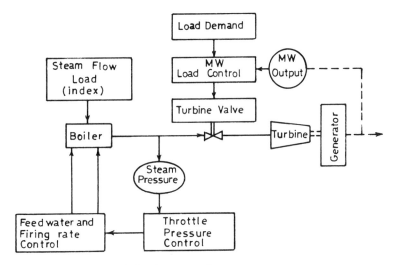

Figure 12.19 Boiler-following control system.

the throttle pressure. In order to maintain a constant throttle pressure, the turbine control valves open and accept the additional boiler output, thus increasing the load. Load response with turbine-following is slow since the turbine-generator must wait for the boiler to change its output before repositioning the control valve to change load.

In the integrated control system (Fig. 12.21), the load control and

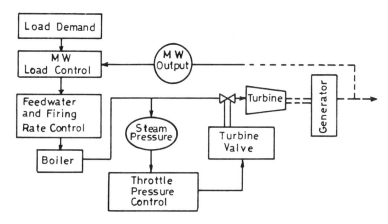

Figure 12.20 Turbine-following control system.

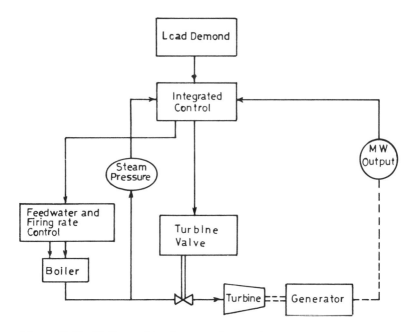

Figure 12.21 Boiler turbine integrated control system.

throttle-pressure control are the responsibility of both the boiler and the turbine generator. The integrated control system was developed primarily for once-through boilers. It can, however, be applied to drum boilers. Distributed digital control algorithms in power stations have been reported (Wallace, 1988).

12.5 DISCUSSIONS

Distributed control systems provide a very convenient method of distributing computing power to its point of use. Hierarchical distributed processing systems, which make use of micro-, mini- and mainframe computers, are particularly suitable for implementation in industrial and power systems. The advantages of this particular approach include progressive build-up of such systems over a period of time and provision

Distributed Digital Control Systems

of computer power for all the users. A prerequisite for the implementation of such systems is a clear formulation of the system development strategy and the specification of the overall system architecture in unambiguous terms. Only then will it be possible to develop integrated systems which make use of a number of computers, with different processing powers, at different levels of the hierarchical computer control system configuration. Such configurations will become particularly desirable when the instrumentation systems or process control systems purchased incorporate microprocessors as just another component or sub-system. It will then be possible to capture the required data at its source and use it for a wide range of other applications. An important consideration in the development of such systems is the need to ensure communications between different elements of an overall integrated system configuration. Such communications can be achieved only by using suitable interfaces and protocols which ensure that different computer systems not only talk but can also understand each other.

It has been shown that applications of the hierarchical control systems provide an organization-type of control by dividing the targeted tasks among various levels. With the recent developments in micros and the increased capabilities they can offer, it is expected that most of the next generation of micro-computers will open new frontiers in the form of sophisticated networking, multiple or dual processors and/or extensive mass storage

NOTES AND REFERENCES

There is increasing motivation on the part of industry to apply advanced computer control methods to its production systems in order to improve productivity, product quality, efficiency in the utilization of resources, and related objectives. The literature on digital control algorithms and their practical implementations is extensive; for a recent account the reader is referred to Warwick (1988), Wallace and Clarke (1983), Horton et al (1982), Hanselmann (1987), Astrom and Wittenmark (1984) and the references cited therein.

13
Advanced Microprocessor Architecture

13.1 OVERVIEW

The evolution of microprocessor technology continues to proceed at an amazing pace. Today commercially available microprocessors have achieved tremendous power and wide functionality to the extent that it becomes difficult to belive that this "computer-on-a-chip" technology is just two decades old or less. In Chapter 9, we looked into the features and characteristics of the early 4-bit and 8-bit generations of microprocessors. This chapter will supplement the knowledge-base about the microprocessor devices and systems by providing relevant information concerning the new 16-bit and 32-bit generations. To appreciate the technological advances in the microprocessor products, it is sufficient to know that over the period 1972–1987, the number of devices per chip has increased by a factor of 500, the clock frequency by a factor of 50, and the overall throughput by three orders of magnitude.

The material covered in this chapter is organized into three major sections; the first deals with 16-bit microprocessors, the second is devoted to 32-bit microprocessors, and the third provides some additional features and potential applications.

Advanced Microprocessor Architectures

13.2 16-BIT MICROPROCESSORS

The era of 16-bit microprocessors began in 1974 with the introduction of the PACE chip by National Semiconductor. The Texas Instruments TMS 9900 was introduced two years later. Subsequently, the Intel 8086 became commercially available in 1978, the Zilog Z8000 in the 1979, and the Motorola MC68000 in 1980. In what follows, we will concentrate on the trends exemplified by the more popular chips. The principal advantages that 16-bit microprocessors offer compared with 8-bit devices are:

1. Faster instruction execution times
2. Wide instruction set (e.g. multiplication and divide instructions)
3. Larger memory addressing range typically, 1 Mbyte (or more) compared with 64 Kbyte
4. Larger integer number range (0–64 K in place of 0–255)
5. More addressing modes to make programs simpler and more efficient
6. The use of co-processors to assist the CPU in executing programs faster

13.2.1 Intel 8086/8088 Family

The 16-bit 8086 represents a major upgrade on the 8-bit 8080 and 8085 devices, whilst the 8088 is virtually identical to the 8086 except that its external data bus is only 8 bits wide. Intel carry the main share of the 16-bit microprocessor market; figures of between 70 and 80% are quoted. The 8086 and 8088 dominate in the 16-bit office computer market, as follows:

Machine	Microprocessor
IBM personal computer (early versions)	8088
ACT Sirius	8088
DEC Rainbow	8088
Apricot	8086

Later versions of the IBM PC and compatibles use enhanced versions of the 8088/8086. These versions are the 80186 and 80286, and the 80386.

Pin functions

Figure 13.1 show the pin functions of the 8086 DIL package which is quite similar to the 8088. It is a 40 pin device with a single +5 V power/ground supply. We notice the following features:

1. The 16-bit address and data buses are multiplexed (share the same pins) on the 8086. The signal ALE (Address Latch Enable) identifies whether a memory address or data is being transferred.
2. There are 20 address bus lines in each case—this gives 1 Mbyte memory addressing range.
3. The 40-pin limitation in DIL size caused Intel to multiplex some signals in the control bus (in addition to multiplexing the address and data buses). The diagrams show that 8 control bus lines (pins 24 to 31) have dual functions, as selected by the setting of the MN/MX signal. This signal selects "Minimum" mode if connected to +5V, and the 8 identities (INTA to HOLD), which are closely related to functions for 8-bit microprocessors, are selected. If MN/MX is connected to Ground, then "Maximum" mode is selected, which causes an expansion of the 8 control signals to facilitate connection to large and complicated circuit configurations. This expansion is created because three signals (S0, S1, and S2) are connected to a 3 to 8 decoder (the Intel 8288 bus controller) to

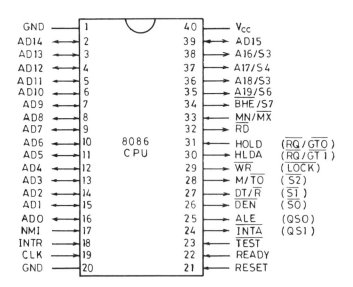

Figure 13.1 8086 pin functions.

provide 8 discrete control bus signals (DEN, DT/R and IO/M are a subset of these 8), and two signals (QS1 and QS0) are connected to a 2 to 4 decoder.

The standard range of 8-bit input/output chips and memory chips are connected to these 16-bit microprocessors in full circuit configurations (examples are given at the end of this chapter). The multiplexed address and data bus PIOs (the Intel 8155 and 8255) can be connected directly, whilst the 8251 UART can be connected by linking the data pins to the multiplexed AD0 to AD7 and its C/D pin to the demultiplexed A0. However, other devices, and in particular memory chips, require that the address and data buses be demultiplexed. This can be achieved using Intel 8282 8-bit latch buffers or Intel 8286 8-bit bi-directional bus transceivers — alternatively the standard TTL SN74245 transceivers can be applied.

One other essential support chip requires mention here. The Intel 8284 clock generator provides the CPU Clock (the input crystal frequency is divided by 3) and Reset signals. A typical Clock rate is 5 MHz.

Processor architecture

The internal operation of the 8086 and 8088 microprocessors is illustrated in Fig. 13.2. The architecture is split into two separate processing units — the execution unit (EU) and the bus interface unit (BIU). The former comprises the main components found in any 8-bit microprocessor — ALU, control unit, instruction register, flags (or status register) and general registers (note that there are eight 16-bit work registers). The bus interface unit processes signal connections to/from the multiplexed address/data bus and control bus, and performs the following functions:

1. Organizes the multiplexing timing
2. Automatically fills an instruction queue with the following instruction/instructions
3. Adds the contents of the four segment registers to memory addresses before they are placed in the address bus, such that memory is accessed in 64 Kbyte "segments"

The last two features are worthy of more detailed descriptions. The instruction queue is a feature that helps to increase the speed of operation of the device. Memory transfers represent a major time delay in the execution of a sequence of program instructions. Therefore the BIU continually attempts to keep the instruction queue full by fetching

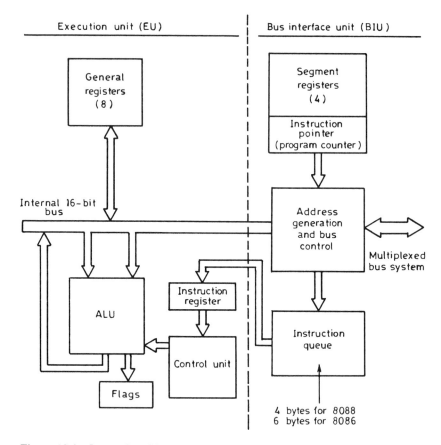

Figure 13.2 Internal architecture of 8086 and 8088.

succeeding instruction bytes from memory when the EU is executing an instruction but does not require use of the buses. In this way when the EU completes an instruction it does not have to perform a memory read to access the next instruction opcode. Instructions are variable in length (1, 2, 3, 4 or more bytes), and the instruction queue is 6 bytes long in the 8086 and 4 bytes long in the 8088. If an instruction (e.g., jump or call) that transfers control to another location in the program is obeyed, the BIU clears the queue and passes the instruction from the new address directly to the EU. It then refills the instruction queue. This queue feature is sometimes called an instruction "pipeline", and it ensures more efficient use of the buses and faster program execution times.

Advanced Microprocessor Architectures

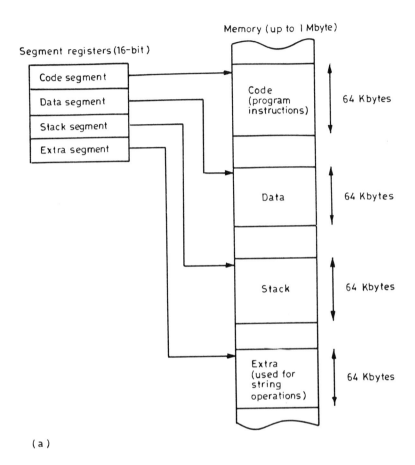

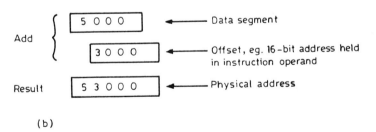

Figure 13.3 Generation of 20-bit addresses using segment registers. (a) Memory segmentation. (b) Calculation of physical address (e.g., accessing data value).

The other facility within the BIU that warrants a more detailed explanation is the use of the segment registers. Figure 13.3 shows the identities of the four segment registers, and how they are applied. The 1 Mbyte address range of memory is divided into 64 Kbyte "segments". Different programs can occupy different segments. Similarly different data lists can be placed in different 64 K segments, and the stack can reside in its own 64 K segment. The contents of a 16-bit segment register are effectively shifted left 4 places and then added to the effective memory address (as used by any 8-bit microprocessor) in the BIU to produce a 20-bit physical memory address that is placed on the address bus. The choice of which segment register is used is implied in the instruction, so that program instructions are fetched from one segment whilst the data values used in that program are accessed from another segment. Segments can overlap, of course, and in a system in which the total amount of memory is less than 64 Kbytes it is possible to set all segment registers to the same value, e.g., zero, and have totally overlapping segments. Segmentation is extremely useful for large multi-tasking systems to provide isolation between program modules and data lists. Additionally programs arc relocatable.

Instruction set

The instruction set is based upon manipulation of data values held in the general registers, and Fig. 13.4 lists these eight 16-bit registers. The first four registers are the principal data manipulation registers, and they can also be accessed in halves (bytes). All four registers act as accumulators, whilst BX can serve as a base register when computing data memory addresses, CX can be used as a counter in multi-iteration

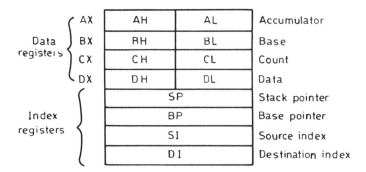

Figure 13.4 8086/8088 general registers.

Advanced Microprocessor Architectures

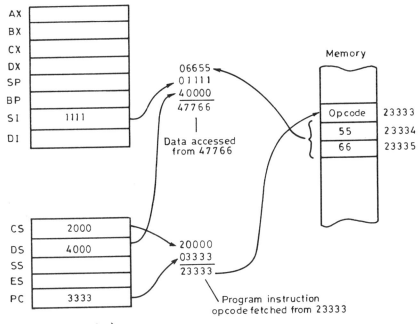

Figure 13.5 Memory address calculations (source is direct indexed addressing).

instructions and DX can be used to transfer data to input/output addresses (ports).

The stack pointer SP operates in the normal way, but the stack segment register is added to it to access the required memory segment, as described in the previous section. The base pointer BP can be used in some instructions to provide a 16-bit base to a computed memory address. The source and destination index registers can be used to add a displacement to a memory address that is applied in either the source or destination within an instruction. Figure 13.5 illustrates in detail the method applied to access a program instruction, and to access the 16-bit data item that is used within that instruction, for "direct, indexed addressing". The instruction set is listed in Appendix E.

The addressing modes available are

1. Immediate
   MOV AX, 29   ; Load AX with 29

2. Register
   MOV AX, BX   ; Load AX with contents of BX
3. Direct
   MOV AX, 1040H   ; Load AX with contents of memory address at offset 1040
4. Indirect, with base register
   MOV AX, (BX)   ; Load AX with contents of memory address held in BX
5. Indirect, with index register
   MOV AX, (SI)   ; Load AX with contents of memory address held in SI
6. Indirect, with base register plus index register
   MOV AX, (BX) (SI)   ; Load AX with contents of
   or BP or DI   memory address computed by summing BX and SI
7. Indirect, with base register or index register, plus offset
   MOV AX, 4000H (BP)   ; Load AX with contents of memory address computed by summing BP and 4000
8. Indirect, with base register plus index register plus offset
   MOV AX, 2100H (BX) (SI) ;Load AX with contents of memory address computed by summing BX, SI and 2100
9. Relative
   JMP -23   ; Jump (unconditional) to PC-23, i.e. back 23 bytes

Additional normal absolute address jump/call instructions are available, e.g.

   JMP 6500H   ; Jump (unconditional) to 6500

Naturally AX in these examples could be replaced with any of the other seven general registers, and the source and destination could be interchanged.

Whilst Table 1 in Appendix E lists the main instructions that are available, the reader is referred to the bibliography at the end of the chapter for a comprehensive list of the instruction set for the 8086 and 8088. The following short program examples illustrate simple input/output operations and the application of a multiply instruction:

Program Example 1.

MOV AX, 1000H   ; Set data segment
MOV DS, AX   ;   to 1000

Advanced Microprocessor Architectures

```
IN AL,2           ; Input from port address 2
MOV 0000H, AX     ; Store in data segment (offset 0000) — memory
                    address 10000
OUT 3,AL          ; Output to port address 3
HLT               ; Halt
```

The program reads in a byte from an input port into register AL, stores that value in memory and then sends that byte to an output port.

An example of a program written for an 8086/8088 assembler is as follows

Program Example 2.

```
; Program multiplies two 16-bit values and stores
; answer in 32-bit form in memory
DATA SEGMENT
   NUM1 DW 1234H
   NUM2 DW 4321H
   ANS1 DW ?
   ANS2 DW ?
DATA ENDS
CODE SEGMENT
   ORG 0000H
   ASSUME CS:CODE, DS:DATA    ; Pseude, to set up segments
   MOV AX, DATA               ; Set up
   MOV DS, AX                 ;    data segment
   MOV AX, NUM1               ; Fetch one number
   MUL NUM2                   ; Multiply by other number
   MOV ANS1, AX               ; Store least significant word
                                of answer

   HLT                        ; Halt
CODE ENDS
END
```

Memory connections

Standard 8-bit memory devices, e.g., 2316 ROM, 2716 EPROM and 6116 static RAM, are connected to the 8088 as if it is a normal 8-bit microprocessor. The only requirement is that the address and data buses must be demultiplexed, as for the 8085. However, the 8086 possesses a full 16-bit external data bus, and although it is a 16-bit word device it addresses memory in byte form. Intel could have designed the 8086 to use a 16-bit opcode and to lose the ability to access byte data values. This would have restricted instruction lengths to 2, 4 or 6 bytes. However, the 8086 maximizes memory usage by using byte opcodes and byte memory addressing. The memory connection arrangement is shown in Fig. 13.6. The bottom address line A0 is not taken to memory devices. Instead it is used as a bank select for lower-order bytes, whilst BHE (set by CPU) selects higher-order bytes. In this way both odd and even address memory bytes can be accessed, whilst words (two bytes) are accessed on even address boundaries (top line of truth table) and odd address boundaries.

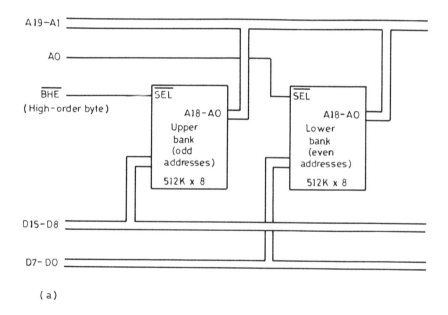

Figure 13.6 8086 memory connection. (a) Circuit arrangement (assuming demultiplexed address and data buses). (b) Truth table for BHE and A0.

Advanced Microprocessor Architectures

Supporting chips (8087/8089)

Two extremely powerful 40-pin supporting devices are offered by Intel to aid the 8086 and 8088 microprocessors in number processing applications and large circuit configurations. These devices are the 8087 numeric data processor, which offers hardware floating point arithmetic operations, and optimizes other input/output operations. Both chips act as co-processors to the CPU, and are connected to the multiplexed address/data bus as well as to selected control bus lines. The CPU must be set to "maximum" mode when these devices are included.

The 8087 monitors the instruction sequence that is being fetched and obeyed by the CPU. If a reserved opcode (11011XYZ) is detected by the 8087, the CPU ignores the instruction and the 8087 executes it in place of the CPU. The 8087 may need to perform memory transfers in order to execute the instruction, and handshaking with the CPU is performed to enable it to "borrow" the CPU buses. The programmer may need to insert CPU wait instructions to ensure that the 8087 has completed its numeric computation task before the result is utilized in the program (there is no hardware indication of task completed). Numeric functions that can be performed are:

1. High-precision fixed-point arithmetic
2. Floating-point arithmetic
3. Complex mathematical functions, e.g., square root, tangent, exponentiation

The 8089 does not monitor instructions to determine if it should replace the CPU for certain instructions, like the 8087. Instead the 8089 input/output processor is triggered when the CPU sets a signal pin on the 8089, and a "mailbox" arrangement is applied to enable the CPU to specify the input/output task required. The mailbox is an area of memory that is shared by the CPU and the 8089, and the CPU (by means of a section of program) must load this memory block with both details of the required input/output transfer and the sequence of 8089 instructions. The 8089 executes this instruction block and notifies the CPU, by interrupt, when it has finished the task. The 8089 possesses nearly 50 instructions.

The 8089 input/output processor possesses two "channels", i.e., it can perform the transfer of two separate streams of data. Normally these channels are used for DMA transfers, e.g., to floppy disk and hard disk, but they can be used for a wide variety and flexible range of tasks, e.g., graphics CRT display refresh, data manipulation during block transfers, spooling messages to printers and bus width matching (byte memory for 8088, and 16-bit memory for 8086 with odd-even banks).

Circuit arrangement

A typical circuit arrangement based on an 8086 CPU is shown in Fig. 13.7. This representation shows the principal components in the Apricot computer. The 8086 is supported by an 8089 input/output

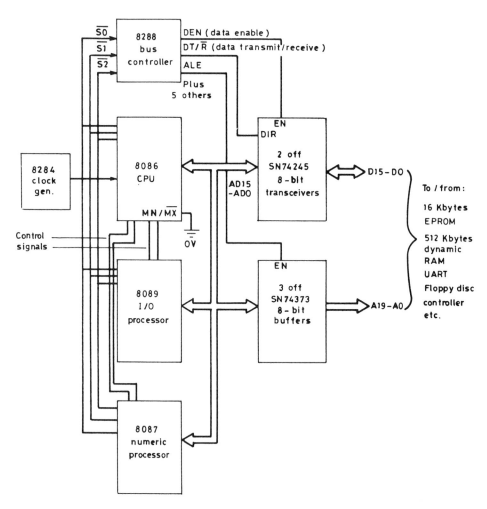

Figure 13.7 A typical 8086 system (Apricot computer).

Advanced Microprocessor Architectures

processor and an 8087 numeric processor. The CPU is applied in maximum mode, i.e., the control bus is expanded (3 to 8) using the 8288 bus controller. The address/data bus is demultiplexed using standard TTL chips, and connects to a large range of memory and input/output devices.

80186 CPU

Intel named their updates on the standard 8088 and 8086 the iAPX188 (or 80188) and the iAPX186 (or 80186) when they were launched in 1982. These devices offered some extra instructions and several on-chip facilities that reduced the need for supporting ICs. In

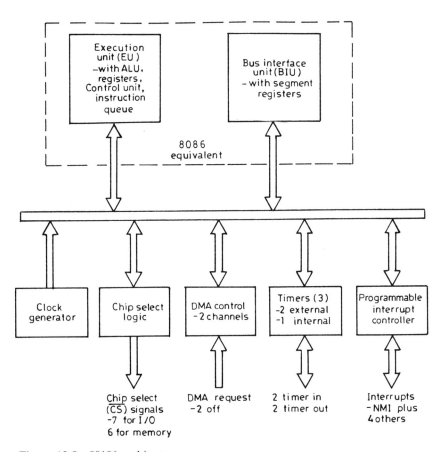

Figure 13.8 80186 architecture.

this way a microcomputer circuit requiring perhaps ten ICs can be supplied by just one of these devices.

The more popular of these devices is the 80186, which is illustrated in Fig. 13.8. The 8086 architecture is retained as the execution unit (EU) and bus interface unit (BIU) (refer to Fig. 13.2).

Additional modules are:

1. Clock generator, with programmable wait-state generation logic
2. Chip select logic, avoiding the need for external address decoding circuitry eliminating decoder propagation delays
3. DMA logic for two channels, with interrupt facilities
4. Programmable timers/event counters, two external and one internal (e.g., for real-time clock)
5. Interrupt controller, which is programmable

The chip packaging technique is the first one encountered in this book to move away from the standard DIL (dual-in-line) method. The 80186 is packaged in a 68-pin JEDEC type A "chip carrier", and this package arranges its pins on all four sides, allowing a smaller package. The device is targeted at small stand-alone applications employing a low chip-count, but it has not achieved as widespread an application as the device described in the following section.

The 80286 CPU

Intel introduced the iAPX286 (better known as the 80286) in 1983. Although this device retains compatibility with the 8086 in the same way as the 80186, it offers a radically different range of additional features. It does not offer the on-chip functions of clock generator, chip select logic, DMA control, timers and interrupt controller; instead it provides memory management and virtual memory facilities. Whilst IBM guaranteed widespread application for the 8088 in their first entry into the microcomputer market with the IBM PC (personal computer), they additionally consolidated the popularity of the 80286 by choosing it as the CPU for the updated IBM PC "AT" version.

The concept of virtual memory is that programs and data files are addressed as if they are held in memory, but in fact they could be held on backing store (disk). Thus backing store is treated as if it is an extension of main memory. In a multi-programming system therefore, the master program ("operating system") can switch from one program task to another, perhaps using "time-slicing" (programs are run in short bursts to avoid a lengthy task dominating machine time), without the

Advanced Microprocessor Architectures

need to implement disk transfers — transfers are performed automatically by the CPU hardware.

The term "memory management" refers to the process of address translation from the addresses used by programs into addresses applied to the memory hardware. The 80286 can be set to operate in either "real address" mode (as used in the 8086) or "virtual address" mode — the setting of a bit in the status register selects the mode. In the latter mode the memory management characteristic of the 80286 uses the segment registers (CS, DS, SS, ES) in a different way to that of the 8086. The contents of these registers are not interpreted by the CPU as 16-bit components of the 20-bit physical address (see Fig. 13.3), instead they are treated by the CPU as memory pointers to 48-bit segment extension registers, as shown in Fig. 13.9. When a task (program) change instruction is executed, the CPU uses the contents of the segment registers to point to memory locations which contain the 48-bit segment descriptor registers. These 48-bit values are automatically transferred to the "cache" (cache means on-chip memory) registers shown in the diagram. The program is then implemented using these 48-bit registers to provide physical memory addressing; note the "access rights", which provides features such as write-protect. If the address specified is on disk (virtual memory), then the following procedure is followed:

1. Program requests access to segment currently on disk
2. CPU checks descriptor table (transferred to cache on CPU)
3. Segment not present causes an interrupt
4. Operating system triggers CPU to transfer segment from disk
5. CPU performs DMA transfer from disk
6. Operating system changes descriptor table
7. Operating system returns to trapped instruction

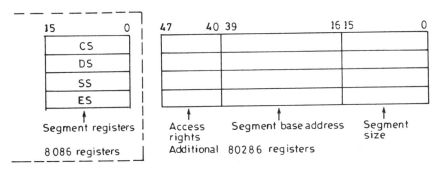

Figure 13.9 80286 segment descriptor cache registers.

A protection system is implicit when the 80286 is in virtual addressing mode, such that one program cannot modify another program, e.g., the operating system. This protection is provided by the 80286 system's four privilege levels. For example, an application program (running in the lowest privilege level) can execute a subroutine CALL instruction into the kernel of the operating system (set at the highest privilege level), but protection is provided to ensure that the operating system, although resident in the same sector of memory, cannot be over-written.

Intel offers a floating-point processor (the 80287) to support the 80286. There is no doubt that the operation of an 80286 in virtual address mode, and supported by an 80287, places such a microcomputer system into the same performance category as most 16-bit minicomputers, and possibly even beyond.

13.2.2 MOTOROLA MC68000 Family

Motorola introduced the 16-bit MC68000 family of microprocessors to succeed the MC6809 8-bit microprocessor, but the devices are not program compatible. The MC68000 series is arguably 32-bit, because the devices possess 32-bit work registers. There are three main processors in the MC68000 series:

1. MC68000: 24-bit address bus and 16-bit data bus; used in the Apple Macintosh computer
2. MC68008: 20-bit address bus and 8-bit data bus; used in the Sinclair QL computer
3. MC68010: as MC68000 with "virtual machine/memory" characteristic (i.e., extends physical main memory to include backing store); used in the Hewlett Packard 9000 computer

Pin functions

The pin functions of the MC68000 are shown in Fig. 13.10. The use of a 64-pin DIL package enables the address and data buses to be nonmultiplexed. Additionally a large number of control bus signals can be accommodated. Therefore the 8288 bus controller and demultiplexing chips required with the Intel 8086/8088 are unnecessary with the Motorola device. Additionally the MC68000 does not possess a "maximum/minimum" option which is selected by the setting of a control pin as with the Intel devices; the 64-pin package enables the "maximum" configuration to be permanently selected.

Pins UDS and LDS indicate whether data are being transferred on the most significant (upper) byte, least significant (lower) byte or both

Advanced Microprocessor Architectures

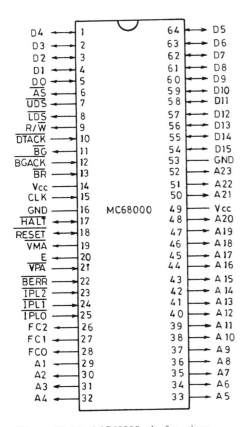

Figure 13.10 MC68000 pin functions.

bytes of the data bus. FC0, FC1 and FC2 represent the function code (type of bus activity) of the device, e.g., user program access, supervisor data memory access, etc. Seven interrupt levels are coded on IPL0, IPL1, and IPL2. Three bus arbitration signals BR, BG, and BGACK are employed when other processors or DMA controllers require to utilize the MC68000's buses. Three further signals E, VPA, and VMA are provided to enable interfacing to standard Motorola MC6800 devices, e.g., the MC6821 PIO (or PIA); the E signal is a synchronizing clock signal which synchronizes data transfers.

Process architecture

The work register block for the MC68000 series is shown in Fig. 13.11. There are seventeen 32-bit data and address registers, and a 32-

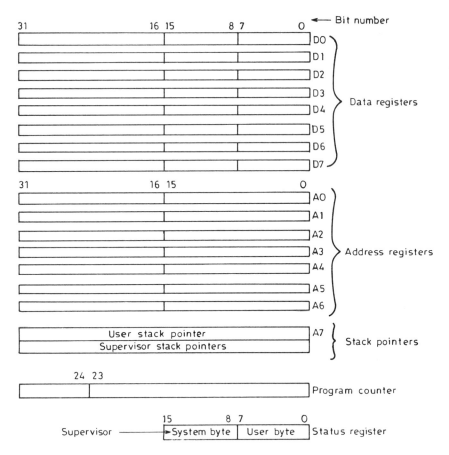

Figure 13.11 Work registers for MC 68000.

bit program counter (only 24 bits are used). The eight data registers can handle data values as bytes, words (16 bits) or long words (32 bits). All of the registers can function as index registers.

Although 32-bit registers are applied, the ALU is only 16 bits wide. The MC68000 utilizes a "pipeline" system in a similar but more limited way to the 8086/8088; it fetches an instruction code during the execution of the two prior instructions. The 16 Mbytes of memory are effectively expanded to 64 Mbytes by utilizing the control bus lines FC0, FC1, and FC2.

The status register is detailed in Fig. 13.12, and is divided into two bytes; the system byte and user byte. The user byte contains the normal

Advanced Microprocessor Architectures

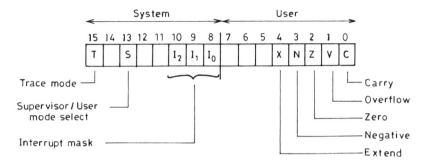

Figure 13.12 Status register.

CPU status indicator flags (the Extend bit is used in multi-precision arithmetic operations), but the device can operate in the "supervisor" or "system" mode as selected by the Supervisor status bit. In supervisor mode, certain privileged instructions can be used and a separate stack pointer is applied. There is no similar system with the Intel 8086/8088. The "trace" mode can be entered when in supervisor mode if the trace bit is set, such that after each instruction is executed a trap is forced. In this way a debugging program can monitor the results of the execution of that instruction.

Motorola call their interrupt system the "exception" system. There is some justification for this nomenclature because the mC68000 exception system handles a much broader range of events than the normal interrupt system in a typical microprocessor. Exceptions can be generated by external events or internal events. Externally generated exceptions are:

1. Interrupts (via IPL0, IPL1, and IPL2)
2. Reset (the RESET signal)
3. Bus errors

Internally generated interrupts are:

1. Instruction traps
2. Trace (as described above)
3. Privilege violations (when an instruction that is reserved for supervisor mode is used in user mode)
4. Illegal opcodes
5. Addressing errors (when an odd address is specified — all memory accesses must be on even address boundaries)

The MC68000 processes 256 exceptions, and the start addresses of the programs for these 256 events are contained in the Exception Vector Table, which is shown in Fig. 13.13. The table occupies the first 1024 bytes in the memory address range, and each 4 bytes contain the 32-bit address (called "vector") which is loaded into the program counter as part of the exception processing sequence. The only variation is that the Reset exception uses two vectors — one to load the program counter and the other to load the supervisor stack pointer. Notice that vectors 64 to 255 allow for the interrupting device (several of which may be daisy-chained on one of the seven interrupt levels) to generate part of the exception vector, e.g., the interrupting device places 8 bits on the data bus lines D7 to D0. The sequence of operations that occurs when an external interrupt signal is set (assuming that its priority is higher than that contained in the interrupt mask) is:

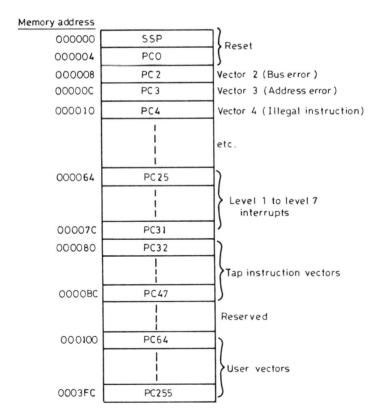

Figure 13.13 Exception vector table.

Advanced Microprocessor Architectures

1. The contents of the status register are stored in an internal register, supervisor mode is entered (S bit is set) and trace is disabled (T bit is cleared).
2. The interrupt mask in the status register is updated with the new priority setting.
3. The exception vector byte is read in (on D7 to D0) from the interrupting device — this byte is multiplied by 4 to generate the full interrupt vector.
4. The program counter and status register are stored on the supervisor stack.
5. The program counter is loaded with the four bytes from the appropriate location in the exception vector table, and instruction execution commences at the start address of the interrupt servicing program.

Instruction set

Within each instruction the opcode is word (16 bits) length; 3 bits within the 16-bit word are used to specify register number (when relevant) and a further 3 bits specify the addressing mode (when relevant). There are only 56 mnemonics (see Appendix E) in the instruction set, and this makes the instruction set an attractively simple one to master. There are no input/output instructions so input/output must be memory mapped. Both signed and unsigned multiply and divide instructions are available.

The length of an instruction varies from 1 to 5 words (2 to 10 bytes) depending on which addressing mode is being used and which data length is required (byte, word or long word). The 14 different addressing modes can be grouped into 6 basic types, as follows:

1. Direct register
   MOVE.W A4, D2   ; Load D2 with contents of register A4
   (Note: W denotes Word, B denotes Byte, L denotes Long Word)
2. Direct memory
   MOVE.W $500200, D2   ; Load D2 with contents of memory location hex. 500200
3. Indirect
   MOVE.W (Al), D2   ; Load D2 with contents of memory address held in A1
   Possible variations of this mode are auto increment/decrement, indexing and displacement.

4. Immediate
   MOVE.W 3, D2   ; Load D2 with 3 (using 16-bit word length)
5. Implied
   PEA A6   ; Push effective address (contents of A6) to stack
6. Program counter relative
   BNE $07   ; Branch (jump) if not equal forward 7 bytes

Once again indexing and displacement can be applied to this mode. Additionally absolute address jump addressing is available, e.g.

   JMP $200000   ; Jump to address hex.200000

Notice that the source and destination are specified in this assembly language mnemonic form in the reverse order to that applied for the Intel 8088/8086 instructions.

The following program examples demonstrate use of the instruction set.

Program Example 3.

```
MOVE.W #5,D3          ; Move 5 (16-bit word) into data register 3
ADD.w $1B0043,D3      ; Add word from memory address hex.1B0043
MOVE.B D3,$200000     ; Output lower byte to PIO
```

This program uses the simplified circuit configuration of Fig. 13.14, which utilizes a standard Motorola MC6821 PIO (called PIA Peripheral Interface Adapter) and illustrates the manner in which 8-bit input/output devices are applied with the 16-bit MC68000 processor. The program section adds two bytes and outputs the result to an output port.

Program Example 4.

; This program (using circuit of Fig. 13.14) initializes a UART to required baud
; rate, number of data bits and parity, then outputs 10 "Line Feed" ASCII
; characters to the transmit pin Tx. The UART (MC6850 ACIA —
; Asynchronous Communications Interface Adapter) has 2 addresses —

Advanced Microprocessor Architectures

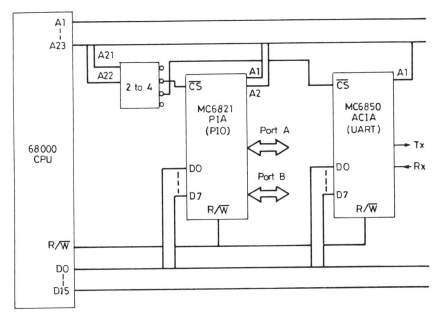

Figure 13.14 Simplified MC6800 input/output circuit.

; hex.400000 (control register output and status register input registers) and
; hex.400002 (Tx output and Rx input data transfer registers).

PROGRAM	EQA	S3000	; Start address of program
UARTCON	EQA	$400000	;Address of UART control and status registers
UARTTX	EQA	S400002	; Address of UART Tx (transmit) register
RESET	EQA	$03	; Control word for master reset
INIT	EQA	$49	; Control word for 4800 baud, 7 data bits, even parity and 1 stop bit
NUMBER	EQA	9	; Number of characters
LINEFEED	EQA	&0A	; ASCII for Line Feed
BUSYBIT	EQA	$1	; Status register transmit busy bit
	ORG	PROGRAM	

	MOVE.B #RESET,UARTCON	; Reset UART (control register)
	MOVE.B #INIT,UARTCON	; Initialize UART
	MOVE.B #NUMBER,D1	; Place number of characters in D1
TRANSMIT	BTST.B #BUSYBIT,UARTCON	; Test if UART busy
	BEQ.S TRANSMIT	; Poll until not busy
	MOVE.B #LINEFEED,UARTTX	; Output Line Feed to Tx
	DBRA D1,TRANSMIT	; Repeat loop until 10 operations
	END	

Notice that the last instruction (DBRA) decrements the data register D1 and branches until the result in D1 is -1. The # symbol denotes that immediate addressing is used.

13.2.3 Zilog Z8000 Family

The Zilog Z8000 series of 16-bit devices includes several CPUs, with differing capacities for memory support (including virtual memory), and several support devices. These support devices include powerful input/output handling chips and co-processors, e.g., floating point hardware chip.

The two principal CPUs, which are almost identical except for memory addressing capacity, are:

1. Z8001: 8 Mbytes memory (16 + 7 address lines) and 16-bit data bus in a 48-pin DIL package
2. Z8002: 64 Kbytes memory (16 address lines) and 16-bit data bus, in a 40-pin DIL package.

The Z8010 memory management unit is used in conjunction with the Z8001 in order to manage the 8 Mbyte address space (providing segment relocation as well as memory protection), whilst the Z8003 and Z8004 are CPUs that are virtually identical to the Z8001 except that they offer the facility of virtual memory.

Pin functions

Figure 13.15 shows the pin functions of the Z8001; the 40-pin Z8002 does not possess the segment number signals (SN0 to SN6) and the

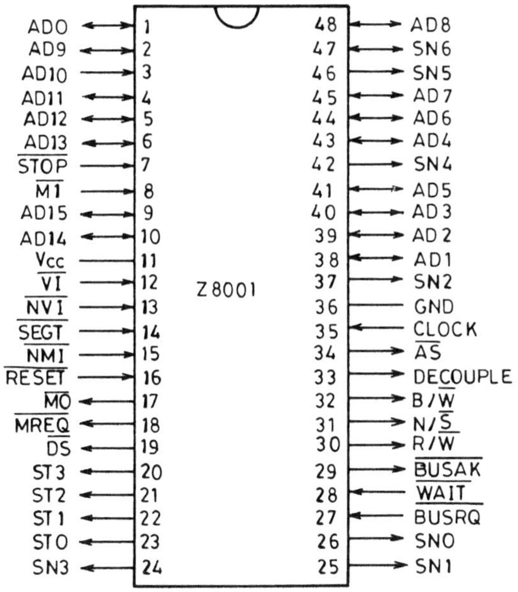

Advanced Microprocessor Architectures

```
ADO   <-> 1          48 <-> AD8
AD9   <-> 2          47 <-  SN6
AD10   -> 3          46 <-  SN5
AD11  <-> 4          45 <-> AD7
AD12  <-> 5          44 <-> AD6
AD13  <-> 6          43 <-> AD4
STOP   -> 7          42 <-  SN4
M1     -> 8          41 <-> AD5
AD15  <-> 9          40 <-> AD3
AD14  <-> 10         39 <-> AD2
Vcc    - 11          38 <-> AD1
VI     -> 12  Z8001  37  -> SN2
NVI    -> 13         36  -  GND
SEGT   -> 14         35 <-  CLOCK
NMI    -> 15         34  -> AS
RESET  -> 16         33  -> DECOUPLE
M0    <-  17         32  -> B/W
MREQ  <-  18         31  -> N/S
DS    <-  19         30  -> R/W
ST3   <-  20         29  -> BUSAK
ST2   <-  21         28 <-  WAIT
ST1   <-  22         27 <-  BUSRQ
ST0   <-  23         26  -> SN0
SN3   <-  24         25  -> SN1
```

Figure 13.15 Z8001 pin functions.

segment trap signal (SEGT). Each Z8000 CPU generates the "Z-bus", which consists of address, data, and control signals which enable data transfers between CPU and memory or input/output. All Z800 CPUs possess multiplexed address and data pins — the setting of the signals AS (Address Strobe) indicates that address information is present on these lines. Signals SN0 to SN6 on the Z8001 are memory segment number signals and they act as additional address lines — increasing binary counts on these signals switch to different 64K "segments". The four status signals ST0 to ST3 can be decoded to produce 16 discrete signals to assist in allocating separate memory spaces for programs, data and stack. This is particularly useful with the Z8002 because it allows the 64K addressing range to be expanded. Notice the following additional signals:

1. Apart from RESET there are four interrupt signals: NMI (non-maskable interrupt), NVI (nonvectored interrupt), VI (vectored interrupt) and SEGT (segmentation trap, used when the memory management unit is used with the Z8001)
2. DMA signals BUSREQ and BUSAK
3. Status signals B/W (byte/word), N/S (normal/system mode) and R/W (read/write)

4. Daisy-chained multi-micro control signals MI and MO to allow one CPU to access a shared device in a multi-processor system

Processor architecture

The work register block for the Z8001 is shown in Fig. 13.16. There are sixteen 16-bit registers, all of which can be used as accumulators and all except R0 can be used as index registers. R0 to R7 can process data in byte form as well as word form, whilst the sixteen registers can be used in pairs to offer 32-bit working. It is also possible to perform some 64-bit (4 word) working. The program counter consists of a 16-bit "offset" and a 7-bit segment number. The CPU can be operated in "system" or "normal" mode (analogous to the supervisor and user modes with the Motorola MC68000), and a bit in the status register

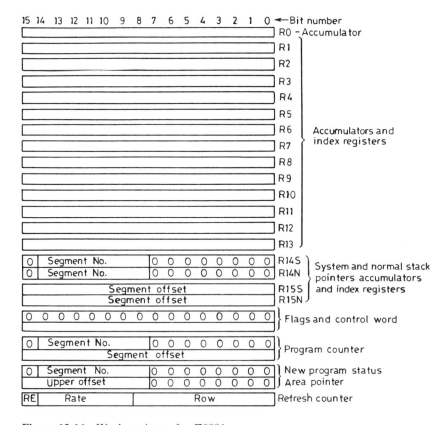

Figure 13.16 Work registers for Z8001.

Advanced Microprocessor Architectures

marks the current mode. There are two stack pointers, one for each mode, and one of the two R15 registers (plus 7 bits in each R14) is used to represent the stack pointer (S = System, N = Normal). Additional instructions are available when in system mode (as with the MC68000). In particular the input and output instructions fall into this category, so that in a multi-programming application a user program (operating in normal mode) must demand that the operating system (operating in system mode) performs input/output tasks. Zilog have maintained the use of a refresh counter (to refresh dynamic RAM) in line with the operation of the 8-bit Z80.

There is no pipeline system with an on-chip instruction queue, as employed by the Intel 8086/8088, but the Z8000 CPUs employ an early instruction decode system. This allows decoding to occur independently of the addressing mode chosen.

Memory addressing is arranged in bytes, but A0 is utilized as with the Intel 8086, i.e., to select one 8-bit bank in lower (even) or upper (odd) address banks. Each opcode is word length, and opcode and data words always start at an even memory address.

Instruction set

There are nine addressing modes, as follows:

1. Register
   LD R4,R7   ; Load R4 with content of R7 (register 7)
2. Immediate
   LD R4,5   ; Load 16-bit word 5 into R4
3. Indirect register
   LD R4,(R2)   ; Load R4 with contents of memory location held in R2
4. Direct address
   LD R4,%1800   ; Load R4 with contents of memory location hex. 1800
5. Index
   LD R4,%4000(R1)   ; Load R4 with contents of memory location computed by adding hex. 4000 to the contents of R1
6. Base
   LD R4,(R6)(4)   ; Load R4 with contents of memory location computed by adding the displacement 4 to the contents of R6

7. Base indexed
   LD R4,(R5)(R12)   ; Load R4 with contents of memory location computed by adding the index value in R12 to the base address in R5

Note: In all these examples, except (2), the source and destination can be interchanged, of course.

8. Relative
   JR + 17   ; Jump (unconditional) to PC + 17

Normal absolute address jump/branch addressing mode is available, e.g.,

   JP %0400   ; Jump (unconditional) to hex. 0400

9. Implied
   LDCTLB FLGR, %31   ; Load 31 into control byte register

The instruction set is summarized in Appendix E, and it uses similar mnemonics to the Z80, as the following Z8001 program example shows

Program Example 5.

   LD R1, #%643D   ! Load 643D into R1 !
   LD R2, ⟨⟨3⟩⟩%5000   ! Load R2 with contents of memory location offset 5000, segment 3 !
   ADD R2,R1   ! Add R1 to 2 !
   OUT 3,RL2   ! Output lower half of R2 to I/O port address 3 !

This program adds a constant to the contents of a memory location, and sends one half of the result to an output port.

Memory management and virtual memory

Zilog offer an extremely useful device to support the Z8001 CPU in order to manage its 8 Mbyte memory addressing space. This is the Z8010 memory management unit, as illustrated in Fig. 13.17. The device converts the 23-bit "logical" address set by the CPU into a 24-bit "physical" address that is applied to the memory circuit. A conversion is performed in the MMU (memory management unit), such that the seven segment registers (SN6 to SN0) are used in a look-up table to generate a 16-bit block identifier (each block is 256 bytes). The low-order 8 address lines bypass the MMU. The physical addresses therefore can access 64 segments in memory, and each segment can have from

Advanced Microprocessor Architectures

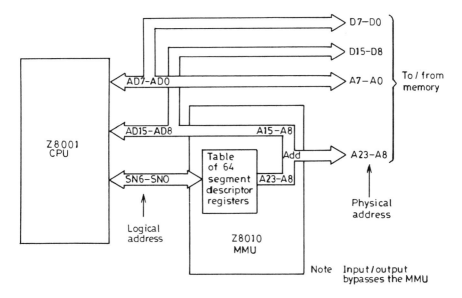

Figure 13.17 Z8010 memory management unit.

256 to 64 Kbytes. The value of this arrangement is:

1. The amount of physical memory in a system may be smaller than the required logical memory for all the programs. The MMU can process two different programs with the same logical address ranges, and route them to different areas of physical memory. A program can therefore be relocated anywhere in physical memory by the MMU. An operating system can re-schedule the physical memory for a program to enable it to be run in a free area of memory in a multi-programming situation.
2. Several attributes can be assigned to each segment within the MMU to provide memory protection, e.g., read only, DMA accesses inhibited, system mode only.

A memory management system gives provision for one more feature that is helpful in multi-programming situations. "Virtual memory" is an arrangement in which backing store is treated as if it is an extension of main memory; it is a common feature with mainframe computers and minicomputers. The operating system, memory management unit and disk combine to make physical addressable memory appear larger to users' programs. A flag can be set in the memory management unit for a particular program/segment such that the CPU is notified (by interrupt) and the MMU is re-programmed to transfer an additional segment from disk into memory. The MMU may have to make space in memory to accommodate this additional segment (by temporarily dumping another segment to disk). This memory manipulation is transparent to the user program. Zilog offer the Z8003 and Z8004 CPUs to replace the Z8001 and Z8002, respectively, so that virtual memory operations can be implemented.

Supporting chips

The Z80 8-bit support chips (PIO, SIO, CTC) can be utilized with Z8000 microprocessors if the address/data bus is de-multiplexed. However, Zilog offer a new range of supporting chips to their 16-bit CPUs. These devices are available in two series:

1. Z-BUS peripherals, which connect to the Z8000 CPU buses, including the multiplexed address/data bus—identified by the number Z80XX
2. Universal peripherals, which connect to conventional non-multiplexed non-Zilog CPUs—identified by the number Z85XX

Most devices in these two ranges are compatible in function, as Table 13.1 shows.

Whilst the Z8X31 is basically a dual-channel UART, the Z8X30 offers asynchronous and synchronous modes as well as byte and bit protocols. The latter enables the bit-oriented protocols such as IBM SDLC (synchronous data link control) and HDLC (high-level data link control), which are more sophisticated than the simple asynchronous RS232-C links, to be implemented. The Z8X36 offers 3 counter/timers (16-bit) and 2 ports, whilst the Z8X90 offers 2 counter/timers, 2 ports

Table 13.1 Zilog Z8000 peripherals

Z-BUS	Universal	Function
Z8016	Z8516	DMA controller
Z8030	Z8530	Serial communications controller
Z8031	Z8531	Asynchronous serial communications controller
Z8036	Z8536	Counter/timer and parallel I/O unit
Z8038	Z8538	FIFO input/output interface unit
Z8060	Z8569	FIFO buffer unit and Z-FIO expander
Z8068	–	Data ciphering processor
Z8070	–	Floating point unit
Z8090/4	Z8590/4	Universal peripheral controller

Advanced Microprocessor Architectures

and interrupt handling facilities. The FIFO devices facilitate buffering between CPUs (or between a CPU and a peripheral device).

The Z8070 floating point unit (FPU) is an example of what Zilog term their "extended processing architecture" the FPU is described as an EPU (extended processing unit). It connects to the CPU buses and performs floating point arithmetic operations while operating in parallel with the CPU. It monitors the same instruction stream as the CPU, and identifies and executes those instructions intended for it. Although the FPU operates internally using an 80-bit floating point format, data transfer between registers in the FPU and CPU is in byte or 16-bit form.

Zilog employ their standard "daisy-chain" system to enable several peripheral devices to share the same interrupt signal, as shown in Fig. 13.18. In this circuit example, three standard Z-BUS peripheral devices are included, and each receives the least significant half of the multiplexed address/data bus. The devices share the same interrupt signal VI (vectored interrupt), such that each device must generate its own identifying 8-bit code, which is used to establish the interrupt vector, when it requires to interrupt the CPU. The hardware priority system is established using the IEI (interrupt enable in) and IEO (interrupt enable out) daisy-chain. This interrupt system is unnecessary, of course, if the devices are to be polled only.

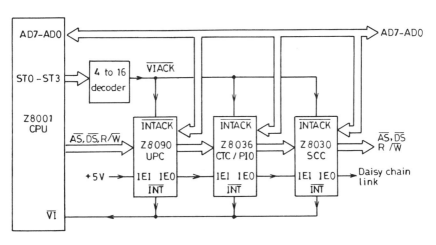

Figure 13.18 A simplified Zilog Z8000 input/output interrupt system.

13.2.4 Comparisons and Evaluation

In this world of increasing variety of 16-bit microprocessor chips, the task of selecting a particular chip is a difficult one. Apart from hardware and architectural details which we have discussed so far, it becomes essential to evaluate system performance at multiple levels (Toong and Gupta, 1982) as follows:

1. *Instruction level*: At this level one compares timings on an instruction-by-instruction basis. This level is especially relevant when the chip is not physically available such as is the case of newly announced or newly released processors.
2. *Routine level*: Sample user tasks coded in assembly language are often represented as competitive analyses between processors by manufacturers. The application programs frequently reflect advantages unique to the "winning" microprocessor.
3. *Support task level*: The sophistication of operating systems support is evaluated. An example of a support task would be a multi-level queue management used by the process scheduling algorithm of the operating system.
4. *System functions level*: Specific operating system and user tasks are encoded, e.g., memory management algorithms for segmentation.
5. *User job level*: This benchmark spawns many processes, thereby exercising the operating system and all lower level benchmarks.

Sixteen-bit microprocessors have been compared and evaluated by Heering (1980), Grappel and Hemenway (1980), Rajulu and Rajaraman (1982), Toong and Gupta (1982) and Prycher (1983). Table 13.2, based on information from Toong and Gupta (1981), Orlando and Anderson (1981), compares the performance at the instruction level. Comparison at higher levels of the evaluation hierarchy are summarized in Table 13.3 based on the same clock frequency.

It is readily seen that the performance is heavily dependent on the choice of algorithm. One would expect that at higher levels comparison is even more application dependent. All general comparisons must be used with extreme caution.

13.3 32-BIT MICROPROCESSORS

13.3.1 Introduction

Following the success of 16-bit microprocessors throughout the 1980s it was inevitable that the principal manufacturers should extend the

Table 13.2 Execution speeds (in microseconds) of 16-bit microprocessors (note that different chips use different clock frequencies)

Operation	Data type	TI9900 at 3 MHz	Intel 8086 at 5 MHz	Zilog Z8000 at 5 MHz	Motorola MC68000 at 8 MHz	National NS16032 at 10 MHz	TI99110 at 6 MHz
Register-to-register move	Byte/word	4.60	0.40	0.75	0.50	0.30	0.50
	Double-word	9.80	0.80	1.25	0.50	0.30	1.00
Memory-to-register move	Byte/word	7.30	3.40	3.50	1.50	1.00	0.83/0.67
	Double-word	14.60	6.80	4.25	2.00	1.50	1.33
Memory-to-memory move	Byte/word	9.90	7.00	7.00	2.50	1.70	1.00/0.83
	Double-word	19.80	14.00	8.50	3.75	2.50	1.67
Add memory to register	Byte/word	7.32	3.60	3.75	1.50	1.20	0.83
	Double-word	21.30	7.20	5.25	2.25	1.60	2.00
Compare memory to memory	Byte/word	9.90	7.00	7.25	3.00	1.70	1.00
	Double-word	19.80	14.00	9.50	4.00	2.50	2.00
Multiply	Byte	21.90	13.00	20.25	N/A	3.50	4.17
	Word	21.90	23.00	16.00	8.75	5.10	4.17
memory-to-memory	Double-	180.64	115.20	85.75	43.00	8.30	26.38
Conditional Branch	Branch taken	3.60	1.60	1.50	1.25	1.60	0.50
	Branch not taken	2.90	0.80	1.50	1.00	0.80	0.50
Modify index branch if zero	Branch taken	7.60	2.20	2.75	1.25	1.30	1.00
Branch to sub-routine		7.90	3.80	3.75	2.25	2.00	1.00

Table 13.3 Comparison of performance of 16-bit microprocessors (n, in the case of the hash algorithm, denotes the number of searches before an open slot is found)

Level	Example used	Performance Intel 8086	Z8000	MC68000
Routine level	(a) Booth's multiplication algorithm	1.00	1.41	1.26
	(b) Polynomial evaluation	1.00	1.80	1.97
Support task level	(a) Stack exerciser	1.00	8.42	3.49
	(b) Hash algorithm	144 + 48*n	188 + 66*n	141 + 54*n

evolutionary process into 32-bit devices. The principal advantages offered by 32-bit devices over their 16-bit predecessors are:

1. 32-bit data manipulation — larger integer numbers
2. Larger memory addressing range — normally 4 gigabytes (2^{32})
3. Faster operation — clock speeds of 16 MHz or more
4. Additional instructions and addressing modes — upwards compatible with their predecessors
5. Intrinsic memory management features
6. Instruction cache — on-chip memory holding most frequently used instructions and data items
7. Approximately 2 to 3 times improvement in processing speed for standard programming benchmark tests.

Manufacturers claim that 32-bit microprocessors rival traditional minicomputers, e.g., the DEC VAX machines, in computing power. Although this is debatable (in terms of instruction speed, co-processor, e.g., floating-point, speed and complex operating system support), 32-bit microcomputer systems are achieving widespread application in the field of engineering workstations, speech recognition, robotic systems, office automation, and large multi-user and multi-processing situations.

The four 32-bit devices that have been developed, and are described in later sections, are:

1. Intel 80386
2. Motorola MC68020
3. Zilog Z8000
4. Inmos T424 transputer (plus several other devices)

Advanced Microprocessor Architectures

Whilst the first three devices represent natural progressions from their 16-bit counterparts, and use conventional CPU architectures, the Inmos transputer is a completely novel approach to machine architecture. It is an example of a RISC (reduced instruction set computer) compared with the traditional CISC (complex instruction set computer), and is designed to operate in a multi-processor configuration, i.e., several transputer CPUs execute a program task in parallel. The concept of RISC represents an attempt to diverge from the evolutionary manner in which CPU design has become increasingly complex with increasing instruction set features. Several research organizations and universities have attempted to produce CPUs with far smaller numbers of instructions offering increased performance and speed of operation. The essential features of a true RISC processor are single-cycle operation (multiple memory transfers are to be avoided) and hardwired control (instruction execution is implemented using fast hardwired logic rather than microcode — microprocessors employ the slow "microcode" technique of a look-up process to determine CPU operations to implement each instruction). Only the Acorn ARM (Acorn RISC machine) and (arguably) the Inmos transputer are commercial 32-bit RISC processors currently available, but it is possible that future computer architectures might lean more to this approach to achieve higher performance.

The 80386, MC68020, and Z80000 32-bit processors employ instruction cache and memory management facilities, and both features deserve more detailed descriptions here. A cache memory is a high-speed memory that is either contained on the CPU itself, or is placed between the CPU and main memory. Large main memory systems are invariably dynamic RAM, which is cheap but slow compared with static RAM. If the most frequently addressed instructions and data are held in fast static RAM cache memory, program execution speeds can be enhanced. Most programs tend to re-access the same memory addresses, and the cache memory holds the contents of these addresses, together with the address itself. When program execution demands the contents of one of these addresses, e.g., the reading of a program instruction, the cache performs a high-speed comparison to determine if the "tag" address requested by the CPU matches one of the stored items within the cache. If it is, a "hit" occurs, and the instruction can be read from cache instead of being transferred from memory with the associated inherent time delay. The hit rate must be high (typically greater than 80%) to make the cache system worthwhile. Typical cache sizes are 4 Kbytes; the larger the cache size, the higher is the hit rate.

The second advanced feature that 32-bit processors employ is memory management. Memory management is applied to allocate different areas of memory to different programs (and data areas) as efficiently as possible, and also to provide access protection to these programs. Once again memory management can be built onto the CPU chip itself, or it may require an additional component. An MMU (memory management unit) translates the "logical" memory address generated by the CPU into a "physical" address that is applied to the memory circuit. Thus, for example, an operating system can transfer control from one program to another, whilst both programs may share the same logical address range but are located separately in physical memory. Additionally the MMU can provide protection of program or data, e.g., read-only and privilege levels.

Thirty two bit microprocessors give provision for co-processors, the most common of which is a floating-point arithmetic processor. All such floating-point co-processors adhere to the IEEE P754 standard, namely 80-bit double extended precision.

Fabrication technologies are either NMOS or CMOS, and each 32-bit microprocessor possesses between 200,000 and 300,000 transistors. Numbers of interconnection pins are far too great, e.g., 84-pin, to enable the conventional DIL package to be used, and so chip carrier packaging (four sides of interconnection pins) is used.

13.3.2 Intel 80386

The Intel 80386 represents the state of the art in high-performance, 32-bit microprocessors. It features absolute object code compatibility with previous members of the iAPX 86 family of microprocessors, including the 80286, 80186, 80188, 8086, and 8088. This protects major investments in application and operating systems software developed for the iAPX 86 family, while offering a significant enhancement in performance. The 80386's architecture and performance should allow it to be used in a wide range of demanding applications, e.g., in engineering workstations, office systems, robotic and control systems, and expert systems.

The 80386 implements a full 32-bit architecture with a 32-bit-wide internal data path including registers, ALU, and internal buses; it provides 32-bit instructions, addressing capability, and data types, and a 32-bit external bus interface. It extends the iAPX 86 family architecture with additional instructions, addressing modes, and data types. It incorporates a complete memory management unit. The 80386 extends the 80286 segmentation model to support four-gigabyte segments and to provide a standard two-level paging mechanism for physical memory management. System designers can use segmentation or paging or both,

Advanced Microprocessor Architectures

without performance penalties, to meet their memory management requirements.

The 80386 architecture is complemented by a bus interface that uses only two clocks per bus cycle; this allows efficient interfacing to high-speed as well as low-speed memory systems. At 16 MHz, the bus can sustain a 32-megabyte-per-second transfer rate. Other bus features include dynamic bus sizing to support mixed 16/32-bit port interfacing and a dynamically selectable pipelined mode to facilitate high-speed memory interleaving and allow longer access times.

Base architecture

Different microprocessor applications require different types of architectural support. Some applications such as those running under Berkeley UNIX, may prefer a linear address space. Others that manage a multitude of dynamic data structures may require hardware-enforced rules to protect the visibility of the dynamically created objects. The 80386 architecture supports these diverse requirements by providing the user with several memory management and addressing models. Further, its repertoire of addressing modes, data types, instructions, and special constructs make it well suited to modern high-level languages.

The base architecture of the 80386 encompasses the register model, data types, addressing modes, and instruction set. It forms the basis for high-level-language compiler code generation and for assembly-language-level application programming.

The CPU architecture is similar to the 8086/8088 model, and Fig. 13.19 illustrates the 80386 register set.

There are eight 32-bit work registers, which can be used to perform calculations and to generate memory addresses. The six segment registers are applied to separate programs, data areas and the stack into different memory segments. Data types that can be processed are integer (two's complement: 8-bit, 16-bit, and 32-bit operations can be implemented), ordinal (unsigned integers: again 8-bit, 16-bit, and 32-bit), BCD, string, bit, and floating-point (if a floating-point co-processor is added). The exponential increase in integer number range achieved by a 32-bit processor compared with its 16-bit and 8-bit predecessors is illustrated in Table 13.4.

Some 64-bit operations, e.g., shift, are possible with the 80386. A wide range of addressing modes is available, including those offered on the 8088/8086, and several modes can be combined within a single instruction. An address can be computed within an instruction as follows:

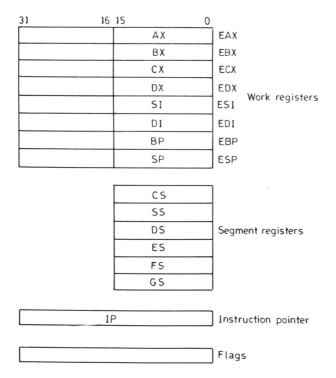

Figure 13.19 80386 register set.

(Base register) + (Index register) * (Scale) + (Displacement)
  Any register   Any register   1, 2, 4, or 8   Two's complement
                 (except ESP)

The manner in which the 8086 uses its segment registers can be emulated by the 80386 if a bit in the Flags register is set. However, the on-chip memory management features of the 80386 provide far more powerful operating system support using memory address translation

Table 13.4 Integer number range of 8-, 16- and 32-bit CPUs

CPU	Integer range
8-bit	-128 to $+127$ (-2^7 to $+2^7 - 1$)
16-bit	$-32,768$ to $+32,767$ (-2^{15} to $+2^{15} - 1$)
32-bit	$-2,147,483,648$ to $+2,147,483,647$ (-2^{31} to $2^{31} - 1$)

Advanced Microprocessor Architectures

and protection. The 80386 converts a logical address that is applied within a program instruction to a physical address, that appears on the pins of the device, using a system of on-chip tables. Firstly a base register is applied to select an entry in a "segment descriptor table" to generate a linear address, and secondly another base register is applied to select an entry in a "page table directory" to convert the linear address to the physical address. Information held in these tables also provides facilities for protection, e.g., read-only. To assist the page translation, the 80386 possesses an on-chip "translation lookaside buffer", which contains the 32 most recent conversions from linear address to physical address.

Both the 80287 and the 80387 floating-point co-processors can support the 80386. Additionally, Intel have introduced the 82258 advanced DMA chip in order to transfer data directly between main memory RAM and peripherals, e.g., floppy disk, and the 82586 local area network co-processor.

Data types

The 80386 directly supports the fundamental data types found in most high-level languages. The basic operations provided by the 80386 for each of these data types are shown in Table 13.5. Most of these operations execute in two clocks when register or immediate operands

Table 13.5 Data types supported by the 80386 instruction set (floating point is available when a numeric co-processor is added)

Operation	Data type					
	Ordinal	Integer	BCD	Floating point	String	Bit string

Move
  to/from memory
  convert precision
	×	×	×	×	×	×

Arithmetics
  add, subtract,
  multiply, divide,
  negate
	×	×	×	×		

Logicals
  AND,OR,XOR,
  shift
	×	×				

Compare	×	×	×	×	×	×
Transcendentals				×		

are used. Furthermore, because of pipelining and the two-clock memory bus, stores to memory also execute in two clocks.

The basic unit of storage is a byte; a 16-bit quantity is a word, and a 32-bit quantity is a double word, or d-word. Words are divined as having a length of 16 bits so that notational compatibility with the other members of the iAPX 86 processor family will be retained. In the 80386, most data types are represented in the form of bytes, words, or d-words, or combinations thereof.

Words comprise two consecutive bytes in memory, with the low-order byte at the lower-numbered address. D-words comprise four consecutive bytes in memory, with the low-order byte at the lowest address and the high-order byte at the highest address. The address of a word or d-word is the address of the low-order byte. Hence, the 80386 utilizes the little-endian storage scheme.

Ordinal. An ordinal is an unsigned number. If it is in the range 0 through 4,294,967,295, it corresponds to a d-word value. If it has a magnitude of less than zero, it corresponds to a word or byte value. An example of an ordinal operation is the instruction sequence

```
MUL EBX, vec[EDX*4]    ; EBX: = EBX * vec[EDX]
INTO                   ; Generate an exception if
                       ; overflow
```

Here, the content of EBX is multiplied by the EDXth element of the d-word-sized ordinal array vec, and the product is stored in EBX. An overflow exception is generated if the product exceeds 4,294,967,295.

Integer. An integer is a signed number in the range $-2,147,483,648$ through $+2,147,483,647$. As with ordinals, d-word, word, and byte integers are supported. Integers are represented in two's-complement notation. This allows a common set of instructions for addition and subtraction. For example,

```
SUB ESP.5
```

subtracts five from ESP whether ESP stores an integer or an ordinal. The settings of the overflow, sign, zero, and carry flags allow a program to determine whether a signed or an unsigned overflow has occurred. However, special instructions are provided for determining overflow in multiply and divide operations involving integers, since an integer multiply has its own rules for overflow and an integer divide produces its own unique bit patterns.

Advanced Microprocessor Architectures

Implementation

The 80386 is organized as eight logical units, with each unit assigned a task or step in the fetching and execution of each instruction. This arrangement allows as much parallel execution of the instruction stream as possible. The units are pipelined and, for the most part, operate autonomously. The units and their interconnections are shown in Fig. 13.20 in the functional block diagram of the 80386.

The eight units are the bus interface unit, the prefetch unit, the instruction decode unit, the control unit, the data unit, the protection test unit, the segmentation unit, and the paging unit. The control, data, and protection test units comprise the execution section of the CPU.

The bus interface unit interfaces the CPU to the external system bus and controls all address, data, and control signals to and from the CPU. The prefetch unit is responsible for fetching instructions from memory.

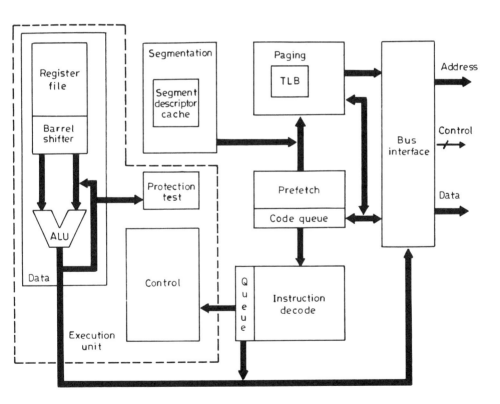

Figure 13.20 Block diagram of 80386.

It uses an advance-instruction-fetch pointer to prefetch code from memory and store it in a temporary code queue. This queue also acts as a buffer between the prefetch unit and the instruction decode unit. Since addresses generated by the prefetch unit are linear, they must be translated to physical addresses by the paging unit before the prefetch bus cycle request can be sent to the bus interface unit.

The instruction decode unit prepares and decodes instructions for immediate execution by the execution unit. It does this by fetching bytes of code from the prefetcher's code queue, transforming them into a fully decoded instruction, and then storing that instruction in a three-level decoded instruction queue. The execution unit then operates on the decoded instruction, performing the steps needed to execute it.

Instructions requiring memory references send their requests to the segmentation unit for logical address computation and translation and segment protection violation checking. The segmentation unit produces a translated linear address which the paging unit then translates into a physical address. The paging unit also checks for paging violations before it sends a bus request and the address to the bus interface unit and external bus.

Pipelining and parallelism. Advanced microprocessors are normally pipelined by overlapping the fetching, decoding, and execution of instructions. In the 80386 microarchitecture, however, the operations of all eight of the logical units are overlapped. This allows the parallel and autonomous operation of the units. They can simultaneously operate on different instructions, thereby significantly boosting the overall instruction processing rate of the CPU. For example, while the bus interface unit is completing a data write cycle for one instruction, the instruction unit can be decoding another, and the execution unit processing a third.

In practice, typical 80386 systems for engineering applications may comprise:

1. 80386 CPU
2. 80287 or 80387 numeric co-processor (floating-point)
3. Memory cache system (e.g., 16 Kbytes static RAM)
4. Main memory (e.g., 4 Mbytes dynamic RAM)
5. 82258 advanced DMA
6. 8272 floppy disk control
7. 82062 fixed disk control
8. 82786 graphics co-processor
9. 8259A interrupt controller
10. 82530 serial control
11. 82586 LAN control

Backplane connections for multi-board systems follow the Intel "Multibus" standard.

13.3.3 Motorola MC68020

The MC68020 represents the first successful extension of a 16-bit microprocessor into the 32-bit world; it provides a full 32-bit data and address bus as well as a 32-bit internal architecture. A natural and elegant 32-bit implementation of the M68000 architecture, it represents the state of the art in high-performance microprocessor design. It is designed to execute all user object code written for previous members of the M68000 family of processors and to execute those programs much faster. Because of the value of the large software base available for M68000 processors, the MC68020 maintains compatibility with previous members of the family.

Although all MC68000 family members have 32-bit user register, operands, and internal registers, the MC68020 adds full 32-bit data paths (internal and external), two 32-bit internal address paths, a 32-bit execution unit, three 32-bit arithmetic units, and an on-board instruction cache. The design goal was to provide a four- to six-fold performance improvement for the 16-MHz MC68020 over the 8-MHz MC68000. This performance increase is realized by a combination of the higher clock frequency at which the MC68020 operates, the 32-bit data bus width, the presence of the on-chip instruction cache, and the addressing mode enhancements.

Register set

The MC68020 offers more addressing modes, a small on-chip cache and external memory management support. It is mounted in a 114-pin grid array package. Unlike the 16-bit family, the MC68020 possesses a small instruction cache and instruction pipeline system within the processor in order to speed up program execution times. The register set is similar, except that several additional registers are employed to handle the more powerful features that give multi-tasking operating system support. Figure 13.21 details the CPU register set, and the close similarity of this set to that of the MC68000 (see Fig. 13.11) is highlighted. Bit, BCD, byte, word (16-bits), long-word (32-bits), and quad word (64-bits) operations are available. The eight data registers act as accumulators, and the seven address registers can be used for base or indexed addressing. Only one of the three stack pointers can be active at one time; the setting of status bits in the status register determines which is active. The supervisor/user mode system is retained in the

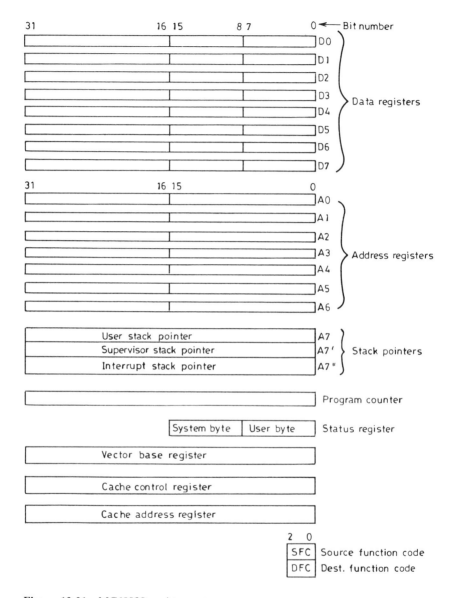

Figure 13.21 MC68020 register set.

Advanced Microprocessor Architectures

MC68020. For example, if the CPU is operating in supervisor (operating system) mode, as indicated by the setting of the appropriate status bits, and an interrupt occurs, the contents of the program counter, status register, and exception vector offset are stored on the master stack automatically.

The vector base register is used to relocate the 1 Kbyte exception vector table anywhere in memory. The two cache registers are used to control the instruction cache, e.g., enable the cache, clear the cache. The cache has 64 entries, and each entry contains one long word (64 bits), which includes a 32-bit tag field. The contents of one of the two function code registers appears on the interconnecting pins and are used to indicate one of the following: user program, user data, supervisor program, supervisor data.

Instruction set enhancements

The MC68020 instruction set includes and extends the MC68010 instruction set and adds new instructions, new data types, and new addressing modes.

Data types. The M68000 family supports data types of bits, bytes, words, long words, and binary-coded decimal (BCD). Variable-width bit fields and packed BCD, quad-word (8-byte), and variable-byte-length operands are four new data types supported by the MC68020. The bit field instructions operate on any bit field from 1 to 32 bits in length regardless of its location in register or memory. Quad-word operands are necessary for long divide and multiply instructions which handle the 64-bit dividend and product, respectively. Variable-byte-length operands are provided to support the co-processor interface; co-processors can define operand lengths suitable for the application.

Addressing modes. The M68000 family supports 14 addressing modes including register direct, register indirect (with predecrement, postincrement, displacement, and indexing), absolute, immediate, program-counter-relative, and implied effective. The MC68020 provides improved support for arrays and lists by expanding the indexed addressing modes to include memory indirection, full 32-bit displacements, sign extension of displacements and index operands, index scaling, pre-indexing and postindexing, and suppression of base register or index value or both. These extensions are available for both data and program address space accesses. Table 13.6 gives the format of the new addressing modes.

Table 13.6 New addressing modes provided by the MC68020

Data Space[a]	Program Space
(bd,an,Xn · sz*scl)	(bd,PC,Xn.Sz*scl)
([bd,An],Xn · sz*scl,od)	([bd,PC],Xn · sz*scl,od)
([bd,An,Xn · sz*scl],od)	([bd,PC,Xn · sz*scl],od)

[a]bd = base displacement (0, 16, 32-bits), An = one of eight 32-bit address registers, Xn = index register, one of 16 data/address registers sizes (sz) and scaled (scl), PC = program counter, od = outer displacement (0, 16, 32-bits), [ ] = memory indirection.

Instruction extensions. The instruction set of the M68000 family required only minor modification to make it consistently 32-bit. Table 13.7 details the additions to the M68000 instruction set that were involved in this extension. The multiply and divide instructions now operate on 32-bit operands. It is now possible to multiply two 32-bit operands to form either a 32- or 64-bit product. A 64/32-bit and a 32/32-bit divide operation is also available. The displacements associated with instructions were expanded from 16 to 32 bits. This is an important capability; it allows branch and link operations to take place without restrictions across the full four-gigabyte address space. This is a significant improvement for compiler support. The move control register (MOVEC) instruction has been expanded to accommodate the new control registers in the MC68020, and the breakpoint (BKPT) instruction has been enhanced to provide more real-time program debug support.

New instructions. Over 20 new instructions have been added to provide new functionality (see Table 13.8). Although most of them will be utilized in general-purpose processing, some are more significant for special-purpose processing.

A large group of the new instructions provides manipulation of variable-width bit fields. Here, a bit field of variable length (up to 32

Table 13.7 MC68020 instruction enhancements

Instructions	Enhancements
MULS, MULU, DIVS, DIVU	Operations extented to 32-bit operands
Bcc, BRA, BSR, LINK	Displacements extended to 32 bits
MOVEC	New control registers may be accessed
BKPT	Opcode substitution supported

Advanced Microprocessor Architectures

Table 13.8 New instructions provided by the MC68020

BFCHG	Bit field change
BFCLR	Bit field clear
BFEXTS	Bit field signed extract
BFEXTU	Bit field unsigned extract
BFFFO	Bit field find first one set
BFINS	Bit field insert
BFSET	Bit field set
BFTST	Bit field test
CALLM	Call module
CAS	Compare and swap
CAS2	Compare and swap (two-operand)
CHK2	Check register against upper and lower bounds
CMP2	Compare register against upper and lower bounds
cpBcc	Co-processor branch on co-processor condition
cpDBcc	Co-processor test condition decrement, and branch
cpGEN	Co-processor general function
cpRESTORE	Co-processor restore internal state
cpSAVE	Co-processor save internal state
cpSETcc	Co-processor set according to co-processor condition
cpTRAPcc	Co-processor trap on co-processor condition
PACK	Pack BCD
RTM	Return from module
UNPK	Unpack BCD

bits), either in a data register or spanning up to five bytes in memory, may be cleared, set, complemented, extracted, inserted, scanned, or tested. These instructions are efficient at manipulating packed data and for communications and graphics applications. One feature of these instructions is that they make it possible, when dealing with a bit field in memory, to specify any 1- to 32-bit field regardless of its orientation in memory. There are no restrictions as to how that field is aligned in memory. As cited above, if a bit field spans five bytes, the processor can internally recognize that condition and make the appropriate adjustments.

On-chip cache

Caches have been used for years in large machines to increase performance without greatly increasing system cost and complexity. It is possible to design a cache that is only a fraction of the size of the main store and that yet provides a significant decrease in the average access time to the main store. This concept has been incorporated into the

MC68020; it contains a 256-byte-on-chip instruction cache which is used to obtain a significant increase in performance by reducing the number of fetches required to external memory. The reduced bus utilization by the MC68020 also increases system performance by providing more bus bandwidth for other bus masters such as DMA devices. The MC68020 only caches instruction stream accesses, to avoid stale data and other data-cache-related problems. The MC68020's cache configuration is shown in Fig. 13.22. The cache interface to the processor data paths allows complete overlap of instruction fetches with data operand accesses, and thus provides a significant increase in performance. If simultaneous instruction and data operand requests are generated by the micromachine, a hit in the instruction cache allows concurrent fetches to take place.

A cache control register is provided to allow the operating system to maintain and optimize the cache. This register is shown in Fig. 13.23.

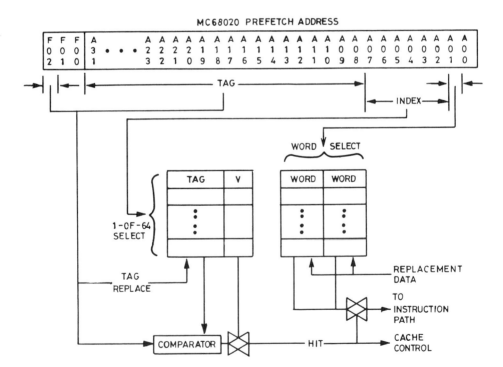

Figure 13.22 MC68020 on-chip cache.

Advanced Microprocessor Architectures

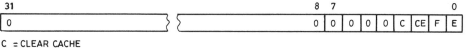

C = CLEAR CACHE
CE= CLEAR ENTRY
F = FREEZE CACHE
E = ENABLE CACHE

Figure 13.23 MC68020 cache control register.

Functional blocks

Figure 13.24 is a block diagram of the MC68020. The processor can be divided into three main sections: the bus controller, the micromachine, and the miscellaneous area. This division reflects the autonomy with which the sections operate. The bus controller comprises the address and data pads and multiplexers required to support dynamic bus sizing, a macro bus controller which schedules the bus cycles on the basis of priority, two micro bus controllers, one to control the bus cycles for operand accesses and the other to control the bus cycles for instruction accesses, and the instruction cache, with its associated control. The micromachine consist of an execution unit, ROM control stores, decode PLAs, an instruction pipe, and miscellaneous control sections. The execution unit has three sections: the instruction address section, the operand address section, and the data section. Microcode control is provided by a modified two-level store of microrom and nanorom. Decode PLAs are used to provide sequencing information. The instruction pipe and other miscellaneous control sections provide the secondary decode of instructions and generate the actual control signals that result in the decoding and interpretation of the control store.

The MC68020 possesses the feature of "dynamic bus sizing". A 32-bit instruction is transferred from four consecutive memory byte locations, which must begin with an even-number address. However, data values can be transferred in byte, word (16-bits) or long word form, and the length of the data value is specified by the memory (or input/output) device itself. This is performed by the support device signaling to the CPU through two control signals (DSACK0 and DSACK1) the size of the data item. For example, whilst RAM is invariably 32 bits wide, e.g., using four 8-bit devices, it is sometimes easier to avoid placing a program or data list into four separate EPROMs. If the EPROM is erased and re-blown, it may be preferable

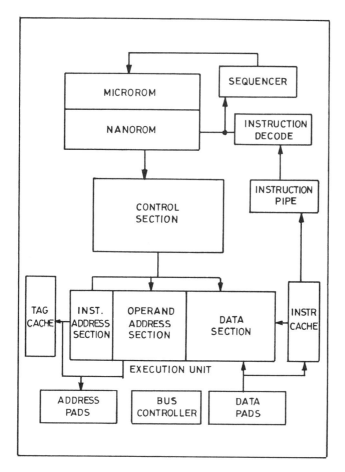

Figure 13.24 Block diagram of the MC68020.

to place the whole program into one EPROM and transfer it one byte at a time when it is executed.

The small (64-entry) on-chip instruction cache of the MC68020 inevitably produces a low "hit" rate. It is advantageous therefore to augment this with an external static RAM cache memory, as applied with 80386 systems.

The MC68020 possesses co-processor control signals to enable interfacing between itself and the MC68881 floating-point processor. The normal IEEE P754 numeric operations, e.g., 80-bit floating-point operations, are available on this device. Motorola have also designed their MC68851 paged memory management unit to act as a co-processor.

Advanced Microprocessor Architectures

This unit provides address translation tables in memory. Additionally the memory management unit provides protection facilities, e.g., user access cannot be made to the operating system, user areas can be made read-only. A further feature included is that, if the CPU encounters a breakpoint instruction, the memory management unit provides the 32-bit instruction that is to be executed in its place. This is useful in program testing situations.

Multi-board configurations based on the MC68020 normally follow the Versabus or VME bus standards.

13.3.4 Zilog Z80000

The 32-bit Z80000 is, as expected, upwards-compatible with the 16-bit Z8000 processors (Z8001 and Z8002). In terms of its on-chip facilities it is probably the most advanced of the 32-bit microprocessors. It possesses an on-chip cache and on-chip memory management unit, and it will run at a clock speed of 25 MHz. A six-stage instruction pipeline is applied, and typical program execution speeds are 4–5 MIPS (million instructions per second).

Register set

The register set is illustrated in Fig. 13.25. As with the Intel and Motorola devices, data operations can be performed on bytes, words (16-bit), long words (32-bit) and quad words (64-bit). One bit in the flag and control word (Zilog nomenclature for "status registers") indicates system or normal mode; some instructions, e.g., input/output, are only allowed when the CPU is set to system mode (when the operating system is running). The contents of the program status register indicate the memory address of the values which are loaded into the program counter and flag and control word when an interrupt occurs. One of the four translation table descriptor registers is used by the memory management unit during memory addressing (to be described later). The overflow stack pointer is used if an address calculation error occurs when an interrupt is set. The system configuration control long word holds bits that control the on-chip cache and memory management functions.

Instruction set

The instruction set is comprehensive, and retains the Zilog features applied on earlier microprocessors, e.g., block transfer, string search. Nine addressing modes are available, and an address calculation can

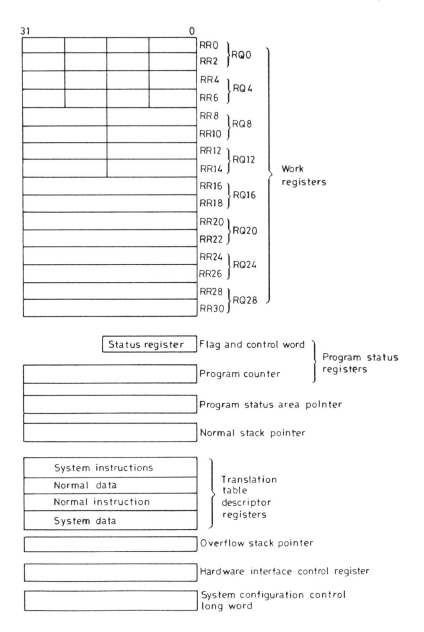

Figure 13.25 Z80000 register set.

Advanced Microprocessor Architectures 577

involve a base register, index register and a displacement. A memory address specified within an instruction is interpreted in one of three ways, as determined by the settings of two bits in the flag and control word. These three representations are:

1. Compact: 16 bits (gives 64 Kbytes)
2. Segmented: 32 bits (15-bit segment with 16-bit offset, for first 2G bytes) (or, 7-bit segment with 24-bit offset, for next 2G bytes)

Note that only the offset field is affected by address calculations in this representation.

3. linear: 32 bits (gives 4 Gbytes)

This memory address specified in the instruction is termed the "logical address", and it is converted into the "physical address" in the memory management unit. The conversion process is performed in one of the following two ways:

1. Using the "translation lookaside buffer" within the CPU; this stores tag addresses and conversion information for the 16 most recently referenced pages in the same manner as an instruction/data memory cache.
2. If the translation lookaside buffer does not produce a tag match, the CPU references translation tables in memory, using one of the four table descriptor registers (see Fig. 13.25) and transfers the required information into the least recently used entry in the translation lookaside buffer.

Memory management

The major function of an operating system is to manage available resources so that system activities and application tasks can be made to accomplish their defined objectives in a timely and efficient manner. A major aspect of this task is memory management. The Z80000 incorporates an on-chip memory management unit (MMU) that support a paged virtual memory addressing scheme with a page size of $1 K = 1024$ bytes.

There are two major concepts involved in memory management that lend themselves to direct hardware support. These are address representation and address translation. Address representation refers to the format used in doing address arithmetic to compute a logical effective address. Address translation refers to the action of converting this logical effective address into an address that can access a physical memory location in a particular hardware configuration. Thus the

CPU's address representation and addressing modes determine the logical effective address, and address translation maps that address onto a physical memory location.

Two types of address representation, usually referred to as linear and segmented, are used on microprocessors. With linear addressing, addresses have a simple integer representation, and address arithmetic is based on the fundamental operation of addition modulo address size. With segmented addressing, computation of an effective logical address involves two separate values. One is referred to as the segment number and is unaffected by address arithmetic. Conceptually the segment number is a pointer identifying an object or block in memory. The other value is commonly referred to as the segment offset. Arithmetic is performed on this offset field only, and is based on the operation of addition modulo segment size. The segment number and offset are then combined to generate the logical effective address.

Cache memory

A cache memory is a rapid-access local storage buffer that stands between the processor and main memory. Some memory accesses cause values to be stored in cache so that subsequent read operations to those same locations will be able to access cache only and will not require an external bus cycle. When the Z80000 fetches from cache, only one system clock cycle is required; when it fetches from off-chip memory, two or more cycles are required. Thus cache can decrease the time required to execute memory cycles, and it can free the bus for DMA or multi-processing transfers, increasing the effective bus bandwidth of the system.

Cache memories are organized in multi-byte blocks (called tag lines) with associated tag fields. The bus structure and internal organization of the processor determine the best block size. When cache is accessed, a comparison is made between valid tag fields and a subset of the address bits. If a match is found, additional bits in the address identify an individual entry (byte, word, or long word) in the tag line.

As shown in Fig. 13.26, the Z80000 has a 256-byte, fully associative cache organized as 16 tag lines, each containing 16 bytes managed as eight words. "Fully associative" means that any tag line can be associated with any address block. Since the Z80000 supports burst-mode transfers, an entire tag line of 16 bytes can be loaded by the four consecutive 32-bit transfers of a burst read, making the 16-byte block size a logical choice. The cache can be configured to hold instructions only, data only, or both instructions and data. The cache can also be

Advanced Microprocessor Architectures

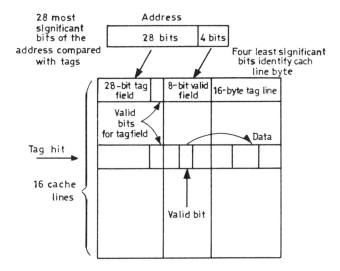

Figure 13.26 The 256-byte cache is organized as 16 tag lines of 16 bytes each.

frozen to hold values from a dedicated memory range, to provide very fast access to either a critical subroutine or important data.

The addresses represented by the tag line are determined by the value stored in the tag field. When a cache access is attempted, the 28 most significant bits of the physical address are simultaneously compared with all valid tag fields. If a match is found, then the least significant bits point to the target word in the cache line.

In addition to the 28-bit tag field, cache tag line contains a validity but indicating whether the address tag field is valid. It also contains an eight-bit field that indicates whether the individual words of the cache line are valid. The 16 bytes of a tag line are aligned on a 16-byte boundary and addressed sequentially. Thus a cache line consists of a 28-bit tag field and a validity bit, and 8-bit validity field, and a 16-byte (8-word) storage block. When a memory reference is made, the 28 most significant bits of the 32-bit physical memory address are simultaneously compared with all cache line tags. If a match is found and the valid bit is set, then bits one to three of the address identify one of the eight words in the block. The validity bit for that word is then checked. If it is set and the operation is a read, then the access is from cache. Otherwise memory is accessed and the cache line can be simultaneously and automatically updated.

Pipelining

Pipelining increases performance by allowing a processor to execute multiple instructions in parallel. This feat is accomplished by dividing each instruction into basic operations and dedicating individual execution units to each one. As shown in Fig. 13.27, the Z80000 divides each instruction into six stages of execution: instruction fetch, instruction decode, operand address calculation, operand fetch, execution, and operand store. No instruction can advance to a stage in the pipe until its immediate predecessor has completed that stage and released the execution unit. When an instruction completes, the instruction just succeeding it may also be partially completed. Under optimal conditions, an instruction will complete as a concurrent set of pipeline stages completes, so that one instruction will enter the pipe as another exits. Under these conditions, processor performance may increase by a factor of six. Unfortunately, the following factors work against this performance increase:

1. Events such as jump or call instructions, interrupts, and traps that cause a new program counter value to be loaded may cause the pipe to be partially or fully flushed.
2. Some instructions do not use all stages of the pipeline; hence, execution units will occasionally be idle.
3. Some stages of the pipe require more time than others, and the stage that requires the longest time to complete determines the rate at which instructions move through the pipe.

These factors make predicting the actual performance increase that will be realized quite difficult. However, nonsequential events are typically infrequent enough that the pipelining scheme of the Z80000 should produce a significant performance increase.

As with the 16-bit Z8000 processors, the Z80000 refers to its coprocessors as extended processing units (EPUs). Zilog apply the same floating-point co-processor (the Z8070) as used with their 16-bit processors. Other 16-bit support devices, e.g. the Z8016 DMA controller, can

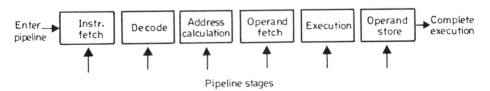

Figure 13.27 The Z80000 user a six-stage pipeline for instruction execution.

Advanced Microprocessor Architectures

be applied with the Z80000. Zilog apply their own backplane interfacing standard ("Z-bus") to large microcomputer configurations.

13.3.6 Inmos T424 Transputer

The Inmos transputer is an exciting and unique approach to computer design. Conventional computer architecture is based on a single CPU. Although the transputer is a single processor, it is not designed to operate in a uni-processor configuration, but instead is intended to operate in an arrangement in which several transputers operate in parallel. This multi-processor configuration produces a theoretical linear increase in performance as the number of processors increase, e.g., doubling the number of transputers doubles the number of instructions per second.

The name "transputer" is an amalgam of "transistor" and "computer". The execution of a program within several parallel processors is termed "concurrency". A large array of transputers is the first realistic solution to the much-vaunted Japanese concept of a fifth-generation computer, which must perform approximately 1000 million instructions per second.

The first transputer developed by Inmos was the T414, which is a 32-bit processor with 4 serial links and 2 Kbytes of on-chip static RAM. This was followed by:

1. T424: 32-bit, 4 serial links, 4K RAM
2. T212: 16-bit, 4 serial links, 2K RAM
3. T800: 32-bit, 4 serial links, on-chip floating point processor

The T414 is fabricated using CMOS technology and is mounted in a 84-pin chip carrier package. At a clock speed of 5 MHz the device can perform 10 MIPS. Its internal organization is shown in Fig. 13.28. In addition to a 32-bit CPU and 2 Kbytes of RAM, it includes four fast serial data links. These duplex links enable messages to be passed to other transputers which contribute to the concurrency of a programming task. Link adapter chips are available to convert from serial to parallel data (bytes) to enable conventional input/output systems to be connected. Up to 4 Gbytes of memory can be addressed using the 32-bit address bus.

The processor module within the transputer is an example of a RISC machine (see Section 13.3.1). It possesses an extremely simple instruction set, and most instructions execute in a single internal clock cycle. Figure 13.29 illustrates the restricted register set and simple byte instruction format.

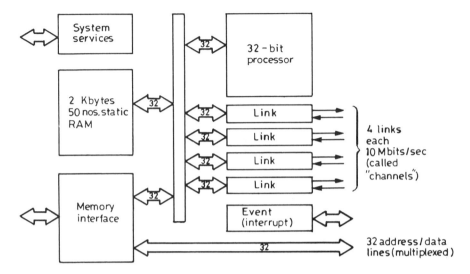

Figure 13.28 Inmos T414 transputer architecture.

It is not intended that the transputer is to be programmed at machine code level; indeed Inmos restrict information concerning the instruction set. A special high-level language named "occam" (after a fourteenth-century Oxford philosopher, who exhorted simplicity in problem-solving techniques) has been developed by Inmos to exploit the concurrency feature of transputer systems. An occam compiler is supplied by Inmos to produce highly efficient machine code for the transputer, and the programmer must specify which parts of the program are to run concurrently in other parallel transputers and which channels are to be used to communicate information. If a multi-transputer network is used, as shown in Fig. 13.30, different parts of the "process" (pro-

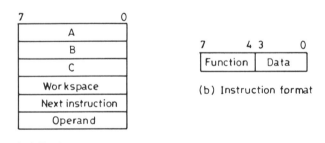

(a) Register set

Figure 13.29 Transputer CPU detail.

Advanced Microprocessor Architectures

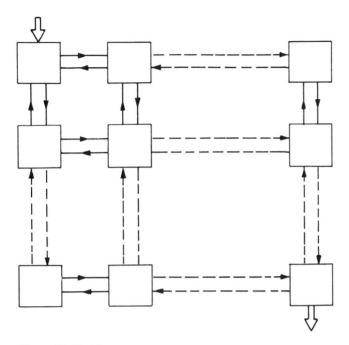

Figure 13.30 Transputer array.

gram) run in different transputers. The larger the array, the greater is the software throughput of the machine. Results quoted by Inmos for large arrays are exciting, e.g., a mainframe performance exceeded by a 256-transputer system.

Support devices produced by Inmos include:

1. F424 floating-point transputer
2. G412 graphics controller transputer
3. M212 disc controller transputer
4. C00l and C002 link adapters (to produce 8-bit parallel link)

13.4 CONCLUSIONS

Broad trends in the microprocessor industry can be characterized along several dimensions as follows.

1. *Chip Technology.* In view of the large number of transistors, most manufacturers have opted to fabricate microprocessors using MOS (metal oxide semiconductor) technology rather than bipolar transistor technology. N-channel MOS (NMOS) technology has been the most

popular technology, by virtue of its high packing density and fast switching speeds. CMOS (complementary MOS) circuits provide lower power consumption than circuits implemented with traditional PMOS and NMOS technology. The disadvantage of CMOS technology lies in its lower packing density; nevertheless, it is becoming increasingly popular for fabricating microprocessors.

2. *Word size.* Word size reflects the basic unit of work for the microprocessor. A larger word size implies more processing power and addressing capabilities. In the early years of microprocessors, the size of registers, the size of internal instruction paths and data paths, and the size of external instruction and data paths were all identical. This is rarely true now. Large external data paths require the chip package to have a high number of pins, which implies high packaging and production costs. Thus, chips nowadays tend to have larger internal paths than external paths. A true 32-bit microprocessor has external paths and all internal units designed to communicate or process at least 32 bits in parallel.

3. *Type of microprocessor.* Some microprocessors process all bits of one word in parallel. Other work with "slices" of data and/or instruction words. In the latter case, called "bit-slice" architecture, several identical chips can be used to process different slices in parallel. In the past, the inability to fabricate a large number of transistors on the same chip made it essential to use multiple bit-slice chips to obtain large word widths. Today bit-slice architecture offers the potential for creating a customized CPU with a word length that is an integral multiple of the bit-slice width. However, for most applications, word sizes of 32 bits are more than adequate, permitting the use of standard (non-bit-slice) 32-bit microprocessors. The advent of 32-bit (and even larger word size) microprocessors makes it less advantageous to employ the bit-slice architecture approach.

4. *Clock frequency* (*hertz*). This is the number of clock cycles per second of the fundamental driving clock circuit. Two clocks of the same frequency, but phase shifted with respect to each other, can be used to generate a clock of higher frequency. An increase in the clock frequency results in a proportionate decrease in the execution time of an instruction.

5. *Number of instructions.* The enhanced ability to fabricate large numbers of transistors on a chip facilitates implementation of a larger instruction set. These instructions are frequently of varying sizes, depending on instruction type, size of data, and addressing mode used. As the instruction size has increased, the complexity of the chip has increased, and so has the design effort. This effort has increased from under one person-year at the dawn of the microprocessor era to over

Advanced Microprocessor Architectures 585

100 person-years of engineering time today. To reduce this massive effort, reduced instruction set architectures have been implemented. A single instruction on a traditional architecture may require several instructions to be executed on a reduced instruction set computer (RISC). A reduction in the total number of instructions can potentially degrade the overall performance of the system.

6. *Addressing capability.* Early microprocessors could reference only limited memory space. Larger word sizes enable direct addressing of larger memory space. In addition, a number of auxiliary addressing modes (such as indirect, indexing, and autodecrementing) have become popular on microprocessors. Specialized memory management chips provide even more enhanced capabilities for efficient memory management and support of virtual memory.

7. *Number of registers.* Registers are required for arithmetic operations, stack operations, storing base and index values, and a variety of other operations, depending on the architecture. General-purpose registers can have multiple uses. Some microprocessors offer general-purpose registers only, others dedicated only, but most offer some combination of the two.

8. *Data types.* All microprocessors support data in the form of bytes and words. However, only some support data in the form of bits, binary coded decimal words, floating-point numbers, words longer than 4 bytes, and character strings. Floating-point capabilities are useful for scientific work. Character string manipulation capability is required for text editing applications. Auxiliary chips, co-processors, or slave processors are sometimes used to perform these functions. As technology improves, more functions are being incorporated on the main chip itself.

9. *Direct memory access (DMA) capability.* DMA capability enables a processor to offer a higher overall performance by allowing input and/or output to proceed concurrently with processing. In most cases, an auxiliary chip is used to control input and output operations, leaving the microprocessor free to process instructions.

10. *Software.* Just as the Burroughs 5000 series introduced a trend in the mainframe computer industry of architectures designed to support high-level languages alone, new microprocessors have established a trend toward providing high-level language-oriented instruction sets and enhanced support for switching from one process to another. This reduces user programming costs.

11. *Multiprocessor capabilities.* In order to increase computational bandwidth and/or system resilience, it becomes necessary to integrate several microprocessors in a single system. The overall throughput and efficiency of such systems are directly dependent on the hardware and

software interconnection mechanisms supported by the basic microprocessor chips. Although all chips offer some facilities for use in a multi-processor environment, it is essential to examine exact features to determine overall maximum efficiency of multi-processor configurations and to estimate software overheads.

12. *Number of chips.* Any computer system, irrespective of size, power, and capabilities, combines three classes of subsystems: CPUs (for arithmetic, logic, and control functions), memories (read/write (RAM) and/or read-only (ROM)), and input/output interfaces for peripheral control. Early microprocessors performed the basic CPU functions only. Additional chips were required to generate timing signals, to provide primary memory for program and data storage, and to interface with peripheral units. As better technology became available to enable integration of a larger number of devices on the same chip, it became feasible to implement an increasing number of auxiliary functions on the microprocessor chip itself.

Single-chip micros constitute an important subset of microprocessors in which all functions, including program and data memory, are implemented on the same chip. In view of the powerful features of the 32-bit microprocessors, the terms "mainframe on a chip" and "micro-mainframe" were used to emphasize the capabilities of their microprocessors and to set them apart from previous 8- and 16-bit models. Designers have borrowed extensively from the concepts and technologies behind much larger systems as VLSI advances have steadily increased the capabilities of microprocessor chip sets. In the latest generation of 32-bitters, this influence is apparent in several ways, including pipelining, memory management, floating-point arithmetic, and processing rates.

NOTES AND REFERENCES

Looking at the wide array of sophisticated (16-bit and 32- bit) microprocessors available today to support virtually all conceivable applications, it is reasonable to anticipate the impact of micros on our engineering systems. We have presented an overview of the features and characteristics of advanced microprocessors. To probe further for more information, the reader is advised to consult the User's Manuals of manufacturers including Intel, Motorola, Zilog, National Semiconductors, Texas Instruments, VAX, and Fairchild among many others. The books of Ciminiera and Valenzano (1987), Gupta (1987) contain an excellent survey of the subject. Other relevant research papers and reports are contained in the bibliography.

Appendix A
Laplace Transforms and Properties

The Laplace transform of the continuous-time signal $f(t)$ is designated hereafter by the symbol $F(s)$ and is formally defined by the integration operation

$$F(s) = \int_{-\infty}^{+\infty} f(t)e^{-st}\, dt \qquad (1)$$

This integral specifies a function of s for every s for which it converges. The variable s appears in (1) is generally complex-valued and is therefore often expressed in terms of its rectangular coordinates

$$s = \sigma + j\omega \qquad (2)$$

where $\sigma = \mathrm{Re}(s)$ and $\omega = \mathrm{Im}(s)$ are referred to as the real and imaginary components of s, respectively. For physical situations, the lower limit in (1) is taken 0 instead of $-\infty$ such that

$$F(s) = \int_0^{\infty} f(t)e^{-st}\, dt \qquad (3)$$

defines the unilateral Laplace transform. We shall omit the term unilateral and consider (3) as the Laplace integral form. The notation $\{f(t), F(s)\}$ will mean that the functions $f(t)$ and $F(s)$ form a Laplace transform pair, that is, $F(s)$ is the Laplace transform of $f(t)$.

Before embarking on a general discussion of the Laplace transform and its properties, it is advisable to take note of the following result:

$$\lim_{T \to \infty} \int_0^T e^{-pt}\, dt = \lim_{T \to \infty} \frac{1 - e^{-pT}}{p}$$

$$= \frac{1}{p}, \quad p > 0 \qquad (4)$$

which is frequently employed in evaluating the Laplace transform integral. In the sequel we shall assume that for all functions under consideration

$$f(t)e^{-st} \to 0 \quad \text{as } t \to \infty \qquad (5)$$

for every s for which the integral converges. By virtue of (4), it is readily seen that the Laplace transform of $f(t) = 1$ is

$$F(s) = \int_0^\infty e^{-st}\, dt = \frac{1}{s} \qquad (6)$$

Consider next the function $f(t) = e^{ct}$ where c is positive or negative constant. Applying (3), it yields

$$F(s) = \int_0^\infty e^{ct} e^{-st}\, dt$$

$$= \int_0^\infty e^{-(s-c)t}\, dt \qquad (7)$$

The last integral is the same as the integral in (4) if $p = s - c$; hence, it equals $1/(s - c)$ provided that $s - c > 0$. We thus conclude that $F(s)$ in (7) is given by

$$F(s) = \frac{1}{s - c}, \quad s > c \qquad (8)$$

To evaluate the Laplace transforms of unit-impulse-type signals that arise when differentiating discontinuous continuous-time signals, we consider $f(t) = \delta(t)$ given by

$$\delta(t) = \begin{cases} 1, & t = 0 \\ 0, & \text{otherwise} \end{cases} \qquad (9)$$

Laplace Transforms and Properties

Then from (3) we obtain

$$F(s) = \int_0^\infty \delta(t) e^{-st} dt = 1, \quad \text{for all } s \tag{10}$$

It should be clear from the above treatment that the Laplace transform provides a systematic procedure for converting a function of a real variable t into a function of a complex variable s. We now present properties that indicate how certain operations effected in one domain manifest themselves in the other domain.

Linearity

For

$$f(t) = \alpha_1 f_1(t) + \alpha_2 f_2(t) \tag{11}$$

direct application gives

$$F(s) = \alpha_1 F_1(s) + \alpha_2 F_2(s) \tag{12}$$

where the pairs $\{f_1(t), F_1(s)\}$, $\{f_2(t), F_2(s)\}$ and $\{f(t), F(s)\}$ are Laplace transformable.

First Translation

Given the pair $\{f(t), F(s)\}$ then

$$\int_0^\infty [e^{at} f(t)] e^{-st} dt = \int_0^\infty f(t) e^{-(s-a)t} dt$$

$$= F(s - a) \tag{13}$$

that is, the pair $\{e^{at} f(t), F(s - a)\}$ is Laplace transformable.

Second Translation

If $\{f(t), F(s)\}$ is a Laplace transform pair and

$$g(t) = \begin{cases} f(t - a), & t > a \\ 0, & t < a \end{cases} \tag{14}$$

then

$$\int_0^\infty g(t) e^{-st} dt = \int_0^\infty f(t - a) e^{-st} dt$$

Let $\tau = t - a$, $dt = d\tau$ thus we have

$$\int_{-0}^{\infty} f(\tau)e^{-s\tau}e^{-as}\,d\tau = e^{-as}\int_{0}^{\infty} f(\tau)e^{-s\tau}\,d\tau = e^{-as}F(s) \qquad (15)$$

Therefore $\{g(t), e^{-as}F(s)\}$ is a Laplace transform pair.

Change of Scale

If $\{f(t), F(s)\}$ is a given Laplace pair, then application of (3) results in

$$\int_{0}^{\infty} f(at)e^{-st}\,dt = \frac{1}{a}F\left(\frac{s}{a}\right) \qquad (16)$$

which adds up the pair $\{f(at), F(s/a)/a\}$, that is a stretch in the time-domain causes a contraction in the frequency-domain and vice versa.

Time Derivative

Given the pair $\{f(t), F(s)\}$, then it is easy to show that the pair $\{df(t)/dt, sF(s) - f(0)\}$ is Laplace transformable. Continuing further, we obtain the pair $\{d^2f(t)/dt^2, s^2F(s) - sf(0) - sdf(0)/dt\}$ and more generally

$$\int_{0}^{\infty}\left[\frac{d^n f(t)}{dt^n}\right]e^{-st}\,dt = s^n F(s) - s^{n-1}f(0)$$

$$- s^{n-2}\frac{dt(0)}{dt} - \cdots - \frac{d^{n-1}f(0)}{dt^{n-1}} \qquad (17)$$

Integration

It is easy to prove that

$$\int_{0}^{\infty}\left[\int_{0}^{t} f(u)\,du\right]e^{-st}\,dt = \frac{F(s)}{s} \qquad (18)$$

which means that the integration operation in the time-domain corresponds to the algebraic division in the frequency-domain.

Multiplication by t^n

Given $\{f(t), F(s)\}$ then

$$\int_{0}^{\infty}[t^n f(t)]e^{-st}\,dt = \frac{(-1)^n d^n F(s)}{ds^n} \qquad (19)$$

or the pair $\{t^n f(t), (-1)^n d^n F(s)/ds^n\}$ is a Laplace transformable.

Laplace Transforms and Properties

Table A.1 Laplace Transform Pairs

Time Signal $f(t)$	Laplace Transform $F(s)$
$af_1(t) + bf_2(t)$	$aF_1(s) + bF_2(s)$
$af(at)$	$F(s/a)$
$e^{at}g(t)$	$F(s-a)$
$g(t) = \begin{cases} f(t-a), & t > a \\ 0, & t < a \end{cases}$	$e^{-as}F(s)$
$df(t)/dt$	$sF(s) - f(0)$
$d^2f(t)/dt^2$	$s^2F(s) - sF(0) - df(0)/dt$
$-tf(t)$	$dF(s)/ds$
$t^2 f(t)$	$d^2F(s)/ds^2$
$\int_0^t f(u)\,du$	$F(s)/s$
$\int_0^t \cdots \int_0^t f(u)\,du^n = \int_0^t [(t-u)^{n-1}/(n-1)!]F(u)\,du$	$F(s)/s^n$
$\int_0^t f(u)g(t-u)\,du$	$F(s)G(s)$
$f(t)/t$	$\int_s^\infty F(u)\,du$
$f(t) = f(t+T)$	$\int_0^T e^{-su}F(u)\,du/(1 - e^{-sT})$
t	$1/s^2$
$t^{n-1}/(n-1)!$	$1/s^n$
$t^{n-1}e^{at}/(n-1)!$	$1/(s-a)^n$
$\sin(at)/a$	$1/(s^2 + a^2)$
$\cos(at)$	$s/(s^2 + a^2)$
$e^{bt}\sin(at)/a$	$1/(s-b)^2 + a^2$
$e^{bt}\cos(at)$	$(s-b)/(s-b)^2 + a^2$
$\sinh(at)/a$	$1/(s^2 - a^2)$
$\cosh(at)$	$s/(s^2 - a^2)$
$e^{bt}\sinh(at)/a$	$1/(s-b)^2 - a^2$
$e^{bt}\cosh(at)$	$(s-b)/(s-b)^2 - a^2$
$e^{bt} - e^{at}/(b-a)$	$1/(s-a)(s-b), \quad a \neq b$
$be^{bt} - ae^{at}/(b-a)$	$s/(s-a)(s-b), \quad a \neq b$
$(1/2a^3)[\sin(at) - (at)\cos(at)]$	$1/(s^2 + a^2)^2$
$t\sin(at)/2a$	$s/(s^2 + a^2)^2$
$[\sin(at) + (at)\cos(at)]/2a$	$s^2/(s^2 + a^2)^2$
$\cos(at) - (at/2)\sin(at)$	$s^3/(s^2 + a^2)^2$
$t\cos(at)$	$(s^2 - a^2)/(s^2 + a^2)^2$
$[(at)\cosh(at) - \sinh(at)]/(2a^3)$	$1/(s^2 - a^2)^2$
$\cosh(at) + (at/2)\sinh(at)$	$s^3/(s^2 - a^2)^2$
$t\cosh(at)$	$(s^2 + a^2)/(s^2 - a^2)^2$
$t^2\sin(at)/2a$	$3s^2 - a^2/(s^2 + a^2)^3$
$(t^2/2)\cos(at)$	$s^3 - 3a^2 s/(s^2 + a^2)^3$
$t^3\sin(at)/24a$	$s^3 - a^2 s/(s^2 + a^2)^4$
$t^2\sinh(at)/2a$	$3s^2 + a^2/(s^2 - a^2)^3$
$(t^2/2)\cosh(at)$	$s^3 + 3a^2 s/(s^2 - a^2)^3$
$t^3\sinh(at)/24a$	$s^3 + a^2 s/(s^2 - a^2)^4$

Division by t

We can show that

$$\int_0^\infty [f(t)/t] e^{-st} dt = \int_s^\infty F(u)\, du \tag{20}$$

The last two properties disclose time-operations and their corresponding Laplace transforms. In view of assumption (5), it is straightforward to see that

$$\lim_{s \to \infty} F(s) = 0 \tag{21}$$

which reflects the asymptotic behavior in the frequency-domain.

Initial-Value Theorem

If the indicated limit in (21) exists, then

$$\lim_{t \to 0} f(t) = \lim_{s \to \infty} sF(s) \tag{22}$$

Final-Value Theorem

If the limit shown in (21) exists, then

$$\lim_{t \to \infty} f(t) = \lim_{s \to 0} sF(s) \tag{23}$$

The results in (22) and (23) give the interlink between the asympotic behavior in the *t*- and *s*-domains. As we have seen from the foregoing analysis, all the Laplace transform pairs have been derived using the basic integrals (3) and (4). For completeness, it is convenient to display the result of our efforts in a Laplace transform pair table. Thus, almost all standard signals and their transform are documented in addition to other useful results in Table A.1.

The Inverse Laplace Transform

Given the pair $\{f(t), F(s)\}$, we denote $f(t)$ as the inverse Laplace transform of the *s*-function $F(s)$. From the previous Laplace transform table, the time function $f(t)$ of some $F(s)$ can be easily read out. The inverse Laplace transform has properties which are analogous to those of the Laplace transform. To determine the inverse Laplace transform, we can use one of the following methods:

1. *Partial fraction method.* Any rational function $P(s)/Q(s)$ where $P(s)$ and $Q(s)$ are polynomials, with the degree of $P(s)$ less than that of $Q(s)$, can be written as the sum of rational functions (called partial

Laplace Transforms and Properties

fractions) having the form

$$A/(as + b)^r, \quad (As + B)/(as^2 + bs + c)^r, \quad r = 1, 2, 3, \ldots$$

By finding the inverse Laplace transform of each of the partial fractions, we can find the desired inverse Laplace transform. The constants A, B, C, ... can be found by clearing of fractions and equating of like powers of s on both sides of the resulting equation.

2. *Series methods.* In the case that $F(s)$ has a series expansion in inverse powers of s given by

$$F(s) = \sum_{j=0}^{\infty} \frac{a_j}{s^{j+1}} \tag{24}$$

then under suitable conditions we can invert term by term to obtain

$$f(t) = \sum_{j=0}^{\infty} \frac{a_j t^j}{j!} \tag{25}$$

3. *The complex inversion formula.* Given the Laplace integral (3), then

$$f(t) = \frac{1}{2\pi j} \int_{\alpha - j\infty}^{\alpha + j\infty} e^{st} F(s) \, ds \tag{26}$$

for $t > 0$ and $f(t) = 0$ for $t < 0$. This result provides a direct means for obtaining the inverse Laplace transform. The integration is to be performed along a line $s = \alpha$ in the complex plane where $s = x + jy$. The real number α is chosen so that $s = \alpha$ lies to the right of all the singularities (poles, branch points or essential singularities) but is otherwise arbitrary. In general, the use of partial fractions along with standard table is quite common.

Appendix B
Elements of Matrix Algebra

Our objective here is to introduce matrices and elementary matrix algebra. Matrices provide a convenient method of dealing with systems of many variables. Additional treatment and applications are found in Noble (1977), Bellman (1968), Gantmakher (1959) and Strang (1976).

B.1 NOTATION

Matrices are rectangular arrays of elements usually referred to as scalars. From now onwards, these scalars will be denoted by lower-case letters a, b, x, p, whereas the matrices are represented by uppercase letters A, B, D, etc. For example

$$A = \begin{bmatrix} 1 & 15 & 3 & 5 \\ 2 & 3 & 4 & 12 \\ -2 & 1 & 2 & 7 \end{bmatrix}$$

is a matrix where the horizontal sets of entries such as (1 15 3 5) and (−2 1 2 7) are called row vectors or rows, while vertical sets of entries such as (1 2 −2) are called columns or column vectors. Vectors will be treated as column vectors and denoted by bold lower-case letters

Elements of Matrix Algebra

such as $\mathbf{x}$, $\mathbf{a}$. Row vectors will be written as $\mathbf{x}^t$ where superscript t signifies transpose. If A has m rows and n columns, it is said to be an $m \times n$ matrix with a_{ij} referring to the element in the ith row and jth column. Sometimes the shorthand notation $A = [a_{ij}]$ will be used. Whenever $m = n$ the matrix is called a square matrix.

B.2 BASIC OPERATIONS

Matrix Equality

The $m \times n$ matrices A and B are equal, written $A = B$, if and only if their corresponding elements are equal, that is, $a_{ij} = b_{ij}$, $1 \leq i \leq m$ and $1 \leq j \leq n$.

Matrix Addition and Subtraction

If $A = [a_{ij}]$ and $B = [b_{ij}]$ are both $m \times n$ matrices then

$$C = A + B = [c_{ij}] \quad \text{and} \quad D = A - B = [d_{ij}]$$

are both $m \times n$ matrices defined by

$$c_{ij} = a_{ij} + b_{ij}, \quad \text{for} \quad i = 1, 2, \ldots, m$$
$$d_{ij} = a_{ij} - b_{ij}, \quad j = 1, 2, \ldots, n$$

Matrix Multiplication

Consider the relation

$$y_1 = a_{11}x_1 + a_{12}x_2 + a_{13}x_3$$
$$y_2 = a_{12}x_1 + a_{22}x_2 + a_{23}x_3$$
$$y_3 = a_{31}x_1 + a_{32}x_2 + a_{33}x_3$$

or $\mathbf{y} = A\mathbf{x}$. This explains the usual rule for matrix multiplication. More generally, the product $C = AB$ where A is $n \times m$ and B is $p \times q$ is defined when $m = p$ or when A and B are conformable. Other forms of multiplication are

1. Multiplying a matrix by scalar

$$xA = Ax = [x \; a_{ij}]$$

2. Schur product is

$$[A \circ B]_{ij} = a_{ij}b_{ij}$$

3. Lie product is

$$[A, B]_{ij} = \sum_{k=1}^{n} [a_{ik}b_{kj} - b_{ik}a_{kj}]$$

4. Kronecker product is

$$A \times B = \begin{bmatrix} a_{11}B & \cdots & a_{1m}B \\ a_{m1}B & \cdots & a_{mn}B \end{bmatrix}$$

B.3 SPECIAL TYPES OF MATRICES

A matrix which has all of its elements equal to zero is called the null matrix, represented by O. The identity or unit matrix I is a square matrix with all elements zero except those on the main diagonal ($i = j$ positions) are 1s. Note that $AI = IA = A$ and $OA = AO = O$.

If $A = [a_{ij}]$ then A^t, the transpose of A, is $[a_{ji}]$. The matrix A is said to be symmetric if $A = A^t$. If $A = -A^t$, then A is skew-symmetric. An important property of matrix transposition of products is illustrated by

$$(AB)^t = B^t A^t, \qquad (A^t)^t = A$$
$$(ABC)^t = C^t B^t A^t$$

but $(A + B)^t = A^t + B^t$.

The conjugate of A, written A^*, is the matrix formed by replacing every element in A by its complex conjugate. Thus $A^* = [a_{ij}^*]$. If all elements of A are real, then $A^* = A$. If all elements are purely imaginary, then $A^* = -A$. The associate matrix of A is the conjugate transpose of A. The order of these two operations is immaterial. Matrices satisfying $A = A^{*t}$ are called Hermitian matrices. Skew–Hermitian matrices satisfy $A = -A^{*t}$. For real matrices, symmetric and Hermitian mean the same thing.

Note that if A is any square matrix then $(A + A^t)$ is symmetric but $(A - A^t)$ is skew-symmetric.

B.4 DETERMINANTS

The determinant of an $n \times n$ matrix A, written $|A|$ or $\det(A)$, is a scalar-value of A. If A and B are square matrices of order n, then $|AB| = |BA| = |A||B|$. Note that $|A| = |A^t|$. If all the elements in any row or in any column are zero, then $|A| = 0$. By the same token, if any two rows (columns) of A are proportional, or if a row is a linear combination of any number of other rows (columns), then $|A| = 0$.

The rank of A, designated as r_A or $\text{rank}(A)$, is defined as the size

Elements of Matrix Algebra

of the largest nonzero determinant that can be formed from A. The maximum possible rank of an $m \times n$ matrix is obviously the smaller of m and n. If A is $n \times n$ and has its maximal rank n, the matrix is said to be nonsingular. If r_A and r_B are the ranks of A and B, then $C = AB$ has rank r_C satisfying.

$$0 \leq r_C \leq \min\{r_A, r_B\}$$

Let A be an $n \times n$ matrix. Then the trace of A, denoted $\text{Tr}(A)$, is the sum of the diagonal elements of A.

$$\text{Tr}(A) = \sum_{j=1}^{n} a_{ij}$$

If A and B are conformable square matrices, then $\text{Tr}(A + B) = \text{Tr}(A) + \text{Tr}(B)$ and $\text{Tr}(AB) = \text{Tr}(BA)$. It is obvious that $\text{Tr}(A^t) = \text{Tr}(A)$ and consequently $\text{Tr}(AB) = \text{Tr}(B^t A^t)$.

Finally

$$\text{Tr}[A^t A] = \sum_{i=1}^{n} \sum_{j=1}^{n} a_{ij}^2$$

B.5 MATRIX INVERSION

For any square $n \times n$ nonsingular ($|A| \neq 0$) matrix A the inverse of A, denoted by A^{-1}, is defined as

$$AA^{-1} = A^{-1}A = I$$

From numerical mathematics, A has a unique inverse given by

$$A^{-1} = C^t/|A|$$

where C is the matrix formed by the cofactors C_{ij}. The matrix C^t is called the adjoint matrix, $\text{Adj}(A)$, and thus $A^{-1} = \text{Adj}(A)/|A|$.

If $A - 1 = A$ then A is said to be involuntary and when $A^{-1} = A^t$ then A is called orthogonal. For an orthogonal matrix A, $\det(A) = \pm 1$. Note that the inverse of a symmetric matrix is symmetric. If A and B are $n \times n$ orthogonal matrices, then so are A^{-1}, A^t, and AB.

Let $A, B, C, \ldots, W$ be any number of conformable nonsingular matrices. Then

$$(ABC \cdots W)^{-1} = W^{-1} \cdots C^{-1} \, B^{-1} \, A^{-1}$$

Note that $(kA)^{-1} = A^{-1}/k$, $|AA^{-1}| = |A^{-1}A| = 1$ and $|A^{-1}| = 1/|A|$.

B.6 PARTITIONED MATRICES

Any matrix A can be subdivided or partitioned into a number of small submatrices which are conformable. We now summarize the important results when dealing with partitioned matrices

1. $\det \begin{bmatrix} A & B \\ 0 & D \end{bmatrix} = \det \begin{bmatrix} A & 0 \\ C & D \end{bmatrix} = |A||D|$

2. $\det \begin{bmatrix} A & B \\ C & D \end{bmatrix} = |A||D - CA^{-1}B|$ (if $|A| \neq 0$)

   $\phantom{\det \begin{bmatrix} A & B \\ C & D \end{bmatrix}} = |D||A - BD^{-1}C|$ (if $|D| \neq 0$)

3. $\det[I + BA] = \det[I + AB]$

4. If $A = \begin{bmatrix} A_1 & A_2 \\ A_3 & A_4 \end{bmatrix}$ then

$$A^{-1} = \begin{bmatrix} (A_1 - A_2 A_4^{-1} A_3)^{-1} & -A_1^{-1} A_2 (A_4 - A_3 A_1^{-1} A_2) \\ -A_4^{-1} A_3 (A_1 - A_2 A_4^{-1} A_3)^{-1} & (A_4 - A_3 A_1^{-1} A_2)^{-1} \end{bmatrix}$$

5. $(A + BDC)^{-1} = A^{-1} - A^{-1}B(D^{-1} + CA^{-1}B)^{-1}CA^{-1}$

which is frequently referred to as the *matrix inversion lemma*.

Different versions can be derived from this lemma the important among those is when $D = I$ leading to

$$(A + BC)^{-1} = A^{-1} - A^{-1}B(I + CA^{-1}B)CA^{-1}$$

For the special case when $B = \mathbf{b}$ (n − vector) and $C = \mathbf{c}^t$ (n − vector), then

$$(A + \mathbf{bc}^t)^{-1} = A^{-1} - \frac{A^{-1}\mathbf{bc}^t A^{-1}}{1 + \mathbf{c}^t A^{-1} \mathbf{b}}$$

6. $\begin{bmatrix} A & B \\ 0 & D \end{bmatrix}^{-1} = \begin{bmatrix} A^{-1} & -A^{-1}BD^{-1} \\ 0 & D^{-1} \end{bmatrix}$

and

$\begin{bmatrix} A & 0 \\ C & D \end{bmatrix}^{-1} = \begin{bmatrix} A^{-1} & 0 \\ -D^{-1}CA^{-1} & D^{-1} \end{bmatrix}$

B.7 RANGE AND NULL SPACES

Let A be an $m \times n$ matrix. The range space of A, written $\mathbb{R}[A]$, is the set of all vectors $A\mathbf{x}$, where $\mathbf{x}$ ranges over the set of all n-vectors. The

Elements of Matrix Algebra

range space has dimension equal to the rank of A, that is, the maximal number of linearly independent vectors in $\mathbb{R}[A]$ is rank A. The null space of A, written $\mathbb{N}[A]$, is the set of vectors $\mathbf{y}$ for which $A\mathbf{y} = \mathbf{O}$. A useful property is that $\mathbb{R}[A^t]$ and $\mathbb{N}[A]$ are orthogonal, that is, if $\mathbf{y} = A^t\mathbf{x}$ for some $\mathbf{x}$ and if $\mathbf{z}$ is such that $A\mathbf{z} = \mathbf{O}$ then $\mathbf{y}^t\mathbf{z} = O$. If A and B are two matrices with the same number of rows, then $\mathbb{R}[A] \subset \mathbb{R}[B]$ if and only if $\mathbb{N}[A^t] \supset \mathbb{N}[B^t]$.

B.8 THE PSEUDO-INVERSE OF A MATRIX

The pseudo-inverse $A^\#$ of a square matrix A is a useful generalization of the inverse of a matrix. There are actually a number of different pseudo-inverses (Barnett, 1971); here, we shall describe the Moore–Penrose pseudo-inverse. The key is to make $A^\# A$ act as the identity matrix on as large a set of vectors as is practicable.

Let A be an $n \times n$ matrix. Its pseudo-inverse $A^\#$ is uniquely defined by

$$A^\# A\mathbf{x} = \mathbf{x}, \quad \text{for all } \mathbf{x} \in \mathbb{R}[A^t]$$

$$A^\# \mathbf{x} = \mathbf{O} \quad \text{for all } \mathbf{x} \in \mathbb{N}[A^t]$$

It is observed that $(A^\# A)$ is the identity on $\mathbb{R}[A^t]$. We have the following properties:

1. $\mathbb{R}[A^\#] = \mathbb{R}[A^t]$
   $\mathbb{N}[A^\#] = \mathbb{N}[A^t]$
2. $(A^\#)^\# = A$
3. $A^\# A A^\# = A^\#$
4. $A A^\# A = A$
5. $A^\# A$ is the orthogonal projection onto $\mathbb{R}[A^t]$
6. $A A^\#$ is the orthogonal projection onto $\mathbb{R}[A]$
7. $(A^\#)^t = (A^t)^\#$

B.9 FUNCTIONS OF A SQUARE MATRIX

For positive m, A^m for a square matrix a is defined as $AA \ldots A$, there being m terms in the product. For negative m, let $m = -n$, where n is positive; then $A^m = (A^{-1})^n$. It follows that $A^p A^q = A^{p+q}$ for any integers p and q, positive or negative, and likewise that $(A^p)^q = A^{pq}$.

A polynomial in A is a matrix $p(A) = \Sigma_{i=0}^r \alpha_i A^i$, where α_i are scalars. Any two polynomials in the same matrix commute, i.e., $P(A)q(A) = q(A)p(A)$, where p and q are polynomials. It follows that

$p(A)q^{-1}(A) = q^{-1}(A)p(A)$, and some rational functions of A also commute.

Let A be a square matrix. Then it can be shown that the series

$$I + A + \frac{1}{2!}A^2 + \frac{1}{3!}A^3 + \cdots$$

converges, in the sense that the ij entry of the partial sums of the series converges for all i and j. The sum is defined as e^A. It follows that

$$e^{At} = I + At + \frac{1}{2!}A^2t^2 + \cdots$$

Other properties are: $p(A)e^{At} = e^{At}p(A)$ for any polynomial, and $e^{-At} = [e^{At}]^{-1}$.

B.10 THE CAYLEY–HAMILTON THEORY

Let A be a square matrix, and let

$$|sI - A| = s^n + \alpha_1 s^{n-1} + \cdots + \alpha_n$$

Then

$$A^n + \alpha_1 A^{n-1} + \cdots + \alpha_n I = 0$$

The Cayley–Hamilton theorem is often stated, rather ambiguously, as "the matrix A satisfies its characteristic equation".

From the Cayley–Hamilton theorem, it follows that A^m for any $m > n$ and e^A are expressible as a linear combination of $I, A, \ldots, A^{n-1}$.

For square A, the minimum polynomial is the unique monic polynomial $m(\cdot)$ of least degree such that $m(A) = 0$. If $p(\cdot)$ is any polynomial for which $p(A) = 0$, then $m(\cdot)$ divides $p(\cdot)$; in particular, $m(\cdot)$ divides the characteristic polynomial.

B.11 DIFFERENTIATION AND INTEGRATION

Suppose an $n \times m$ matrix A is a function of a scalar variable t, in the sense that each entry of A is a function of t. The derivative of A with respect to t is defined by the matrix whose (i,j)th element of the original matrix. Thus

$$\frac{dA(t)}{dt} = \begin{bmatrix} da_{11}(t)/dt & \cdots & da_{1m}(t)/dt \\ \vdots & & \vdots \\ da_{n1}(t)/dt & \cdots & da_{nm}(t)/dt \end{bmatrix}$$

Elements of Matrix Algebra

It follows that:
1. $d(AB)/dt = dA/dt \cdot B + A \cdot dB/dt$
2. $d(A + B)/dt = dA/dt + dB/dt$
3. $d[Ak(t)]/dt = dA/dt \cdot k(t) + A \cdot dk(t)/dt$
4. $dA^{-1}/dt = -A^{-1}[dA/dt]A^{-1}$

If $J(\mathbf{x})$ is a scalar function of a vector $\mathbf{x}$, then

$$\frac{\partial J}{\partial \mathbf{x}} = [\partial J/\partial x_1 \ldots \partial J/\partial x_n]^t$$

$$\frac{\partial^2 J}{\partial \mathbf{x}^2} = \begin{bmatrix} \partial^2 J/\partial x_1^2 & \partial^2 J/\partial x_1 \partial x_2 & \ldots & \partial^2 J/\partial x_1 \partial x_n \\ \ldots & & & \ldots \\ \partial^2 J/\partial x_n \partial x_1 & & \ldots & \partial^2 J/\partial x_n^2 \end{bmatrix}$$

For a scalar function $V[\mathbf{x}(t)]$, we have

$$\frac{dV[\mathbf{x}(t)]}{dt} = \left(\frac{\partial V}{\partial \mathbf{x}}\right)^t \frac{d\mathbf{x}}{dt}$$

Given an $m \times 1$ matrix $\mathbf{f}(\mathbf{x})$ of an n-vector $\mathbf{x}$, then

$$\frac{\partial \mathbf{f}}{\partial \mathbf{x}} = \begin{bmatrix} \partial f_1/\partial x_1 & \partial f_2/\partial x_1 & \ldots & \partial f_m/\partial x_1 \\ \partial f_1/\partial x_2 & \partial f_2/\partial x_2 & \ldots & \partial f_m/\partial x_2 \\ \vdots & \vdots & & \vdots \\ \partial f_1/\partial x_n & \partial f_2/\partial x_n & \ldots & \partial f_m/\partial x_n \end{bmatrix}$$

which is often termed the Jacobian. Note that

$$\frac{d(e^{At})}{dt} = Ae^{At} = e^{At}A$$

In view of the above we have:
1. $\partial[A\mathbf{x}]/\partial \mathbf{x} = A^t$
2. $\partial[\mathbf{x}^t \, A\mathbf{x}]/\partial \mathbf{x} = A\mathbf{x} + A^t\mathbf{x}$
   which reduces to $(2A\mathbf{x})$ if A is a real symmetric matrix
3. $\partial[\mathbf{x}^t \, A\mathbf{y}]/\partial \mathbf{x} = A\mathbf{y}$ and $\partial[\mathbf{x}^t \, A\mathbf{y}]/\partial \mathbf{x} = A^t\mathbf{x}$

By similarity, the integral of an $n \times m$ matrix $A(t)$ with respect to t is defined by the matrix whose (i,j)th element is the integral of the (i,j)th element of the original matrix, or

$$\int A(t)\,dt = \begin{bmatrix} \int a_{11}(t)\,dt & \cdots & \int a_{1m}(t)\,dt \\ \vdots & & \\ \int a_{n1}(t)\,dt & \cdots & \int a_{nm}(t)\,dt \end{bmatrix}$$

B.12 EIGENVALUES AND EIGENVECTORS

Let A be an $n \times n$ matrix. Construct the polynomial $|sI - A|$. This is termed the characteristic polynomial of A; the zeros of this polynomial are the eigenvalues of A. If λ is an eigenvalue of A, there always exists at least one vector x satisfying the equation

$$A\mathbf{x} = \lambda^i \mathbf{x}$$

The vector $\mathbf{x}$ is termed an eigenvector of the matrix A. If λ is not a repeated eigenvalue, i.e., if it is a simple zero of the characteristic polynomial, to within a scalar multiple, x is associated with λ. If λ_i is real, the entries of $\mathbf{x}$ are real, whereas if λ_i is complex, the entries of $\mathbf{x}$ are complex.

If A has zero entries everywhere off the main diagonal, i.e., if $a_{ij} = 0$ for all i,j, with $i \neq j$, then A is termed diagonal. (Note: zero entries are still permitted on the main eigenvalue that the diagonal entries of the diagonal A are precisely the eigenvalues of A.

It is also true that for a general A,

$$|A| = \prod_{i=1}^{n} \lambda_i$$

If A is singular, A possesses at least one zero eigenvalue.

The eigenvalues of a rational function $r(A)$ of A are the numbers $r(\lambda_i)$, where λ_i are the eigenvalues of A. The eigenvalues of e^{At} are $e^{\lambda_i t}$.

If A is $n \times m$ and B is $m \times n$, with $n \geq m$, then the eigenvalues of AB are the same as those of BA together with $(n - m)$ zero eigenvalues. An important property is that

$$\mathrm{Tr}[A] = \sum_{j=1}^{n} \lambda_j$$

B.13 NORMS

A norm is a function which assigns to every vector $\mathbf{x}$ in a given space a real number denoted by $\|\mathbf{x}\|$ such that

Elements of Matrix Algebra

1. $\|\mathbf{x}\| > 0$ for $\mathbf{x} \neq 0$
2. $\|\mathbf{x}\| = 0$ if and only if $\mathbf{x} = 0$
3. $\|k\mathbf{x}\| = |k|\,\|\mathbf{x}\|$
   where k is a scalar and $|k|$ is the absolute value of k
4. $\|\mathbf{x} + \mathbf{y}\| \leq \|\mathbf{x}\| + \|\mathbf{y}\|$ for all $\mathbf{x}$ and $\mathbf{y}$
5. $|\mathbf{x}^t\mathbf{y}| \leq \|\mathbf{x}\|\,\|\mathbf{y}\|$ (Schwarz inequality)

Some of the widely used norms are

1. $\|\mathbf{x}\| = \sqrt{x_1^2 + x_2^2 + \cdots + x_n^2}$
2. $\|\mathbf{x}\| = \sum_{j=1}^{n} |x_j|$
3. $\|\mathbf{x}\| = \max_{j}\{|x_j|\}$

Extending the concept of norms to matrices we have:

1. A norm $\|A\|$ of an $n \times n$ matrix A may be defined by

$$\|A\| = \min k$$

such that

$$\|A\mathbf{x}\| \leq k\|\mathbf{x}\|$$

From the norm of $\|\mathbf{x}\|$, the above relation is equivalent to

$$\|A\|^2 = \max_{\mathbf{x}}\{\mathbf{x}^t A^t A \mathbf{x}; \mathbf{x}^t\mathbf{x} = 1\}$$

which means that $\|A\|^2$ is the maximum of the "absolute value" of the vector $A\mathbf{x}$ when $\mathbf{x}^t\mathbf{x} = 1$

2. Another norm may be defined by

$$\|A\| = \left(\sum_{i=1}^{n}\sum_{j=1}^{n}|a_{ij}|^2\right)^{1/2}$$

3. A third norm is given by

$$\|A\| = \max\left(\sum_{k}^{n}|a_{jk}|\right)$$

4. A different norm still has the form

$$\|A\| = \max_j \sqrt{\lambda_j(A^t A)}$$

which is often called the spectral norm.

All the above norms have the following properties

1. $\|A\| = \|A^t\|$
2. $\|A + B\| \leq \|A\| + \|B\|$
3. $\|AB\| \leq \|A\|\,\|B\|$
4. $\|A\mathbf{x}\| \leq \|A\|\,\|\mathbf{x}\|$

Appendix C
A Derivation of the Riccati Equation

The optimal state regulator problem has been studied thoroughly by Kalman (1960), who has established that a state feedback solution can be obtained for linear plants with quadratic performance indices. Our purpose here is to provide a procedure of deriving the Riccati equation.

Consider a plant modeled by

$$\dot{\mathbf{x}} = A\mathbf{x} + B\mathbf{u}, \qquad \mathbf{x}(0) = \mathbf{x}_0 \qquad (1)$$

and a performance index

$$I = \int_0^\infty [\mathbf{x}^t Q \mathbf{x} + \mathbf{u}^t R \mathbf{u}]\, dt \qquad (2)$$

where $Q = Q^t \geq 0$ and $R = R^t > 0$.

In terms of the output $\mathbf{y} = C\mathbf{x}$, we let Q be factored as $C^t C$. We require that the plant (1) be completely controllable and observable. By restricting $\mathbf{u}$ to the form

$$\mathbf{u} = K\mathbf{x} \qquad (3)$$

the problem, then, is to find K to minimize (2) subject to (1) and the above conditions. We apply the *calculus of variations* which permits writing

$$\mathbf{u} = \mathbf{u}^* + \eta \Delta \mathbf{u} \qquad (4)$$

where **u** is the optimal **u**, $\Delta\mathbf{u}$ is an arbitrary vector time function and η is a scalar parameter. Observe that the substitution of **u** from (4) into (2) with the aid of (3) yields the index I, a function of η. In addition, I has a minimum for $\eta = 0$ since $\mathbf{u}^*$ is the optimal value. Thus

$$\frac{dI(\eta)}{d\eta} = 0 \quad \text{at } \eta = 0 \tag{5}$$

To use (5), we substitute (3) and (4) into (1) to obtain

$$\dot{\mathbf{x}} = A\mathbf{x} + BK\mathbf{x} + B\eta\Delta\mathbf{u}$$
$$= [A + BK]\mathbf{x} + B\eta\Delta\mathbf{u} \tag{6}$$

whose solution is $\mathbf{x}(t, \eta)$, a function of both t and η. By differentiating (2) with respect to η along with (3) and (4) we obtain

$$\frac{dI}{d\eta} = \int_0^\infty \frac{d}{d\eta}[\mathbf{x}^t Q_0 \mathbf{x} + 2\eta \mathbf{x}^t K^t R\Delta\mathbf{u} + \eta^2 \Delta\mathbf{u}^t R\Delta\mathbf{u}]\, dt \tag{7}$$

where

$$Q_0 = Q + K^t RK \tag{8}$$

Note that $Q_0 = Q_0^t$. It is easy to show that

$$\frac{d}{d\eta}[\mathbf{x}^t Q_0 \mathbf{x}] = 2\mathbf{x}^t Q_0 \frac{\partial \mathbf{x}}{\partial \eta} \tag{9}$$

$$\frac{d}{d\eta}[2\eta \mathbf{x}^t K^t R\Delta\mathbf{u}] = 2\mathbf{x}^t KR\Delta\mathbf{u} + 2\eta\left[\frac{\partial \mathbf{x}^t}{\partial \eta}\right] K^t R\Delta\mathbf{u} \tag{10}$$

since $\Delta\mathbf{u}$ is independent of **u** and

$$\frac{d}{d\eta}[\eta^2 \Delta\mathbf{u}^t R\Delta\mathbf{u}] = 2\eta\Delta\mathbf{u}^t R\Delta\mathbf{u} \tag{11}$$

In view of condition (5) the derivatives (9)–(11) have to be evaluated at $\eta = 0$. The result is that (9) remains intact, but (11) vanishes and (10) becomes

$$\frac{d}{d\eta}[2\eta \mathbf{x}^t K^t R\Delta\mathbf{u}] = 2\mathbf{x}^t KR\Delta\mathbf{u} \tag{12}$$

Substituting (9) and (12) back into (7) yields

$$\left.\frac{dI}{d\eta}\right|_{\eta=0} = \int_0^\infty \left[2\mathbf{x}^t Q_0 \frac{\partial \mathbf{x}}{\partial \eta} + 2\mathbf{x}^t KR\Delta\mathbf{u}\right]_{\eta=0} dt = 0 \tag{13}$$

A Derivation of the Riccati Equation

Using the results established in Chapter 3, the solution to (6) has the form

$$\mathbf{x} = e^{[A+BK]t}\mathbf{x}_0 + \int_0^t e^{[A+BK](t-\tau)} B\eta \Delta \mathbf{u}(\tau)\, d\tau \tag{14}$$

whose derivative at $\eta = 0$ is given by

$$\left.\frac{\partial \mathbf{x}}{\partial \eta}\right|_{\eta=0} = \int_0^t e^{[A+BK](t-\tau)} B\Delta \mathbf{u}(\tau)\, d\tau \tag{15}$$

Also from (14) we have

$$\mathbf{x}|_{\eta=0} = e^{[A+BK]t}\mathbf{x}_0 \tag{16}$$

On substituting (15) and (16) into (13)

$$\int_0^\infty \Bigg[\mathbf{x}_0^t e^{[A+BK]^t t} Q_0 \int_0^t e^{[A+BK](t-\tau)} B\Delta \mathbf{u}(\tau)\, d\tau$$

$$+ \mathbf{x}_0^t e^{[A+BK]^t t} K\, R\Delta \mathbf{u}(t) \Bigg] dt = 0 \tag{17}$$

Rearranging the order of integration in (17) yields

$$\mathbf{x}_0^t \int_0^\infty e^{[A+BK]^t \tau} \Bigg[\int_0^\infty e^{[A+BK]^t \delta} Q_0 e^{[A+BK]\delta} B\, d\delta$$

$$+ K^t R \Bigg] \Delta \mathbf{u}(\tau)\, d\tau = 0 \tag{18}$$

Since (18) is valid for all $\mathbf{x}_0$ and $\Delta \mathbf{u}(\tau)$, it is necessary and sufficient that the expression in the parentheses be zero. The result is

$$RK + B^t \int_0^\infty e^{[A+BK]^t \delta} Q_0 e^{[A+BK]\delta}\, d\delta = 0 \tag{19}$$

Define

$$P = \int_0^\infty W\, d\delta \tag{20}$$

$$W = e^{[A+BK]^t \delta} Q_0 e^{[A+BK]\delta} \tag{21}$$

When the matrix $[A + BK]$ is asymptotically stable, it turns out that P satisfies

$$P[A + BK] + [A + BK]^t P + Q_0 = 0 \tag{22}$$

From (20) in (19) we arrive at
$$K = -R^{-1}B^t P \qquad (23)$$
Using (8) in (22) and rearranging we reach the matrix Riccati equation
$$PA + A^t P - PBR^{-1}B^t P + Q = 0 \qquad (24)$$
It is obvious from (20)–(24) that $P = P^t > 0$ provided the conditions of controllability, observability, and asymptotic stability are met.

Appendix D
Introduction to Random Variables and Gauss–Markov Processes

Our purpose here is to introduce some fundamental notions of probability theory, random variables and Gaussian processes. The material covered will be sufficient for understanding the techniques of estimation and related topics. More advanced treatment can be found in Sage and Melsa (1971), Astrom (1970), Anderson and Moore (1979), Meditch (1969), and Jazwinski (1970).

D.1 BASIC CONCEPTS OF PROBABILITY THEORY

We start by considering an experiment with a number of possible outcomes, examples of which are the throwing of a dice, the drawing of a card from a card deck or the picking of a colored ball from a basket of balls. Three fundamental concepts are introduced. The first is that of *sample space* Ω, the set of possible outcomes of an experiment. We will call any particular member of this set ω. As an example, for the case of throwing of a dice $\Omega = \{1, 2, 3, 4, 5, 6\}$.

The second fundamental concept is that of an *event*, which is defined as any subset of the sample space. For example, "obtain the number 5" or "obtain a red ball" are events.

The third fundamental concept is that of *probability*. A probability measure $P(\cdot)$ is a mapping from events into the real line satisfying the following axioms:

1. $P(A) \geq 0$
2. $P(\Omega) = 1$
3. For a countable set $\{A_j\}$ of mutually disjoint events, that is $A_j \cap A_m = \emptyset$ for all j, m, $P(\cap A_j) = \Sigma_j P(A_j)$. Here $\emptyset$ denotes the empty set

Some important formulae which arise from these axioms are:

1. $P(\emptyset) = 0$
2. $P(A) \leq 1$
3. $P(A^*) = 1 - P(A)$; A^* is the complement of A, since $A^* \cap A = \emptyset$ and $A^* \cup A = \Omega$.
4. $P(A \cap B^*) = P(A) - P(A \cap B)$. This is true in view of the fact that the events $(A \cap B^*)$ and $(A \cap B)$ are mutually disjoint and their union is A

If we write the event $(A \cap B)$ as the union of two mutually exclusive events, we obtain

$$P(A \cap B) = P(A) + P(A^* \cap B)$$

In view of the previous result, (4), we have

$$P(A^* \cap B) = P(B) - P(A \cap B)$$

Combining the above two relations yields:

5. $P(A \cap B) = P(A) + P(B) - P(A \cap B)$. We note that result (5) reduces to the third axiom when A and B are mutually disjoint.

Suppose A and B are two events and an experiment is conducted with the result that event B occurs. The probability that event A has also occurred, *the conditional probability of A given B*, written as $P(A|B)$ is given by

$$P(A|B) = P(AB)/P(B) \qquad (1)$$

where $P(AB) = P(A \cap B)$ is the joint probability of A and B. We point out that

1. $P(B) = 0$ in (1), otherwise the definition of $P(A|B)$ would be meaningless.
2. $P(A|B)$ for fixed B and variable A satisfies the probability measure axioms.

We now consider the notion of independence. Events $A_1, A_2, \ldots, A_n$

are *mutually independent* if and only if

$$P(A_{j_1} \cap A_{j_2} \cap \cdots \cap A_{j_m}) = P(A_{j_1})P(A_{j_2})\ldots P(A_{j_m}) \quad (2)$$

for all integers $j_1, j_2, \ldots, j_m$ selected from the set of integers $[1, 2, \ldots, n]$ where no two are the same. We caution the reader to distinguish between the notions of independence and of mutually disjoint events.

In the case of two independent events A and B, then (2) becomes

$$P(AB) = P(A)P(B) \quad (3)$$

which, when used in (1), yields

$$P(A|B) = P(A)$$

This result agrees with our intuitive idea of independence and conditional probability in that, since B and A are independent, we do not need to know B to arrive at the probability $P(A|B)$. Consider the situation of three events A, B, C such that each pair is mutually independent, that is

$$P(AB) = P(A)P(B)$$
$$P(BC) = P(B)P(C)$$
$$P(CA) = P(C)P(A)$$

It is easy to show that these conditions do not imply that A, B, C are mutually independent.

We say that two events A and B are *conditionally independent given an event* C when

$$P(AB|C) = P(A|C)P(B|C) \quad (4)$$

If $A_j, j = 1, 2, \ldots, n$ are mutually disjoint and $UA_j = \Omega$, then for arbitrary B we have

$$P(B) = \sum_j P(B|A_j)P(A_j) \quad (5)$$

An important consequence of (1) is *Bayes' Rule*:

$$P(A|B) = P(B|A)P(A)/P(B) \quad (6)$$

provided that $P(B) = 0$. Again consider n mutually disjoint events A_j with $U_j A_j = \Omega$. By virtue of (5) and (6) we have

$$P(A_j|B) = P(B|A_j)P(A_j)/\left\{\sum_j P(B|A_j)P(A_j)\right\} \quad (7)$$

We now proceed to consider random variables and examine their mathematical properties.

D.2 MATHEMATICAL PROPERTIES OF RANDOM VARIABLES

It is often desirable to have a procedure by which one can evaluate the output records of an experiment. A suitable way would be to measure quantities associated with the outcome of an experiment. Such a quantity is called *a random variable*. Strictly speaking, a random variable X is a real valued function from the ω in a sample space Ω to the set of real numbers. A value of the random variable X is the number $X(\omega)$ when the outcome ω occurs. When X takes on a discrete value, it is called a discrete random variable.

Since by definition a random variable is a function on a probability space, it is often of interest to be able to know the probability that a certain value of the random variable $X(\cdot)$ occurs in a given set. We adopt the notation $P(X - \alpha)$ to mean $P([\omega | X(\omega) = \alpha])$, that is the probability of the subset of Ω consisting of those outcomes ω for which $X(\omega) = \alpha$. In a similar way, $P(X > 0)$ means $P([\omega | X(\omega) > 0])$. It is required for X to be a random variable that

1. $P(X = -\infty) = P(X = +\infty) = 0$
2. For all real β, the quantity $[\omega | X(\omega) \leq \beta]$ is an event which implies that

$$P([\omega | X(\omega) \leq \beta]) = P(X \leq \beta)$$

Distribution Functions

One way of describing random variables is in terms of their distribution functions. Given a random variable X, *the distribution function $F(x)$ is a mapping from the reals to the interval $[0, 1]$*:

$$F(x) = P(X \leq x) \tag{8}$$

where the argument x is a typical value. The distribution function is monotonously increasing in the sense that

$$\lim_{x \to \infty} F(x) = 1 \quad \text{and} \quad \lim_{x \to \infty} F(x) = 0$$

Another way of describing random variables is in terms of their density functions. When $F(x)$ is continuous and differentiable everywhere, *the probability density function $p(x)$ associated with the random variable X*

Random Variables and Gauss–Markov Processes

is

$$p(x) = dF(x)/dx \tag{9}$$

From (8) and (9), it is readily seen that $p(x)\,dx$ to first order is $P(x < X < x + dx)$.

A random vector $\mathbf{X}$ of order n consists of n random variables $X_1, X_2, \ldots, X_n$ with distribution and probability density functions defined by

$$F(\mathbf{x}) = P[(X_1 \leq x_1) \cap \cdots \cap (X_n \leq x_n)] \tag{10}$$

$$p(\mathbf{x}) = (\partial^n / \partial x_1 \ldots \partial x_n) F(\mathbf{x}) \tag{11}$$

Consider the events $[X \leq x]$ and $[Y \leq y]$ associated with the random variables X and Y, respectively. If these events are independent for all x and y, then it follows from (3) and (8) that the joint distribution function is

$$F(x, y) = F(x)F(y) \tag{12}$$

and correspondingly the joint probability density function is

$$p(x, y) = p(x)p(y) \tag{13}$$

Let $h(\cdot)$ be a well-behaved scalar valued function of a scalar variable and X a random variable. If X and Z are independent random variables, so are $h(X)$ and $g(Z)$.

Mathematical Expectations

We now define mathematical expectation. The *mathematical expectation* or *mean* of a random variable X, written as $E[X]$, is the number defined by

$$E[X] = \int_{-\infty}^{+\infty} x p(x)\,dx \tag{14}$$

where the integral is assumed absolutely convergent. In the same way, a function $g(X)$ of the random variable X will have the mathematical expectation

$$E[g(X)] = \int_{-\infty}^{+\infty} g(x)p(x)\,dx \tag{15}$$

As an operator, the mathematical expectation has the following properties:

1. For a constant β, $E[\beta] = \beta$.

2. It is a linear operator. More precisely, if $g_1(X)$ and $g_2(X)$ are two functions of the random variable X and α and β are two constants, then
$$E[\alpha g_1(X) + \beta g_2(X)] = \alpha E[g_1(X)] + \beta E[g_2(X)]$$

3. If $X_1, \ldots, X_n$ denote mutually independent random variables, then
$$E[X_1 X_2 \ldots X_n] = E[X_1]E[X_2] \ldots E[X_n] \tag{16}$$

The *variance* σ^2 of a random variable provides a measure of the dispersion around the mean value and is defined by
$$\sigma^2(X) = E[(X - E[X])^2]$$
$$= \int_{-\infty}^{+\infty} (x - E[X])^2 p(x)\, dx \tag{17}$$

An alternative form of (17) is
$$\sigma^2(X) = E[X^2 - 2E[X]X + (E[X])^2]$$
$$= E[X^2] - 2(E[X])^2 + (E[X])^2$$
$$= E[X^2] - (E[X])^2 \tag{18}$$

where we have made use of the properties of the expectation operator. Form (18) is easy to remember. We note that the definition of the mean generalizes in an obvious way to a vector. Let $\mathbf{X} = [X_1 X_2 \ldots X_n]^t$. Thus,
$$E[\mathbf{X}] = [E[X_1]E[X_2] \ldots E[X_n]]^t \tag{19}$$

For random n-vector $\mathbf{X}$, the variance is now replaced by the $(n \times n)$ *covariance matrix* $\text{Cov}(\mathbf{X})$ given by

$\text{Cov}(\mathbf{X}) = E[(\mathbf{X} - E[\mathbf{X}])(\mathbf{X} - E[\mathbf{X}])^t]$

$$= \begin{bmatrix} \sigma^2(X_1) & \text{Cov}(X_1, X_2) & \cdots & \text{Cov}(X_1, X_n) \\ \text{Cov}(X_1, X_2) & \sigma^2(X_2) & \cdots & \vdots \\ \vdots & & & \vdots \\ \text{Cov}(X_1, X_n) & & \cdots & \sigma^2(X_n) \end{bmatrix} \tag{20}$$

where the superscript t denotes matrix transpose. From (18) and (19) we note that the variance is always nonnegative, and the covariance matrix is nonnegative definite and symmetric.

Two Random Variables

For two random variables X and Y we summarize some important relations:

Random Variables and Gauss–Markov Processes

1. The *conditional probability density of X given Y*. $p(x|y)$ is given by Bayes' Rule.
$$p(x|y) = p(x, y)/p(y) \tag{21}$$
from which one obtains the important formula
$$p(x) = \int_{-\infty}^{+\infty} p(x|y)p(y)\,dy \tag{22}$$

Also, if X and Y are independent, then using (13) in (21), it reduces to
$$p(x|y) = p(x) \tag{23}$$

2. By definition
$$E[X] = \int_{-\infty}^{+\infty}\int_{-\infty}^{+\infty} xp(x, y)\,dx\,dy \tag{24}$$

$$E[Y] = \int_{-\infty}^{+\infty}\int_{-\infty}^{+\infty} yp(x, y)\,dx\,dy \tag{25}$$

$$E[X^2] = \int_{-\infty}^{+\infty}\int_{-\infty}^{+\infty} x^2 p(x, y)\,dx\,dy \tag{26}$$

$$E[Y^2] = \int_{-\infty}^{+\infty}\int_{-\infty}^{+\infty} y^2 p(x, y)\,dx\,dy \tag{27}$$

$$E[XY] = \int_{-\infty}^{+\infty}\int_{-\infty}^{+\infty} xyp(x, y)\,dx\,dy \tag{28}$$

and
$$\mathrm{Cov}(X, Y) = E[(X - E[X])(Y - E[Y])] \tag{29}$$

The quantity $E[XY]$, defined by (28), is often called *the correlation of X and Y*. As a consequence, we define *the coefficient of correlation* between X and Y by:
$$\rho(X, Y) = \mathrm{Cov}(X, Y)/\sqrt{\sigma^2(X)}\sqrt{\sigma^2(Y)} \tag{30}$$
provided that the variances of X and Y are finite and strictly positive.

3. The random variables X and Y are said to be *uncorrelated* if $E[X^2]$ and $E[Y^2]$, as defined by (26) and (27), respectively, are finite and
$$\mathrm{Cov}(X, Y) = 0 \tag{31}$$
From (30) this implies that
$$\rho(X, Y) = 0 \tag{32}$$

4. Suppose X and Y are two independent random variables. Then

it is easy to show that they are uncorrelated. Starting from (29), expanding and using (18) we obtain

$$\begin{aligned} \text{Cov}(X, Y) &= E[XY] - E[X]E[Y] - E[X]E[Y] + E[X]E[Y] \\ &= E[X]E[Y] - E[X]E[Y] \\ &= 0 \end{aligned}$$

which agrees with (31). Therefore, an alternative way to define two uncorrelated random variables is when $E[XY] = E[X]E[Y]$. If $E[XY] = 0$, the random variables X and Y are termed *orthogonal*.

We caution the reader that two uncorrelated random variables need not necessarily be independent. The absence of correlation implies that the general condition

$$E[h(X)g(Y)] = E[h(X)]E[g(Y)]$$

is only satisfied for $h(X) = X$, whilst independence requires that this condition be satisfied for all functions $h(\cdot)$ and $g(\cdot)$. The conditionally expected value of a random variable X, given that Y has taken the value y, is

$$\begin{aligned} E[X|Y=y] &= E[X|Y] \\ &= \int_{-\infty}^{+\infty} x p(x|y)\, dx \end{aligned} \qquad (33)$$

Note that the result of integration will be a number, depending on y. But since y is the outcome of a random experiment, the conditional expectation is a random variable. To calculate its expected value, we proceed as follows: from (16) and (33) we obtain

$$\begin{aligned} E[E[X|Y]] &= \int_{-\infty}^{+\infty} p(y) \int_{-\infty}^{+\infty} x p(x|y)\, dx\, dy \\ &= \int_{-\infty}^{+\infty} x \left\{ \int_{-\infty}^{+\infty} p(x|y) p(y)\, dy \right\} dx \end{aligned}$$

using (22) it simplifies to

$$\begin{aligned} &= \int_{-\infty}^{+\infty} x p(x)\, dx \\ &= E[X] \end{aligned} \qquad (34)$$

If the random variables X and Y are independent, then it follows from (23) and (33) that

Random Variables and Gauss–Markov Processes

$$E[X|Y] = E[X] \tag{35}$$

and more generally

$$E[h(X)|Y] = E[h(X)] \tag{36}$$

for any function $h(\cdot)$. We can generalize (33) by using $g(X, Y)$ in place of X to obtain

$$E[g(X, Y)|Y] = \int_{-\infty}^{+\infty} g(x, y) p(x, y) \, dx \tag{37}$$

and again the result is a random variable which is a function of the random variable Y. To emphasize this point, let $g(X, Y) = g_1(X)g_2(Y)$, then (37) becomes

$$E[g(X, Y)|Y] = g_2(Y)E[g_1(X)|Y] \tag{38}$$

which represents a useful formula.

Since the various notions of random variables can be easily extended from random scalars to random vectors, with the notation (19) in mind, we next go on to consider Gaussian random vectors since most of our analysis for state and parameter estimation will assume that the probability distributions are Gaussian.

D.3 STOCHASTIC PROCESSES

Definition and Properties

Hitherto, most of our discussions have been centered around an experiment with a number (or an n-tuple of numbers) of possible outcomes and the time factor has been set aside. In this section we extend the previous analysis to the case where the outcome is a function mapping an underlying time set (commonly nonnegative integers) into the reals. Thus, we will deal with a random process rather than a random variable. More precisely, *a discrete-time random process* results in a function mapping from $\omega \in \Omega$ to a set of values $x_\omega(k)$ for $k \in I_t \triangleq \{0, 1, 2, \ldots\}$, the discrete-time set. Looked at in this light, a scalar discrete-time random process behaves like an infinite-dimensional random vector. We adopt the notation $\{\mathbf{x}(k)\}$ to denote $\{(\mathbf{x}_\omega(\mathbf{k}), \mathbf{k}) | \mathbf{k} \geq \mathbf{0}, \omega \in \Omega\}$, that is a particular sequence of vectors taken as a result of an experiment. The quantity $\mathbf{x}(m)$ will then denote the random vector obtained by looking at the process at time m, as well as the value taken by that vector.

From the above discussion, it is readily seen that a random process is just a generalization of the concept of a random variable. Hence, most of the properties presented in Section D.1 will carry over here. As an example, let m be an arbitrary integer and $(k_1, k_2, \ldots, k_m)$ be arbitrary instants in the underlying time set I_t. Then the set of all probability densities

$$p\{x(k_1), x(k_2), \ldots, x(k_m)\}$$

or the corresponding distribution functions can serve to define the probability structure of the random process. In what follows, we will provide some fundamental properties of random processes.

The *mean* $m(k)$ of a random process is simply the time function $E[x(k)]$. Given two discrete-time instants j and $r \in I_t$ and let

$$\mathbf{x}(\cdot)[x_1(\cdot) x_2(\cdot) \ldots x_n(\cdot)]^t$$

then the *autocorrelation matrix* $R(j, r)$ is the set of quantities $E[\mathbf{x}(j)\mathbf{x}(r)^t]$, written in full as

$$R(j,r) = \begin{bmatrix} E[x_1(j)x_1(r)] & E[x_1(j)x_2(r)] & \ldots & E[x_1(j)x_n(r)] \\ E[x_2(j)x_1(r)] & E[x_2(j)x_2(r)] & \ldots & \vdots \\ \vdots & \vdots & & \vdots \\ E[x_n(j)x_1(r)] & \ldots & & E[x_n(j)x_n(r)] \end{bmatrix}$$
(39)

In a similar way, *the covariance matrix* $W(j, r)$ is the set of quantities $E[\{\mathbf{x}(j) - \mathbf{m}(j)\}\{\mathbf{x}(r) - \mathbf{m}(r)\}^t]$ for all j and r. Its full description takes a form similar to (39) with appropriate changes. When $j = r$, the covariance matrix $W(j, j)$ becomes a nonnegative definite symmetric matrix. Thus, we see that a random process is entirely characterized by the properties of the random variable (or vector) at different discrete time instants.

The *first order densities* of a process are the set of densities $p\{\mathbf{x}(j)\}$ for all $j \in I_t$. The *second order densities* of a process are the set $p\{\mathbf{x}(j), \mathbf{x}(r)\}$ for all $j, r \in I_t$. Given these densities, we can apply the rules of the previous section like (14), (17), to obtain the mean and variance of a process.

Define $[j_m, j_{m+1}]$ as a set of nonintersecting intervals in the discrete-time set I_t. Then a process is said to have *uncorrelated* (*orthogonal* or *independent*) increments if the quantity $[\mathbf{x}(j_m) - \mathbf{x}(j_{m+1})]$ is a sequence of *uncorrelated* (*orthogonal or independent*) random vectors.

A process $\{\mathbf{x}(k)\}$ is said to be *strict-sense stationary*, or simply *stationary* if its associated probability densities are unaffected by time translation; that is, for arbitrary integer m and discrete times $j_1, \ldots, j_m$ and s,

Random Variables and Gauss–Markov Processes

$$p\{\mathbf{x}(j_1), \mathbf{x}(j_2), \ldots, \mathbf{x}(j_m)\} = p\{\mathbf{x}(j_{1+s}), \mathbf{x}(j_{2+s}) \ldots, \mathbf{x}(j_{m+s})\} \quad (40)$$

If we consider two first order densities, we have

$$p\{\mathbf{x}(j)\} = p\{\mathbf{x}(j+s)\} \quad (41)$$

which implies that the first order probability density is, in this case, independent of j. Consequently, the mean $\mathbf{m}(j)$ of the process $\{\mathbf{x}(j)\}$ is a constant,

$$E[\mathbf{x}(j)] = \mathbf{m}(j) = \mathbf{m} \quad (42)$$

For the second order density we have

$$R(j, r) = E[\{\mathbf{x}(j) - \mathbf{m}\}\{\mathbf{x}(r) - \mathbf{m}\}^t]$$
$$= R(j - r) \quad (43)$$

that is the autocorrelation function depends only on the difference $(j - r)$.

We now move a step forward and consider pairs of random processes. In view of the above discussions, we summarize the important properties:

1. Two random processes $\{\mathbf{x}(k)\}$ and $\{\mathbf{y}(k)\}$ are said to be *uncorrelated* if

$$E[\mathbf{x}(j)\mathbf{y}^t(r)] = E[\mathbf{x}(j)]E[\mathbf{y}^t(r)] \quad (44)$$

   for all $j, r \in I_t$.

2. Two random processes $\{\mathbf{x}(k)\}$ and $\{\mathbf{y}(k)\}$ are said to be *orthogonal* if

$$E[\mathbf{x}(j)\mathbf{y}^t(r)] = [0] \quad (45)$$

   for all $j, r \in I_t$.

3. Two random processes $\{\mathbf{x}(j)\}$ and $\{\mathbf{y}(k)\}$ are said to be *independent* if for any sets $\{j_i\}$ and $\{r_i\}$ the vector random variable $[\mathbf{x}^t(j_1)\mathbf{x}^t(j_2) \ldots \mathbf{x}^t(j_n)]^t$ is independent of the vector random variable $[\mathbf{y}^t(r_1)\mathbf{y}^t(r_2) \ldots \mathbf{y}^t(r_m)]^t$.

4. Two random processes $\{\mathbf{x}(k)\}$ and $\{\mathbf{y}(k)\}$ are *jointly stationary* if the combined process $\{[\mathbf{x}^t(k)\mathbf{y}^t(k)]^t\}$ is *stationary*.

Gauss and Markov Processes

Having presented the description and mathematical properties of a random process, we now direct attention to a particular class of stochastic processes called *Markov processes*. Consider a set of ordered parameters $j_0 < j_1 < j_2 < \cdots < j_n$. A stochastic process $\{\mathbf{x}(j)\}$ is called a

Markov process if we can write

$$p\{\mathbf{x}(j_n)|\mathbf{x}(j_{n-1}), \ldots, \mathbf{x}(j_0)\} = p\{\mathbf{x}(j_n)|\mathbf{x}(j_{n-1})\} \qquad (46)$$

which means that the entire past history of the process is contained in the last state.

We now develop an expression for the joint probability density function for a Markov process. Using Bayes' theorem, similarly to (21), we have

$$p\{\mathbf{x}(j_n), x(j_{n-1}), \ldots, \mathbf{x}(j_0)\} =$$
$$p\{\mathbf{x}(j_n)|\mathbf{x}(j_{n-1}), \ldots, \mathbf{x}(j_0)\} xp\{\mathbf{x}(j_{n-1}), \ldots, \mathbf{x}(j_0)\} \qquad (47)$$

If the process is Markovian, then from (46) and (47) we obtain

$$p\{\mathbf{x}(j_n), \mathbf{x}(j_{n-1}), \ldots, \mathbf{x}(j_0)\} = p\{\mathbf{x}(j_n)|\mathbf{x}(j_{n-1})\}$$
$$\times p\{\mathbf{x}(j_{n-1}), \ldots, \mathbf{x}(j_0)\} \qquad (48)$$

Doing the same operation on $p\mathbf{x}(j_{n-1}), \ldots, \mathbf{x}(j_0)$ and repeating, we finally obtain

$$p\{\mathbf{x}(j_n), \mathbf{x}(j_{n-1}), \ldots, \mathbf{x}(j_0)\} = p\{\mathbf{x}(j_n)|\mathbf{x}(j_{n-1})\}$$
$$\times p\{\mathbf{x}(j_{n-1})|\mathbf{x}(j_{n-2})\} \ldots p\{\mathbf{x}(j_1)|\mathbf{x}(j_0)\} p\{\mathbf{x}(j_0)\}$$

This means that we can describe a Markov process completely in terms of its *transition probability densities* $p\{\mathbf{x}(j_m)|\mathbf{x}(j_{m-1})\}$ and the distribution of the initial state.

Another important class of stochastic processes is *white noise*. Recall that a stationary discrete-time stochastic process $\{\mathbf{x}(k)\}$ with zero mean is one whose autocorrelation function is

$$R(s) = E[\{\mathbf{x}(j+s)\}\{\mathbf{x}(j)\}^t] \qquad (49)$$

The *power spectrum* of this random process is given by

$$\phi(z) = \sum_{k=-\infty}^{+\infty} z^{-k} R(k) \qquad (50)$$

that is, the power spectrum is the Z-transform of the autocorrelation function of a stationary random process. In the important special case, where $\phi(z)$ is constant, we obtain white noise. Note the analogy with white light which contains all frequencies in its spectrum. Constancy of the power spectrum is equivalent to

$$E[\mathbf{x}(j)\mathbf{x}^t(r)] = C\delta_{jr} \qquad (51)$$

for some constant matrix C. The discrete-time (Kronecker) delta func-

Random Variables and Gauss–Markov Processes

tion δ_{jr} is 0 for $j \neq r$ and 1 for $j = r$. In this case, the vectors $\mathbf{x}(j_n), \mathbf{x}(j_{n-1}), \ldots, \mathbf{x}(j_0)$ constitute a sequence of uncorrelated random variables.

Next, we consider Gaussian random variables, vectors and processes. a random variable X is a *Gausian random variable* if its probability density is of the form

$$p(x) = (1\sqrt{2\pi\sigma^2}) \exp[-(x-m)^2/2\sigma^2] \qquad (52)$$

Simple evaluation of $E[X]$ in (14) using (52) shows that

$$E[X] = m \qquad (53)$$

and the variance $E[(X - E[X])^2]$ is σ^2.

Sometimes the notation "X is $N(m, \sigma^2)$" is used to denote that X is Gaussian (normal) with mean m and variance σ^2.

Let $\mathbf{X}$ be a random m-vector. If its covariance matrix $\mathrm{Cov}(\mathbf{X})$ in (20) is nonsingular then we say that $\mathbf{X}$ is *Gaussian* if and only if its probability density is of the form

$$p(\mathbf{x}) = (1/[2\pi]^{n/2})(1/\det[W]^{1/2}) \exp[-\tfrac{1}{2}(\mathbf{x}-\mathbf{m})^t W^{-1}(\mathbf{x}-\mathbf{m})] \qquad (54)$$

for some vector m and matrix W. By similarity to the scalar case, we can easily show that

$$E[\mathbf{X}] = \mathbf{m} \qquad (55)$$

and

$$E[(\mathbf{X} - \mathbf{m})(\mathbf{X} - \mathbf{m})^t] = W \qquad (56)$$

Likewise, we write "$\mathbf{X}$ is $N(\mathbf{m}, W)$" to denote that $\mathbf{X}$ is Gaussian with mean $\mathbf{m}$ and covariance W.

From these definitions, we see why Gaussian distributions are so attractive. The Gaussian probability density function can be described uniquely in terms of the two quantities: mean and covariance so that we do not have to worry about higher order moments. This is fortunate, since many physical phenomena can be described in terms of Gaussian distributions.

Due to the importance of Gaussian random variables in our work on estimation, we prove below some of its basic properties

1. Given a Gaussian random vector $\mathbf{X} = [X_1 \ldots X_n]^t$, any vector formed by some of its components is also Gaussian, that is $\mathbf{Y} = [X_j \ldots X_m]^t$ is Gaussian for all $j \neq m \leq n$.
2. When e random variables $X_1, \ldots, X_n$, which comprise a Gaussian random vector $\mathbf{X}$, are uncorrelated then they are independent. This

is easy to see, since in this case the covariance matrix W and its inverse are diagonal implying that

$$p(\mathbf{x}) = p(x_1)p(x_2) \ldots p(x_n) \tag{57}$$

where A is a constant matrix and $\mathbf{b}$ is a vector. Then by simple evaluation we can show that

$$E[\mathbf{Y}] = A\mathbf{m} + \mathbf{b} \tag{58}$$

and

$$\text{Cov}[\mathbf{Y}] = AWA^t \tag{59}$$

that is $\mathbf{Y}$ is $N(A\mathbf{m} + \mathbf{b}, AWA^t)$.

Finally, we conclude this section by defining Gaussian processes. Let $j_1, \ldots, j_n$ be any selection of points in the discrete-time set I_t. A random process is a *Gaussian random process* if the random variables $x(j_1)$, $m(j_2), \ldots, x(j_n)$ are jointly Gaussian, that is

$$p\{x(j_1), \ldots, x(j_n)\} = (1/[2\pi]^{n/2})(1/\det[W]^{1/2})$$
$$\times \exp[-\tfrac{1}{2}(\mathbf{x} - \mathbf{m})^t W^{-1}(\mathbf{x} - \mathbf{m})] \tag{60}$$

where

$$x = [x(j_1)x(j_2) \ldots x(j_n)]^t$$
$$m = E[x(j_i)]$$
$$\mathbf{m} = [m_1 m_2 \ldots m_n]^t$$
$$w_{ir} = E[\{x(j_i) - m_i\}\{x(j_r) - m_r\}]$$
$$W = [w_{ir}] \tag{61}$$

A complete probabilistic description of the process is thus provided by $E[x(j_i)]$ and $\text{Cov}[x(j_i), x(j_r)]$ for all j_i and j_r.

D.4 LINEAR DISCRETE MODELS WITH RANDOM INPUTS

We shall now consider the application of the previous concepts to linear discrete-time dynamical models described by

$$\mathbf{x}(k + 1) = A(k)\mathbf{x}(k) + G(k)\mathbf{w}(k) \tag{62}$$

$$y(k) = H(k)\mathbf{x}(k) \tag{63}$$

$$\mathbf{z}(k) = \mathbf{y}(k) + \mathbf{v}(k) \tag{64}$$

where the integer k denotes the discrete-time instant, and $\mathbf{x}(k)$ is the

Random Variables and Gauss–Markov Processes

system state, $\mathbf{y}(k)$ is the system output when the output disturbance is absent, $\mathbf{z}(k)$ is the measured system output, $\mathbf{w}(k)$ is the random input to the system, and $\mathbf{v}(k)$ is the output disturbance.

We have $\{\mathbf{x}(k)\}$, $\{\mathbf{y}(k)\}$, $\{\mathbf{z}(k)\}$, $\{\mathbf{w}(k)\}$, and $\{\mathbf{v}(k)\}$ representing the system process, output process, measurement process, input noise process and output noise process, respectively. Certain assumptions are needed:

1. The processes $\{\mathbf{v}(k)\}$ and $\{\mathbf{w}(k)\}$ are each white noise processes. This means that the random vectors $\mathbf{v}(j)$, $\mathbf{v}(r)$ are independent for any j and r with $j \neq r$. Similarly, $\mathbf{w}(j)$, $\mathbf{w}(r)$ are independent random vectors for any j and r with $j \neq r$.
2. The processes $\{\mathbf{v}(k)\}$ and $\{\mathbf{w}(k)\}$ are individually zero mean, Gaussian random processes with known covariances. This implies that

$$E[\mathbf{v}(k)] = \mathbf{0} \tag{63}$$

$$E[\mathbf{w}(k)] = \mathbf{0} \tag{64}$$

$$E[\mathbf{v}(k)\mathbf{v}^t(j)] = R(k)\delta_{kj} \tag{65}$$

$$E[\mathbf{w}(k)\mathbf{w}^t(j)] = Q(k)\delta_{kj} \tag{66}$$

with $R(k)$ and $Q(k)$ being nonnegative definite matrices for all k.
3. The processes $\{\mathbf{v}(k)\}$ and $\{\mathbf{w}(k)\}$ are independent processes. In view of the above assumption and the zero mean assumption, we have

$$E[\mathbf{v}(k)\mathbf{w}^t(j)] = 0 \tag{67}$$

for all k and j.
4. The initial state $\mathbf{x}(0)$ is a Gaussian random vector with a known mean $\mathbf{m}_0$ and known covariance W_0, that is

$$E[\mathbf{x}(0)] = \mathbf{m}_0 \tag{68}$$

$$E[(\mathbf{x}(0) - \mathbf{m}_0)(\mathbf{x}(0) - \mathbf{m}_0)^t] = W_0 \tag{69}$$

5. The noise processes $\{\mathbf{v}(k)\}$ and $\{\mathbf{w}(k)\}$ are independent of $\mathbf{x}(0)$, that is

$$E[\mathbf{x}(0)\mathbf{v}^t(k)] = 0 \quad \text{for all } k \tag{70}$$

$$E[\mathbf{x}(0)\mathbf{w}^t(k)] = 0 \quad \text{for all } k \tag{71}$$

For convenience, we can sum up the above assumptions as follows. The noise processes $\{\mathbf{v}(k)\}$ and $\{\mathbf{w}(k)\}$ are zero mean, independent Gaussian processes with covariances given by (65) and (66). The initial state is $N(\mathbf{m}_0, w_0)$ which is independent of the processes $\{\mathbf{v}(k)\}$ and $\{\mathbf{w}(k)\}$.

We now provide some important properties of the random process $\{\mathbf{x}(k)\}$ of the system (62)–(64). The first property is that $\mathbf{x}(k)$ *is a Gaussian random vector*. To show this we use (62) iteratively to yield

$$\mathbf{x}(k) = \Gamma(k, 0)\mathbf{x}(0) + \sum_{j=0}^{k-1} \Gamma(k, j+1)G(j)\mathbf{w}(j) \qquad (72)$$

where the transition matrix $\Gamma(s, r)$,

$$\Gamma(s, r) = A(s-1)A(s-2)\ldots A(r), \quad s > r, \Gamma(s, s) = I$$

$$\Gamma(s, r)\Gamma(r, m) = \Gamma(s, m) \quad \text{for all } s, r \text{ and } m$$

$$\text{with } s \geq r \geq m \qquad (73)$$

Since the random vectors $\mathbf{x}(0), \mathbf{w}(0), \mathbf{w}(1), \ldots, \mathbf{w}(k-1)$ are individually Gaussian and independent, then they are jointly Gaussian. By virtue of the fact that linear transformations of Gaussian random vectors preserve their Gaussian character, it follows that $\mathbf{x}(k)$ is a Gaussian random vector.

Recall the property that a random vector derived from the entries of a Gaussian random vector is also Gaussian. Then, for arbitrary s and j_i, $i = 1, \ldots, s$, the set of random vectors $\mathbf{x}(j_i)$ is jointly Gaussian. This leads to the second property: $\{x(k)\}$ *is a Gaussian random process*.

A little reflection on (72) will show that $\mathbf{x}(m)$ depends on the first m values of the noise process $\{\mathbf{w}(k)\}$ and on the initial state $\mathbf{x}(0)$. Therefore, $\mathbf{x}(m)$ is independent of $\mathbf{w}(m)$ and we can show in the same way that $\mathbf{x}(m-1), \ldots, \mathbf{x}(0)$ are all independent of $\mathbf{w}(m)$. We could thus write

$$p\{\mathbf{x}(k)|\mathbf{x}(k-1), \ldots, \mathbf{x}(0)\} = p\{\mathbf{x}(k)|\mathbf{x}(k-1)\} \qquad (74)$$

Because of (46) and (74), the process $\{\mathbf{x}(k)\}$ is *Markovian*. To summarize, we say that the state of the system (62) is a *Gauss–Markov process*.

For essentially the same reasons, one can show that the measurement process $\{\mathbf{z}(k)\}$ *is a Gaussian process*. In view of the assumptions made on the noise processes, we see that $\{x(k)\}$ *and* $\{z(k)\}$ *are jointly Gaussian processes*. But, since the output process $\{\mathbf{y}(k)\}$ is not white and the random variables $\mathbf{y}(s)$ and $\mathbf{y}(r)$ for $|s - r| > 1$ may be correlated, we conclude that the process $\{\mathbf{z}(k)\}$ *is a non-Markov process*. To elaborate more, it may be possible that $\mathbf{y}(m-2)$ and $\mathbf{y}(m-1)$ jointly convey more information about $\mathbf{y}(m)$ than $\mathbf{y}(m-1)$ alone.

Random Variables and Gauss–Markov Processes   625

Hence, the past history is not necessarily included in the last value and one may need more records to recover the present.

Since the processes $\{x(k)\}$ and $\{z(k)\}$ are jointly Gaussian, their probabilistic properties are entirely determined by their means and covariances. Our purpose now is to determine the behavior of these means and covariances along the discrete time set. We start by evaluating the evolution of the means.

From (72), the linearity of the expectation operator, and making use of (63), (68), we obtain

$$E[x(k)] = \Gamma(k,0)m_0 \tag{75}$$

Equivalently,

$$\begin{aligned} E[x(k+1)] &= m(k+1) \\ &= A(k)E[x(k)] \\ &= A(k)m(k) \end{aligned} \tag{76}$$

Similarly, from (64) we have

$$\begin{aligned} E[z(k)] &= H(k)E[x(k)] \\ &= H(k)m(k) \end{aligned} \tag{77}$$

Note that (76) is a recursive relationship which enables us to compute the mean at each discrete-instant.

Next consider the covariance functions. The covariance matrix at the $(k+1)$th instant is defined by

$$W(k+1) = E[\{x(k+1) - m(k+1)\}\{x(k+1) - m(k+1)\}^t] \tag{78}$$

To calculate the inner term in (78) we combine (62) and (76) to obtain

$$\begin{aligned} x(k+1) - m(k+1) &= A(k)[x(k) - m(k)] \\ &\quad + G(k)w(k) \end{aligned}$$

Substituting the above expression into (78), it becomes

$$\begin{aligned} W(k+1) = E[\{A(k)[x(k) - m(k)] + G(k)w(k)\} \\ \times \{[x(k) - m(k)]^t A^t(k) + w^t(k)G^t(k)\}] \end{aligned} \tag{79}$$

We have already seen that the random sequences $x(k)$ and $w(k)$ are independent so that

$$E[x(k)w^t(k)] = E[w(k)x^t(k)] = 0$$

In addition

$$E[\mathbf{m}(k)\mathbf{w}^t(k)] = \mathbf{m}(k)E[\mathbf{w}^t(k)] = 0$$

and

$$E[\mathbf{w}(k)\mathbf{m}^t(k)] = 0$$

since the noise is assumed to have a zero mean, see (64). The use of the above expressions in (79) with some manipulations, reduces it to

$$W(k + 1) = A(k)W(k)A^t(k) + G(k)Q(k)G^t(k) \tag{80}$$

where we have used (66) and the linearity property of the expectation operator. This constitutes a difference equation for $W(k)$, allowing computation of this matrix recursively, starting with the known matrix W_0. Given that the process is Gaussian, the two quantities $\mathbf{m}(k + 1)$ and $W(k + 1)$, defined by (76) and (80), are sufficient to determine the probabilistic structure of the state $\mathbf{x}(k + 1)$.

We can equally determine a general expression for $W(k)$. To accomplish this we start from the definition

$$W(k, j) = E[\{\mathbf{x}(k) - \mathbf{m}(k)\}\{\mathbf{x}(j) - \mathbf{m}(j)\}^t] \quad \text{for } k \geq j \tag{81}$$

Following a parallel development, it can be shown that

$$W(k, j) = \Gamma(k, j)\{\Gamma(j, 0)W_0\Gamma^t(j, 0) + \sum_{m=0}^{j-1} \Gamma(j, m + 1)$$
$$\times G(m)Q(m)G^t(m)\Gamma^t(j, m + 1) \tag{82}$$

where $\Gamma(s, r)$ is defined in (73). It should be noted that (75) and (82) together provide an alternative form for all the probabilistic information there is to know about the Gaussian process $\{\mathbf{x}(k)\}$. Specializing (82) to the case $k = j$, we arrive at

$$W(k, k) = W(k)$$
$$= \Gamma(k, 0)W_0\Gamma^t(k, 0) + \sum_{m=0}^{k-1} \Gamma(k, m + 1)$$
$$\times G(m)Q(m)G^t(m)\Gamma^t(k, m + 1) \tag{83}$$

Since (82) is valid for $k \geq j$, then the comparison of (82) and (83) reveals that

$$W(k, j) = \Gamma(k, j)W(j, j)$$
$$= \Gamma(k, j)W(j), \quad k \geq j \tag{84}$$

To obtain the corresponding expression when $k \leq j$, we note from (81)

Random Variables and Gauss–Markov Processes

that $W(k, j) = W^t(j, k)$; therefore,

$$W(k, j) = W(j)\Gamma^t(j, k), \quad k \leq j \tag{85}$$

Turning to the measurement process $\{z(k)\}$ for which the mean has already been defined by (74). The covariance essentially follows from that for the process $\{x(k)\}$. Using (63), (64), and (77), it follows that

$$\begin{aligned}\operatorname{Cov}[z(k), z(j)] &= E[\{z(k) - H(k)m(k)\}\{z(j) - H(j)m(j)\}^t] \\ &= E[H(k)\{x(k) - m(k)\}\{x(j) - m(j)\}^t H^t(j)] \\ &\quad + E[H(k)\{x(k) - m(k)\}v^t(j)] \\ &\quad + E[v(k)\{x(j) - m(j)\}^t H^t(j)] \\ &\quad + E[v(k)v^t(j)]\end{aligned}$$

which can be simplified into

$$\begin{aligned}\operatorname{Cov}[z(k), z(j)] &= H(k)W(k)\Gamma^t(j, k)H^t(j) + R(k)\delta_{Kj} \\ &\quad \text{for } k \leq j \\ &= H(k)\Gamma(k, j)W(j)H^t(j) + R(k)\delta_{Kj} \quad \text{for } k \geq j\end{aligned}$$

where we have made use of (65), (81) and the fact that the process $\{v(k)\}$ is independent of $\{x(k)\}$.

Appendix E
Instruction Sets

In what follows, the instruction sets of three 16-bit microprocessors are presented. They are given in the form of tables for convenience.

Table E.1 8086/8088 instruction set (main instructions)

Mnemonic		Description
Data move		
MOV	dest, source	Move data value (byte or 16-bit word) from source to destination (both can be a register or memory location, or immediate value for source)
LDS	reg, source	Load pointer using data segment
LEA	reg, source	Load effective address into register
LES	reg, source	Load pointer using extra segment
XCHG	dest, source	Exchange contents of register, or of register with memory
XCHG	AX (or AL), reg	Exchange register and accumulator
IN	AL, port	Input from port address to AL (or AX)
IN	AL	Input from port address held in DX to AL (or AX)
OUT	port, AL	Output to port address from AL (or AX)

Instruction Sets

Table E.1 *Continued*

Mnemonic		Description
OUT	AL	Output to port address held in DX from AL (or AX)

Usual register PUSH and POP instructions

Data modify

ADD	dest, source	Add source to destination
ADC	dest, source	Add with carry
SUB	dest, source	Subtract with borrow
AND	dest, source	AND
TEST	dest, source	AND with no result obtained
OR	dest, source	OR
XOR	dest, source	Exclusive-OR
INC	dest	Increment (register or memory)
DEC	dest	Decrement
AAA		ASCII adjust for add
DAA		Decimal adjust for add
NEG	dest	Negate (two's complement)
AAS		ASCII adjust for subtract
DAS		Decimal adjust for subtract
MUL	source	Multiply accumulator (by register or memory)
DIV	source	Divide accumulator (by register or memory)
IMUL	source	Signed multiply
IDIV	source	Signed divide
AAM	dest	ASCII adjust for multiply
AAD	dest	ASCII adjust for divide
CBW		Convert byte to word
CWD		Convert word to double-word (in AX or DX)
NOT	dest	Invert (one's complement)
SHL(SAL)	dest	Shift left by count in CL register
SHR	dest	Shift logical right by count in CL register
SAR	dest	Shift arithmetic right by count in CL register
ROL	dest	Rotate left by count in CL register
ROR	dest	Rotate right by count in CL register
RCL	dest	Rotate left through carry
RCR	dest	Rotate right through carry

Branch/jump

Jcc	offset	Jump conditional relative to PC (with variety of conditions, e.g., JLE – 63)

Table E.1 *Continued*

Mnemonic		Description
JMP	address	Jump to 16-bit address within segment or intersegment
JMP	offset	Jump relative to PC (-128 to $+127$ bytes)
JMP	source	Jump using indirect addressing within segment or intersegment
CALL	address	Call within segment or intersegment
CALL	source	Call using indirect addressing within segment or intersegment
RET		Return
RET	data	Return and add immediate data to SP
LOOP	offset	Loop CX times
LOOPZ	offset	Loop whilst CX is nonzero and zero flag is not set
LOOPNZ	offset	Loop whilst CX is nonzero and zero flag is set

Control/miscellaneous

(a) String operations instructions, e.g.

REP		Repeat string operation decrementing CX until CX is zero
MOVSB		Move source string to destination string (source is addressed by SI, destination by DI)

(b) Various control instructions, e.g.

INT	data	Software interrupt (data represents type)
CLC		Clear carry flag
WAIT		Wait until TEST pin is set

Table E.2 Z8000 instruction set

Mnemonic		Description
Data move		
LD	dest, source	Move data value (byte, word or long word) from source to destination — various addressing modes
LDM	reg, source, n	Load multiple (consecutive words) from memory to registers
LDM	dest, reg, n	Load multiple (consecutive words) from registers to memory

Instruction Sets

Table E.2 *Continued*

Mnemonic		Description
EX	reg, source	Exchange words (EXB – exchange bytes)
IN	reg, source	Input to register from source (INB – input byte)
OUT	dest, reg	Output from register to source (OUTB – output byte)

Additional range of input/output instructions, e.g.

INIR	dest, source, reg	Input from source to destination, autoincrement destination address, decrement register and repeat until register contents are zero
PUSH	index reg, source	Push
POP	dest, index reg	Pop

Data modify

ADC	reg, source	Add with carry
ADD	reg, source	Add (ADDB – add bytes; ADDL – add long words)
CP	reg, source	Compare
DAB	dest	Decimal adjust
DEC	dest, n	Decrement by n
DIV	reg, source	Divide (signed)
EXTS	dest	Extend sign (from lower half through to higher half)
INC	dest, n	Increment by n
MULT	reg, source	Multiply (signed)
NEG	dest	Negate
SBC	reg, source	Subtract with carry
SUB	reg, source	Subtract
AND	reg, source	AND
COM	dest	Complement
OR	reg, source	OR
TCC	cond, dest	Test condition code (set LSB)
TEST	dest	Test
XOR	reg, source	Exclusive-OR
BIT	dest, b	Test bit static (b can be replaced by register)
SET	dest, b	Set bit static
TSET	dest	Test and set
RL	dest, n	Rotate left (RLB – rotate byte)
RLC	dest, n	Rotate left through carry
RLDB	reg, source	Rotate digit left

Table E.2 *Continued*

Mnemonic		Description
RR	dest, n	Rotate right
RRC	dest, n	Rotate right through carry
RRDB	reg, source	Rotate digit right
SDA	dest, reg	Shift dynamic arithmetic (by contents of register)
SDL	dest, reg	Shift dynamic logical
SLA	dest, n	Shift left arithmetic
SLL	dest, n	Shift left logical
SRA	dest, n	Shift right arithmetic
SRL	dest, n	Shift right logical
Branch/jump		
CALL	dest	Call subroutine
CALR	dest	Call relative
DJNZ	reg, dest	Decrement register and jump if nonzero
IRET		Interrupt return
JP	cond, dest	Jump conditional
JR	cond, dest	Jump conditional relative
RET	cond	Return conditional
SC	source	System call
Control/miscellaneous		
COMFLG		Complement flag
DI		Disable interrupts
EI		Enable interrupts
HALT		Halt
LDCTL	CTRL, source	Load into control register
NOP		No operation

Range of multi-micro control operations, and block transfer and string manipulation instructions, e.g.

CPSD	dest, source, reg, cond	Compare string and decrement (register)

Instruction Sets

Table E.3 MC68000 instruction set

Mnemonic		Description
Data move		
MOVE	source, dest	Move data value (byte, word or double-word) from source to destination (all addressing modes are allowed)
MOVE	register, dest	Move multiple registers (the contents of the specified registers are transferred to contiguous memory locations)
MOVEP	data reg, addr reg, displ	Move peripheral data (transfer contents of data register to address computed by adding displacement to contents of address register)
EXG	reg, reg	Exchange the contents of the specified registers
LEA	source, addr reg	Load source address into address register
SWAP	data reg	Exchange the 16-bit halves of a data register
PEA	address	Push effective address
Data modify		
ABCD	source, dest	Add decimal (BCD) with extend bit
ADD	source, dest	Add binary
AND	source, data reg	AND
ASL	data reg	Arithmetic shift left (variable shift count)
ASR	data reg	Arithmetic shift right (variable shift count)
BCHG	bit no, dest	Test a bit and change
BCLR	bit no, dest	Test a bit and clear
BSET	bit no, dest	Test a bit and set
BTST	bit no, dest	Test a bit
CLR	dest	Clear an operand (register or memory)
CMP	source, data reg	Compare (and set flags)
DIVS	source, data reg	Signed divide (16-bit)
DIVU	source, data reg	Unsigned divide (16-bit)
EOR	data reg, dest	Exclusive-OR
EXT	data reg	Sign extend data register (byte to word to double-word)
LSL	data reg	Logical shift left
LSR	data reg	Logical shift right
MULS	source, data reg	Signed multiply (32-bit result)
NBCD	dest	Negate decimal with extend

Table E.3 *Continued*

Mnemonic		Description
NEG	dest	Negate (two's complement)
NOT	dest	Logical complement (one's complement)
OR	source, data reg	OR
ROL	dest	Rotate left without extend (variable rotate count)
ROR	dest	Rotate right without extend
ROXL	dest	Rotate left with extend
ROXR	dest	Rotate right with extend
SBCD	data reg,	Subtract decimal with extend bit
SUB	source, dest	Subtract binary
Branch/jump		
Bcc	displ	Branch conditionally by displacement relative to PC (with variety of conditions, e.g. BLE − 24)
BRA	displ	Branch always (relative to PC)
BSR	displ	Branch to subroutine
DBcc	data reg, displ	Test condition, decrement and branch
JMP	address	Jump to address
JSR	address	Jump to subroutine
RTE		Return from exception
RTR		Return and restore (status register)
RTS		Return from subroutine
Control/miscellaneous		
CHK	source, data reg	Check register against bounds (and generate exception)
LINK	address reg, displ	Link stack and allocate (address register to stack)
NOP		No operation
RESET		Reset external devices
Scc	address	Set (byte) according to condition
STOP		Stop
TAS	dest	Test and set an operand (to synchronize other CPUs)
TRAP	vector	Trap
TRAPV		Trap on overflow
TST	dest	Test an operand (set flags)
UNLK	addr reg	Unlink (load SP from address register, and then address register from stack)

Bibliography

S. I. Ashon, *Microprocessors*, Tata McGraw-Hill, India, 1984.

S. T. Allworth, *Introduction to Real-time Software Design*, Macmillan, London, 1979.

H. Amrehn, Computer control in the polymerization industry, *Automatica*, *13*, 533 (1977).

D. K. Anand, *Introduction to Control Systems*, Pergamon Press, Oxford, 1984.

F. Anceau, *The Architecture of Microprocessors*, Addison-Wesley, U.K., 1985.

B. D. O. Anderson and J. B. Moore, *Linear Optimal Control*, Prentice Hall, Englewood Cliffs, NJ, 1971.

B. D. O. Anderson and J. B. Moore, *Optimal Filtering*, Prentice Hall, Englewood Cliffs, NJ, 1979.

B. D. O. Anderson and S. Vongpanitlrd, *Network Analysis and Synthesis: A Modern Systems Theory Approach*, Prentice Hall, Englewood Cliffs, NJ, 1973.

M. Andrews, *Programming Microprocessor Interfaces for Control and Instrumentation*, Prentice Hall, Englewood Cliffs, NJ, 1982.

E. S. Armstrong, *ORACLS, A Design for Multivariable Control*, Marcel Dekker, New York, 1980.

B. A. Artwick, *Microcomputer Interfacing*, Prentice Hall, Englewood Cliffs, NJ, 1980.

K. J. Astrom, *Introduction to Stochastic Control Theory*, Academic Press, New York, 1970.

K. J. Astrom, Computer-aided modeling, analysis and design of control systems—A perspective, *IEEE Control Systems Mag.*, *3*, 4 (1983).

K. J. Astrom and B. Wittenmark, *Computer-Controlled Systems-Theory and Design*, Prentice Hall, Englewood Cliffs, NJ, 1984.

P. C. Badavas, Direct synthesis and adaptive controls, *Chem. Eng.*, *116*, 99 (1984).

A. V. Balakrishnan, A Martingale approach to linear recursive state estimation, *SIAM J. Control*, *10*, 754 (1972).

S. Barnett, *Matrices in Control Theory*, Van Nostrand Reinhold, UK, 1971.

S. Barnett, Matrices polynomials and linear time-invariant systems, *IEEE Trans. Autom. Control*, *18*, 10 (1973).

G. C. Barney, *Intelligent Instrumentation*, Prentice Hall, Englewood Cliffs, NJ, 1985.

R. V. Bartman, Dual composition control in a C3/C4 splitter, *Chem. Engineering Progress*, *114*, 58 (1980).

N. H. Beachley, and H. Harrison, *Introduction to Dynamic System Analysis*, Harper and Row, New York, 1979.

R. E. Belman, *Matrix Analysis*, McGraw-Hill, New York, 1968.

S. Benett, *Real-Time Computer Control: An Introduction*, Prentice Hall, Englewood Cliffs, NJ, 1988.

G. J. Bierman, A comparison of discrete linear filtering algorithms, *IEEE Trans. Aerospace Electronic Systems*, *9*, 28 (1973).

P. F. Blackman, *Introduction to State Variables Analysis*, Macmillan, London, 1977.

K. L. Bowles, *Microcomputer Problem Solving Using Pascal*, Springer-Verlag, Berlin, 1977.

E. H. Bristol, Organization and discipline for distributed process control, *Instrum. Technol.*, *26*, 41 (1979).

E. H. Bristol, On a new measure of interaction for multivariable process control, *IEEE Trans. Autom. Control*, *31* (1986).

W. L. Brogan, Applications of a determinant identity to pole-placement and observer problems, *IEEE Trans. Autom Control*, *19*, 612 (1974).

W. L. Brogan, *Modern Control Theory*, Prentice Hall, Englewood Cliffs, NJ, 1985.

G. Brookes, G. A. Manson, and J. A. Thompson, *CP/M 80 System Programming*, Blackwell, U.S.A., 1985.

R. S. Bucy, Optimum finite time filters for a special nonstationary class of

Bibliography

inputs, Internal Memo BBD-600, Johns Hopkins University, Baltimore, MD, 1959.

J. P. Buzen, I/O subsystem architecture, *Proc. IEEE*, 63, 871 (1975).

D. A. Cassell, *Microcomputers and Modern Control Engineering*, Reston Publishing Co., U.S.A., 1983.

J. L. Casti, *Dynamical Systems and Their Application*, Academic Press, New York, 1977.

C. T. Chen, *Introduction to Linear System Theory*, Holt, Rinehart and Winston, U.S.A., 1970.

L. Ciminiera and A. Valenzano, *Advanced Microprocessor Architectures*, Addison-Wesley, Reading, MA, 1987.

C. M. Clare and D. K. Frederick, *Modeling and Analysis of Dynamic Systems*, Houghton Mifflin, CA, 1978.

A. Clements, *Microcomputer Design and Construction*, Prentice Hall, Englewood Cliffs, NJ, 1982.

T. C. Coffery, Automated frequency domain synthesis of multiloop control systems, *AIAA J.*, 8, 48 (1970).

J. W. Coffron, *Using and Troubleshooting the MC68000*, Reston, U.S.A., 1983.

D. Comer, *Operating Systems Design*, Prentice Hall, Englewood Cliffs, NJ, 1985.

H. Cox, On the estimation of state variables and parameters for noisy dynamic systems, *IEEE Trans. Autom. Control*, 9, 5 (1964).

J. F. Craine and G. R. Martin, *Microcomputers in Engineering and Science*, Addison-Wesley, Reading, MA, 1985.

J. B. Cruz, Jr., *Feedback Systems*, McGraw-Hill, New York, 1972.

J. J. D'Azzo and C. H. Houpis, *Linear Control System Analysis and Design*, McGraw-Hill, New York, 1981.

E. J. Davidson, On pole assignment in linear systems with incomplete state feedback, *IEEE Trans. Autom. Control*, 15, 348 (1970).

P. M. DeRusso, R. J. Roy, and C. M. Close, *State Variables for Engineers*, Wiley, New York, 1965.

C. A. Desoer, *Notes for A Second Course On Linear Systems*, Van Nostrand Reinhold, New York, 1970.

C. A. Desoer and M. Vidyasagar, *Feedback Systems: Input Output Properties*, Academic Press, New York, 1975.

R. C. Dorf, *Time-Domain Analysis and Design of Control Systems*, Addison-Wesley, Reading, MA, 1965.

R. F. Dorf, *Modern Control Systems*, Addison-Wesley, Reading, MA, 1980.

H. M. J. M. Dortmans, Application of microprocessors, *J. Phys. E. Sci. Instrum.*, 14, 777 (1981).

T. E. Dyliacco, The adoptive reliability control system, *IEEE Trans. Power Apparatus Systems*, 86, 517 (1967).

O. I. Elgerd, *Control Systems Theory*, McGraw-Hill, New York, 1967.

M. E. El-Hawary, *Control System Engineering*, Reston, U.S.A., 1984.

C. Evans, *The Making of the Micro*, Gollancz, U.K., 1981.

V. W. Eveleigh, *Introduction to Control Systems Design*, McGraw-Hill, New York, 1972.

P. Eykoff, *System Identification*, Wiley, New York, 1969.

M. M. Fahmy and J. O'Reilly, On eigenstructure assignment in linear multivariable systems, *IEEE Trans. Autom. Control*, 27, 1982.

M. M. Fahmy and J. O'Reilly, Eigenstructure assignment in multivariable systems — A parametric solution, *IEEE Trans. Autom. Control*, 28, 990 (1983).

P. L. Falb and W. A. Wolovich, Decoupling in the design and synthesis of multivariable control systems, *IEEE Trans. Autom. Control*, 12, 651 (1967).

B. K. Fawcett, *The Z8000 Microprocessor*, Prentice Hall, Englewood Cliffs, NJ, 1984.

W. Finbisen, Multilevel systems control, *Automation Remote Control*, 9, 1447 (1970).

L. H. Fink, Concerning power system control structures, *Proceedings of the 26th Annual ISA Conference*, USA, pp. 1–11 (1971).

W. I. Fletcher, *An Engineering Approach to Digital Design*, Prentice Hall, Englewood Cliffs, NJ, 1984.

G. F. Franklin and R. D. Powell, *Digital Control of Dynamical Systems*, Addison-Wesley, Reading, MA, 1981.

F. R., Gantmakher, *Theory of Matrices*, Chelsea Publishing Co., U.S.A., 1959.

A. Ghosh, Checklist for batch process computer control, *Chem. Engng.*, 112, 88 (1980).

G. A. Gibson and Y. Lin, *Microcomputers for Engineers and Scientists*, Prentice Hall, Englewood Cliffs, NJ, 1981.

E. Gilbert, Controllability and observability in multivariable control systems, *SIAM J. Control*, 1, 128 (1963).

R. Gilbert, The general-purpose interface bus, *IEEE Micro*, 2, 41 (1982).

B. E. Gladstone, Comparing microcomputer development system capabilities, *Computer Design*, 18, 83 (1979).

J. L. Goodrich, Interfacing microcomputers for control and data acquisition, *AIIE Trans.*, 12, 106 (1980).

L. A. Gould, *Chemical Process Control*, Addison-Wesley, Reading, MA, 1969.

R. Grappel and J. Hemenway, Evaluating the 16-bit chips, *Mini-Micro Systems*, Dec. 152 (1980).

Bibliography

R. Grappel and J. Hemenway, A tale of four µPs: benchmarks quantify performance, *EDN*, April 1, 179 (19781).

A. Gupta ed., *Advanced Microprocessors, II, IEEE Press*, New York, 1987.

S. C. Gupta and L. Hasdorff, *Fundamentals of Automatic Control*, Wiley, New York, 1970.

H. Hanselmann, Implementation of digital controller—A survey, *Automatica*, *23*, 7 (1987).

P. M. Hansen, et al., A performance evaluation of the Intel IAPX 432, *Computer Architecture News*, *10*, 17 (1982).

R. J. L. Hanus, A new technique for preventing control winup, *Journal A*, *21*, 1 (1980).

M. J. J. Hautus, *Controllability and Observability Conditions of Linear Autonomous*, Ned. Akad.

J. Heering, The Intel 8086, the Zilog Z8000 and the Motorola MC68000 microprocessors, *Euromicro J.*, *6*, 135 (1980).

J. L. Hilburn and P. M. Julich, *Microcomputers/Microprocessors: Hardware, Software Applications*, Prentice Hall, Englewood Cliffs, NJ, 1976.

M. Hordeski, Interfacing microcomputers in control systems, *Instrum. Control Systems*, *51*, 56 (1978).

J. S. Horton, et al., The Baltimore Gas and Electric Company energy control system—An overview, *IEEE Power Apparatus Systems*, *101*, 26 (1982).

R. Imriale, System family covers all MC development needs, *Electron Design*, *28*, 141 (1980).

R. Iserman, *Digital Control Systems*, Springer-Verlag, New York, 1981.

R. G. Jacquot, *Modern Digital Control*, Marcel Dekker, New York, 1981.

A. H. Jazwinski, *Stochastic Processes and Filtering Theory*, Academic Press, New York, 1970.

C. D. Johnson, *Microprocessor-Based Process Control*, Prentice Hall, Englewood Cliffs, NJ, 1984.

B. E. Jones, *Instrumentation, Measurement and Feedback*, McGraw-Hill, New York, 1979.

E. I. Jury, *Sampled-Data Control Systems*, Wiley, New York, 1958.

T. Kailath, *Linear Systems*, Prentice Hall, Englewood Cliffs, NJ, 1980.

S. H. Kaisler, *The Design of Operating Systems for Small Computer Systems*, Wiley, New York, 1982.

S. A. Kallis, Jr., Developing software for micro applications, *Machine Design*, *51*, 66 (1979).

R. E. Kalman, On the general theory of control systems, *Proc. First Int. Congress on Automatic Control*, London, pp. 481–493 (1960).

R. E. Kalman, A new approach to linear filtering and prediction problems, *J. Basic Engng*, *82*, 34 (1960).

R. E. Kalman, Mathematical description of linear dynamical systems, *J. Soc. Ind. Appl. Math. Ser. A. Control*, *1*, No. 2 (1963).

R. E. Kalman, Mathematical description of linear systems, *SIAM J. Control*, *1*, 152 (1963).

R. E. Kalman, When is a linear control system optimal, *Trans SME J. Basic Engng*, *86D*, 51 (1964).

R. E. Kalman, *Lectures on Controllability and Observability*, C.I.M.E., Bologna, 1968.

R. E. Kalman and J. E. Bertram, General synthesis procedures for computer control of single-loop and multi-loop systems, *Trans. AIEE*, *77*, 602 (1958).

R. E. Kalman and J. E. Bertram, A unified approach to the theory of sampling systems, *J. Franklin Inst.*, *267*, 405 (1959).

R. E. Kalman and J. E. Bertram, Control system analysis and design via the second method of Lyapunov, *Trans. ASME J. Basic Engng*, *82*, 371 (1960).

R. E. Kalman and R. S. Bucy, New results in linear filtering and prediction theory, *ASME, Ser. D. J. Basic Engng*, *83*, 95 (1961).

R. E. Kalman and R. W. Kopcke, Optimal synthesis of linear sampling control systems using generalized performance indexes, *Trans. ASME Ser. D. J. Basic Engng*, *80*, 1812 (1980).

A. Kanunker et al., A 32 bit microprocessor with virtual memory support, *IEEE J. Solid State Circuits*, *16*, 548 (1981).

D. C. Karnopp and R. C. Rosenberg, *System Dynamics: A Unified Approach*, Wiley, New York, 1975.

P. Katz, *Digital Control Using Microprocessors*, Prentice Hall, Englewood Cliffs, NJ, 1981.

M. B. Kendler and N. P. Lutte, Microprocessors applied to industrial control systems, *Electron. Engng*, *51*, 37 (1979).

A. J. Khambata, *Microprocessors/Microcomputers: Architecture, Software and Systems*, Wiley, New York, 1982.

A. K. Kochhar, Distributed real time data processing for manufacturing organizations, *IEEE Trans. Engng, Management*, *24*, 119 (1977).

A. K. Kochhar and N. D. Burns, *Microprocessors and their Manufacturing Applications*, Pitman, London, 1983.

A. K. Kochhar and N. D. Burns, *Microprocessors and their Manufacturing Applications*, Edward Arnold, London, 1983.

A. K. Kochhar and J. Parnaby, Dynamical modeling and control of plastics extrusion processes, *Automatica*, *13*, 177 (1977).

E. Kriendler and A. Jameson, Optimality of linear control systems, *IEEE Trans. Autom. Control*, *17*, 349 (1972).

B. C. Kuo, *Automatic Control Systems*, Prentice Hall, 1982.

B. C. Kuo, *Digital Control Systems*, Prentice Hall, Englewood Cliffs, NJ, 1984.

Bibliography

H. Kwarkernaak and R. Sivan, *Linear Optimal Control Systems*, Wiley-Interscience, New York, 1972.

L. Lapidus and R. Luus, *Optimal Control of Engineering Processes*, Blaisdell, U.S.A., 1967.

S. C. Lee, *Microcomputer Design and Applications*, Academic Press, New York, 1977.

I. Lefkowitz, Multilevel approach to control system design, *Trans. ASME J. Basic Engng*, 88, 392 (1966).

I. Lefkowitz, Integrated control of industrial systems, *Trans. Royal Soc. London*, 287, 443 (1977).

I. Lefkowitz and J. D. Schoeffler, Multilevel control structures for three discrete manufacturing processes, *Proc. Fifth IFAC Congress*, Paris (1972).

S. Lehnigk, *Stability Theorems for Linear Motions*, Prentice Hall, Englewood Cliffs, NJ, 1966.

J. R. Leigh, *Applied Control Theory*, IEE Control Engineering Series, No. 18, Peter Peregrinus, U.K. 1982.

J. R. Leigh, *Applied Digital Control*, Prentice Hall, Englewood Cliffs, NJ, 1985.

L. A. Leventhal, *Introduction to Microprocessors: Software, Hardware, Programming*, Prentice Hall, Englewood Cliffs, NJ, 1978.

L. Levine, *Methods for Solving Engineering Problems Using Analog Computers*, McGraw-Hill, New York, 1964.

A. M. Lister, *Fundamentals of Operating Systems*, Macmillan, London, 1979.

B. Litman, An analysis of rotating amplifiers, *Trans. AIEE*, 68, Pt. III, 1111 (1949).

D. G. Luenberger, An introduction to observers, *IEEE Trans. Autom. Control*, 16, 596 (1971).

D. G. Luenberger, *Introduction to Dynamic Systems*, Wiley, New York, 1979.

W. L. Luyben, *Process Modeling, Simulation and Control for Chemical Engineers*, McGraw-Hill, New York, 1974.

A. G. J. MacFarlane, A survey of some recent results in linear multivariable feedback theory, *Automatica*, 8, 455 (1972).

A. G. J. MacFarlane, *Frequency-Response Methods in Automatic Control*, IEEE Press, New York, 1979.

M. S. Mahmoud, Multilevel systems control and applications: a survey, *IEEE Trans. Systems, Man Cybernet.*, 7, 125 (1977).

M. S. Mahmoud and M. G. Singh, *Discrete Systems: Analysis, Optimization and Control*, Springer-Verlag, Berlin, 1984.

J. M. Malcolm, Microprocessor-based controllers as applied to multiple cascade control systems, *J. Inst. Measurement Control*, 14, 459 (1981).

Mayer, O., *The Origins of Feedback Control*, MIT Press, Cambridge, MA, 1970.

MC68000 16/32 Bit Microprocessor Programmer's Manual, Motorola, Prentice Hall, Englewood Cliffs, NJ, 1984.

MC68020 32-Bit Microprocessors User's Manual, Prentice Hall, Englewood Cliffs, NJ, 1985.

D. R. McGlynn, *Distributed Processing and Data Communications*, Wiley Interscience, New York, 1978.

T. C. McIntyre, *Software Interpreters for Microcomputers*, Wiley, New York, 1978.

J. S. Meditch, *Stochastic Optimal Linear Estimation and Control*, McGraw-Hill, New York, 1969.

G. A. Mehta, The benefits of batch process control, *Chemical Engineering Progress*, *117*, 47 (1983).

J. L. Melsa and D. G. Schultz, *Linear Control System*, McGraw-Hall, New York, 1969.

M. D. Mesarovic, Multilevel systems and concepts in process control, *Proc. IEEE*, *59*, 111 (1970).

H. J. Mitchell, *32-bit Microprocessors*, Collins, U.S.A., 1986.

C. L. Morgan and M. Waite, *8086/8088 16-bit Microprocessor Primer*, McGraw-Hill, New York, 1982.

M. Mori, Root-locus method of pulse transfer function for sampled-data control systems, *I.R.E. Trans. Autom. Control*, 13 (1957).

S. P. Morse and D. J. Albert, *The 80286 Architecture*, Wiley, New York, 1986.

A. S. Morse and W. A. Wonham, Triangular decoupling of linear multivariable systems, *IEEE Trans. Autom. Control*, *15*, 447 (1970).

A. S. Morse and W. M. Wonham, Status of noninteracting control, *IEEE Trans. Autom. Control*, *16*, 568 (1970).

Motorola Semiconductors, *Microcomputer Development Systems and Subsystems*, USA, 1979.

I. J. Nagrath and M. Gopal, *Control Systems Engineering*, Wiley, New Delhi, 1982.

G. Newton, L. Gould, and J. Kaiser, *Analytical Design of Linear Feedback Controls*, Wiley, New York, 1957.

B. Noble, *Applied Linear Algebra*, Prentice Hall, Englewood Cliffs, NJ, 1977.

K. Ogata, *System Dynamics*, Prentice Hall, Englewood Cliffs, NJ, 1978.

K. Ogata, *Modern Control Engineering*, Prentice Hall, Englewood Cliffs, NJ, 1979.

C. A. Ogdin, *Microcomputer Design*, Prentice Hall, Englewood Cliffs, NJ, 1978.

C. A. Ogdin, *Software Design for Microcomputers*, Prentice Hall, Englewood Cliffs, NJ, 1978.

Bibliography

H. Olson, Interfacing micros with AC/DC loads, *Instruments Control Systems*, 51, 79 (1978).

R. V. Orlando and T. L. Anderson, An overview of the 9900 microprocessor family, *IEEE Micro*, 1, 38 (1981).

R. Ortega, Experimental evaluation for four microprocessor based advanced control algorithms, *Microprocessing/Microprogramming*, 10, 229 (1982).

J. Parnaby, A. K. Kochhar, and B. Wood, Development of computer control strategies for plastics extruders, *Polymer Engng Sci.*, 15, 594 (1975).

W. R. Perkins and J. B. Cruz, Jr., *Engineering of Dynamic Systems*, Wiley, New York, 1969.

W. R. Perkins and J. B. Cruz, Jr., Feedback properties of linear regulators, *IEEE Trans. Autom. Control*, 16, 659 (1971).

L. Pipes, *Matrix Methods for Engineering*, Prentice Hall, Englewood Cliffs, NJ, 1963.

E. P. Popov, *The Dynamics of Automatic Control Systems*, Addison-Wesley, Reading, MA, 1962.

M. Prycher, A performance comparison of three contemporary 16-bit microprocessors, *IEEE Micro*, 3, 26 (1983).

J. R. Ragazzini and G. F. Franklin, *Sampled-Data Control Systems*, McGraw-Hill, New York, 1958.

R. G. Rajulu and V. Rajaraman, Execution-time analysis of process control algorithms on microprocessors, *IEEE Trans. Ind. Electron.*, 29, 312 (1982).

F. H. Raven, *Automatic Control Engineering*, McGraw-Hill, New York, 1978.

B. Reynolds, Guide to analog I/O boards, *Instruments Control Systems*, 51, 53 (1978).

R. J. Richards, *An Introduction to Dynamics and Control*, Longman, London, 1979.

B. Roffel and J. E. Rijnsdorp, *Introduction to Process Dynamics, Control and Protection*, Ann Arbor, MI, 1981.

B. Roffel, P. A. Chin, and R. D. Chung, Control of an in-line blending process, *Proc. 36th C.S. Ch.E. Conf.*, Sarnia, Canada (1986).

R. A. Rohrer, *Circuit Theory: An Introduction to the State-Variable Approach*, McGraw-Hill, New York, 1970.

H. H. Rosenbrock, Multivariable and State-Space Theory, Wiley, New York, 1970.

K. Roud, Design of control systems using microelectronics, *Design Engng*, 36 (1980).

A. P. Sage and J. L. Melsa, *Estimation Theory with Applications to Communications and Control*, McGraw-Hill, New York, 1971.

A. P. Sage and C. C. White, *Optimum Systems Control*, Prentice Hall, Englewood Cliffs, NJ, 1977.

A. F. Shackil, *Microprocessors, IEEE Spectrum,* Jan. 32 (1982).
J. L. Shearer, A. T. Murphy, and H. H. Richardson, *Dynamic Systems,* Addison-Wesley, Reading, MA, 1967.
S. M. Shinners, *Modern Control System Theory and Application,* Addison-Wesley, Reading MA, 1978.
F. G. Shinskey, *Process-Control Systems,* McGraw-Hill, New York, 1981.
M. E. Sloan, *Introduction to Minicomputers and Microcomputers,* Addison-Wesley, Reading MA, 1980.
H. W. Sorenson, Least-squares estimation from Gauss to Kalman, *IEEE Spectrum,* 7, 63 (1970).
G. Stephanapoulos, *Chemical Process Control,* Prentice Hall, Englewood Cliffs, NJ, 1984.
S. M. Stinners, How to approach the stability analysis and compensation of control systems, *Control Engng.,* 62 (1978).
H. S. Stone, *Microcomputer Interfacing,* Addison-Wesley, Reading, MA, 1982.
G. Strang, *Linear Algebra and Its Applications,* Academic Press, New York, 1976.
F. G. Stremler, *Introduction to Communication Systems,* Addison-Wesley, Reading, MA, 1982.
P. H. Sydenham, *Transducers in Measurement and Control,* Adam Hilger, U.S.A., 1980.
Y. Takahashi, M. J. Rabins, and D. M. Auslander, *Control and Dynamic Systems,* Addison-Wesley, Reading, MA, 1970.
G. J. Thaler, *Design of Feedback Systems,* Dowden, Hutchinson and Ross, U.S.A., 1973.
H.-M. D. Toong and A. Gupta, An architectural comparison of contemporary 16-bit microprocessors, *IEEE Micro,* 1, 26 (1981).
J.-M. D. Toong and A. Gupta, Evaluation kernels for microprocessor performance analysis, *Performance Evaluation,* 2, 1 (1982).
Transputer Reference Manual, Inmos, U.S.A., 1987.
W. Twaddell, General-purpose microprocessors: performance and features, *Electron. Design,* Oct. 14, 118 (1982).
S. G. Tzafestas ed., *Microprocessors in Signal Processing, Measurement and Control,* Reidel, Dordrecht, 1983.
H. Unbehauen, C. Schmid and F. Bottiger, Comparison and application of DDC algorithms for a heat exchanger, *Automatica,* 12, 393 (1976).
Universal development system, *Microprocessors Microsystems,* 4, 190 (1980).
P. Uronen and L. Yliniemi, Experimental comparison and application of different DCC algorithms, *Proc 5th IFAC/IFIP Conference on Digital Computer Applications to Process Control,* North-Holland, Amsterdam, 1977.
J. N. Wallace, Application of distributed digital control algorithms in power

Bibliography

stations, Chap 14, *Industrial Digital Control Systems* (Warwick, ed.) 1988, pp. 335–363.

J. N. Wallace and R. C. Clarke, The application of Kalman filtering estimation techniques in power station control systems, *IEEE Trans. Autom. Control*, 28, 416 (1983).

Warwick ed., *Industrial Digital Control Systems*, 1988.

T. W. Weber, *An Introduction to Process Dynamics and Control*, Wiley, New York, 1973.

P. E. Wellsted, *Introduction to Physical System Modelling*, Academic Press, London, 1979.

T. Westerlund et al., Some simple facts a chemical engineer should know about stochastic control, *Technical Publication of Abo Akademi*, Report 85-1, Abo, Finland, 1985.

H. Wetenschappen, *Proc. Ser. A.*, 72, 443 (1969).

I. R. Whitworth, *16-Bit Microprocessor*, Granada, U.S.A., 1984.

D. M. Wiberg, *State Space and Linear Systems*, Schaum's Outline Series, McGraw-Hill, New York, 1973.

P. Wintz, Fundamentals of microprocessor development systems, *Digital Designs*, 70, 30 (1980).

P. M. Wolfe, Data input/output using a microprocessor system, *AIIE Trans.*, 12, 114 (1980).

W. M. Wonham, On pole assignment in multi-input, controllable linear systems, *IEEE Trans. Autom. Control*. 12, 660 (1967).

G. A. Woolvet, *Transducers in Digital Systems*, Petregrinus, U.K., 1977.

G. A. Woolvet, Transducers in digital systems, *IEE Control Engineering Series*, No. 3, Peter Peregrinus, U.K. 1979.

J. G. Ziegler and N. B. Nichols, Optimum settings for automatic controllers, *Trans. ASME*, 64, 759 (1942).

Index

Accelerometer, 41–43
Actuators, 474–477
Advanced control (*see also* Control)
 adaptive, 459–461
 constraint, 451–459
 dead-time compensator, 427–437
 feedforward, 445–451
 inferential, 437–445
 multivariable, 461–462
 self-tuning, 462–463

Batch process control
 software support of, 402–404
Block diagram
 algebra, 24–25
 rules, 26–29
Bode
 logarithmic plots, 137–143

Canonical forms
 controllability, 79
 controller, 79
 observability, 80
 observer, 80
 phase-variable, 76
Capacity storage
 high, 337
 medium, 336
Cauchy theorem, 145–146
Cayley-Hamilton method, 84–86
Characteristic equation, 103–108
Compensation
 design, 160
 methods, 161–175
 cascade, 161
Compensator
 lag, 169–172
 lead, 165–169
 lead-lag, 172–174

    proportional derivative,
      162–164
    proportional integral,
      164–165
Control
    decoupling, 187–188
    feedback, 7
    feedforward, 8
    inferential, 8
    linear-quadratic, 280–284
    linear-quadratic-Gaussian,
      309–312
    on-off, 257
    systems, 2
Controllability
    concept, 181

Data acquisition, 344–346
    modules, 393–398
Debugger, 383
Difference equations, 230–236
Digital
    algorithms, 250–256
    controller design, 249–250
    design realization, 265–268
Discretization
    principles of, 236
    schemes, 238–241
Dynamic
    systems, 4–9
    variable, 2

Editors, 382
Equilibrium point, 129
Error
    generalized coefficient, 56
    series, 57
    steady state, 53–59
Estimation
    problem, 289–291

    error, 291
Estimate
    maximum *a posteriori*, 293
    maximum likelihood,
      292–293
    minimum variance, 291–292

Feedback
    design, 159–162
    observer-based, 188–191
    output, 185–187
    state, 182–184
File
    handling and management,
      384–386
Filtering problem, 289
    optimal, 294–301

Gaussian processes, 288–293,
    609–627

Hardware
    devices, 315–344
Heating system, 3
Hierarchical processing system,
    483–485
Horizontal processing systems,
    485–487
Hurwitz criterion, 113–115
Hydraulic systems
    linear actuator, 40–41
    transmitter, 38–40

Integrated systems control
    controller, 507–508
    information processor, 508–
      509
    plant, 506–507

Index

Intel 8086/8088 family
    instruction set, 530–533
    memory connections, 533–538
    pin functions, 526–527
    processor architecture, 527–530
Intel 80386, 560–567
Inmos T424 transputer, 581–583
Interface
    components, 343–344
    microcomputer, 387–389
I/O
    common, 341
    methods, 337–339
    parallel ports, 390–392
    serial, 339–341
    serial ports, 392–393
    transfer, 342–343

Jury test, 246–248

Kalman filter
    equations, 294–301
    properties, 301–302

Laplace transforms, 587–593
Loaders, 381–382
Loop
    closed, 3
    open, 2
Linkers, 382–383
Lyapunov
    equation, 131
    function, 131
    method, 129–131
    theorem, 130

Matrix algebra, 594–604
Matrix inversion lemma, 294
Memory
    erasable programmable read only, 11
    hierarchies, 333–334
    random access, 11
    read only, 11
    programmable read only, 11
Microprocessors
    architecture, 317–326
    families, 326–332
    functions, 315–317
Monitor, 383–384
Multi-layer hierarchy
    control functions of, 509–512
    functional levels of, 513–515
Multi-level hierarchy
    control levels, 515–518
    in power systems, 517–520
Multi-processor systems, 490–506
Motorola MC68020, 567–575
Motorola MC68000 family
    instruction set, 545–548
    pin functions, 540–541
    process architecture, 541–545

Nyquist
    contour, 147
    stability criterion, 146–149

Observability
    concept, 181
Operating systems
    classification, 360–363
    real-time, 377–381
    single foreground, 373–377

single user, 363-373
Optimization
   dynamic, 280-282
   parameter, 194-196
Optimum
   output regulator, 202-204
   state regulator, 198-202

Parseval's theorem, 196-198
Pneumatic flapper valve, 45-47
Polar plot, 133-137
Positive definite function, 130-132
Prediction problem, 290
Process control
   actions, 406-414
   concepts, 404-406
   software for, 424-425
Processing
   analog, 9
   digital, 9-10
   distributed, 482-490
Programmable logic controllers (PLC), 400-402
Programming
   cascade, 266-268
   direct, 265-266
   parallel, 268

Response
   forced, 81
   free, 80
   frequency, 132-134
Riccati equation, 200, 605-608
   discrete, 283-286
Root-locus method, 119-129
Routh criterion, 108-113

Sensitivity
   analysis, 149-152

root, 149-150
   logarithmic, 150
Sensors
   acceleration, 473-474
   displacement, 472-973
   flow, 471-472
   level, 469-471
   pressure, 464-466
   temperature, 466-469
   velocity, 473-474
   vibration, 473-474
Servomotors
   ac, 32-33
   dc, 33-36
Smoothing problem, 290-291
Software support
   control, 348-352
   development tools, 381-390
Spectral factorization, 204-206
Spring-mass-dashpot system, 22-23
Stability
   asymptotic, 129-131
   bounded-input bounded-output, 103
   concept of, 101
   criteria, 102
   input-output, 116-117
   margins, 143-145
   relative, 115-116
State space
   equations, 75-80
   formulation, 72-75
State transition matrix, 82-86
Stirred tank heater, 4
Supervisory control
   basics, 414-416
   cascade control, 420-422
   program control, 418-419
   ratio control, 416-418
   sequence control, 419-420
System sensitivity, 53-55

Index

Systems
  control, 1
  liquid-level, 5–6

Tachometer, 36–38
Thermal system, 47–49
Transducers (*see* Sensors)
Transfer function
  all-pass, 143
  minimal polynomial, 105
  minimum phase, 143
  nonminimum phase, 143
  poles and zeros, 227–229
  pulse, 220–223
  single-input single-output, 102–103

Ward-Leonard system, 5–7
Working store, 334–336

Z-transform
  definition, 216
  inverse, 223–226
  properties, 217–220
Zilog Z8000 family
  instruction set, 551–552
  memory management, 552–556
  pin functions, 548–550
  process architecture, 550–551
Zilog Z80000, 575–581